Nanoparticles Reinforced Metal Nanocomposites

Santosh K. Tiwari · Vijay Kumar · Sabu Thomas
Editors

Nanoparticles Reinforced Metal Nanocomposites

Mechanical Performance and Durability

Springer

Editors
Santosh K. Tiwari
Department of Chemistry
NMAM Institute of Technology
Nitte (Deemed to be University)
Karnataka, India

Faculty of Chemistry
Warsaw University
Warsaw, Poland

Vijay Kumar
Department of Physics
National Institute of Technology
Srinagar, Jammu and Kashmir, India

Department of Physics
University of the Free State
Bloemfontein, South Africa

Sabu Thomas
School of Energy Materials
Mahatma Gandhi University
Kottayam, India

ISBN 978-981-19-9728-0 ISBN 978-981-19-9729-7 (eBook)
https://doi.org/10.1007/978-981-19-9729-7

This Springer imprint is published by the registered company Springer Nature Singapore Pte Ltd.
The registered company address is: 152 Beach Road, #21-01/04 Gateway East, Singapore 189721, Singapore

Contents

Metal and Materials Engineering: Historical Prospect

Raunak Pandey, Nannan Wang, Bibhuti B. Sahu, Srikanta Moharana, and Santosh K. Tiwari

Abstract Materials are the backbone of our society and have become a functional part of the engineering process for the formation and modifications of technologies that have intrigued humankind. It is impossible to think about the next generation of technologies without discovering new materials. However, for the investigation of new materials, knowledge of the historical prospects of materials is very crucial. This chapter is dealing with the historical prospects of different kinds of materials, including ceramics, glasses, metals, non-metals, alloys, plastics, composites, and nanomaterials. The chapter deals with the different ages of material development, including stone, metal, copper, bronze, iron, and modern ages. Furthermore, the material engineering task at hand required the selection or development of a unique material, which in turn allowed for novel approaches to designing the final product. The materials way of thinking requires a system perspective or the abandonment of the linear concept of innovation.

Keywords Historical prospective · Materials · Engineering nanomaterials · Composites · Polymers

---

R. Pandey
Department of Chemical Science and Engineering, Kathmandu University, Dhulikhel, Kavre, Nepal

N. Wang
Key Laboratory of New Processing Technology for Nonferrous Metals and Materials, Guangxi Institute Fullerene Technology (GIFT), Ministry of Education, School of Resources, Environment and Materials, Guangxi University, Nanning 530004, China

B. B. Sahu (✉)
Department of Physics, Veer Surendra Sai University of Technology, Burla, Odisha, India
e-mail: bibhubhusan78@gmail.com

S. Moharana
School of Applied Sciences, Department of Chemistry, Centurion University of Technology and Management, Burla, Odisha, India

S. K. Tiwari
Department of Chemistry, NMAM Institute of Technology, Nitte (Deemed to Be University), Karnataka 574 110, India

S. K. Tiwari et al. (eds.), *Nanoparticles Reinforced Metal Nanocomposites*,
https://doi.org/10.1007/978-981-19-9729-7_1

1 Introduction

Our universe is made up of different kinds of materials, and the entire phenomenon that is happening in this universe has somewhat of a connection with the type and properties of materials. Since the prehistoric era, our ancestors have made progress and developed their lives through the utilization of various kinds of materials. They were using different kinds of materials as survival tools in terms of foods, habitats, protection, and migration from one place to another [1, 2]. The real progress in the use of materials by our ancestors from 2.6 million years ago to the present has been accounted for in Table 1, along with the important milestones achieved in the particular time period [1–4]. Indeed, with time, the use of materials by mankind has changed as per their requirements, especially, for the improvement of life quality and sustainability. Similarly, with time, the scalability of the use of different kinds of materials has also changed. For example, over the last 50 years, silicon (liquid silicone, fluorinated silicone, high-consistency silicon rubber, silicon foam, and silicon emulsions) and carbon (such as carbon black, graphite, carbon felt, carbon sheets, carbon foam, carbon paper, carbon tapes, carbon cloth, carbon brushes, carbon resins, fullerenes, carbon nanotubes, carbon nanoplates, carbon nanopowder, and carbon nanopowder) have been widely used in different industries dealing with computers, automobiles, electronics, clothing, waterproof devices and textiles, construction, health care, etc.

Similarly, during the last two centuries, different kinds of metallic and non-metallic materials like aluminum, iron, copper, zinc, gold, ceramics, polymers, and their derivatives (metal alloys, hybrid material composites, and blend composites) have been innovated as per the requirements of engineering and industries [1, 2, 4]. These materials have drastically changed in terms of properties and applications due to the advanced cerebral features of human civilization and the highly advanced lifestyle of human beings. For example, just a hundred years ago, it was impossible to replace the bones in our body with any artificial material owing to adoptability, toxicity, and durability issues. However, at present, due to the rapid progress in materials processing, we have different kinds of excellent artificial materials that can be easily used as an alternative for the bones. Thus, materials have endured drastic measures to develop and modify them for exceptional uses in human civilization, and we have tried to provide an overview of how these materials were developed to this stage from the earliest known age of the Stone Age [1, 2]. The study of the link between the processing of materials, their structure, their qualities, and their performance is what material science is all about, which is given a concise idea by the schematic as shown in Fig. 1.

In Western Europe, the Iron Age started around 3000 B.C. and is still going strong now. Iron and steel, being both stronger and more cost effective, had a profound impact on people's day-to-day lives [1, 2]. In the modern materials age and throughout the Iron Age, several novel materials (ceramic, semiconductors, polymers, composites, etc.) have been developed and offered to the market. In order to become familiar with the interconnections between the structure of the material and its processing and performance, a synthesis of advanced materials with intelligence has been done.

Table 1 Important developments and achievements that human civilization has achieved in terms of material processing from the Stone Age to the Advanced Ages [1–3]

S. no	Ages	Key development and notable achievements
1	Stone age (2.6 M years ago)	(a) Wood, bone, fibers, feathers, and animal skins were used as materials for building weapons, ornaments, shelters, utensils, etc (b) Earlier, they were handheld, but in later years, the stone flakes were attached to wooden handles (c) The Egyptians used to make pottery, and glazed minerals were also used to decorate it
2	Copper–Stone age (3000 BCE–1000)	(a) Carving or hammering copper led to the start of the copper age (b) After the discovery of its manufacturing, copper and stones became inseparable parts of the copper age (c) Copper smelting was considered to be borrowed from pottery manufacturing (d) Copper utensils, weapons, and ornaments were found in the Egyptian graves. The use of copper in various activities in China through the epics of Shu Chiang was also found (e) Weapons such as a copper axe with wooden handles were utilized (f) Egyptians also manufactured glasses and glazes for utensils and decorations (g) The glassblowers did not figure out how to pick a specific color until they had worked with a lot of different ones. Oxides of manganese, copper, cobalt, iron, tin, antimony, and lead were used as pigments. The opacifiers, which were admixtures used to make glass opaque, were composed of Cu_2O and Sb_2O_3, along with CaO, PbO, and SnO_2
3	Bronze age (3300 to 1200 B.C)	(a) When human beings discovered the ways to craft bronze by adding tin and arsenic to copper, this led to the start of the Bronze Age (b) At first, the alloy of copper and arsenic was fabricated, but arsenic was later substituted by tin to form bronze (c) Thought to have originated in the Middle East or the Mediterranean region, bronze was also found in Thailand, India, and China during these times (d) Bronze was utilized in crafting weapons, utensils, bracelets, ornaments, two-tone bells, human and animal figurines, etc (e) The Chinese were masters of producing bronze tools and equipment, as no traces of hammering for the fabrication of the appliances were found. Also, beautiful figures were carved in the appliances (f) The Ueticians were skilled bronzesmiths who crafted safety pins, tools, weapons, and jewelry that they traded with the Middle East and Western Europe (g) The Indus Valley Civilization mastered the use of bronze in the creation of human-like objects such as figurines, vases, arrowheads, spearpoints, knives, and axes (h) The use of gold, fur, amber, glass beads, bracelets, and necklaces in application as well as currency was also found

(continued)

Table 1 (continued)

S. no	Ages	Key development and notable achievements
4	Iron age (600–1200 B.C)	(a) The discovery of meteoritic iron and its use in tools shaped the formation of the Iron Age (b) Pure iron was achieved by hammering the slag when copper was smelted from malachite in the presence of iron oxide. Today, this type of iron is known as wrought iron (c) The discovery of steel was also made when iron was heated with charcoal, which led to the combination of iron and carbon, forming steel. Romans, Hittites, and some ancient Greek texts from the first millennium B.C. suggest different procedures for the formation and improvement of carbonized iron or steel (d) Chinese people improved the steel by removing excess carbon from iron, allowing the fabrication of a metal jacket (e) The mass production of steel and iron materials by the Chinese people was done through casting; however, the Western people relied on shaping and carbonizing their goods individually by hammering (f) Structures, monuments, utensils, wootz steel, weapons, ships, helmets, glass cups, glass windows, glass flowers, fruits, etc., were built using iron materials (g) Iron production, copper-based alloy manufacturing, lead, and silver manufacturing were enhanced during these times (h) Steel manufacturing through the casting process and puddling process yielded appliances such as plowshares, hoes, cart bearings, and harness buckles used in agriculture in China, while the Western people formed weaponry, utensils, and other utilities from steel using the hammering process of wrought iron in the presence of carbon to form steel
5	Medieval age (500 to 1400 CE)	(a) The beginning of steel manufacturing was done in the Iron Age, but this got more attention in the medieval age (b) Manufacturing of swords and weaponry, utensils, megastructures, ships, other transport vehicles, etc., through brass, steel, iron, copper-based alloys, etc., was conducted (c) Production of transparent glasses in most of Europe and smalt glass, mosaics, and art glass production in Roman civilizations and other parts of Europe were observed (d) The post-medieval age was the foundation stone of the industrial revolution in Europe, with improvements in the metallurgy and manufacture of steel, iron, copper, lead, silver, gold, etc

(continued)

Table 1 (continued)

S. no	Ages	Key development and notable achievements
6	Industrial revolution (1760–1840)	(a) The industrial revolution shaped modern metallurgy and material sciences (b) With advancements in the production and manufacturing of iron, steel, lead, copper, silver, alloys, and so on, charcoal to coke was the primary fuel for factories (c) Different techniques for the fabrication of these materials were also discovered, and the ones that were already discovered were enhanced for more efficient production
7	After the revolution (1840–1945)	(a) Large-scale production of metals started with metals such as wrought iron, mild steel, gold, silver, brass, bronze, and other metals that are useful in buildings, weaponry, vehicles, utensils, etc. The addition of various reagents to the metals, like oxides, acids, and bases, and the removal of phosphorous were done for better performances, efficiency in production, the achievement of high-quality materials, etc (b) The development of enhanced furnaces and manufacturing units for the production of these metals was also done
8	Advanced ages (1950–present)	(a) The study of composites, plastics, metals, metalloids, non-metals, etc., for the production of electronics, energy, materials, automobiles, industries, and other applications that have leaped mankind into a more advanced age was done

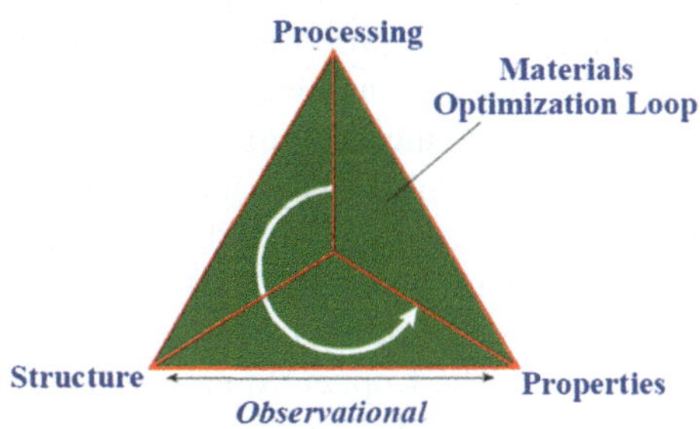

Fig. 1 Schematic illustration of the processing, structure, quality, and performance of materials science

The characteristics of materials have come a long way, increasing our knowledge of the correlation between their structure and composition [1, 3, 5]. The tremendous improvement in the strength to density ratio of the materials has allowed for the creation of many novel items, from dental materials to tennis racquets, among many others, as has been well demonstrated in the graph given in Fig. 2. From the literature, the most current chapter provides an overview of the historical development of metal working and metal engineering at a high level. Furthermore, this chapter will offer a broad view of the history of materials, starting from the Stone Age to the contemporary age of diverse materials and engineering [1, 5].

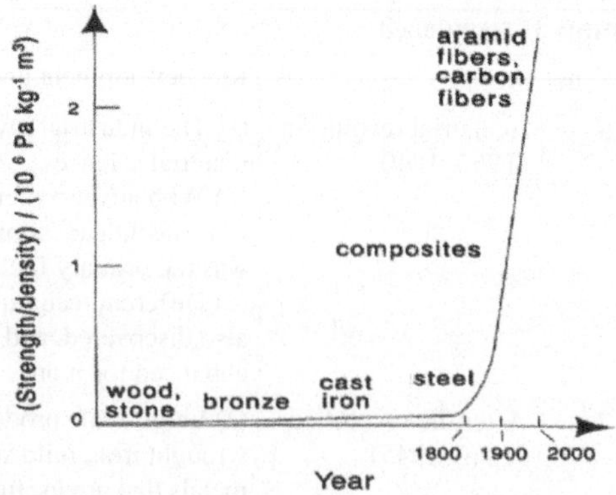

Fig. 2 Presentation of strength/density ratio with respect to the years

2 Historical Panorama in Metals and Materials

2.1 *Stone Age*

Materials have constrained mankind essentially from the very beginning of its existence. Stone and wood were predominantly used by mankind as the tools for survival, but other materials such as fibers, bone, clay, animal skin, and shells also had very specific purposes for the people of the Stone Age [1, 2, 6]. Materials in the Stone Age served specific purposes in weapons for hunting, utensils, tools, and for creating jewelry or decorations, which was the means of self-expression. A rise in human consciousness coincided with greater usage and the production of ever-more complex materials. In other words, it appears that more developed civilizations invented and utilized more complex materials, whose origins can also be found in the Stone Ages. This remark still holds true today, most likely. Historical periods have been labeled by researchers according to the materials that were most often employed during that time period. The Copper–Stone Age (also known as the Chalcolithic period), the Bronze Age, and the Iron Age are all periods that fall under this category. It is estimated that the Stone Age began around 2.5 million years ago. There are three periods that make up this era: the Paleolithic, the Mesolithic, and the Neolithic (New Stone Age). The development or vast use of materials started mostly during the Neolithic and Chalcolithic periods. So, we have focused on the discussion of these periods in the case of the Stone Age. Surprisingly, despite pottery's significant importance over long stretches of time, these classifications do not contain a "ceramics age." Certain linguistic usages have adopted the names of specific metals. For instance, the Greeks distinguished between the Golden Age and the Silver Age, the former of which was said to be characterized by peace and happiness. These disparities have more figurative connotations than literal descriptions of the materials that were employed. In particular, the human race has always placed a high value on gold. Medals are awarded in gold, silver, or bronze for exceptional performances (in sporting events, etc.). Gold, silver, and iron are used to classify specific wedding anniversaries. Up until very recently, the mastery of materials has been primarily attained by empirical methods or, at its best, through a type of alchemy. Systematic investigation only produced the interdisciplinary field of study known as materials science in the nineteenth and twentieth centuries. Later chapters will go into more detail on this and provide examples. Before materials take on their ultimate shape and designation, they frequently need to be cut, shaped, or smoothed. To accomplish this, a tool that is more powerful than the work item must be used. For example, wood can be shaped into different pieces by using flint stones that have a sharp edge. In order to obtain stool tools from big and bulky rocks, percussion flaking has been done. Striking a lump stone with another helps achieve the rocks' specific shapes as desired. It is a common practice to employ flaky edges as blades for the purpose of hunting and survival. To provide a more secure grasp and a more robust application of force while using stone tools, people throughout the world from 5,000 to 10,000 years ago began engineering these flakes of stone with wooden handles derived from vegetable resins or fibers [1, 4–8].

2.2 *Metal Ages*

It is not immediately apparent why stone-based societies finally gave way to metal-based ones. As a side note, this change did not occur in all regions of the world at the same time. If it did happen at all, the introduction of metals took place over a period of about 5,000 years and seems to have started separately in a number of different places. For instance, metals were utilized quite early on in Anatolia, which is now a part of Turkey and serves as a bridge between Asia and Europe. Once upon a time, this area was home to a sophisticated culture that cultivated grain-bearing grasses (such as wheat and barley) and raised cattle, sheep, and goats as livestock. The transition from a nomadic to a settled civilization allowed individuals to devote more time to activities other than the necessity of finding and preparing food for the next meal. So, it is not strange that people would be curious about things like copper, gold, silver, mercury, and lead concentrations in their immediate area. Metals in their pure state (not combined with other elements as they are in ores) can be worked into different shapes and hardnesses by hammering and heating, respectively. Pieces of native metals likely commanded a high price due to their limited availability and rarity. Stone was still the material of choice for most tools and weapons, since pure metals were often too pliable to be a suitable substitute. As a result, pure metals, notably copper, silver, and gold, were primarily put to use in the creation of ornaments and decorations as well as for the performance of rituals. Figure 3 displays an example of one of the very oldest copper artifacts, which is an oval-shaped pendant that is 2.3 cm in length and 0.3 cm in thickness. It was discovered in a cave in the northeastern region of Iraq. As a result, the beginning of the metal ages coincided with the beginning of the Copper Age, and the conclusion of the metal ages coincided with the end of the Iron Age [1, 9, 10].

Fig. 3 Copper pendant that was discovered in the Shanidar Cave of North-Eastern Iraq. It was thought be completely mineralized and have been built at about 9500 B.C. by carving the malachite ore and finished with abrasive [11]

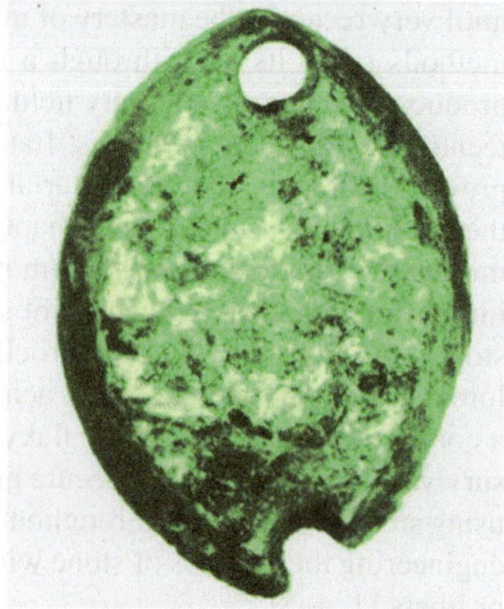

2.3 *Copper Ages*

Sometime around 9500 B.C., people probably started working with local copper by pounding it or maybe even cutting out copper ore, marking the beginning of the Copper Age. Copper was used in each of these procedures. Having metal cooking implements must have given their owner a sense of pride. In particular, copper's availability and striking appearance made a tremendous impact (especially after men learned how to smelt it). In conclusion, there was a long period of overlap between the Stone Age and the Copper Age. This is why this time period is called the Chalcolithic or the Copper–Stone Age. The earliest known usage of copper by human's dates back to the Neolithic era, around 8000 B.C., yet it will be difficult to pinpoint an exact date for this discovery. Copper tools and weaponry dating back to roughly 5000 B.C. were unearthed from ancient Egyptian tombs. The epics of Shu Ching claim that copper was in use in China much before 2500 B.C. Native copper, which was discovered in large quantities around Lake Superior in Michigan, USA, at the turn of the Common Era (100–200 AD), was likely used for decoration. Some scholars have suggested that indigenous people in the Americas were already making use of copper by 4000 B.C. Once upon a time, the native copper and other metals were likely mined to near exhaustion. This led Neolithic man to look elsewhere for metals, specifically at the minerals themselves. Malachite is a very popular copper resource. It is possible that Neolithic man encountered it in significant numbers in places like Anatolia and the Sinai Peninsula. Cyprus, which is abundant in chalcopyrite, is only one example of a potential copper ore deposit (a copper-iron sulfide). Now, after the discovery of various copper ores, the process of smelting copper ore in order to remove oxygen, sulfur, and carbon was not and is not an easy task. Extreme heat, more than the melting point of pure copper (1084 degrees Celsius), and a "reducing atmosphere" (an atmosphere devoid of oxygen and rich in carbon monoxide) are both required.

Charcoal is created by burning wood or other materials. When the moment is right and the conditions are ideal, oxygen is drawn out of the copper ore, where it interacts with carbon monoxide to generate gaseous carbon dioxide, which is then released from the ore. The last stage of the reduction process requires a fluxing agent, such as iron ore. Additionally, it aids in the last step of separating the molten copper from the slag once the melt has cooled. This takes place after the melt has cooled. Iron ore, in particular, combines with the unwanted sand particles already present in the ore. Massive amounts of heat were generated by burning charcoal, and then air was blown onto the burning charcoal either by physically pushing the air into the furnace through bellows and/or blow tubes (called tuyères) or by situating the furnace on the peak of a mountain and taking advantage of the updraft winds. To this day, nobody knows how Neolithic men could have figured out this sequence of actions without the help of initiates or at least some degree of intuition. Recent research by archeometallurgists has disproven the hypothesis that copper may have been accidentally generated in campfires whose surroundings would have consisted of rocks containing copper ore. This was done to disprove the theory that copper was created accidentally in campfires. It is impossible to melt copper in a campfire because the temperatures involved (600–700 °C) are not high enough and the reducing atmosphere does not stay long enough. However, lead may be recovered from its ores and smelted using this process due to its lower melting temperature [1, 9, 10].

The "technique" of copper smelting is now widely believed to have been adapted from the skill of pottery production, which originated around 9000 BCE (or earlier in some regions). The contention relies on the assumption that pottery production and copper smelting happened simultaneously. A fertility statue known as the "Venus of Vestonice" was unearthed in the Czech Republic. It was made of baked clay and dates back to around 23,000 B.C. It is believed that this figure was made from the first cooked clay object. However, it appears that the development of both pottery making and copper smelting occurred simultaneously. For instance, Neolithic man learned that clay bricks hardened when dried in the sun but softened when treated with rain, a second time, allowing for more flexibility. Intentionally exposing the mud bricks to the heat of a fire to hasten the drying process is probably what led to the discovery of an irreversible hardening process. At temperatures around 500 °C, clay undergoes a chemical transformation that gives it a permanent consistency and renders it impenetrable in terms of water storage. This discovery likely prompted the methodical growth of pottery making and the construction of kilns to replace traditional methods of drying clay in the open air. These two changes occurred gradually over time. Surely Neolithic man knew that a mound of wood fuel on top of shattered pottery and dirt would cause a temperature increase [1, 9, 10, 12, 13].

The oldest known account of glass fabrication was known to have occurred in Egypt and Mesopotamia during the sixteenth–fifteenth centuries B.C., but the oldest known glass was discovered in Egypt around the thirty-fifth century B.C. Royal families exhibited the ornaments and beads that were manufactured using the glass, along with a beautiful vessel made of hollow glasses. Three methods were known to produce glasses in the presence of heat: consistent hammering, polishing, drying, cleaning, etc. Most ancient Egyptian spectacles, even those of similar design, were opaque and came in a rainbow of colors. The low temperatures of the first glass-making pit kilns and the use of a wide range of admixtures—typically classified as either dyes, decolorizers, or opacifiers—were to blame. As far as can be told, the initial translucence and color palette of ancient glass were determined by random admixtures. The glassmakers did not figure out how to select a certain color until they had already amassed a great deal of experience. Dyes were made from the oxides of several metals, including manganese (Mn), copper (Cu), cobalt (Co), iron (Fe), and antimony (Sb). Moreover, lead (Pb) oxide was used. Opacifiers included copper oxide (Cu_2O), antimonous oxide (Sb_2O_3), calcium oxide (CaO), lead oxide (Pb_2O_5), and tin oxide, which made the glass behave as see-through. Tin oxide and lead oxide (PbO) were also used as opacifiers (SnO_2). As a result, the metal ages began with the Copper Age. That was back in the day when people first started working with copper to make tools, artifacts, utensils, weapons, and so on [2, 14, 15].

2.4 *Bronze Age*

Without a doubt, one of the most important discoveries ever made was that adding tin to copper while casting could provide a large amount of extra strength in the cast form without the need for additional cold working. However, it is possible that the idea developed slowly through time and that there was a time when arsenic and very small amounts of tin were used together in the Near East. Sometimes, the Bronze Age is broken up into early, middle, and late segments based on the typology of the metals used during the era, but this is far from the rule. Arsenical coppers and straight tin-bronzes, neither of which contain arsenic or lead, were the metals of choice in prehistoric Britain and Ireland. Straight tin-bronze, with tin concentrations commonly exceeding 10%, was a common alloy throughout the Middle Bronze Age (Sn). Despite the fact that the alloy often contained 10% tin, lead was first used in castings at the end of the Bronze Age. Lead's usage did not explode the moment it was introduced; in fact, it appears to have been confined to the southeast of the country in Britain. However, in other regions, lead is frequently found in bronzes without any attempt being made to date when lead was used in the creation of the bronzes. Because of this, the period is best divided into three parts: the early, exploratory age; the middle, complete Bronze Age; and the late, full Copper Age [3, 16, 17]. Clearly, the people of the Chalcolithic era knew that copper had a number of advantages over other materials like stone and organic matter that made it the material of choice for a variety of tasks. Copper's malleability and ductility made it possible to shape the metal into useful shapes, while the metal's slicability made it possible to shape it into sheets and other shapes without sacrificing its ductility. Copper was employed by Chalcolithic people because of its durability, which may be enhanced by plastic deformation during hammering. Now that molten copper can be poured into molds and allowed to cool slowly, more complex forms of copper may be made. However, Chalcolithic man probably worried at least a little bit about porosity due to surface oxidation and gases trapped during the melting and casting processes. On the other hand, cast copper tends to be somewhat pliable, making it unsuitable for use in the creation of long-lasting tools or weapons. Ultimately, it was obvious that a different strategy was needed, one that placed a premium on imagination. A different material was required. It was determined that this material was bronze. It is unclear if Chalcolithic man discovered the toughness of cast alloys could be greatly improved by adding different metals to copper through experimentation or by sheer luck. No matter the method of discovery, it is known that the hardness of the cast alloy was significantly increased (an alloy is a combination of several metals). Cast bronze is harder than pure copper without any further hammering, thus it may be used right out of the mold. As an added bonus, it is possible that Chalcolithic man knew that some copper alloys had a substantially lower melting temperature compared to pure copper (by roughly 100 °C) and that molten alloys flowed more smoothly throughout the casting process. As expected, some of the impurities in the copper ore made their way into the molten metal. Sulfur, phosphorus, and iron were among the contaminants found. Rarely detected elements included arsenic, antimony, silver, lead, iron, bismuth, and even tin. To a lesser extent than in an alloy, these impurities were not present in the final product. Except for bismuth, which may make copper brittle even at extremely low concentrations, these impurities often do not have much of an influence on copper's features. It appears that the first important and intentional addition to copper was arsenic (at least in the Middle East). However, alloying may be achieved there by melting together arsenic-containing ores and copper ores. Easily accessible copper-arsenic ores might be found there. Metal objects from the Middle East that date back to 3000 B.C. often included copper, but they also contained arsenic at levels of up to 7%. There is a compelling case to be made for highlighting a piece of archeological evidence that was discovered in the year 1961 A.D. A total of 429 artifacts, the great majority (428) of which were fashioned of a copper-arsenic alloy, were discovered in the almost inaccessible "Cave of the Treasure," which is located near the cave in which the Dead Sea Scrolls were unearthed. They probably belonged to a temple or shrine and were brought there by refugees around 3000 B.C. There were 240 intricately decorated mace heads discovered, as well as other treasures such as chisels and axes of various sizes and shapes. On the other hand, copper-arsenic alloys saw little application. There was probably an investigation into the deaths of metalworkers at some point when it was discovered that they had been caused by arsenic fumes emitted during the melting process. Finally, the ideal alloy mass percentage of 10% was reached after determining that tin was the best component to mix with copper. The copper-tin alloy in question is better known by its common name, bronze. Copper with 10% tin has a melting temperature of roughly 950 °C, while pure copper's melting point is 1084 °C. The melt can be poured into the molds without any problem, and there are no difficulties with porosity (air pockets). Most importantly, the alloy is already rather tough after casting and the subsequent cooling process, but it may be worked even further to increase its hardness by hammering. Overall, the copper-tin alloy outlasts the copper-arsenic alloy and is far less brittle. The reader is invited to weigh in on a few intriguing issues. One is the debate over whether bronze was "made" in a single location (the Middle East, as many experts once believed) or separately at a variety of places across the world. There has not been a final word on this matter yet. However, recent archeological evidence suggests that in addition to the Mediterranean region, considered by many Westerners to be the "cradle of civilization," independent bronze-producing centers existed in Northern Thailand (Ban Chiang) in the third or fourth millennium B.C. and also in the isolation of China during the Shang dynasty beginning around 1400 B.C. These geographically isolated areas seem to have experienced a transition into the Bronze Age in ways that were inconsistent with one another [1–3, 10–18]. For instance, Indo-China produced a great deal of bronze, and it is thought that many staple crops, such as rice, bananas, coconuts, yams, taro, and sugarcane, were first brought to humankind from that region. The locals built their dwellings from bamboo, made a living as potters, and raised pigs, chickens, and cows, among other domesticated animals. Bronze axes, spearheads, socket tools, bronze bracelets, clay crucibles, and sandstone molds have all been uncovered in this region, dating back to between 3000 and 2300 B.C. The most intriguing find was made when archeologists realized the people of Ban Chiang appeared to have skipped the stages of copper manufacturing and arsenical bronze, going straight to the tin-bronze period. There was no doubt that everything needed to cast bronze could be found in one place (in contrast to the Near East, as we shall elucidate below). Rich alluvial concentrations of tin and copper ores have been discovered in locations ranging from Southern China to Thailand and Indonesia. One of the most intriguing findings by archeologists is that the Thai people appeared to have existed in a "calm bronze period." This conclusion is based on the lack of discoveries of bladed weapons such as swords, battle axes, daggers, and mace heads. Rather, bronze was mostly used for decorative purposes. The fact that so many infants were buried with bronze bracelets suggests that they were not a sign of social status [18–20].

In contrast, during the Shang dynasty (1600–1122 B.C.), early Chinese bronze was mostly employed for ceremonial containers to offer food and drink to ancestral spirits. The Shang dynasty saw the construction of these ships [1–3]. Somewhere between 3 and 30% tin and 3% and 5% lead can be found in bronze (which makes the melt flow easier). Numerous animals, like elephants, water buffaloes, tigers, and even fabled dragons, are depicted in the relief patterns on bronze artifacts from the Shang era. The Chinese have a lot of experience with casting. This was accomplished by first cutting the desired patterns into already-burnt clay molds. No more forging or pounding of metal was performed. A massive cauldron that weighs 875 kg in its entirety and was cast in one piece stands as proof of their skill. (Local villagers used it in 1939 A.D. to store food for their pigs; it was discovered in Anyang.) Another recent astounding find was a set of bronze bells from 433 B.C. that were buried in a tomb belonging to the Marqui Yi. The design of these instruments allows for a lower pitch to be produced when hit in the middle as opposed to the edges. According to modern academics, the two tones represent the Chinese concept of a universe ruled by two opposing forces, Yin and Yang, who coexist in harmony (such as day and night, heaven and earth, sun and moon). Supposedly, the two-tone bells represented the harmonious coexistence of opposing forces. Music was highly valued by the Chinese of that time since it was a means by which they could express their appreciation to the generations gone before them for blessing them with health and wealth. This was not a party with music provided by bells and drums, but rather a ritual meal shared with and celebrated by their forefathers [20, 21].

The sum of the reputed archeological evidence seems to indicate that during the Chalcolithic period, there were no major known tin sources located in the Near East, with the possible exception of some native tin in the Zagros Mountains, which are situated on the eastern edge of the Mesopotamian plain. However, if they were real, they vanished almost immediately after being uncovered. Instead, documents detailing massive caravans used to transport tin have been uncovered. It is also possible that tin was transported all the way from the Middle East to the Far East by coastal ships. Recently, a shipwreck carrying tin was discovered during excavations off the coast of Israel. There was no readily available tin for experimentation or the accidental discovery of bronze; therefore, it is still unclear how the Bronze Age may have begun in the Mediterranean region about 2000 B.C. Due to the lack of evidence to the contrary, it is important to explore the possibility that bronze technology was exported from the Far East or another place to the West. Malachite and cassiterite, tin and copper ores, are found on the southern slopes of the Caucasus in what is now Armenia. One theory puts this out as a possible tin source. It is likely that tin-bronze was made when these minerals were accidentally heated together [22, 23].

Some scholars believe that trade connections linking the Middle East and Eastern Europe (where tin may be found in Bohemia, Saxony, and other locations) date back to 2500 B.C. Because of this, we may now focus on the Europeans, and more specifically, the Ueticians, who became the dominant population in Europe around 1500 B.C. [1–3]. The Unitarians were named after a small town near Prague that influenced a large area, specifically the Rhine Valley and Ukraine. The Ueticians were highly proficient bronzesmiths who mass-produced a wide range of goods. Jewelry, implements (such as axes and plowshares), weapons, and clothing pins were among the items produced. One product, a neck ring, was produced in such quantity that it was used as a kind of money, meaning that it could be traded for more valuable commodities like gold, furs, amber, and glass beads. The high volume of manufacturing made this a realistic option. However, these neck rings appeared to be very similar to those discovered in Syria [2, 3, 19, 22]. Results of unethical labor have been found in many tombs across the British Isles, Scandinavia, and Ireland, indicating that they traded with more than just the southern areas. They were original minds that developed innovative uses for existing ideas or made copies of popular items. Actually, it was the Unitarians who thought of using a safety pin.

Archeologists excavating sites linked to the Unetic civilization have found knitting needles, the remnants of a complicated loom, and a strainer used in the production of cheese [3, 4]. A Bronze Age settlement wonderfully preserved in a peat swamp near the Federsee (a lake in southern Germany) yielded a large number of bronze-metal artifacts. These included axes, chisels, spears, knives, bracelets, pins, and a chain. Aside from the region around the Indus River in northwest ancient India, this was another crucial area for the advancement of bronze technology in this part of the country. The Harappan culture, which flourished there between 7000 B.C. and 1500 B.C., was an extremely sophisticated civilization (until the Aryans invaded the land). Excavations at Mehrgarh, which took place in what is now Pakistan, indicate that the Harappans were skilled bronze artists as early as 2300 B.C. In the case of bronze, the Harappans employed lost-wax casting, annealing, and riveting. During this time period, they also developed the ability to manufacture human figures, containers, arrowheads, spearheads, knives, and axes. Evidence from bronze sickles found at the site suggests that the area was used for agricultural purposes. Huge mountains of copper slag indicate that the copper ores used in these processes came from the plains around the southern end of the Indus River (Mohenjodaro) and from areas to the northwest of the Indus valley, in what is now called Afghanistan [3, 18–22]. The copper ingots discovered in this area have ashapem like a half circle. Similar to the Near East, though, tin's history is clouded by mystery. Researchers believe the middle and Western Indian Deccan Plateau was the point of origin [24, 25]. Copper smelting occurred across the Harappan civilization, not just in urban centers. However, various Indian communities spread out over the continent have the technology to work with copper and bronze, but not always at the same degree of sophistication. Greek mythology contains a written record devoted to the metal bronze. The Greek deity of fire, Hephaestus, crafts a superior shield for Achilles out of copper, tin, silver, and gold in Homer's Iliad, which was likely written between 800 B.C. and 700 B.C.C [18, 19, 22–24].

Finally, it is theorized that the indigenous inhabitants of the central highlands of Peru used bronze technology both before and during the Inca Empire. Around 1450 A.D., or maybe even earlier, this technology is said to have first appeared. In contrast to European and Asian standards, the maximum allowable arsenic content in the products (including pins, chisels, and axes) was only 1.5%, while the allowable tin content was 3% or less. Most of the objects unearthed here lacked even trace levels of alloy components, much less enough to noticeably change their mechanical qualities. Therefore, the deliberateness of the alloying process is up for debate. Regardless, potential tin resources would have been nearby in Northern Bolivia. As opposed to popular belief, bronze technology was developed not just in the Middle East but also in other parts of Europe and Asia [1, 3]. In conclusion, several parts of the ancient globe had intricate bronze technology.

2.5 Iron Age

Historians have suggested a time range of 1500–1000 B.C. for the start of the Iron Age (at least in some parts of the world). This does not mean that people did not know about iron before that time; on the contrary, this was the case. Meteoritic iron, which includes a substantial quantity of nickel, was likely used by ancient people as early as 4000 B.C. They hammered and shaped it into various tools and weapons. It is easy to see why iron was called "metal from the skies" in so many languages as it is believed that meteoric iron was used, hence his name. Until this reason, stone, copper, and bronze were widely used in construction for at least the first two millennia BCE [3, 9, 10]. However, iron ores served a few essential functions far into the Chalcolithic period and beyond the Bronze Age. In the analysis of the manufacture of copper, fluxing agents are necessary during the smelting process for copper when malachite is present. When melting the metal, iron oxide was added to get rid of the unwanted sand particles that are found in malachite. Over time, slag formed, and as the melt cooled to the right temperature, the copper was easily separated from the slag [2, 3, 10]. Iron, however, has a high melting point of 1538 °C; hence, there has been much debate in the academic community over whether or not prehistoric humans could have obtained this metal from on-Earth resources. For the Western (or eastern) half of the planet, such a high temperature was unthinkable at that age. But the answer to how humans obtained iron is taken from the previously mentioned slag as an example, where large quantities of it have been uncovered in areas that were once home to extensive copper smelting activity [1, 3, 9, 25].

It was discovered that this slag included some reduced iron, but in a porous form now known as "sponge iron" or "bloom." Once the slag was studied, the findings became clear. Slag can be broken up and removed, and the iron can become more compacted when bloom is hammered at high temperatures for an extended period of time [3, 16, 17]. With this technique, you may get iron that is nearly pure. Modern metallurgists use the term "worked iron" to describe the final product (also written as wrought iron). It takes a temperature around a thousand degrees Celsius lower than what is needed to melt pure iron in order to reduce iron ore into spongy bloom. The use of this technique to produce iron seems like a logical beginning point for the manufacture of iron in that period. This technique reveals that the heat was never turned up high enough to melt anything. But iron in its purest form is extremely malleable, much more so than bronze. Furthermore, pure iron corrodes extremely rapidly when exposed to air with a high humidity content. This suggests that, prior to the knowledge of how to produce "good iron," as it was characterized in ancient writings, prehistoric humans probably were not too interested in the metal itself. There is widespread agreement among archeometallurgists that the Hittites, or more likely Hittite subjects (known as the Chalybes), who resided in what is now Turkey but was then known as Anatolia-Mesopotamia, were responsible for the finding of high-quality iron. The Hittites were effective colonizers; they eventually came to rule over large swaths of the Mediterranean, including Assyria, Babylon, and parts of Northern Palestine. It is generally agreed that the Hittite administrative system was more advanced than that of its neighbors and that the Hittite legal system prioritized reparation above punishment. The Indo-Germanic Hittite language was recorded using hieroglyphics or cuneiform, a syllabic writing method borrowed from the Mesopotamians. As a global people, the Hittites communicated with one another using Akkadian. Iron swords, spears, and arrows were their most effective weapons [3, 26, 27], with the mythology claiming that they could easily pierce their enemies' bronze shields.

Although the speed and agility of their nimble chariots certainly helped them win, their superior design was also likely a factor. For two centuries, from around 1400 B.C. [1, 3] to around 1200 B.C., the Hittites allegedly kept their method of making high-quality iron a secret. Hammering the bloom to remove the slag and compress it after many heating cycles in a charcoal furnace at temperatures close to 1200 °C with the goal of softening the bloom produced good iron. During the heat treatment, the bloom and, eventually, the iron were often exposed to carbon monoxide gas, which was created by the burning charcoal. According to previous statements, this technique aids in facilitating carbon diffusion into the iron's surface. This process results in an iron-carbon alloy known as steel, which, while having a carbon content of just about 0.5% (as in the preceding example with bronze, Cu-10% Sn), is far more resistant to abrasion and corrosion (steel is defined as iron that contains up to 2.11 mass percent carbon). The carbon content of steel ranged between 0.3% and 0.6% of the original steel. Cold work helps to increase strength. The inhabitants of the Iron Age must have also realized that limiting carbonization to an object's surface (like the edge of a blade or the tip of a tool) produced a balance between great surface hardness and outstanding interior ductility. Iron with a surface carbon concentration of 1.5%, the result of a process called "selective steeling," has been found as early as 1200 B.C. The manufacturing of modern case-hardened iron is quite similar to this technology. But for a long time, no one realized the role carbon played in establishing the tenacity of iron and steel. Actually, Aristotle (384–322 B.C.), an ancient Greek scholar and philosopher, was one of many who believed (incorrectly) that steel was a purer form of iron due to the "purifying impact of charcoal fire." When cast iron was dissolved in acid, a "graphite-like residue" was left behind that was not spotted until 1774 A.D. by a Swedish metallurgist named S. Rinman. A further seven years passed before Bergman and Gadolin released their results on the varied carbon levels in the various irons and steels. Two more discoveries, probably made in the first millennium BCE, improved the quality of the carbonized iron even further. One such method is called "quenching," and it involves plunging a piece of red-hot carbonized iron into a tub of ice water to cool it down quickly. To our surprise, Homer describes this procedure in great detail in The Odyssey. By increasing its strength, the material being worked on might become so tough that it cracks easily when handled. Blades of quenched swords, tools, and other equipment may have developed fractures or even shattered as a direct result of this. The second discovery was made at the end of the first millennium B.C. and entailed briefly heating the previously quenched steel to temperatures of around 600 °C. This treatment, now known as tempering, restores some of the material's ductility and reduces some of its brittleness, but at the expense of some of its hardness [25, 26].

A high-carbon percentage makes the iron in a structure very brittle. The substance easily breaks or shatters when struck, leaving it nearly unusable for tools and weapons. This means that cast iron needs further processing. The Chinese probably invented this novel treatment around 500 B.C. Surface carbon from high-carbon iron was removed as part of the process. This procedure led to the creation of a steel coat with properties on par with those of steel made in the west by carbonizing wrought iron [1–6]. To accomplish the intended outcome of decreasing the carbon content of the cast iron, the iron was heated in the presence of air at temperatures ranging from 800 to 900 °C. Carbon monoxide gas is released into the environment when oxygen from the air combines with some of the carbon already there. The Chinese and Mediterranean peoples both arrived at quite similar solutions, albeit independently. Nonetheless, they ended up at the same place in the end. While the Western world had to form and carbonize its items one at a time by hammering, the Chinese could form their products via casting, allowing for simple mass manufacturing. The ability to mass-produce their goods was, however, the greatest benefit of Chinese technology. Another major technological innovation originated in China during the first century of the Common Era (A.D.). Carbon-rich iron has to be agitated in order to stimulate a reaction between carbon in the melt and the air. In this way, the carbon content of the melt was already low enough to allow steelmaking. This technique, often known now as puddling, was found in England in 1784 A.D. Massive industrial complexes were active around Zheng-Zhou and other locations throughout the latter half of the Han period (202 B.C.–A.D. 220). Several gigantic furnaces, each measuring about 4 by 3 m in size and towering about 3 m tall, were housed in these structures. The iron from these furnaces might have weighed several tons per day. Stack casting, in which many molds are piled on top of one another, was also pioneered in China. Up to 120 distinct spells might be cast at once. Production of plowshares, hoes, cart bearings, and harness buckles was so extensive that each item could be made at a cheap per-unit cost. As a result, it is possible that agriculture improved, leading to larger production and a larger population [9, 26–28]. This was because it facilitated the widespread availability of tools essential for agricultural labor, such as plows, cultivators, and diggers. Thus, the Iron Age paved the way for the discovery of iron and steel and the mass manufacture of these materials, laying the framework for the progressive development of structure in the contemporary world, whether consciously or not. Since this era marked the beginning of the efforts to create the modern world, its time range is considered to be the basis for the modern age as a whole.

3 History of Glass Materials

Glass has been used as a separate material since around 2500 BC, mostly as beads. It may have originated in India or Mesopotamia and was then introduced to Egypt and other nations in Europe. Glassware first emerged in 1450 B.C., under the rule of Thutmose III, an Egyptian pharaoh from the eighteenth dynasty. The Crystal Palace, erected by Joseph Paxton in 1851 A.D. to hold the Great Exhibition, served as a precursor to the use of glass as a construction material. The public began using glass as a construction material for residential and horticultural architecture as a result of Paxton's groundbreaking new structure. With the assistance of renowned French glassmaker Georges Bontemps, the British Crown Glass Company (later Chance Brothers) was the first business to use the cylinder process to create sheet glass in 1832 [1, 2].

4 Brief History of Polymeric Materials

The original meaning of the term "plastic" was "pliable and readily moldable." It was only recently given a name for the class of materials known as polymers. Polymers are composed of lengthy chains of molecules, and the term "polymer" implies "of many pieces." In nature, polymers are abundant. The component of plant cell walls known as cellulose is a widely used natural polymer. John Wesley Hyatt created the first synthetic polymer in 1869 A.D. after being motivated by a New York company's $10,000 reward for anybody who could come up with an alternative to ivory. By combining cellulose, which is made from cotton fiber, and camphor, Hyatt discovered a plastic that could be molded into a variety of shapes and made to resemble natural materials such as tortoise shell, horn, linen, and ivory [29, 30].

5 Brief History of Carbon Materials

Carbon has been used since at least 3750 B.C., when the Egyptians and Sumerians reduced copper (Cu), zinc (Zn), and tin (Sn) ores to produce bronze. As early as 157 A.D. [38, 39], animal and plant-based carbons were used to treat a variety of illnesses. The absorptive capacity of carbon-derived compounds from various sources was first observed in 1773 A.D. by Car Wilhelm, a chemist from Pomerania, a region of Europe on the Baltic coast that was governed by Sweden. However, Eponit (trade name), the first activated carbon manufactured industrially, was initially marketed by the Austrian Fanto Works in 1911. The municipal water treatment sector is now the greatest market for activated carbon [31, 38, 39].

6 Brief History of Nanomaterials

One of the most intriguing instances of nanotechnology in the ancient world was presented by the Romans in the fourth century AD, who employed nanoparticles and structures. One of the most remarkable works of ancient glass art is the Lycurgus cup, which is part of the British Museum collection [32–35]. It is the first well-known instance of dichroic glass. Dichroic glass refers to two distinct glass kinds that, depending on the illumination, may change color. The Advanced Research Project Agency (ARPA) of the USA initiated the fabrication of material science during the advanced modern ages [32–35]. ARPA funded a project in material science with five esteemed universities in the USA in 1960 A.D., which led to the initiation of the formation of a new branch of science and technology called "Material Science and Engineering." As for the modern world, extensive use of nanomaterials began in that year [36, 37, 40].

7 Conclusions

The historical perspective provides an overview of how the modifications have been made to achieve the goal with improved inherent properties. However, the historical prospect of material and material engineering will provide a brief and detailed idea about the modifications that have taken place since the primitive age. This alteration of improvement is still in the process of evolving, although with the industrial revolution, there has been a significant acceleration in the development process. The "materials method of thinking," which led to the unlikely development of a generic concept of materials, emerged just recently. However, in the twentieth century, they were reimagined as a generic entity to be examined in a multidisciplinary techno-scientific paradigm rather than as separate entities that posed challenges to scientific reason in the context of the contemporary scientific paradigm. The significant position of materials has been profoundly reshaped in this historical process, where science, technology, and society are interconnected; they are no longer seen as preconditions placing restrictions and limitations on engineering, but rather as objects of design. This chapter will provide the connecting link between the past, present, and future of materials and material engineering with the knowledge of historical perspectives and statistics regarding the development of materials over the course of history.

Acknowledgements We sincerely acknowledge the use of data about historical prospects, mainly from the books titled "Understanding Materials Science: History, Properties, and Applications" and "Revisiting the History of Materials Science Glass, Glaze, and Enamel over the Millennia," along with ideas about different engineering approaches in ancient times from the different articles cited in this chapter. We are very grateful to the authors directly or indirectly cited in this chapter.

References

1. Hummel RE (1998) Understanding materials science: history, properties, applications. Springer, New York, p 407
2. Gnesin GG (2016) Revisiting the history of materials science glass, glaze, and enamel over the millennia. I. Glass. Powder Metallurgy Metal Ceramics 54(9):624–630
3. Tylecote RF (1977) A history of metallurgy. Br Corros J 12(3):137–140
4. Cahn RW (2001) The coming of materials science. New York

5. Hill JD (1994) Barry Cunliffe (ed.). The Oxford illustrated prehistory of Europe. xii+ 532 pages, 46 colour plates, 19 maps. 1994. Oxford: Oxford University Press; ISBN 0–19–814385–0 hardback£ 30. Antiquity 68(260):668–670
6. Mehl RF (1984) Brief history of the science of metals. AIME
7. Wilson A (1994) The living rock: the story of metals since earliest times and their impact on developing civilization. Woodhead Publishing
8. Lessem D (1994) The iceman. Crown, New York
9. Raymond R (1984) Out of the fiery furnace: the impact of metals on the history of mankind. Penn State Press
10. Smith CS (1977) Metallurgy as a human experience. ASM International (formerly American Society of Metals), Materials Park, OH
11. Smith CS (1977) Metallurgy as a human experience: an essay on man's relationship to his materials in science and practice throughout history. American Society for Metals
12. Harrison RJ (1980) The beaker folk: copper age archaeology in western Europe, vol 97. Thames and Hudson
13. Parr JG (1958) Man, metals, and modern magic. Published jointly by American Society for Metals [and] Iowa State College Press, Cleveland, Ohio; Ames, Iowa
14. Scheel B (1989) Egyptian metalworking and tools. Shire Publications, Aylesbury, UK
15. Spindler K (1994) The man in the ice. Harmony, New York
16. Craddock PT, Gale D (1987) Evidence for early mining and extractive metallurgy in the British Isles: problems and potentials. In: Science and archaeology Glasgow 1987. Proceedings of a conference on the application of scientific techniques to archaeology, Glasgow, pp 167–191
17. Stickland P (1975) The recovery of tin into copper by surface additions of tin-bearing minerals. Under graduate dissertation, Department of Metallurgy, Cambridge
18. Dickinson O, Dickinson OTPK (1994) The Aegean bronze age. Cambridge University Press
19. Cunliffe B (ed) (2001) The Oxford illustrated history of prehistoric Europe. Oxford Illustrated History
20. Fong W (1980) The great bronze age of China. Knopf, New York
21. Chase WT (1991) Ancient Chinese Bronze Art, China House Gallery, China Institute in America, New York
22. Higham C (1996) The bronze age of Southeast Asia. Cambridge University Press
23. Mellaart J (1966) The chalcolithic and early bronze ages in the Near East and Anatolia
24. Langmaid NG (1976) Bronze age metalwork in England and Wales, shire archaeology series. Shire Publications, Aylesbury, UK
25. Pigott VC (1992) Iron versus bronze. J Metals 42ff
26. Macqueen JG (1986) The Hittites and their contemporaries in Asia Minor, vol 83. Thames and Hudson
27. Pleiner R, Wertime TA, Muhly JD (1980) The coming of the age of iron. Yale University Press, Newhaven and London, p 40
28. Schmidt PR, Childs ST (1995) Ancient African iron production. Am Sci 83(6):524–534
29. Nicholson JL, Leighton GR (1942) Plastics come of age. Harper's Magazine, p 306
30. Freinkel S (2011) Plastics: a toxic love story. New York, Henry Holt, p 4
31. Gupta T (2018) Historical production and use of carbon materials: the activated carbon. In: Carbon. Springer, Cham, pp 47–70
32. Bayda S, Adeel M, Tuccinardi T, Cordani M, Rizzolio F (2019) The history of nanoscience and nanotechnology: from chemical–physical applications to nanomedicine. Molecules 25(1):112
33. Geiger RL (1992) Science, universities, and national defense, 1945–1970. Osiris 7:26–48
34. Choi H, Mody C (2013) From materials science to nanotechnology: Institutions, communities, and disciplines at Cornell University, 1960–2000. Hist Stud Nat Sci 43(2):121–161
35. Bensaude-Vincent B (2001) The construction of a discipline: Materials science in the United States. Hist Stud Phys Biol Sci 31.2:223–248
36. Tiwari SK, Kumar V, Huczko A, Oraon R, Adhikari AD, Nayak GC (2016) Magical allotropes of carbon: prospects and applications. Crit Rev Solid State Mater Sci 41(4):257–317

37. Tiwari SK, Mishra RK, Ha SK, Huczko A (2018) Evolution of graphene oxide and graphene: from imagination to industrialization. Chem Nano Mat 4(7):598–620
38. Inagaki M (2006) Carbon materials science and engineering: from fundamentals to applications. 清华大学出版社有限公司
39. Inagaki M, Kang F (2014) Materials science and engineering of carbon: fundamentals. Butterworth-Heinemann
40. Sudha PN, Sangeetha K, Vijayalakshmi K, Barhoum A (2018) Nanomaterials history, classification, unique properties, production and market. In: Emerging applications of nanoparticles and architecture nanostructures. Elsevier, pp 341–384

Introduction of Metal Nanoparticles, Dental Applications, and Their Effects

Md. Alamgir, Manoj Panchal, Ashis Mallick, G. C. Nayak, and Santosh Kumar Singh

Abstract Investigation and evolution in the involved sciences at the molecular or atomic level is the demand of the day under the discipline of nanoscience or nanotechnology, with massive impact on almost all regions of human health and conditioning, including various medical domains such as clinical diagnoses, pharmacological investigations, and additional resistant methods. The domain of nano-dentistry has appeared due to the assorted dental applications and consequences of nanotechnology. This chapter first outlines classifications and categories of nanomaterials established on their source, biochemical arrangement, resources, and measurements. The basic effects of value conversion at the nanoscale and the multiple improved and valuable effects of fabricated nanomaterials, such as incarceration effects, texture influences, mechanical impacts, structural developments, thermal developments, optical effects, magnetic effects, and dental application, are also described. In the final section, we looked at in vivo, in vitro, and the effects of nanomaterials.

Keywords Nanomaterials · Distinct properties · Dental Application · Excellent performance

---

Md. Alamgir (✉)
Department of Mechanical Engineering, Adwaita Mission Institute of Technology, Shivdham, Baunsi-Banka, Bihar 813104, India
e-mail: alam.jugnu@gmail.com

M. Panchal
Department of Mechanical Engineering, Rajeev Gandhi Memorial College of Engineering and Technology (Autonomous), Nandyal, Andhra Pradesh 518501, India

A. Mallick
Department of Mechanical Engineering, Indian Institute of Technology (ISM), Dhanbad, Jharkhand 826004, India

G. C. Nayak
Chemistry and Chemical Biology, Indian Institute of Technology (ISM), Dhanbad, Jharkhand 826004, India

S. Kumar Singh
Department of Mechanical Engineering, Adwaita Mission Institute of Technology, Shivdham, Baunsi-Banka, Bihar 813104, India

S. K. Tiwari et al. (eds.), *Nanoparticles Reinforced Metal Nanocomposites*,
https://doi.org/10.1007/978-981-19-9729-7_2

1 Introduction

The free technological process, which was the fate of the eighteenth century, sparked the advancement of industrial research and the achievement of newly developed materials [1]. At this point, the barriers are the miniaturization of apparatuses as well as the apparatuses themselves: lower magnitude, lower energy consumption, but excellent implementation. The ability to complete minute designs with high precision is required for advancement, as is the exploration of novel, acceptable fabrics. However, the development is not so smooth and effortless. Nanotechnology [2] is one of the most magnificent approaches developed to respond to such an essential. Recently, the investigation using nanoscale materials has gathered a significant amount of concentration from researchers. They accept nanotechnology as the innovative technology of the twenty-first century [3]. The expression “nanotechnology" comes from the Greek word “nano,” which stands for "very little,” and so it refers to materials of minute-size spectra [4, 5]. The interdisciplinary science of nanotechnology is a competent domain wherever collections of particles as well as grains have been controlled at the nanometer scale. It has the formatting of materials, elements, appliances, and procedures at near-atomic classes. Nanomaterials typically have dimensions ranging from 1 to 100 nm (nm). behavior that has been developed, determined, transformed, demonstrated, and contracted with distinct effects at the nanoscale, such as economical, solid, eco-approachable, good, stable, flimsier, and typical for a variety of definitions [6–8]. The purpose of nanotechnology has been separated into two domains: The first is the amount of material around fabricating at sizes of 1–100 nm, and the second is the near attributes of resources at the nanoscale that produce potential for their activity in unexplored applications. Because materials have fundamentally different effects than their majority peers, dimensions content that maintains an excellent arrangement of concentration is typically from 100 nm down to the atomic level [9]. The multiple significant explanations for this revolution in interpretation are the improved importance of the exterior as well as the interfacial area [10]. At this exact moment, nanotechnology is a new-fangled paradigm that is essential for reflecting on and gaining knowledge concerning biological creation, where the bottom-up technique is the control and not a peculiarity. In this unexplored method, one has to imagine periods of particles and the way they act together to construct valuable ingredients, designs, appliances, and procedures [11–13]. For a long time, nanotechnology has progressed from the laboratory environment to applications and client effects [14]. Nanotechnology will change people’s attitudes toward the environment, and their contracts have been found to provide the most sound technical as well as technical advancement in a variety of domains in separate touches: microchip technology, animation, climate, transmission, fitness, and medical care [15]. Nanotechnology also has a broad perspective in the fields of biology, pharmacy, material science, and physics, which could be combined to improve medical care. Even though the perception of nanotechnology has been examined in healthcare analysis for the previous three periods, it is still considered to remain in the early stages of growth as estimated medicinal benefits have not been apprehended [16, 17]. Both academic and industrial groups are devoting time to research the effect of nanotherapeutics on surface experiments and analyzing the theoretically demonstrated benefits of nanoparticulate techniques for dental applications. While nanotechnology is still in its early stages, it is already producing rapid results, opening up new avenues for analytical minds to apply this superior equipment for human well-being [18]. This chapter handles the up-to-date execution of nanomaterials by delivering a comprehensive assessment of the recent improvements in the nanotechnology domain. It draws awareness to the additional characterizations, large-scale types, fundamental effects, synthesis routes, dental application, and effects of nanomaterials.

2 Nanomaterials

Nanomaterials, formerly known by Paul Ehrlich as "magical bullets" [19], have been one of the principal studied materials of the century that provided delivery to a novel constituent of science known as nanotechnology [20]. Nanomaterials are chemically significant materials that are formed or used on a very small scale. The term "material" refers to an infinite number of elements that, when combined, produce an average statistical execution. As a result, the nanomaterials project is concerned with distinct interface outcomes and explains factors related to size and the determined number of constituents [21]. Nanomaterials are various types of imports that have structural components lower than 100 nm to a minimum of one extent. Nanomaterials consist of nanoparticles (NPs), which are particles, with at least two extents between almost 1 and 100 nm [22]. Novel-designed materials presented in the demand as well as the summary of the contribution of dentists to the interpretation of clinical applicability and efficiency of nano-materials is resembled those presently deployed in clinical preparations. Various levels contain disparities in confidence in describing nanomaterials [23]. To be categorized as nanomaterials, the material must be smaller than 100 nm in size in a minimum of one order. The International Standardization Organization (ISO) has described nanomaterials as "materials with any exterior nanoscale measurement or carrying the interior nanoscale texture structure" (ISO/TS 27,687 2008; ISO/TS 80,004–1 2010). The US Food and Drug Administration (USFDA) also represents nanomaterials as "materials that contain at least one dimension in the range of approximately 1 to 100 nm and demonstrate dimension-dependent phenomena." As per the European Union Commission, nanomaterials represent "a manufactured or natural material that achieves untied, aggregated, or agglomerated particles, where exterior dimensions are between 1 and 100 nm size ranges" [24]. The use of various descriptions across numerous domains is regarded as the most powerful impediment to regulatory measures because it demonstrates how to permit delay in involving regulatory strategies for imperceptible nanomaterials. As a result, the main purpose of developing a single transnational definition for nanomaterials is to persuade differing concerns.

3 Category of Nanomaterials

The broad category of nanomaterials, including organic, inorganic, and carbon-based materials, is presented in Fig. 1.

3.1 Organic Nanomaterials

Due to its strength, capability, and delivery approaches, the drunken medication procedure defines the separate kinds of benefits and effectiveness, however, of the material effects, including sizes, arrangements, and exterior morphologies. Organic nanomaterials have general use in biomedicine for targeted medicine delivery [25–29]. Such distinct elements generate an excellent prospect for medication delivery. Polymers or organic nanomaterials generally enclose dendrimers, micelles, liposomes, ferritin, etc. Several of these particles, such as liposomes and micelles, have hollow structures known as nanocapsules that are sensitive to thermal and electromagnetic radiation, including heat and rays, which is an abbreviation for biodegradable and non-toxic [30].

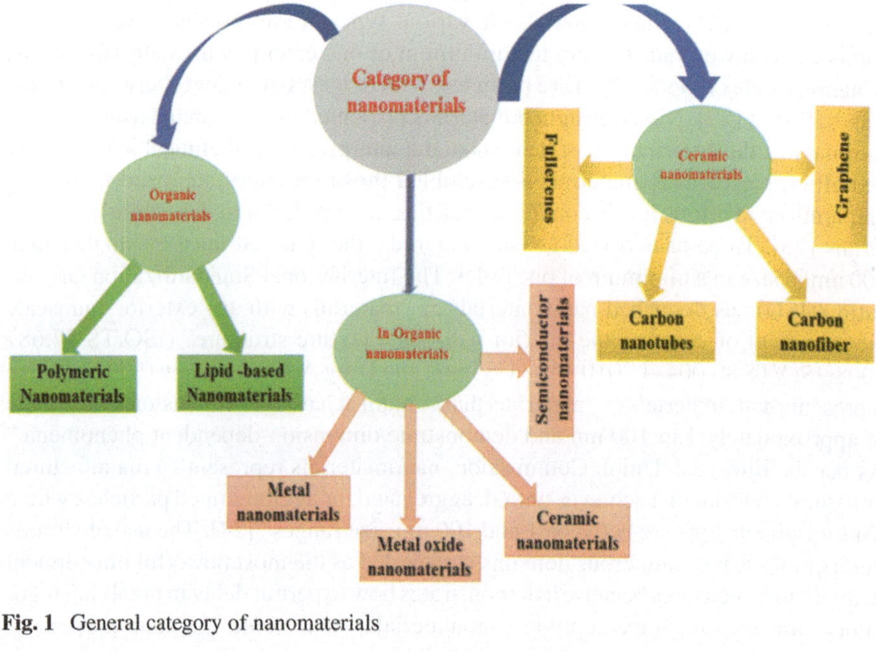

Fig. 1 General category of nanomaterials

3.1.1 Polymeric Nanomaterials

This type is usually for organic-based nanomaterials, which practically have nanosphere or nanocapsule conditions and can be smoothly functionalized. Nanospheres are matrix particles with a wide solid assembly, and additional molecules are immersed in the spheroidal surfaces' external surfaces. Nanocapsules are regular masses that have been encapsulated within a particle [31, 32].

3.1.2 Lipid-Based Nanomaterials

These nanomaterials, varying in diameter between 10 and 100 nm, maintain the fatty acid element and have useful applications in considerable bioinformatics domains. They, like polymeric nanomaterials, consist of a solid substance made of lipids and a matrix containing lipotropic molecules, though the mixing or detergent stabilizes the extreme nature. They locate various applications in medicine carriers, delivery, and RNA release for treating cancer [25].

3.2 *Inorganic Nanomaterials*

Metal and metal oxide-based nanomaterials are typically categorized below.

3.2.1 Metal Nanomaterials

Nanomaterials of decent metals, such as Au, Ag, and Cu, and alkali maintain an extensive adsorption elevation in the detectable area of the electromagnetic solar range; factors, sizes, and shape-observed metal Nanomaterials are highly valued as advanced developed materials [33–35]. These are entirely manufactured from metal prototypes. Given their standard localized exterior plasmon resonance (LSPR) effects, they want distinct optical-electrical features.

3.2.2 Metal Oxide Nanomaterials

Such nanomaterials show amazing characteristics in comparison with their metal analogs [36, 37]. Some samples are titanium oxide (TiO_2), aluminum oxide (Al_2O_3), zinc oxide (ZnO), magnetite (Fe_3O_4), cerium oxide (CeO_2), silicon dioxide (SiO_2), and iron oxide (Fe_2O_3), which are frequently synthesized oxides. Mainly, metal oxide nanomaterials are synthesized because of their higher reactivity and significance.

3.2.3 Ceramic Nanomaterials

Ceramic nanomaterials are inorganic nonmetallic solids that have been synthesized through reheating and successive freezing and exist in polycrystalline, viscous, unstructured, absorbent, or open structures with applications in catalyzed, photolyzed, and stunning applications [38, 39].

3.2.4 Semiconductor Nanomaterials

Semiconductor materials have metal and nonmetal structures, and because of their expansive hole semiconductor laser, their components have changed significantly as the hole semiconductor laser has improved. As a result, they have simply dominant materials in photonics, photocatalysis, and photoelectronic procedure [40].

3.3 Carbon Founded Nanomaterials

Carbon-based nanomaterials can be assembled along a nanometer scale within fullerenes, CNTs, GO, CNFs, black carbon, and periodically actuated carbon [41].

3.3.1 Fullerenes

C60 is a spheroidal carbon particle formed active of carbon particles linked via sp^2 implants with approximately 28–1500 carbon particles, combining globular networks with diameters of 8.2 nm for different coatings and 4–36 nm for multidimensional [42].

3.3.2 Graphene

A sheet of graphene has a consistency of about 1 nm [43, 44]. Graphene has one of the lowest concentrations of carbon. It has a hexadic grid created by a bore lattice of carbon particles on a two-dimensional surface.

3.3.3 Carbon Nanotubes

Carbon nanotubes are graphene nano-foils with a honeycomb lattice of carbon atoms that have been supported by a hollow cylinder, resulting in nanotubes as small as 0.7 nm for a one-layered carbon nanotube and 100 nm for multilayered carbon nanotubes that range in size from a few micrometers to many millimeters. The ends of nanotubes may be unfilled or assembled via a half fullerene molecule [42, 45, 46].

3.3.4 Carbon Nanofiber

Equivalent to graphene, nano-foils have been utilized for making carbon nanofibers and carbon nanotubes. They pivot into a cone structure, ideally a standard conical pipe [47].

3.3.5 Carbon Black

It has an amorphous carbon object that is usually rounded with a diameter in the range of 20–70 nm. The particles have a significant connection that leads to the binding of the wholes, so that about 500 nm-sized agglomerates are designated [48].

4 Synthesis of Nanomaterials

Two primary processes have been arranged for the synthesis of nanomaterials and have large orders in top-down and bottom-up approaches (Fig. 2).

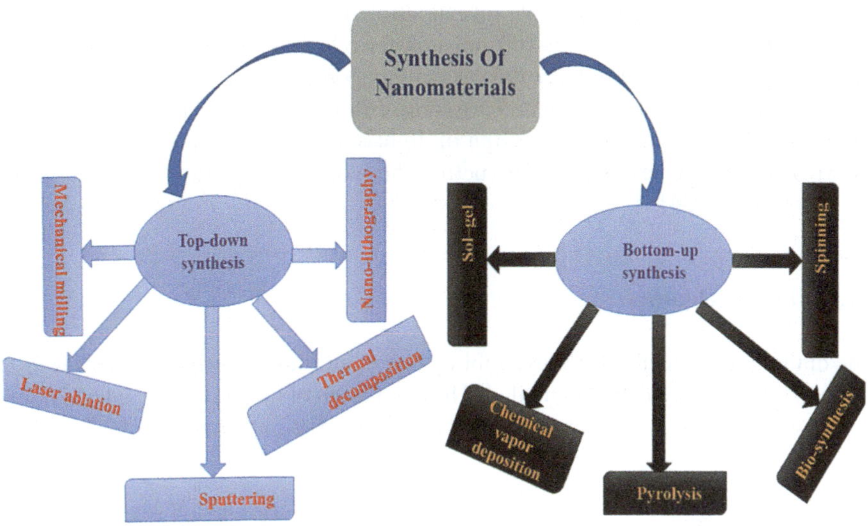

Fig. 2 Traditional synthetic approaches for nanomaterials

4.1 Top-Down Synthesis

The upper-down procedure guides the reduction of bulk materials to nanometer-gauge atoms with mechanical grinding, popping, nano-lithography, laser ablation, and thermal deterioration as the fabricated methods with wide-extent applications.

4.1.1 Mechanical Milling

Among the distinct top-down methods, mechanical milling has been one of the widely habituated approaches for developing additional nanoparticles and it has involved mill and post-anneal nanoparticles during syntheses, in which different elements are milled. Moldable deformation parameter influences the mechanical milling, which results in particle formation and fracturing resulting in the reduction in particle dimensions, and cold-welding, conducting to an addition in the particle size [49–51].

4.1.2 Nano-lithography

The fabrication of nanometer-scale investigating designs with at least one measurement in the size range of 1 to 100 nm is referred to as "nano-lithography," with additional nano-imprint mixtures such as scanning probe, optic, multi-photon, lithography, and electron light [52]. When available, lithography refers to the technique of publishing the planned form or design on light-sensitive materials, which destroys a portion of the materials for constructing the planned formation and design in a distinct way [53, 54].

4.1.3 Laser Ablation

One of the famous methods for assembling nanomaterials in distinct solvents is laser ablation synthesis, which has entailed. The irradiation of metal submerged in a liquid by a laser ray reduces a plasma plume, thus causing the nanomaterials [55].

4.1.4 Sputtering

Sputtering is the process of depositing NPs on the outside of a body by ejecting particles from it during a collision with ions [56]. The NPs' sizes and conditions are specified by the layer consistency, annealing duration, temperature, and kind of substrate [57]. It usually entails depositing a light coating of NPs followed by annealing.

4.1.5 Thermal Decomposition

The thermal decomposition methodology guides the chemical deterioration via heating, which is endothermic and cracks down the chemical adhesives in the mixtures. Nanomaterials are formed by the deterioration of metal at specific temperatures, which causes a chemical reaction, resulting in secondary goods [58, 59].

4.2 Bottom-Up Synthesis

Developing materials from particles toward clumps to nanomaterials is known as a "bottom-up" or "productive" technique with overall applications for creating nanomaterials via chemical vapor deposition (CVD), spinning, sol–gel, pyrolysis, and biosynthesis.

4.2.1 Sol–Gel

One of the overall bottom-up approaches is the sol–gel process because of its straightforwardness. Sol–gel is a wet-chemical method with a chemical resolution that has a prototype for discrete particles. The final system contains a solid and a liquid stage. A separate stage is involved for retrieving the nanomaterials via additional methods, including sedimentation, filtration, and centrifugation. Vapor is frequently further destroyed through drying [60, 61]. Afterward, the prototypes are sprinkled into a host liquid by mixing and shaking. Metal oxide and chloride are two prototypes that are usually deployed in the sol–gel technique [62–65].

4.2.2 Spinning

A spinning disk reactor (SDR) has been utilized to synthesize nanomaterials through spinning, which has involved a spinning disk within a chamber in which material variables, including temperature, may be monitored. In widespread use, nitrogen or different inert gases are filled in the reactor to extract oxygen and avoid the chemical response. The disk rotates at different speeds, during which moisture and a prototype are pumped in. The spinning technique influences the fusion of molecules or particles, which has been observed in showers, supplies, and parching. The effects of the nanomaterials synthesized from SDR are distinguished by different functional variables such as disk characters, disk wheel acceleration, liquid ratio, liquid or precursor ratio, and spread background [66–70].

4.2.3 Chemical Vapor Deposition (CVD)

The benefits of CVD include higher purity, uniformity, and hardness, with the drawback that it requires particular mechanisms and gaseous by-products that would be highly lethal [71, 72]. The response results in the appearance of a light coating of effect on the substrate's exterior, which is recyclable and reusable. The temperature of the substrate is one of the elements exploited by CVD. CVD refers to the deposit of a light coating of gaseous reactants over a substrate. The sedimentation procedure is completed at room temperature in a reaction compartment via a gas-molecule mixture in which the chemical response takes residence when a wild substrate structure with the mixed gas is present [73].

4.2.4 Pyrolysis

The moisture or fluid form of the prototype is delivered into the furnace with improved stress via a small orifice for discharge [58]. Later, the outbreak or product gases are classified in order to save the nanomaterials. Pyrolysis, the burning of a precursor with a flame, is a technique with general application in industries with large-scale production of NPs. Pyrolysis appreciates advantages such as clarity, efficiency, affordability, and a continual methodology with significant results. Several furnaces use lasers and plasma rather than flames to generate high temperatures for easy evaporation [74, 75].

4.2.5 Bio-synthesis

Bio-synthesis is an eco-friendly approach for synthesizing nanomaterials, which is rather less harmful and has the probable benefit of biodegradable materials [76]. It deploys fungi, bacteria, manufacturer extracts, and enzymes with prototypes for designing nanoparticles, preferable to conventional chemicals for bio-reduction and limiting definitions. The biosynthesized nanomaterials have specific improved biocompatibility attributes that are valuable for biomedical applications [77].

5 Nanomaterials Features

In widespread use, nanomaterial topographies are categorized into physical and chemical structures.

5.1 Physical Features

Electrical and magnetic elements, including conductivity, semiconductivity, and resistivity, provide the conditions for utilizing nanomaterials in modern electronics, thermal conductivity, and renewable power applications [33, 78]. Moreover, it concerns mechanical characteristics, including elasticity, flexibility, tensile strength, and flexibility, which contribute significantly to their application. Physical properties include optical aspects such as nanomaterial pigment, light penetration adsorption, consideration capabilities, UV adsorption, and thinking ability in a resolution or painted over a texture. Notably, several current companies use other elements such as hydrophilicity, hydrophobicity, suspension, dispersion, and determining effects.

5.2 Chemical Features

Chemical features affect the reactivity of the nanomaterials with their mark, strength, and sensitivity to variables such as sunlight, atmosphere, humidity, and heat that specify the applications of nanomaterials. Corrosive, anti-corrosive, oxidation, degradation, and flammability properties of nanomaterials distinguish their applications [79, 80]. Antibacterial, anti-fungal, disinfection, and toxicity are excellent nanomaterial attributes for biomedical and environmental uses.

6 Characterizing Nanomaterials

Different description techniques have been designed, i.e., infrared spectroscopy (IR), X-ray diffraction (XRD), X-ray photoelectron spectroscopy (XPS), transmission electron microscopy (TEM), field emission scanning electron microscopy (FESEM), Brunauer–Emmett–Teller (BET), and particle size analyses, to explore distinct physicochemical characteristics of nanomaterials.

6.1 Morphological Properties

FESEM is based on the electron scanning direction, which gives details regarding the materials at the nanoscale [81]. Different characterization techniques have been offered for morphological studies; yet, microscopic methods, including FESEM, polarized optical microscopy (POM), and TEM, are the most famous procedures. The morphological effects of nanomaterials have always been particularly assumed because morphology always involves a maturity of the nanomaterials' characteristics.

6.2 Structural Properties

The structural characteristics are important for analyzing the composition and character of the crucial materials. XRD, EDX, XPS, Raman, IR, BET, and zeta potential and size investigation are the dominant systems utilized for analyzing the structural characteristics of nanomaterials [82].

6.3 Particle Size and Surface Area

It has the potential to operate different techniques for evaluating the size of the nanomaterials, such as TEM, XRD, and FESEM, while the zeta possibility and size estimation by dynamic light scattering (DLS) may be involved in discovering the sizes of excessively minor nanomaterials [83].

6.4 Optical Features

The optical characteristics are essential in photocatalytic applications, and the details of the instruments can be influenced by photochemical methods. Such characteristics are consistent with the well-known Beer-Lambert's law and the essential light regulations. Such approaches feed an understanding of the luminescence, absorption, reflectance, and glowing characteristics of the nanomaterials [84].

7 Applications of Nanomaterials

7.1 Dental Applications of Nanomaterials

An investigation has revealed that different types of nanomaterials imitate the host tissue parts [85, 86], though the facts of such characteristics among dental residents are not known. All of the overhead difficulties may be addressed via restorative interventions and the application of biocompatible manufactured materials. Nanomedicines involved as dental materials maintain specific physicochemical and physical characteristics, which cause them to be outstanding for devastating flank outcomes connected to better traditional dental treatments [87]. Therefore, the current review concentrates on the features of different metal and polymer-based nanomaterials utilized in adhesive and beneficial dentistry, acrylic resins, periodontology, tissue engineering, endodontics, and implant dentistry [88–90] (Fig. 3). Injured dental tissues may result in dental caries, periodontal infections, tooth sensitivity, undesirable puffiness, and oral precancerous and cancerous diseases. Teeth perform in the

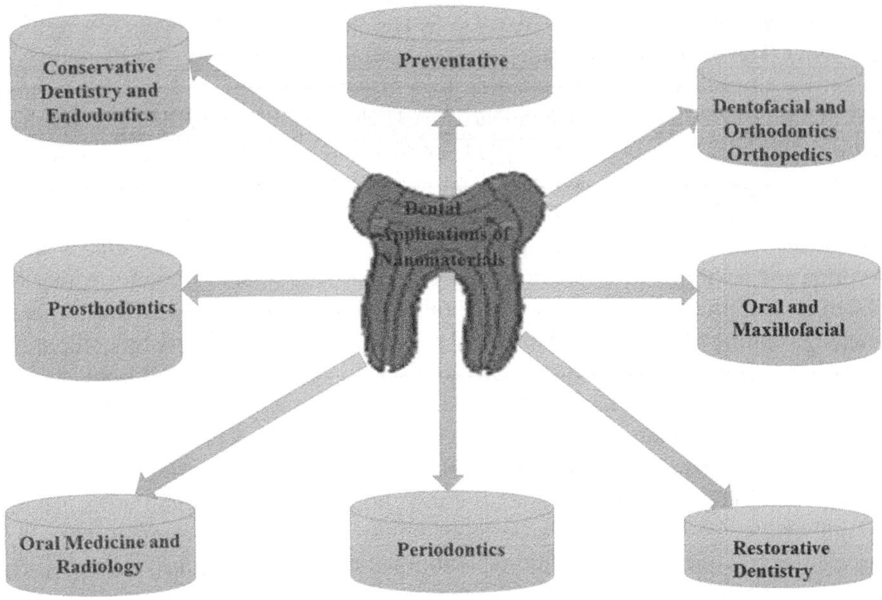

Fig. 3 Dental applications of nanomaterials

active atmosphere of the oral hollow, wherein it has a great challenge to control tooth erosion. Because of the collected details ground on oral infections, preventative dentistry has a crucial and recreates a meaningful role. Nanomaterials are used in preventative dentistry, such as working biofilms at the character of teeth with nano-apatites and demineralizing the early phase of submicron-sized enamel lesions [91, 92]. Periodontal conditions involving the hard and soft tissues near the teeth lead to gum infections, bone failure, and, in extreme situations, tooth failure. Periodontics is the extension of dental science that focuses solely on the tooth-supporting systems such as the gingiva, the periodontal ligament, the cementum, the alveolar bone, and the mucosa, as well as the illnesses that influence them. Diverse medicines are available for the management of periodontal conditions; some of them contain both restorative therapy and surgical interventions. Medicine molecules in medicinal medicines are intrinsically macro-sized particles that seek to penetrate periodontal pockets. Contrarily, the nanoscale measurements of nanoparticles make it easier for them to reach subgingival provinces. For example, medication delivery via nanoparticle inclusion for periodontal medicine therapy with TCS has been used as a suitable sample [93, 94]. It has been employed by inserting a patch within the periodontal pocket, which releases the medication into the manufactured location in a sustained manner [95]. In another investigation, nanoparticles of carbonate and apatite crystals exhibited sufficient sealing of dentinal tubules, which is believed to be important for long-term therapy of dentinal hypersensitivity [96]. Tetracycline nanoparticles (Tet NPs) have been examined for periodontal treatment as well. The commercially available tetracycline-packed microsphere patch was marketed as

Arestin® (Valeant, Bridgewater, MA, USA) and Nanogen® (Orthogen, Springfield, IL, USA). The medicine and management of intraosseous periodontal defects were recently evaluated with nanocrystalline hydroxyapatite (HA), and it demonstrated approving effects [97]. Metal nanoparticles such as hydroxyapatite NPs and titanium oxide NPs have better biological approval than conventional metals for the imitation of prosthodontic dentures. The expansion of nanofillers in polymethylmethacrylate (PMMA) in prosthodontics exhibits a significant increase in transverse resilience, suitable biological compatibility, and exterior hardness and reduces water sorption and solubility [98, 99]. Prosthodontics has dealt with the diagnosis, therapy planning, and protection of the normal oral position of patients with clinical needs related to cutting teeth or enunciating and maxillofacial tissues, using biocompatible expedients. The latest-generation materials, acrylic resins and dental implants, for example, have pleasant effects and are available for extensive prosthodontic restorative approaches. Manufacturing dentures has employed nanoceramic materials that exhibit high resilience, color strength, and lower electrical and thermal conductivity [100]. The biological aging method, oral infections, or trauma can all influence the normal oral environment and cause tooth failure, necessitating a replacement for the regular oral environment. Multiple additional metal varieties have been used for creating a prosthodontic denture with titanium, cobalt-chromium, and molybdenum alloys [101, 102]. Their mixtures have exhibited exceptional mechanical effects and corrosion resistance to stainless steel or gold alloy [103, 104]. Developed nanomaterials exhibit superior biocompatibility and deliver outstanding outcomes over standard therapy possibilities [105]. Materials with low biocompatibility cause facial skin irritation, post-operative condition, and contusion. Localized nano-drug delivery can even assist in maintaining surrounding healthy tissues while targeting malignant tissue. Maxillofacial and oral surgery is the department of dentistry that concentrates on the therapy of oral infections with damages and deficiencies in the hard and soft tissues of the oral (mouth) and maxillofacial (jaws and face) provinces. Some oral conditions or trauma cause facial deformities and bone deficiencies that need surgical intervention to restore the normal facial characteristics. Oral bone implants or bone-forming biocompatible materials have been established to transform the facial formation of the patient. Several types of research have shown that magnetic nanoparticles could be utilized for the delivery of targeted cancer drugs. However, this therapy causes symptoms such as a burning mouth and hair loss. Nanotechnology can also help in decreasing systemic toxicity by lowering the mandated parts of anticancer medicines via specific localization and defeating cancer enclosures by nano-drug delivery [106]. They can also be used as platforms for fresh bone-building due to their ability to promote osteogenic differentiation and biomineralization in enclosures. Oral cancer has evolved into one of the most life-threatening oral conditions and has been considered the main stake to human fitness. The biggest side effect of cancer therapy is the systemic toxicity induced by chemotherapy. They can be directly injected into the tumor location tissue via an intravenous route, and because of their nanoscale size, they require a lower dosage of medications, which reduces systemic toxicity and provides the expected outcome of tumor regression via precise targeting of medicine delivery [107–109]. Endodontics is the department of dentistry associated with the biology of the regular dental pulp and the etiology and therapy of infections and harms of the dental pulp, along with associated periradicular infections. A recently reported study showed the inclusion of biopolymeric NPs in root canal disinfectants provided powerful antibacterial action [110]. Nanotechnology can recreate the important role it played in the development of developed endodontic materials in endodontic therapies. The effects of endodontic materials can be enhanced through the application of nanotechnology by the inclusion of antibacterial nanoparticles, which can control recurrent disease and defeat root canal treatments [111]. There are several types of materials demanded in endodontic therapies, such as dental amalgam, glass ionomer cement (GIC), dental composite, gutta-percha, root channel disinfectant, and sealers. In another investigation, the inclusion of QPEI (quaternary ammonium polyethyleneimine) NPs enhanced the antibacterial action of the root canal sealer against biofilms of Enterococcus faecalis strains [112]. Microorganisms in the oral cavity can cause dental caries that guide various endodontic techniques such as root canal treatment, which is one of the most common causes of deep dental caries (Table 1).

Dentofacial and Orthodontic Orthodontics is the domain of dentistry that mainly trades with the diagnosis, prevention, and modification of malpositioned teeth. The teeth should be in the ideal position for the regular process of filling the hollow, but

Table 1 Role of nanotechnology in conservative dentistry endodontics

S. no	Role of nanotechnology	Benefits	References
1	Therapy of dental hypersensitivity	Carbonate hydroxyapatite nanocrystals in toothpaste for the interception of dentinal tubules to treat dental hypersensitivity	[96]
2	Nano-loaded glass ionomer types of cement	Adequate mechanical and optical effects	[113]
3	Nanoparticle-filled restoring composite resins	Higher filler loading, better mechanical effects, glossy exterior	[113–115]
4	Nano-loaded adhesion mechanisms	More immune to degradation and adequate adhesion properties	[116]
5	Endodontic sealers	Because of their antibacterial activity, Quaternary ammonium polyethylenimine nanoparticles (QPEI NPs) are used as a filler in commercially available endodontic sealers such as Guttaflow, Epiphany, and AH Plus	[117]
6	Remineralization of tooth design	A variety of polyvinylpyrrolidone (PVP) and Amorphous calcium phosphate (ACP) nanofibers for remineralization of demineralized dentine in vitro	[118–120]

orthodontic medicine is required due to insufficient oral soundness, malocclusion, tooth filling, or reaching between teeth. The nanoparticles acted as spacers, reducing texture irregularities, and even in high-load applications, the metal nanoparticle layer served as a solid grease film, allowing for extremely low conflict and comfortable gliding of orthodontic wire over the frame [121]. Nanomaterials with developed effects are constantly being researched and commercialized [122]. Conflict and mechanical resistance between orthodontic wires and frames can be facilitated by a layer of nanoparticles. A nanolayer of antibacterial nanoparticles in orthodontic materials can control dental plaque shape near the orthodontic devices and contain dental caries associated with orthodontic therapies. Kachoei et al. demonstrated in 2006 that nano-coatings of globular metal nanoparticles reduced the conflict significance between arch wires and self-ligated stands. The use of nanotechnology for orthodontic applications is in its early stages of evolution.

7.2 Nanomaterials in Oral Medicine and Radiology

Organic and inorganic nanoparticles of silica, zirconia, HA, and titanium dioxide have been employed in oral treatment for medicinal applications [123]. Nanotechnology-based oral medicinal techniques have several advantages over the traditional process. Nanoparticles have unmatched particles and an extensive exterior area that tends to form powerful chemical or material bonds that develop mechanical and physical effects. Oral medicine is the entity of dentistry that is affected by the clinical diagnosis and non-surgical direction of non-dental pathologies involving the verbal and facial provinces. Inadequate oral health can lead to innumerable oral infections. Keeping a healthy oral situation requires an earlier and more accurate diagnosis of oral infections. Nanotechnology has given us the credentials for the greatest imaging and adequate therapy of oral infections. Nanoimaging is a unique concept for dentistry. Digital imaging is tested with nanophosphor scintillators that rapidly radiate visual dawn when revealed to actually emit a very low amount of ionizing radiation. The nanoimaging process demands an extremely intense portion of the radiation to provide high-quality images corresponding to conventional techniques, which can be very useful for dental applications [124].

7.3 Nanomaterials in Restorative Dentistry

In the last few years, resin-based dental restorative materials have made significant advances. The accumulation of nanoparticles in the dental compound resin matrix can greatly improve its mechanical effects, low polymerization shrinkage, increased abrasion resistance, and exterior hardness [125, 126]. A newly documented analysis indicated that the expansion of fluoro-aluminosilicate glass nanoparticles in GIC enhanced the mechanical and esthetic effects [113]. Newly, Nanoionomers (Ketac™

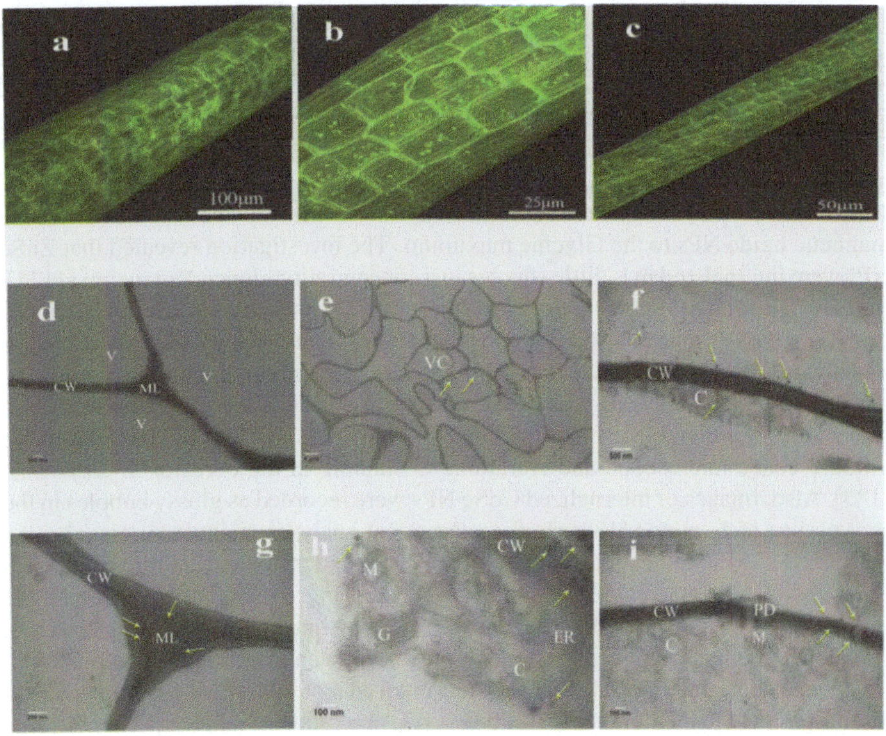

Fig. 4 Fluorescent diagrams of *L. minor* origins. **a** Control, **b**, **c** plants exposed to L-cysteine-capped and tannic acid-capped Se NPs, **d–i** TEM micrographs of *L. minor* root cells, **d** control **e**, **f** plants exposed to L-cysteine-covered Se NPs, and **h**, **i** plants uncovered to tannic acid plugged (Reprinted with permission from Tarrahi et al. (2021) [147])

Nano; 3M ESPE) have been sold for the clinical benefit [115]. Restorative dentistry has dealt with the diagnosis and management of illnesses connected to the teeth, and they are living systems. Advanced methods are requirements for the restoration and relief of injured tooth configurations, the restoration of tooth position, and adequate esthetics [127]. Nanotechnology has aided in the development of biocompatible, non-toxic dental restorative materials such as GIC, dental composite, dental implants, and endodontic materials [114] (Figs. 4 and 5).

7.4 *In Vivo* Toxicological Investigations

Translocation in the plants can be branded by the material properties of NPs, and the opening can be described by the structural characteristics of TiO_2 NPs [128, 129]. In additional research, TiO_2 NPs were found on *S. polyrrhiza* as brilliant spots in the tissues using fluorescence microscopy [130]. On the other hand, opposing statements have been documented [131] and demonstrated that TiO_2 NPs were not toxic to *L. minor* and did not penetrate the chambers, though connected to the cell partition. TEM imaging of ZnSe and CdSe NPs documented their internalization and localization into the enclosure wall, cytoplasm, and multivesicular structures of *L. minor* with the succeeding collapse of organelles (Fig. 5). The same damages, as mentioned for *L. minor* and by Se NPs, have been revealed in these analyses [132, 133] (Fig. 4). Fluorescence microscope analysis confirms the diffusion of awesome magnetic oxide NPs to the Glycine maximum. The investigation revealed that ZnSe NPs were internalized in L. little sources in a concentration-dependent manner [134]. Endocytosis is an available method for absorbing NPs as well as pores and plasmodesmata, which extend ways for NPs to penetrate plants. Larger NPs infiltrate via hydathodes, bloom stigmas, and stomata. Santos et al. reported that mercaptopropionic acid-coated CdSe/ZnS QDs were current in cell recess cultures of Medicago sativa [132]. Further, black holes of CuO NPs were marked in an analysis employing a fluorescence microscope and Auramine O staining in the cores of *S. polyrrhiza* [135]. Also, images of internalized CdSe NPs were recorded as glossy bubbles in the root tissues of *L. minor* [133]. In the subsequent analyses, manufacturers revealed that Se NPs exhibited encouraging ignorant scars under fluorescent microscopy; in other words, NPs were evident in root tissues [136] (Fig. 4). Moreover, a comparable influence has been observed in soybeans affected by magnetite nanoparticles [137]. In the microscopic specimens, nanomaterials have been traced inside the manufacturer's enclosures in different analyses. It is cracked that saturation has been done via a semi-permeable membrane, vessel pressures, or a crude trend [138]. Therefore, *S. polyrhizal* roots revealed to be L-cysteine-limited CdS nanoparticles exhibited identical areas under fluorescence microscopy [139]. The endoplasmic reticulum (ER) and mitochondria have been destroyed. The effects revealed that NPs have infiltrated the cell wall and are carried throughout the cytoplasm. Moreover, cellular remains (disrupted regions) have been detected to be affected by the impact of Se NPs on the organelles (Fig. 5). The existence of ZnO NPs in the core cells of Fagopyrum esculentum as well as the collapse of organelles induced by NPs poisoning are reported in an investigation [140]. The extension of NPs to the exterior of enclosures can generate a shading development that can be described by a decrease in light absorption, thus reducing the rate of photosynthesis. Attaching NPs can also guide material collapses to partitions and nutrient impairment, controlling evolution and expansion [141]. It is proven that ER stress is interconnected with NP cytotoxicity, which eventually guides cells to apoptosis [142]. Moreover, glowing areas of mercaptopropionic acid-coated CdSe/ZNS QDs, as well as Pt NPs, were observed in the cytosol, cell division, and organelles of Medicago sativa. In a work by Tarrahi et al., L-cysteine and tannic acid-checked Se NPs have been shown inside the enclosure wall and cytoplasm of *L. minor*. In other investigations, the internalization of Au and CeO_2 NPs into the core enclosures of *Arabidopsis thaliana* and cotton has been proved by TEM [143, 144]. Moreover, CdSe NPs were evaluated by SEM images in *Chlorella vulgaris* (algae) and demonstrated substantial alterations in the cell morphology in comparison with the sound structure of the algae (prior to the treatment); deformation of the cell exterior of the alga was evident behind the revealing of NPs [145]. One instance is related

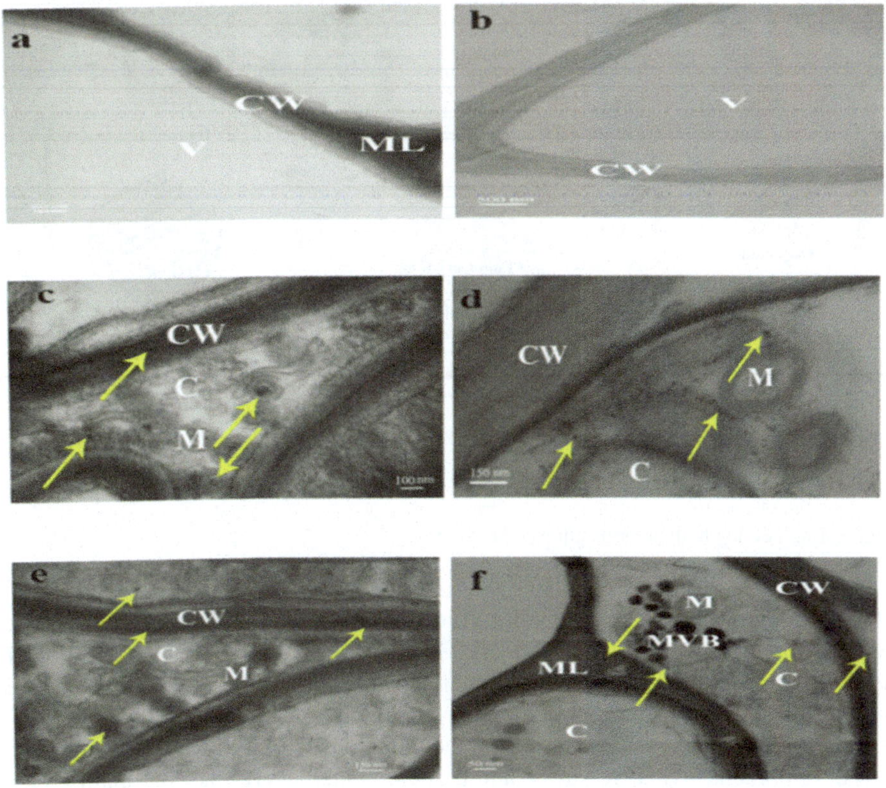

Fig. 5 TEM micrographs of *L. minor* root cells. **a** Control, **b–d** Cells treated with ZnSe NPs. CR: cell remnants (reprinted with permission from Tarrahi et al. (2021) [147])

to Ag NPs that generated different ER stress features in human THP-1 monocytes and a succeeding fast ER stress response [146].

7.5 *Tissue and Cell Culture Nano Toxicological Studies*

The addition of NPs to culture medium has effectively led to a regime of microbial pollutants in plant tissue cultures [148, 149]. They can be an effective additional sample procedure that contributes valuable information to whole-plant investigations. Although nanotoxicology has been broadly analyzed using various entire plants, in vitro cultures have also been used to study the intrinsic genetic and metabolic abilities of plant cells in reaction to NPs. The effects showed the negative effect of modified NPs on metabolism and caused the display of protective blends [150]. Tissue culture is an essential technique involved in different areas of plant biology [151]. Tobacco BY-2 cell culture has been applied as a plant cell model, comparable to the

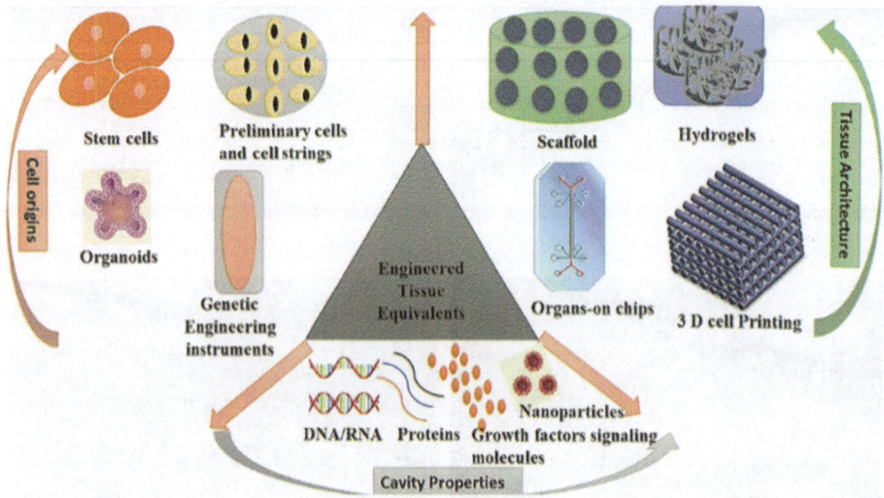

Fig. 6 Engineering human tissue equivalents in vitro

HeLa cells employed for creature cell cultures, to explore the effects of NPs on plants. All medicines except for whole NPs and Fe_2O_3-NH_2 exhibited growth reduction at a lower concentration as compared to maintenance. Revised magnetic NPs caused significant changes in protein quantity, phytochelatin scopes, and glutathione S-transferase activity. Several assay types, including the exhibit of ROS, cell viability, cell development, oxidative stress, and particle uptake assays, are the main issues in such experiments. In one investigation, tobacco BY-2 cells were exposed to 10, 50, and 250 μg mL^{-1} of cerium oxide NPs for 24 h. The results demonstrated a concentration-dependent addition of Ca^{2+} and ROS at all cerium oxide NP therapies. Exposure of BY-2 cells to magnetic NPs (γ-Fe_2O_3) demonstrated no significant impact on cell viability, but the application of modified magnetic NPs (γ-Fe_2O_3-OH and Fe_2O_3-NH_2) led to a significant reduction of cell viability at all concentrations (1, 10, and 100 ng mL^{-1}) after 120 h of cure (Fig. 6).

However, at concentrations of 50 and 250 g mL^{-1}, significant DNA damage and changes in antioxidant defense systems were clearly discernible. Moreover, the effects of this investigation have indicated an unconventional autophagy-mediated, antioxidative activity and the geno-protective prospect of cerium oxide NPs [152]. In addition, a cell biology technique utilizing tobacco cell suspension cultures revealed the various toxic products of chemically and biologically synthesized CuO NPs. According to the physiological studies, chemically synthesized CuO NPs were more toxic than biologically synthesized CuO NPs [153]. But despite the favorable effects, plant tissue culture-based studies have demonstrated that NPs have potentially toxic impacts on plants. In vitro investigations may conceal the complexity of the natural conditions or organismal methods; however, they supply a steady situation to investigate the toxicity mechanisms at a molecular level, which is difficult to perform in in vivo analyses. Nanoparticles have been widely used for distinct objectives such as improving minor metabolite exhibit, callus installation, organogenesis, somatic embryogenesis, somaclonal deviation, and genetic modifications [154]. Several reports indicate the positive results of nanoparticle application in plant tissue cultures. Tissue cultures such as calluses, cell suspensions, and hairy roots propose a mixture of experimental benefits, including practical convenience and speed, besides the results of microflora and translocation barriers compared with whole-plant systems [155]. The treatment of tobacco factory cells with water-soluble carboxy fullerenes demonstrated their existence on the cell wall, which disrupted the cell wall and membrane and, as a result, prevented cell development. Furthermore, a glycosyl peak persists after cell division, and ROS elevation in the treated cells suggested a possible protection mechanism for the manufacturer cells against fullerene stress [156]. Differences in multiple oxidative stress biomarkers, including a significant decrease in cell development and viability, a decrease in H_2O_2 scopes, an increase in ROS scavenging enzyme activities, and a decrease in phenolic and flavonoid compound scopes, confirmed that Ag NPs caused oxidative stress in N cell culture tabacum [157]. Analyzing the cytotoxic impact of SnO_2 and Ag/SnO_2 NPs in tobacco cell cultures as a prototype design explained the significance of structure modification in NPs' toxic effects. SnO_2 NPs were not toxic, but Ag-doping was effective in producing toxicity in tobacco cells via oxidative stress. Microscopic examination with a neutral red (NR) vital stain not only revealed cell death in medicines with high levels of SnO_2 NPs (0.5 mg mL^{-1}) and precise low concentrations of Ag/SnO_2 NPs (0.2 mg mL^{-1}), but also revealed a high concentration of NR in the vacuole of tobacco cells, revealing the stress caused by NPs and the acidification of vacuolar pH [158]. Arabidopsis break cultures have also been utilized as prototype cells to explore the potential toxicities of some NMs. The temperamental reaction was considered a potential toxic agent caused by the MWCNTs [159]. The toxic impacts were severely increased by lowering the size of the agglomerates of the MWCNTs. MWCNTs induced an increase in ROS, followed by a reduction in cell viability in Oryza sativa cell break cultures. The addition of mercaptopropionic acid-coated CdSe/ZnS quantum dots (QDs) to the broken culture of Medicago sativa demonstrated a substantial decrease in cell growth at a concentration of 100 mM after 72 h. The accumulation of QDs in the cytoplasm and middle of the cells resulted in the generation of ROS that was dose- and time-dependent [132]. The toxicity of graphene (0–80 mg^{-1}) on *A. thaliana* (Columbia ecotype) T87 cells was documented as employing fragmented nuclei, membrane deterioration, and mitochondrial dysfunction. Graphene led to a 3.3-fold increase in ROS, which suggests that ROS are the main mediators in the cell death signaling pathway. Moreover, the migration of graphene into the cells and an endocytosis-like system were detected, which proposed graphene entry into the cells by endocytosis [160]. In the case of the tolerant combination, only the non-enzymatic antioxidant approach was stimulated, while in the exposed combination, both enzymatic and non-enzymatic pathways were stimulated for scavenging ROS [161]. Based on the data of cell dryness importance, cell viabilities, cell chlorophyll amounts, and SOD activities, the agglomerates of multi-walled carbon nanotubes (MWCNTs) were discovered to be toxic to the Arabidopsis T87 cells (Fig. 6). The endocytosis-like pathway was hypothesized as a means of cellular uptake for these NPs. The expansion of 20-nm gold and silver NPs to an increased medium of a cell suspension culture of *A. thaliana* (L.) Heynh was accompanied by an increase in the intracellular free amino acid pool (alanine, -aminobutyric acid, and valine), which is a factor in plant cell abiotic pressure reactions. Likewise, the NPS direction altered the design of extracellular proteins in the cell culture [162]. The authors hypothesized that the presence of person nanotubes at the cell wall could trigger a hypersensitive response signaling cascade, resulting in the production of ROS [163]. Multifunctional silica NPs (CMB@SiO_2, with a middle diameter of 47 nm) and SiO_2 NPs (29 nm) had no harmful impact on A. thaliana cell break cultures, despite the considerable toxicity of molybdenum octahedral group bromide (CMB, 1 nm). Exposure of silica NPs (19 nm, 48 nm, and 202 nm) in rice cell cultures revealed that Cd toxicity was noticeably reduced by reducing the size of silica NPs. Silica NPs induced an expansion in the Si uptake ability and a reduction in the Cd uptake ability, which led to the comfort of the toxicity of Cd in the cells [164]. Internalization of gold and silver NPs in *Linum usitatissimum* L. calli and regenerants has been demonstrated. It was suggested that the addition of the vacuum-deposited metal nanolayers on the powdered hormone in the culture medium could be applied for the delivery of metal NPs to plant cells [165]. Impact of Ag NPs and $AgNO_3$ on MDA and H_2O_2 scopes of potato in vitro cultures revealed that the phytotoxicity of Ag NPs was more important than $AgNO_3$, and the toxicity was probably due to the induction of oxidative stress [166]. Also, according to the altered actions of antioxidant enzymes and the deduction of glutathione and ascorbate in potato shoot cultures, it was supposed that Ag NPs included a more harmful development in comparison with the equivalent quantity of Ag ions [167]. Different types of cell cultures have also been utilized to estimate the phytotoxicity of NMs. A study of the interactions of fluorescein isothiocyanate-labeled mesoporous silica NPs (MSNs) with Liriodendron hybrid break cells revealed that MSNs (5–15 nm) were internalized by the plant cells via endocytosis. The cytotoxic and genotoxic outcomes of mercaptopropionic acid-coated CdSe/ZnS QDs (10, 50, and 100 nM) were displayed in *M. sativa* cells by the activation of both the DNA restoration genes and the ROS-releasing enzymes [168]. Because of the outstanding biocompatibility of MSNs, these NPs are suggested to be possible nano-carriers for walled plant cells [169]. The exposure of *Corylus avellana* L. cells to the various attentions of Ag NPs (0, 2.5, 5, and 10 ppm) for 1 week reduced the cell viability and improved the movement of some extreme scavenging enzymes. Again, the amounts of H_2O_2, taxol, and phenolic combinations, as well as the action of phenylalanine ammonia-lyase, were increased [170]. It has been stated that both Ag NPs and silver ions demonstrated similar toxicity impacts on the stress-tolerant and stress-sensitive types of wheat. The obtained results described nanosilver toxicity mechanisms as morphology disorder, membrane damage, oxidative stress, and an increase in non-enzymatic antioxidant production. Callus culture, an undifferentiated tissue, has been utilized as an essential mechanism in plant biotechnology. Investigators have used callus cultures to explore the effect of NMs on plants. Lower groups of CuO NPs (1 and 15 mg L^{-1}) showed poor renewal in Basmati 370 [171]. In vanilla node cultures, Ag NPs induce growth stimulation at 25 and 50 mg/L and excellent deterrence at 100 and 200 mg/L. Production of ROS and mineral nutrition were identified as the main means of AgNPs-induced hormesis for vanilla in vitro cultures [172]. However, the high concentration (500 mg/L) reduced the growth and induced the production of the highest amounts of H_2O_2, with notably more polyphenol oxidase (PPO), peroxidase (POD), and secondary metabolic activities. Interaction of MWCNTs with plant cells can activate specific enzymes such as POD and NADPH oxidases, resulting in the production of ROS. The production of ethylene and jasmonic acid are signaling pathways provoked by carbon nanomaterials. The response of plants to carbon nanomaterials is similar to that of biotic elicitors such as pathogens or herbivore attacks. The biosynthesized CuO NPs (1–20 mg/L) utilizing the Azadirachta indica leaf section revealed a favorable influence on the callogenesis and resurrection of O. sativa L. According to the effects, the best renewal was noticed in Basmati 385 (92%) at a high concentration of 20 mg/L, Basmati 2000 (80%), Super Basmati (52%), and Basmati 370 (32%), respectively. The generated ROS oxidize polyunsaturated fatty acids (PUFA) to polyunsaturated fatty acid hydroperoxide (PUFA-OOH), which are transformed into oxylipins, resulting in the upregulation of genes incorporated in the production of secondary metabolites [173]. The growth of Satureja khuzestanica Jamzad calli was significantly aided by increasing CNT concentrations up to 50 mg L^{-1}.

8 Conclusion

Nanomaterials attach directly to natural, random, or fabricated materials, including particles in unbound, agglomerated, or aggregated conditions. They are materials with essential structural units, particles, fibers, or other essential features smaller than 100 nm in at least one dimension. Nanotechnology can be characterized as the performance, control, and manipulation of materials, having measurements around the 1–100 nm range, where traditional physics breaks down. It has been seen that nanomaterials are diverse from their bulk moieties and cannot be studied as precisely as bulk or small molecules because of their distinct effects at the nanoscale. The properties of nanomaterials depend upon their design, chemistry, particle dimension, and relationships with different materials. The manufacturing of nanomaterials is accomplished mostly via two methods specified as "top-down" and "bottom-up" processes. Researchers consider nanotechnology to be the innovatory technology of the twenty-first century. Nanoscience and nanotechnology can deal with a lot of the universal challenges confronting the community today and enhance the quality of life. The device of nanotechnology continues to make momentous endowments to inventiveness, dental applications, and advantageous effects across wide areas. In fact, nanotechnology aims to develop novel functional smart materials and apparatuses with a broad field of devices, and it is important to put focus on the emergence of recent topics like nanoenergy, nanomedicine, nanoelectronics, and nanofood. The first way stands for cracking down the bulk material into fewer and fewer dimensions, while the second one is based on reducing the small clusters.

Acknowledgements The authors are thankful to Adwaita Mission Institute of Technology, Shivdham, Baunsi-Banka, Bihar, for their financial support.

References

1. Fajardo AR, Pereira AG, Muniz EC (2015) Hydrogels nanocomposites based on crystals, whiskers and fibrils derived from biopolymers. InEco-friendly Polymer Nanocomposites. Springer, New Delhi, pp 43–71
2. Huyen DN (2011) Carbon nanotubes and semiconducting polymer nanocomposites. In: Carbon nanotubes-synthesis, characterization, applications. IntechOpen
3. Royal Society & The Royal Academy of Engineering (2004) Nanoscience and nanotechnologies. https://royalsociety.org/~/media/Royal_Society_Content/policy/publications/2004/9693.pdf. Accessed 18 Aug 2019
4. Nikalje AP (2015) Nanotechnology and its applications in medicine. Med Chem 5:81–89
5. Rai M, Yadav A, Gade A (2008) Current trends in phytosynthesis of metal nanoparticles. Crit Rev Biotechnol 28(4):277–284
6. Asmatulu R, Asmatulu E, Zhang B (2010) Nanotechnology and nanoethics in engineering education. In: Proceedings of the 2010 Midwest section conference of the American Society for Engineering Education. Lawrence, KS, pp 1–11
7. Taniguchi T (1974) On the basic concept of nanotechnology. In: Proceedings of the international conference of production engineering, Tokyo, part II. Japan Society of precision Engineering, pp 18–23
8. Pradeep T (2007) Nano: the essentials—understanding nanoscience and nanotechnology. Tata McGraw- Hill Publishing, New Delhi, pp 3–15
9. Patel JK, Patel A, Bhatia D (2021) Introduction to nanomaterials and nanotechnology. In: Emerging technologies for nanoparticle manufacturing. Springer, Cham, pp 3–23
10. Wardak A, Gorman ME, Swami N, Deshpande S (2008) Identification of risks in the life cycle of nanotechnology—based products. J Ind Ecol 12:435–448
11. Raza H, Raza TZ (2013) Introducing nanoengineering and nanotechnology to the first year students through an interactive seminar course. J Nano Educ. 4:41–46
12. Rocco MC (2007) National nanotechnology initiative—past, present, future. In: Goddard WA, Brenner DW, Lyshevski SE, Iafrate GJ (eds) Handbook on nanoscience, engineering and technology, 2nd edn. Taylor and Francis/CRC Press, Boca Raton, pp 3.1–3.26
13. Rocco MC, Mirkin CA, Hersam MC (2011) Nanotechnology research directions for societal needs in 2020: retrospective and outlook. Springer, Dordrecht
14. Barakat N, Jiao H (2011) Nanotechnology integration to enhance undergraduate engineering education. In: Bernardino J, Quadrado JC (eds) Proceedings of the SEFI annual conference. 1st world engineering education flash week. Lisbon, Portugal, pp 623–630
15. Daryoush B, Darvish A (2013) A case study and review of nanotechnology and nanomaterials in green architecture. Res J Environ Earth Sci. 5:78–84
16. Miyazaki K, Islam N (2007) Nanotechnology systems of innovation—an analysis of industry and academia research activities. Technovation 27(11):661–675
17. Sandhiya S, Dkhar SA, Surendiran A (2009) Emerging trends of nanomedicine an overview. Fundam Clin Pharmacol 23(3):263–269
18. Daniel MC, Astruc D (2004) Gold nanoparticles: assembly, supramolecular chemistry, quantum size related properties and applications towards biology catalysis and nanotechnology. Chem Rev 104(1):293–346
19. Kreuter J (2007) Nanoparticles-a historical perspective. Int J Pharm 331:1–10
20. Nasir Khan M, Mobin M, Abbas ZK, AlMutairi KA, Siddiqui ZH (2017) Role of nanomaterials in plants under challenging environments. Plant Physiol Biochem 110:194–209

21. Guo D, Xie G, Luo J (2014) Mechanical properties of nanoparticles: basics and applications. J Phys D Appl Phys 47:1–25
22. Klaine SJ, Alvarez PJJ, Batley GE, Fernandes TE, Hand RD, Lyon DY et al (2008) Nanoparticles in the environment: behavior, fate, bioavailability and effects. Environ Toxicol Chem 27(9):1825–1851
23. Boverhof DR, Bramante CM, Butala JH, Clancy SF, Lafranconi M, West J et al (2015) Comparative assessment of nanomaterial definitions and safety evaluation considerations. Regul Toxicol Pharmacol 73(1):137–145
24. Potocnik J (2011) Commission recommendation of 18 October 2011 on the definition of nanomaterials. Off J Eur Communities Legis L275:38–40
25. Palazzolo S, Bayda S, Hadla M, Caligiuri I, Corona G, Toffoli G, Rizzolio F (2018) Curr Med Chem 25:4224
26. Broza YY, Vishinkin R, Barash O, Nakhleh MK, Haick H (2018) Chem Soc Rev 47:4781
27. Yu Y, Shi Y, Zhang B (2018) Acc Chem Res 51:1711
28. Oveisi M, Asli MA, Mahmoodi NM (2018) J Hazard Mater 347:123
29. Andre RS, Sanfelice RC, Pavinatto A, Mattoso LH, Correa DS (2018) Mater Des 156:154
30. Harb ME, Ebrahim S, Soliman M, Shabana M (2018) J Electron Mater 47:353
31. Zhu J, Hou J, Zhang Y, Tian M, He T, Liu J, Chen V (2018) J Membr Sci 550:173
32. Makvandi P, Gu JT, Nazarzadeh Zare E, Ashtari B, Moeini A, Tay FR, Niu LN (2020) Acta Biomater 101:69
33. Chen Y, Fan Z, Zhang Z, Niu W, Li C, Yang N, Zhang H (2018) Chem Rev 118:6409
34. Cheng H, Yang N, Lu Q, Zhang Z, Zhang H (2018) Adv Mater 30:1707189
35. Roach KA, Stefaniak AB, Roberts JR (2019) J Immunotoxicol 16:87
36. Zhang D, Zhang C, Liu J, Chen Q, Zhu X, Liang C, Appl ACS (2018) Nano Mater 2:28
37. Aline Pires L, de Azevedo Silva LJ, Ferrairo BM, Erbereli R, Parreira Lovo JF, Ponce Gomes O, Henrique Rubo J, Lisboa-Filho PN, Griggs JA, Fortulan CA, Sanches Borges AF (2020) Dent Mater 36:38
38. Foster EJ, Zahed N, Tallon C (2018) Small 14:1870215
39. Jung DH, Sharma A, Jung JP (2018) J Alloys Compd 743:300
40. Singh M, Goyal M, Devlal K (2018) J Taibah Univ Sci 12:470
41. Cai Q, Subramani K, Mathew RT, Yang X (2019) Nanobiomaterials in clinical dentistry, 2nd edn. Elsevier, p 429
42. Clancy AJ, Bayazit MK, Hodge SA, Skipper NT, Howard CA, Shaffer MS (2018) Chem Rev 118:7363
43. Ghosal K, Sarkar K, Biomater ACS (2018) Sci Eng 4:2653
44. Islam Nizami MZ, Takashiba S, Nishina Y (2020) Appl Mater Today 19:100576
45. Rao R, Pint CL, Islam AE, Weatherup RS, Hofmann S, Meshot ER, Carpena-Nuñez J (2018) ACS Nano 12:11756
46. Laux P, Riebeling C, Booth AM, Brain JD, Brunner J, Cerrillo C, Jungnickel H (2018) Environ Sci 5:48
47. Yang B, Zhang M, Lu Z, Luo J, Song S, Zhang Q (2018) ACS Sustain Chem Eng 6:8954
48. Bott J, Franz R (2019) Appl Sci 9:214
49. Panda NR, Pati SP, Das D (2019) Appl Surf Sci 491:313
50. Khade P, Bagwaiya T, Bhattacharaya S, Singh A, Jhaand P, Shelke V (2018) AIP Conf Proc 1942, 110052
51. Bello SA, Agunsoye JO, Adebisi JA, Hassan SB (2018) Eng Appl Sci Res 45:262
52. Varghese RJ, Parani S, Thomas S, Oluwafemi OS, Wu J (2019) Nanomater Sol Cell Appl 2019:75–95
53. Marcovici A, Le Saux G, Bhingardive V, Rukenstein P, Flomin K, Shreteh K, Schvartzman M (2018) ACS Nano 12:10016
54. Bera A, Bhattacharya A, Tiwari N, Jha SN, Bhattacharyya D (2018) Surf Sci 669:145
55. Davari SA, Gottfried JL, Liu C, Ribeiro EL, Duscher G, Mukherjee D (2019) Appl Surf Sci 473:156
56. Wang X, Li H, Li M, Li C, Dai H, Yang B (2018) Diamond Relat Mater 86:179

57. Thomas N, Blake S, Morris C, Moles DR (2018) Int J Paediatr Dent 28:226
58. Shang W, Cai T, Zhang Y, Liu D, Liu S (2018) Tribol Int 118:373
59. Cazaña F, Latorre N, Tarifa P, Labarta J, Romeo E, Monzón A (2018) Catal Today 299:67
60. Budnyak T, Aminzadeh S, Pylypchuk I, Riazanova A, Tertykh V, Lindström M, Sevastyanova O (2018) Nanomater 8:950
61. Aydin C (2018) J Mater Sci 29:20087
62. Lin J, He Y, Du X, Lin Q, Yang H, Shen H (2018) Curr Comput-Aided Drug Des 8:384
63. Mao H, Li B (2018) NANO 13:1850027
64. Catauro M, Tranquillo E, Dal Poggetto G, Pasquali M, Dell'Era A, Vecchio Ciprioti S (2018) Materials 11:2364
65. Ansari F, Sobhani A, Salavati-Niasari MJ (2018) Colloid. Interface Sci 514:723
66. Molnár G, Rat S, Salmon L, Nicolazzi W, Bousseksou A (2018) Adv Mater 30:1703862
67. Salva JM, Gutierrez DD, Ching LA, Ucab PM, Cascon H, Tan NP (2018) Nanotechnol 29:50LT01
68. Náfrádi NP, Choucair M, Forró L (2018) Nanomater 1:67
69. Tian W, Liu S, Deng L, Mahmood N, Jian X (2018) Composites 149:92–98
70. Tien NA, Diem CH, Linh NTT, Mittova VO, Mittova IY (2018) Nanosystems 9:424
71. Manawi Y, Samara A, Al-Ansari T, Atieh M (2018) Materials 11:822
72. Atchudan R, Edison TNJI, Perumal S, RanjithKumar D, Lee YR (2019) Int J Hydrogen Energy 44:2349
73. Noorani B, Tabandeh F, Yazdian F, Soheili ZS, Shakibaie M, Rahmani S (2018) Int J Polym Mater Polym Biomater 64:754
74. Nizamuddin S, Siddiqui MTH, Mubarak NM, Baloch HA, Mazari SA, Tunio MM, Riaz S (2018) Curr Org Chem 22:446
75. Kalaiselvan S, Balachandran K, Karthikeyan S, Venckatesh R (2018) Silicon 10:211
76. Guo Z, Cui K, Zeng G, Wang J, Guo X (2018) Sci Environ 643:1325
77. Monteiro N, Thrivikraman G, Athirasala A, Tahayeri A, França CM, Ferracane JL, Bertassoni LE (2018) Dent Mater 34:389
78. Yao J, Wang H, Chen M, Yang M (2019) Microchim Acta 186:395
79. Zafar MS, Alnazzawi AA, Alrahabi M, Fareed MA, Najeeb S, Khurshid Z (2019) Adv Dent Biomater 18:477
80. Gao X, Lowry GV (2018) Nano Impact 9:14
81. Wang J, Chen R, Xiang L, Komarneni S (2018) Ceram Int 44:7357
82. Cui ML, Chen YS, Xie QF, Yang DP, Han MY (2019) Coord Chem Rev 387:450
83. Li Y, Xu Y, Fleischer CC, Huang J, Lin R, Yang L, Mao H (2018) J Mater Chem 6:9
84. Manera MG, Colombelli A, Taurino A, Martin AG, Rella R (2018) Sci Rep 8:12640
85. Pokrowiecki R, Pałka K, Mielczarek A (2018) Nanomedicine 13:639
86. Khurshid Z, Zafar M, Qasim S, Shahab S, Naseem M, AbuReqaiba A (2015) Materials 8:717
87. Besinis A, De Peralta T, Tredwin CJ, Handy RD (2015) ACS Nano 9:2255
88. Fukuda H (1992) Dental implant. US Pat., US5174755A
89. Feng X, Chen A, Zhang Y, Wang J, Shao L, Wei L (2015) Int J Nanomed 10:3547
90. Elkassas D, Arafa A (2017) Nanomed 13:1543
91. Hannig M, Hannig C (2019) Nanobiomater. Clin Dent 201:1
92. Goldberg M (2018) Dentistry Oral Health Care 1:19
93. Arcos D, Lopez-Noriega A, Ruiz-Hernandez E, Terasaki O, Vallet-Regi M (2009) Ordered mesoporous microspheres for bone grafting and drug delivery. Chem Mater 21(6):1000–1009
94. Piñón-Segundo E, Ganem-Quintanar A, Alonso-Pérez V, Quintanar-Guerrero D (2005) Preparation and characterization of triclosan nanoparticles for periodontal treatment. Int J Pharm 294(1–2):217–232
95. Khurshid Z, Zafar M, Qasim S, Shahab S, Naseem M, AbuReqaiba A (2015) Advances in nanotechnology for restorative dentistry. Materials 8(2):717–731
96. Vano M, Derchi G, Barone A, Covani U (2014) Effectiveness of nano-hydroxyapatite toothpaste in reducing dentin hypersensitivity: a double-blind randomized controlled trial. Quintessence Int 45(8):703–711

97. Chitsazi MT, Shirmohammadi A, Faramarzie M, Pourabbas R, Rostamzadeh AN (2011) A clinical comparison of nano- crystalline hydroxyapatite (Ostim) and autogenous bone graft in the treatment of periodontal intrabony defects. Medicina Oral, Patologia Oraly Cirugia Bucal 16:448–453
98. Jasim BS, Ismail IJ (2014) The effect of silanized alumina nanofillers addition on some physical and mechanical properties of heat cured polymethyl methacrylate denture base material. J Baghdad College Dent 26(2):18–23
99. Li Y, Zhao B, Xie S, Zhang S (2003) Synthesis and properties of poly (methyl methacrylate)/montmorillonite (PMMA/MMT) nanocomposites. Polym Int 52(6):892–898
100. Wang W, Liao S, Zhu Y, Liu M, Zhao Q, Fu Y (2015) Receptor mediated drug delivery systems targeting to glioma. J Nanomater 6(1):3
101. Tang N, Li YP, Kurosu S, Matsumoto H, Koizumi Y, Chiba A (2011) Interfacial reactions between molten Al and a Co–Cr–Mo alloy with and without oxidation treatment. Corros Sci 53(12):4324–4326
102. Tang N, Li Y, Koizumi Y, Kurosu S, Chiba A (2013) Interfacial reaction between Co–Cr–Mo alloy and liquid Al. Corros Sci 75:262–268
103. Tang N, Li Y, Koizumi Y, Chiba A (2014) Effect of nitriding treatment on corrosion behaviour of Co–Cr–Mo alloy in liquid Al. Corros Sci 78:244–250
104. Lombardo GH, Nishioka RS, Souza RO, Michida SM, Kojima AN, Mesquita AM et al (2010) Influence of surface treatment on the shear bond strength of ceramics fused to cobalt-chromium. J Prosthodont 19(2):103–111
105. Foster BL, Ramnitz MS, Gafni RI, Burke AB, Boyce AM, Lee JS et al (2014) Rare bone diseases and their dental, oral, and craniofacial manifestations. J Dental Res 93(7):7S–19S
106. Huber F-X, Belyaev O, Hillmeier J, Kock H-J, Huber C, Meeder P-J et al (2006) First histological observations on the incorporation of a novel nanocrystalline hydroxyapatite paste OSTIM® in human cancellous bone. BMC Musculoskeletal Disorders 7(1):50
107. Malik A, Tahir Butt T, Zahid S, Zahid F, Waquar S, Rasool M et al (2017) Use of magnetic nanoparticles as targeted therapy: theranostic approach to treat and diagnose cancer. J Nanotechnol 2017:1098765
108. Zhengli X, Jiao S (2009) Application of nanotechniques and nanomaterials in oral medicine. Chin J Dental Mater Dev 4
109. Yu K, Liu M, Dai H, Huang X (2020) Targeted drug delivery systems for bladder cancer therapy. J Drug Delivery Sci Technol 56:101535
110. Fan W, Wu D, Ma T, Fan B (2015) Ag-loaded mesoporous bioactive glasses against enterococcus faecalis biofilm in root canal of human teeth. Dent Mater J 34(1):54–60
111. Shrestha A, Kishen A (2016) Antibacterial nanoparticles in endodontics: a review. J Endodontics 42(10):1417–1426
112. Kesler Shvero D, Zaltsman N, Weiss EI, Polak D, Hazan R, Beyth N (2016) Lethal bacterial trap: cationic surface for endodontic sealing. J Biomed Mater Res Part A 104(2):427–434
113. Moshaverinia A, Ansari S, Movasaghi Z, Billington RW, Darr JA, Rehman IU (2008) Modification of conventional glassionomer cements with N-vinylpyrrolidone containing polyacids, nano-hydroxy and fluoroapatite to improve mechanical properties. Dent Mater 24(10):1381–1390
114. Hua Y, Gu L, Watanabe H (2013) Micromechanical analysis of nanoparticle-reinforced dental composites. Int J Eng Sci 69:69–76
115. Uysal T, Yagci A, Uysal B, Akdogan G (2010) Are nanocomposites and nano-ionomers suitable for orthodontic bracket bonding? Eur J Orthodontics 32(1):78–82
116. Cheng L, Weir MD, Zhang K, Arola DD, Zhou X, Xu HHK (2013) Dental primer and adhesive containing a new antibacterial quaternary ammonium monomer dimethylaminododecyl methacrylate. J Dent 41(4):345–355
117. Abramovitz I, Beyth N, Weinberg G, Borenstein A, Polak D, Kesler-Shvero D et al (2012) In vitro biocompatibility of endodontic sealers incorporating antibacterial nanoparticles. J Nanomater 2012:858073
118. Selwitz RH, Ismail AI, Pitts NB (2007) Dental caries. The Lancet 369(9555):51–59

119. White JM, Eakle WS (2000) Rationale and treatment approach in minimally invasive dentistry. J Am Dental Assoc 131(1):13S–19S
120. Ericson D, Kidd E, McComb D, Mjör I, Noack M (2003) Minimally invasive dentistry—concepts and techniques in cariology. Oral Health Prev Dent 1:59–72
121. Kachoei M, Eskandarinejad F, Divband B, Khatamian M (2013) The effect of zinc oxide nanoparticles deposition for friction reduction on orthodontic wires. Dental Res J 10(4):499–505
122. Aitken RJ, Chaudhry MQ, Boxall ABA, Hull M (2006) Manufacture and use of nanomaterials: current status in the UK and global trends. Occup Med 56(5):300–306
123. Tay CY, Fang W, Setyawati MI, Chia SL, Tan KS, Hong CHL et al (2014) Nano-hydroxyapatite and nano-titanium dioxide exhibit different subcellular distribution and apoptotic profile in human oral epithelium. ACS Appl Mater Interfaces 6(9):6248–6256
124. Sahoo SK, Parveen S, Panda JJ (2007) The present and future of nanotechnology in human health care. Nanomed Nanotechnol Biol Med 3(1):20–31
125. Hojati ST, Alaghemand H, Hamze F, Babaki FA, Rajab-Nia R, Rezvani MB et al (2013) Antibacterial, physical and mechanical properties of flowable resin composites containing zinc oxide nanoparticles. Dental Mater 29(5):495–505
126. Turssi CP, Ferracane JL, Ferracane LL (2006) Wear and fatigue behavior of nano-structured dental resin composites. J Biomed Mater Res B Appl Biomater 78(1):196–203
127. Mount GJ (2003) Minimal intervention dentistry: rationale of cavity design. Oper Dent 28(1):92–99
128. Jia G, Wang H, Yan L, Wang X, Pei R, Yan T, Zhao Y, Guo X (2005) Cytotoxicity of carbon nanomaterials: single-wall nanotube, multi-wall nanotube, and fullerene. Environ Sci Technol 39:1378–1383
129. Zhu H, Wang X, Li Y, Wang Z, Yang F, Yang X (2009) Microwave synthesis of fluorescent carbon nanoparticles with electrochemiluminescence properties. Chem Commun 2009:5118–5120
130. Movafeghi A, Khataee A, Abedi M, Tarrahi R, Dadpour M, Vafaei F (2018) Effects of TiO_2 nanoparticles on the aquatic plant *Spirodela polyrrhiza*: evaluation of growth parameters, pigment contents and antioxidant enzyme activities. J Environ Sci 64:130–138
131. Li L, Sillanpaa M, Tuominen M, Lounatmaa K, Schultz E (2013) Behavior of titanium dioxide nanoparticles in *Lemna minor* growth test conditions. Ecotoxicol Environ Saf 88:89–94
132. Santos AR, Miguel AS, Tomaz L, Malho R, Maycock C, Vaz Patto MC, Fevereiro P, Oliva A (2010) The impact of CdSe/ZnS quantum dots in cells of Medicago sativa in suspension culture. J Nanobiotechnol 8:24–37
133. Tarrahi R, Movafeghi A, Khataee A, Rezanejad F, Gohari G (2019) Evaluating the Toxic impacts of cadmium selenide nanoparticles on the aquatic plant *Lemna minor*. Molecules 24:410–424
134. Tarrahi R, Khataee A, Movafeghi A, Rezanejad F (2018) Toxicity of ZnSe nanoparticles to *Lemna minor*: evaluation of biological responses. J Environ Manag 226:298–307
135. Tarrahi R, Khataee A, Movafeghi A, Rezanejad F, Gohari G (2017) Toxicological implications of selenium nanoparticles with different coatings along with Se^{4+} on *Lemna minor*. Chemosphere 181:655–665
136. Tripathi DK, Shweta SS, Singh S, Pandey R, Singh VP, Sharma NC, Prasad SM, Dubey NK, Chauhan DK (2017) An overview on manufactured nanoparticles in plants: uptake, translocation, accumulation and phytotoxicity. Plant Physiol Biochem 110:2–12
137. Ghafariyan MH, Malakouti MJ, Dadpour MR, Stroeve P, Mahmoudi M (2013) Effects of magnetite nanoparticles on soybean chlorophyll. Environ Sci Technol 47:10645–10652
138. Nowack B, Bucheli TD (2007) Occurrence, behavior and effects of nanoparticles in the environment. Environ Pollut 150:5–22
139. Khataee A, Movafeghi A, Nazari F, Vafaei F, Dadpour MR, Hanifehpour Y, Joo SW (2014) The toxic effects of L-cysteine-capped cadmium sulfide nanoparticles on the aquatic plant *Spirodela polyrrhiza*. J Nanopart Res 16:1–10

140. Lee S, Kim S, Kim S, Lee I (2013) Assessment of phytotoxicity of ZnO NPs on a medicinal plant, *Fagopyrum esculentum*. Environ Sci Pollut Res 20:848–854
141. Wang F, Guan X, Ding M, Ma T (2019) Effects of nanoparticles on algae: adsorption, distribution, ecotoxicity and fate. Appl Sci 9:1534–1548
142. Simard J-C, Vallieres F, de Liz R, Lavastre V, Girard D (2015) Silver nanoparticles induce degradation of the endoplasmic reticulum stress sensor activating transcription factor-6 leading to activation of the NLRP-3 inflammasome. J Biol Chem 290:5926–5939
143. Ma Y, Zhang P, Zhang Z, He X, Li Y, Zhang J, Zheng L, Chu S, Yang K, Zhao Y, Chai Z (2015) Origin of the different phytotoxicity and biotransformation of cerium and lanthanum oxide nanoparticles in cucumber. Nanotoxicology 9:262–270
144. Taylor AF, Rylott EL, Anderson CWN, Bruce NC (2014) Investigating the toxicity, uptake, nanoparticle formation and genetic response of plants to gold. PLoS ONE 9:e93793
145. Movafeghi A, Khataee A, Rezaee A, Kosari-Nasab M, Tarrahi R (2019) Toxicity of cadmium selenide nanoparticles on the green microalga Chlorella vulgaris: inducing antioxidative defense response. Environ Sci Pollut Res 26:36380–36387
146. El-Shahate R et al (2011) Evaluation of the effect of three different pesticides on *Azolla pinnata* growth and NPK uptake. J Am Sci 7:1020–1031
147. Tarrahi R, Mahjouri S, Khataee A (2021) A review on in vivo and in vitro nanotoxicological studies in plants: a headlight for future targets. Ecotoxicol Environ Saf 15(208):111697
148. Helaly MN et al (2014) Effect of nanoparticles on biological contamination of 'in vitro' cultures and organogenic regeneration of banana. Aust J Crop Sci 8:612–624
149. Mandeh M, Omidi M, Rahaie M (2012) In vitro influences of TiO2 nanoparticles on barley (*Hordeum vulgare* L.) tissue culture. Biol Trace Elem Res 150:376–380
150. Krystofova O, Sochor J, Zitka O, Babula P, Kudrle V, Adam V, Kizek R (2013) Effect of magnetic nanoparticles on tobacco BY-2 cell suspension culture. Int J Environ Res Public Health 10:47–71
151. Thorpe TA (2007) History of plant tissue culture. Mol Biotechnol 37:69–180
152. Sadhu A, Ghosh I, Moriyasu Y, Mukherjee A, Bandyopadhyay M (2018) Role of cerium oxide nanoparticle-induced autophagy as a safeguard to exogenous H_2O_2- mediated DNA damage in tobacco BY-2 cells. Mutagenesis 33:161–177
153. Mahjouri S, Movafeghi A, Divband B, Kosari-Nasab M (2018) Toxicity impacts of chemically and biologically synthesized CuO nanoparticles on cell suspension cultures of Nicotiana tabacum. Plant Cell Tissue Organ Cult (PCTOC) 135:223–234
154. Kim DH, Gopal J, Sivanesan I (2017) Nanomaterials in plant tissue culture: the disclosed and undisclosed. RSC Adv 7:36492–36505
155. Doran PM (2009) Application of plant tissue cultures in phytoremediation research: incentives and limitations. Biotechnol Bioeng 103:60–76
156. Liu Q, Zhang X, Zhao Y, Lin J, Shu C, Wang C, Fang X (2013) Fullerene-induced increase of glycosyl residue on living plant cell wall. Environ Sci Technol 47:7490–7498
157. Mahjouri S et al (2018) Assessing the toxicity of silver nanoparticles in cell suspension culture of *Nicotiana tabacum*. Biointerface Res Appl Chem 8:3252–3258
158. Mahjouri S, Kosari-Nasab M, Mohajel Kazemi E, Divband B, Movafeghi A (2020) Effect of Ag-doping on cytotoxicity of SnO_2 nanoparticles in tobacco cell cultures. J Hazard Mater 381:121012
159. Lin C, Fugetsu B, Su Y, Watari F (2009) Studies on toxicity of multi-walled carbon nanotubes on Arabidopsis T87 suspension cells. J Hazard Mater 170:578–583
160. Begum P, Fugetsu B (2013) Induction of cell death by graphene in *Arabidopsis thaliana* (Columbia ecotype) T87 cell suspensions. J Hazard Mater 260:1032–1041
161. Barbasz A, Kreczmer B, O'cwieja M (2016) Effects of exposure of callus cells of two wheat varieties to silver nanoparticles and silver salt ($AgNO_3$). Acta Physiol Plant 38:76–86
162. Selivanov NY, Selivanova OG, Sokolov OI, Sokolova MK, Sokolov AO, Bogatyrev VA, Dykman LA (2017) Effect of gold and silver nanoparticles on the growth of the *Arabidopsis thaliana* cell suspension culture. Nanotechnol Russ 12:116–124

163. Tan X, Lin C, Fugetsu B (2009) Studies on toxicity of multi-walled carbon nanotubes on suspension rice cells. Carbon 47:3479–3487
164. Cui J, Liu T, Li F, Yi J, Liu C, Yu H (2017) Silica nanoparticles alleviate cadmium toxicity in rice cells: mechanisms and size effects. Environ Pollut 228:363–369
165. Kokina I, Gerbreders V, Sledevskis E, Bulanovs A (2013) Penetration of nanoparticles in flax (*Linum usitatissimum* L.) calli and regenerants. J Biotechnol 165:127–132
166. Homaee MB, Ehsanpour AA (2015) Physiological and biochemical responses of potato (*Solanum tuberosum*) to silver nanoparticles and silver nitrate treatments under in vitro conditions. Indian J Plant Physiol 20:353–359
167. Homaee MB, Ehsanpour AA (2016) Silver nanoparticles and silver ions: oxidative stress responses and toxicity in potato (*Solanum tuberosum* L) grown in vitro. Horticult Environ Biotechnol 57:544–553
168. Santos AR, Miguel AS, Macovei A, Maycock C, Balestrazzi A, Oliva A, Fevereiro P (2013) CdSe/ZnS quantum dots trigger DNA repair and antioxidant enzyme systems in Medicago sativa cells in suspension culture. BMC Biotechnol 13:111–121
169. Xia B, Dong C, Zhang W, Lu Y, Chen J, Shi J (2013) Highly efficient uptake of ultrafine mesoporous silica nanoparticles with excellent biocompatibility by Liriodendron hybrid suspension cells. Sci. China Life Sci. 56:82–89
170. Jamshidi M, Ghanati F, Rezaei A, Bemani E (2016) Change of antioxidant enzymes activity of hazel (*Corylus avellana* L.) cells by AgNPs. Cytotechnology 68:525–530
171. Anwaar S, Maqbool Q, Jabeen N, Nazar M, Abbas F, Nawaz B, Hussain T, Hussain SZ (2016) The effect of green synthesized CuO nanoparticles on callogenesis and regeneration of *Oryza sativa* L. Front Plant Sci 7:1330–1338
172. Spinoso-Castillo JL, Chavez-Santoscoy RA, Bogdanchikova N, Pérez-Sato JA, Morales-Ramos V, Bello-Bello JJ (2017) Antimicrobial and hormetic effects of silver nanoparticles on in vitro regeneration of vanilla (Vanilla planifolia Jacks. ex Andrews) using a temporary immersion system. Plant Cell Tissue Organ Cult (PCTOC) 129:195–207
173. Tian L, Ding J, Zhang W, Yang H, Fu W, Zhou X, Zhao W, Zhang L, Fan X (2011) Synthesis and photoelectric characterization of semiconductor CdSe microrod array by a simple electrochemical synthesis method. Appl Surf Sci 257:10535–10538

Recent Progress in the Development of Metallic Composite for Advanced Technologies

Yogesh Kumar Kumawat, Rishabh Sehgal, Irfan Ayoub, Rakesh Sehgal, and Vijay Kumar

Abstract Every day, millions of things are manufactured for the pleasant and delightful existence of humans, and the number is continually expanding with the growth of modern technology. The fascinating characteristics of composite materials have piqued the interest of the research community across the globe. Ever-increasing demands are now being made to develop cost-effective, eco-friendly, and efficient composite materials, thereby replacing the previously used monolithic technology. In the past few years, considerable work has been devoted in this field to designing an enormous variety of metallic composite configurations. Among the developed composite configurations, metal matrix composites are considered to be the most important because of their excellent characteristic features, such as wear, creep, fatigue, lightweight. Besides this, they also find a wide variety of applications in every aspect of life. In this book chapter, a brief introduction to metal matrix composites will be provided, followed by their different classifications based on their properties. Following that, a brief discussion of their significant properties is provided, providing the readers with in-depth knowledge. Besides that, an outlook on the different synthesis techniques as well as their applications in different technological aspects is presented in the chapter.

Keywords Metallic · Composites · Mechanical · Technology · Metal matrix · Ceramic matrix

---

Y. K. Kumawat · I. Ayoub · V. Kumar (✉)
Department of Physics, National Institute of Technology Srinagar, Hazratbal 190006, Jammu and Kashmir, India

R. Sehgal
Department of Electrical and Computer Engineering, University of Texas at Austin, Austin, TX 78751, USA

R. Sehgal
Department of Mechanical Engineering, National Institute of Technology Srinagar, Hazratbal 190006, Jammu and Kashmir, India

V. Kumar
Department of Physics, University of the Free State, P.O. Box 339, Bloemfontein ZA9300, South Africa

S. K. Tiwari et al. (eds.), *Nanoparticles Reinforced Metal Nanocomposites*,
https://doi.org/10.1007/978-981-19-9729-7_3

1 Introduction

Modern technology has increased the demand for materials, and it is the necessity of the hour to create new materials in order to overcome the constraints of the materials that exist now in terms of their applicability in many industries [1]. The materials researchers have a complex design challenge when selecting the proper components, but if done well, it may serve as the foundation for the material's best usage and function [2]. Likewise, metallic composites are now at the forefront of materials technology with efficient characteristic parameters that lead to their high performance, which is suitable for high-demanding application areas. Besides that, they are also cost-effective and eco-friendly in nature [3]. Because of the remarkable features possessed by them, they are currently being used in different applications like lightweight equipment, transportation, sports equipment, aviation and automotive industries, construction, marine and space activities, electronics, computer systems, security, surveillance, and so on [4–9]. Metallic composites originally arose as a unique technology during a period when higher efficiency for sophisticated military systems was a top priority for material development [5–13]. By the late 1970s, the materials development and commercialization orientation had shifted. Even in defense applications, the increasing reliance on price and increased sensitivity to risk made certifying new high-performance materials for applications more difficult [12]. However, functional qualities of metallic composites such as excellent dimensional efficiency, high durability, and appealing thermoelectric characteristics have enabled the transportation services (vehicles and railways), thermal management, aviation, industrial, recreational, and infrastructure industries. The term "metallic composites" itself expounds on the concept of a composite substance, which is a mixture of materials. When two or more constituent materials with considerably different physical or chemical properties get merged, a material with distinctive aspects independent of its constituent parts is produced [14]. In a similar way, metal matrix composites are made up of a continuous matrix of metal or alloy with reinforcement in the form of particles, small fibers or whiskers, or continuous fibers [15]. Aluminum (Al), magnesium (Mg), copper (Cu), and titanium (Ti) must make up the majority of the metallic matrix in metal matrix composites (MMCs), and the reinforcement can either be metallic fibers or be distributed ceramics consisting of oxides and carbides (tungsten, molybdenum, and lead). Nearly 50% of the total volume of the composite material is made up of reinforcement [14]. Excellent examples of composite materials include wood, which is mostly made up of fibrous chains of cellulose molecules in a matrix of the organic polymer lignin [14, 16]. The development of MMCs has made it feasible to combine the pliable, strong, and conductive properties of metals with the wear resistance, toughness, and hardness of ceramic materials involved in the metal composites [1, 12]. The matrix reinforcement involves materials with high strength, stiffness, and low thermal expansion, as it enables the MMCs to stand up to stress conditions. The reinforcement phase in composites is probably stronger than the continuous matrix [5, 14, 17, 18]. Many modern metal composite materials use metallic reinforcements, such as tungsten-reinforced copper, tungsten heavyweight alloys, copper–niobium, superconductors. Furthermore, the use of metallic composites has resulted in a decline in the wastage of food items and other eating products by employing revolutionary food-packaging methods that improve food storage life, safety, and quality, which has piqued academic and industry interest [19, 20]. In order to meet the needs of marine applications, metals have a substantially greater modulus than human hard tissues, while ceramics are both stiffer and brittle than naturally mineralized tissues. However, composite materials are more ductile than hard tissues and are not stiff enough to replace them in load-bearing applications [8, 21]. They are also employed in a variety of marine applications, including hull materials for small boats and impact-resistant structures in the military sector. It is mostly employed in bulletproof vest applications and the bullet-proofing of vehicles used by the defense industry and VIPs [5, 18]. Many researchers have been creating various sorts of modern manufacturing procedures to boost productivity and efficiency [14, 22, 23]. The different components of a metallic composite material have an impact on its qualities; therefore, it is necessary to systematically study their classification and distinguishing properties for future applications. This book chapter focuses on the different aspects of metallic composites. Firstly, an overview of the classification based on the matrix phase is provided, wherein the different classes of composite materials are briefly discussed. This section is followed by a brief discussion on the different properties of these composite materials. Finally, an overview of the various applications of composite materials in various fields is provided.

2 Classification of Composite Materials

Depending on their scale, reinforcement, matrix, and other characteristics, composite materials can be categorized in a variety of ways. Figure 1 provides a visual representation of many classes in the various classification schemes. The primary focus has remained on metallic composites, with polymers, ceramics, and metallic composites explored among the various classification schemes. Metal matrix composites have been briefly mentioned among the metallic composites, including those made of aluminum, magnesium, titanium, and other metals. Further, the classification is based on the various parameters, such as:

i. Scale: Composite materials can be classified based on the size parameter. This can be further subdivided into nano-sized composite materials.
ii. Reinforcement: In this section of classification, the composite materials are classified based on the components of reinforcement materials such as fiber, particles, and many more reinforcing materials.
iii. Matrix materials: The matrix is a monolithic substance that can be chosen according to the intended use of the composite material, which heavily influences the selection of acceptable matrix alloys. They can be further classified into polymers, ceramics, metals, etc.

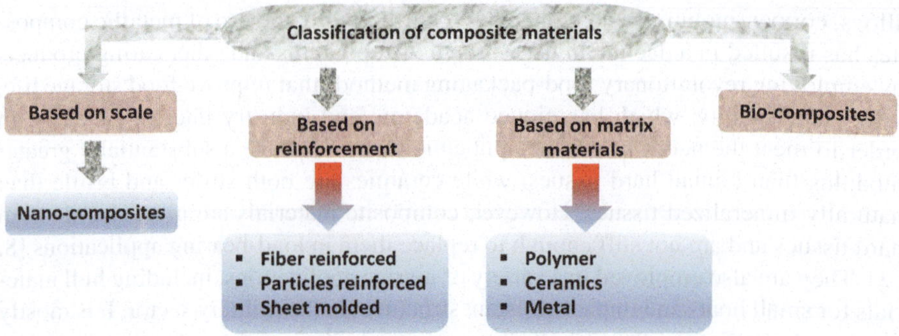

Fig. 1 Pictorial representation of the classification of composite materials

iv. Biocomposites: The materials that can be manufactured using biomaterials are described in this section.

2.1 *Ceramic Matrix Composites*

The matrix is a monolithic substance in which the reinforcement is generally incorporated and must be dispersed equally overall. Ceramic matrix composites (CMCs) are the material structures that generally consist of carbon, silicon carbide (SiC), aluminum oxide (Al_2O_3), and silicon nitride (SiN) fibers incorporated in a ceramic matrix structure [14, 24]. Ceramic materials seem to have substantial physical and mechanical properties, but their implementations in the applied field are limited due to their brittleness. Furthermore, particulate and fiber reinforcement have been used to optimize the strength, particularly the toughness, of brittle ceramics, despite their low efficacy [25]. Ceramic composites are utilized in those situations where reliable operation under harsh circumstances is demanded due to their outstanding characteristics. A few examples of the CMCs used in aerospace rockets and gas turbine engines are wear-resistant elements (e.g., brake systems), heat pumps, furnaces, etc. [25, 26]. Traditional ceramic fabrication techniques include integrating the powder form of the matrix material with the reinforcing agent, then handling at high temperatures, such as hot pressing or sintering [26, 27]. During sintering, the inner phase and matrix are developed around the fibers' gaseous or liquid precursors [14, 26–28]. Generally, ceramic precursors change the composition and microstructure of a ceramic matrix (polymer or gaseous). Hence, it was concluded that the commercialization of a wide range of composites with high ceramic yield and a cheap cost will open up new applications in the electrical, automotive, and aerospace sectors. In the recent years, the Astrium firm has worked closely with several partners to develop equipment for hypersonic engines and liquid rocket propulsion systems. Various hot-firing tests with subscale (scale 1:5) and full-scale nozzle extensions were carried out in the year 2000.

This year saw another significant milestone in the field of tiny thrusters, with long-term testing demonstrating the exceptional stability of the C/SiC material [29]. Moreover, continuous fiber-reinforced SiC ceramic matrix composites (FRCMCs-SiC) have become the dominant material for high-value hot-section components, safety–critical components, and braking components (in aerospace, energy, and transportation), sparking machining demand. DACC (South Korea) developed brake disks for the F16 fighter, which have been widely used in aircraft braking systems. The Ferrari Enzo, Mercedes-Benz AMG, and Corvette ZR1 are among the vehicles included, as are the French TGV NG, the British Heathrow Express train, and the South Korean TCV train [30]. Further, parallel to the development of novel ceramic composites with increased mechanical and thermal characteristics, the cost-effectiveness of material production procedures must be improved.

2.2 *Polymer Matrix Composite*

Polymer matrix composites (PMCs) are made up of a matrix phase of polymer that is typically reinforced with organic or inorganic fibers. They are usually made of woven fabrics linked together by a polymer matrix in a laminate structure [31]. A polymer contributes as the continuous matrix phase in a PMC, while long or short fiber particles or nanoparticles serve as the discontinuous reinforcement [33]. Epoxies, vinyl esters, and polyesters were popular polymer matrices used in structural composites [33]. The fiber/polymer interfacial region's durability is critical to the structural stability and lifespan efficiency of fibrous polymeric composites [34–36]. Polymer composites have several useful properties, such as chemical abrasion resistance, electrical, magnetic, and optoelectronic properties, wear resistance, insulation, high-frequency dielectric capabilities [37]. Polymer matrix composites can be produced using a number of fabrication techniques, as shown in Fig. 2. The kind of reinforcement, relative volume proportion, fiber orientation, stacking direction, and number of layers all have a significant impact on the physicochemical properties of polymer composites [37, 38]. Due to variations in physical and chemical properties, thermoplastic and thermosetting plastic are the two types of manufacturing procedures utilized to create polymer composites. With pultrusion, desired cross-sectional shapes can be produced without length restrictions by dragging layers of cloth or fiber through a heated die while submerged in resin. The process yields high productivity, minimal material waste, excellent quality, and rapid processing of composite materials. However, the drawbacks of this procedure include fiber breakage, die jamming, and poor fiber wetting [38]. Similarly, another prominent method for producing thermosetting polymer composites is resin transfer molding. The pre-shaped dry reinforcements are pressed into a heated die mold. The procedure has the benefit of allowing huge, complicated shapes and curvatures to be created with automation at a low cost and with expert labor [39]. Furthermore, the other widely used techniques for creating thermoplastic composites are injection molding

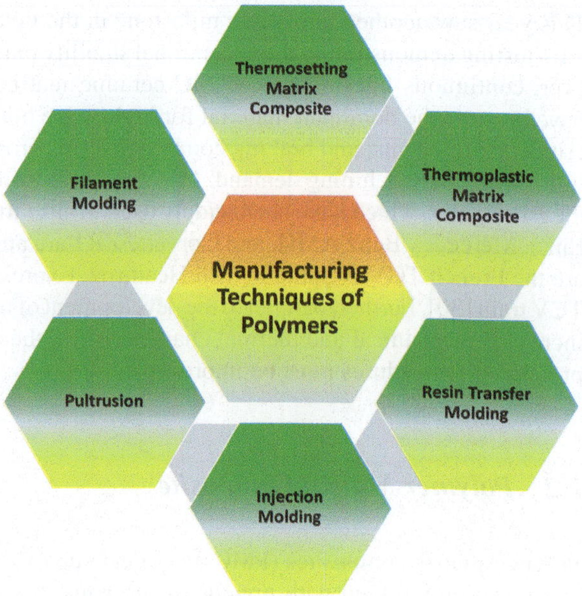

Fig. 2 Pictorial representation of different manufacturing techniques for the polymer matrix composites

and screw extrusion (because of their low cost). Pellets of polymer grains and reinforcing particles are poured into the hopper. The slurry is then pushed into a heated barrel and placed into a split mold by a rotating feeding screw. The generation of feedstock filament for composite material additive manufacturing is the technique's most notable advantage [40]. The different types of materials that are being used in the reinforcement, along with their properties and applications, are given in Table 1.

2.3 *Metal Matrix Composites (MMCs)*

Wrought iron and most ordinary steels can theoretically be considered MMCs because they have a metallic matrix reinforced with dispersoids of oxides, sulfides, carbides, etc. Several standard engineering alloys, including some steels featuring dispersion developed by shifting of the alpha–gamma interface, may also be included even after they are defined by microstructural design [45, 46]. It has been observed that at least one ingredient of MMC is a metal or alloy that produces at the most one circulating network, while the remaining component is encased in the matrix material. This phenomenon is known as reinforcement. To develop a metal matrix composite material, a high-strength substance called reinforcement is spread within the matrix material [47]. Owing to their improved strength-to-weight ratio, aluminum, magnesium, and copper have attained the status of being the most studied matrix materials [48, 49]. Silicon, zinc, magnesium, and copper are endowed with sufficient solubility, thus enabling them to be used as significant alloying elements [49, 50]. Particle size

Table 1 Polymer matrix composites: implementations and qualities

S. no.	Matrix material	Material of reinforcement	Manufacturing methods	Qualities	Implementations	References
1	Epoxy resin	Carbon nanotubes	Resin transfer molding	Superior tensile strength and conductivity	Aviation industry, automobile industry, and sports equipment industry	[37, 41]
2	Plastic	Boron	Pultrusion	Elasticity with a high modulus	Space and aerospace programs	[37, 42]
3	Polyester	Banana	Compression molding	Elastic strength and impact modulus, storage modulus	For construction of materials	[37, 43]
4	Nylon	Iron particle	Injection molding or extrusion	High elasticity with modulus	Tolling at a breakneck pace	[32, 37]
5	Polyethylene	Flax	Extrusion	Exceptional tensile strength	Automobile equipment and construction material	[38, 44]

is a very important property regarding MMCs. Grain refining can minimize thermal effects, resulting in enhanced matrix strength. The matrix's strength improves as grain size decreases, but the overall output does not boost considerably. The reinforcement's structure, size, and volume fraction all play a role in the composite's strength [49, 51]. That is why MMCs have also made their way into the "physical world" of engineering disciplines, varying from well-known engineering materials (such as WC–Co hard metal) to newer specialized materials [10, 15, 52]. Furthermore, MMCs are extensively used in a variety of sectors, notably aerospace, automobiles, and construction, as a prospective class of materials used to produce lightweight components and structures. A strong demand for MMCs in the aircraft industry and automobile sector, in particular, confirms the universal need for these materials. Meanwhile, MMC's development has a multitude of challenges, the majority of which come from its complex treatment and ineffective cost productivity [4, 53]. Metal matrix composites can be divided into several categories. In particle, layer, fiber, and composite materials, one categorization is the kind and contribution of reinforcement components as depicted in Fig. 3. Continuous fiber composite materials (multi- and monofilament) and short fiber composite materials (whisker composite materials) are two types of fiber composite materials [2]. Table 2 lists various MMC

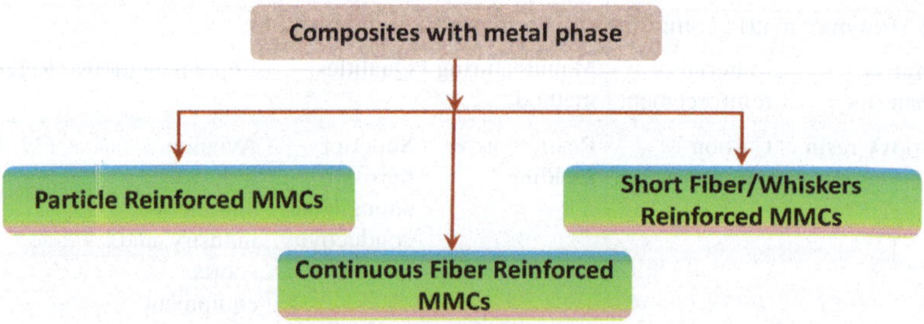

Fig. 3 Classification of metal phase reinforced MMCs

Table 2 Reinforcements commonly used in metal matrix composites

Type	Aspect proportions	Diameter (μm)	Examples	References
Particle	~1–4	1–25	SiC, Al_2O_3, WC, TiC	[15, 54]
Continuous fiber	>1000	0.1–25	C, B, W, NbTi, Nb_3Sn	[54]
Short fiber	~10–1000	3–150	SiC, Al_2O_3, Al_2O_3 + SiO_2, C	[12, 49]

parameters, such as aspect (length/diameter) ratios and diameters of some of the most common reinforcements used in metal matrix composites. MMCs are divided into numerous types based on the matrix materials typically deployed. The following are among the most frequently utilized metallic matrix configurations:

2.3.1 Aluminum-Based Composites

Aluminum and its alloys have a peculiar combination of characteristics that makes them ideal for composite manufacturing. Based on their reaction to precipitation strengthening, aluminum alloys are classified as heat treatable or non-heat treatable [55]. Aluminum alloys are widely used in the manufacturing of MMCs due to their significant properties. Some of these characteristics are high strength, low density, outstanding wear response, fatigue life, good thermal conductivity, corrosion resistance, renewability, acceptable production costs, and flexibility in reshaping, rolling, drawing, extruding, and welding [56–58]. To improve these properties, the volume fraction, sorts, and sizes of the reinforcing particles could all be changed. As a consequence, aluminum metal matrix composites (AMMCs) are employed to produce a variety of aerospace components to ensure increased wear resistance, frictional resistance, and some remarkable mechanical properties [55, 59, 60]. These composites are designed to replace metallic alloys in aerospace applications such as wings and fuselage, as well as automotive components such as brake disks, drums, and pistons, at a cheaper price [55, 61, 62]. The number of studies conducted in the field of AMMCs has increased significantly over the previous decade. Different researchers have studied the different characteristics of the AMMCs. Stojanovic et al. [63] have observed that to increase the characteristics of the basic aluminum metals, there are different reinforcing elements that can be added in the form of particles, whiskers, or fibers in varied proportions, such as SiC, alumina (Al_2O_3), AL–Li, silicon nitride (Si_3N_4), boron carbide (B_4C), titanium dioxide (TiO_2). In this context, Vasudevan et al. [64] revealed that if lithium is employed as a primary alloying element with aluminum, then it has the ability to raise the elastic modulus while reducing the density of the alloy. Consequently, the aviation industry has been a major focus of this growth. Similarly, Ramnath et al. [65] described that silicon carbide is among the most broadly utilized reinforcement materials as it enhances the tensile strength, hardness, density, and wear resistance of aluminum. Furthermore, Cavaliere et al. [66] have employed the spark plasma sintering (SPS) procedure for synthesizing the AL–SiC composites. While analyzing, they observed that by applying the mentioned synthesis route, the hardness and tensile strength of the produced 6 wt% SiC composites were found to have risen by 34 and 26%, respectively. Afterward, Ekambaram et al. [67] applied the stir casting fabrication method for developing with 4, 6, and 8% alumina and found that the composites had fewer casting flaws than the aluminum alloy. Also, the analysis report has shown remarkable improvements in the hardness, ductility, and lightness of composite materials. Similarly, Gomez et al. [68] added that AMCs supplemented with B_4C particle reinforcements can be easily manufactured via a solid-state processing technique. In addition to this study, the tribological properties of B_4C composites have also demonstrated a higher friction coefficient with a lower wear rate than their equivalents, which might be very useful in braking applications. At the early developmental stages, the aluminum-based composites were developed by two methods: solid-state processing (powder metallurgy, mechanical alloying) and liquid-state processing. The combining, degassing, and sintering of metal alloys and reinforcements are part of the powder metallurgy process [58]. Mechanical alloying is a technique in which constituents interact to generate reinforcement. For the production of aluminum composites, stir casting is perhaps the most common and cost-effective approach so far [59, 69, 70]. A few examples of the most commonly used aluminum alloys, along with their properties, manufacturing method, and applications in different fields, are given in Table 3.

2.3.2 Magnesium-Based Composites

Magnesium is the lightest metal on the planet and has great potential as little more than a structural alloy. It is around 35% lighter than aluminum [74]. Lightweight alloys and composites are gaining popularity among aerospace and automotive companies because they can be used in place of heavier metals such as steel, cast iron, zinc alloys, and even aluminum alloys [75]. Due to the expensive price of magnesium, its applicability seems to have been initially confined; however, due to their enhanced solidification properties over cast metals such as aluminum and copper alloys, magnesium alloys are gaining significant attraction. As an outcome, magnesium composite materials have emerged as a cutting-edge study subject in a variety of linked sectors [75, 76]. The studies have revealed that magnesium is not sufficiently strong in its pure

Table 3 Properties and applications of commonly used aluminum alloys

Matrix material	Properties	Manufacturing methods	Application in various sectors	References
Pure Al	Good recyclability, ductility, and workability	Powder metallurgy	Decorative foil for beverages, drinks, and snacks, as well as food-packing trays	[55]
AA2024	High fatigue strength, processability, and surface quality	Powder metallurgy	Automobile wheels, laboratory equipment, sheets, biomedical applications, and aircraft components are just a few examples	[71]
AA3003	Weldability. machinability, castability	Vapor infiltration	Heat exchangers, pressure vessels, and a storage tank	[72]
AA4043	Exceptional corrosion resistance	Accumulative roll bonding	Used as filler material	[73]

state, so it is alloyed with different elements to achieve certain specialized features, specifically a high strength-to-weight ratio. Consequently, the use of varying sorts of reinforcements has addressed the limitations of monolithic magnesium qualities such as low elastic modulus, quick loss of strength with rising temperature, and poor creep resistance [77]. Some of the common reinforcing elements used in magnesium-based composites are silicon carbide (SiC), aluminum oxide (Al_2O_3), titanium carbide (TiC), etc. [78]. Lim et al. [79] found that during the dry sliding process, Mg-based metal matrix composites augmented with silicon carbide particulates (SiC) exhibit substantially improved wear resistance under lower stresses. Squeeze casting methods were used by Jayalakshmi et al. [80] for developing composites from the magnesium alloy AM100 (Mg-9.3 to 10.7Al–0.13Mn) containing alumina (Al_2O_3). Jiang et al. [81] explored the fabrication of B_4C (10, 15, and 20%) particle-reinforced magnesium matrix composites by powder metallurgy. It was found that the toughness and wear resistance were higher than those of as-cast Mg ingots and improved as the percentage of B_4C particles rose from 10 to 20%. Similarly, Poddar et al. [82] used a stir casting technique to make MMC from SiC (15 vol%) reinforced cast magnesium matrix composite (AZ91D). Tang et al. [83] used squeeze casting to investigate the mechanical properties of magnesium matrix composites and discovered that 10% W14Al86 alloy particles improve mechanical strength. Dudina et al. [84] investigated a magnesium alloy (AZ-911) matrix composite combined with metal glass using powder metal working. Compared to Mg alloy alone, the fabricated Mg alloy-metallic glass composite was found to exhibit higher mechanical strength without a significant loss of flexibility. According to the International Magnesium Association's (IMA) projection, the usage of wrought die-cast magnesium alloy in automotive sectors will increase significantly at an unprecedented yearly pace, indicating the massive and long-term expansion of die-cast automotive parts in the upcoming years [85]. Nowadays, industrialists are focusing on weight reduction processes in the automotive sector to increase mileage. For an increase in energy savings, weight reduction technology for automotive and structural materials is desired. As a response, it would also be extremely important to adopt high strength-to-weight ratio materials in large quantities, with the ability to replicate steel products components [75, 86]. Also, magnesium-based composites and alloys are being used as an alternative to steel, zinc, and aluminum in the products and materials of the world's largest automobile manufacturers. Magnesium alloys and composites have superior mechanical and tribological characteristics. Some of the most well-known automobile manufacturers, along with their versions and supplies, are included, like Honda, Ford, Lexus, Toyota, Alfa, Romeo, Volvo, BMW, Honda-City, etc. [75, 87]. Currently, there has been a surge in interest in producing magnesium (Mg)-based biomaterials that are biodegradable. Magnesium-based orthopedic implants have been used in clinical practice for over a century. In 1900, an Austrian physician named Erwin Payr pioneered the use of magnesium implants (plates and sheets) for musculoskeletal problems [88]. The biomaterials community has been placing close emphasis on the development of magnesium-based alloys with improved corrosion resistance and mechanical qualities that can be used in biomedical applications. Zn, Zr, Sr, Ca, Nd, Ce, Y, Gd, Mn, and Al are some common alloying elements used to improve the qualities [89–93]. Hydroxyapatite (HAP), α-tricalcium phosphate, ZnO, MgO, and other bioresorbable ceramic reinforcements in magnesium matrices are among the most frequently documented. Magnesium oxide is a bioactive ceramic that may be used in a variety of applications. MgO also has antibacterial characteristics, and its breakdown products are quite similar to those of calcium. Therefore, when MgO is combined with an Mg matrix, it results in an increase in a variety of mechanical and biological properties due to good interfacial interaction [94]. Similarly, nanozinc (ZnO) particles are widely used in biological domains such as biomedicine, bioimaging, and trauma treatments [88]. When incorporated in the metal matrix, significant improvements in the mechanical and corrosion characteristics have been observed [95]. It is commonly utilized as a food additive and is included as a nutritional supplement in the FDA's database program [96]. Hydroxyapatites are calcium phosphate mineral with the chemical formula $Ca_5(PO_4)_3(OH)$. In addition to the above-mentioned characteristics and properties, the magnesium matrix can also boost the composite's resistance to corrosion, mechanical and physical characteristics, and biocompatibility [97]. Likewise, B-TCP has received a lot of interest in the orthopedic field for bone tissue healing [98]. There are a variety of production methods for manufacturing magnesium-based composites for different applications.

2.3.3 Titanium-Based Composites (TMMCs)

Titanium-based MMCs composites are the subject of significant research and development programs across the world due to their unique features. Despite the fact that titanium-based composites (TMCs) are one of the most researched and sought-after materials, there are still an enormous number of loopholes that need to be addressed. However, constant attempts were made in the early days of TMC research, notably by NASA, to broaden its use for industrial purposes. One such example is the titanium matrix composite developed during the turbine engine consortium program. Titanium and its alloys are now progressively becoming a central concern for a variety of industries, including the automotive and aerospace sectors. These materials are lighter in weight ratio and offer a variety of appealing qualities, including specific strength, chemical stability, and biocompatibility. Besides this, they also act as a decent choice for structural, biochemical, petrochemical, maritime, and biological applications due to their concoction of qualities [99]. It has Young's modulus of 115 GPa and a density of 4.5 g/cm^3. The density of titanium alloys varies from 4.3 to 5.1 g/cm^3, whereas the elasticity can be varied from 80 to 130 GPa. It is necessary to have high strength-to-weight and modulus-to-weight ratios. Titanium has an extremely high melting point (1672 °C) and high thermal conductivity and is also resistant to corrosive environments [15]. Therefore, aluminum alloys are not suitable at supersonic speeds because the aircraft's surface gets enormously hot, and that is why titanium alloys have been used at such high temperatures [75]. Also, titanium is a reactive metal that reacts well with oxygen, nitrogen, and hydrogen. As a natural consequence, every step of the process of welding titanium requires protection from the environment. Due to titanium's strong chemical reactivity, it is difficult to choose it as a reinforcing material and processing method for Ti matrix composites. Thus, mostly SiC and boron act as effective fiber reinforcements for Ti matrix composites. In recent years, coatings, such as boron fibers coated with SiC and amorphous B_4C, have gained the most attention during the manufacturing of these materials. Meanwhile, the best mechanical properties were achieved by using a SiC fiber with an outer layer composition that is initially carbon rich and then returns to stoichiometric SiC (SCS-6, Textron) [74]. Titanium composites have been produced with a 1–5 μm carbon-rich layer coating on the SiC fibers to reduce and regulate the interfacial area during manufacturing and service. Recently, TiC particles, Ti_3SiC_2 bars, and ultrafine Ti_5Si_3 needle reinforcements for Ti64 matrix composites were successfully generated in situ by Liu et al. [100]. Particularly those composites comprising 5.0 vol% reinforcements and made using 0.5 m SiC, which resulted in a UTS of 1171 MPa and elongation of 5.3%, demonstrated a good blend of strength and endurance as compared to monolithic Ti64 alloy. In addition, Kim et al. [101] looked at the wear and frication behavior of titanium composites (TiB + TiC). They combined granular B_4C with practically pure (CP) Ti (Grade 2) and used vacuum induction melting to create TMCs. They discovered that increasing the frication characteristic at a 20% reinforcement content is the best course of action. As the reinforcement content is increased, the wearing loss also decreases as a result. In a different investigation, An et al. [102] created in-situ TiBw/Ti64 composites with a network architecture using a powder metallurgical process. According to their findings, the TiBw network border functioned as a "barrier wall" that successfully resisted abrasion, leading to a notable improvement in hardness and wear characteristics over the Ti64 alloy. They also found that the network's size has a significant impact on the wear attributes and process. The wear process switches from micro-cutting to brittle debonding as the network size is increased from 60 to 200 m. They concluded that the best composite had a network size of 60 and contained 8.5% TiBw. Stanley et al. [103] recently examined the resistance to fatigue fracture growth from unbridged faults at room temperature from 300 and 450 °C in air and vacuum, respectively. Due to its effect on fiber failure, they found that the initially applied stress intensity factor range (delta Kapp) value significantly affects the composite's crack arrest/catastrophic failure (CA/CF) transition. Modern jet engines (turbine and compressor shafts), fuselage sections, and other aviation parts are some of the applications for the titanium alloys [15]. TMCs are now more commonly acknowledged as potential biomaterials for use in the field of biomedical implants and devices. The utilization of these composites has risen for biomedical applications due to the proper balance of hydroxyapatite bioactivity and the advantageous mechanical properties of titanium. Furthermore, because hydroxyapatite (HA) is harmless, bioactive, and bioadaptable, implants and bones osseointegrate better as a result [104]. Powder metallurgy has been found to make it simple to create TiAl and HA composites. In this method, the sintered composites are made from powders of Ti and HA, with the HA particles acting as "islands" inside the Ti matrix. Over 40% of US combat aircraft, including the F-22, use titanium composites, according to a research conducted by ex-NASA engineers [105]. It is reasonable to infer that this number has increased given the development of cutting-edge, high-tech aircraft like the F-35. It has been discovered that the TMC replacement weighs 40% less than the traditional version, which was made of high-strength steel. In terms of corrosion and fatigue resistance, metallic composite materials perform better than steel and aluminum. Rolls-Royce has successfully conducted testing on a carbon/titanium composite for advanced ultra-fan engine designs. Fan blades made of these composite materials can save up to 1500 pounds per aircraft. This weight reduction corresponds to free transportation for an additional seven individuals. Additionally, the Trent engine will use 20% less fuel and release 20% less CO_2 than the previous generation [99]. Hence, from the above discussion, it can be concluded that the TMCs are being used in almost every industrial sectors.

2.3.4 Copper-Based MMCs

Copper has strong electrical, thermal, and corrosive qualities; however, its uses at high temperatures are limited because of its weak mechanical properties [106]. Copper and its alloys are widely used in various technical and industrial sectors, such as autos, buildings, sanitary systems (underground or undersea drinking water pipelines), electrical equipment, and lead-frame substances for integrated circuit design [107]. Copper-based MMCs have been gaining popularity in the manufacturing area for some years because of their low density, improved fatigue strength, and specific strength. Nowadays, copper is used as a metal matrix in niobium-based superconductors, which is one of the most primary applications for the metallic composites [15]. The development of copper-based materials with both high tensile strength and good electrical conductivity is required for a wide range of technological purposes. The accessibility of Cu-based alloys that accomplish the distinguishable profile is extremely significant for the production of high field resistive and pulse magnets. As a result, efforts have been made to design materials with minimal Joule heating and outstanding strength to endure macroscopic Lorentz forces [108]. Numerous efforts have been made to enhance the mechanical and other characteristics of copper matrix using various reinforcements such as carbon nanotube (CNT), MoS_2, SiC, Al_2O_3, TiO_2, B_4C, WS_2, graphite, $MoSe_2$, carbon fiber (CF) [106, 109]. Alloying is one of the most successful strategies for improving copper's mechanical strength and corrosion resistance. As a result of the alloying action, the timeframe for safe operation rises. The reinforced hybrid composite shows improved friction and wear metrics in addition to a higher degree of hardness. Kavalchenko et al. [110] improved the tribological properties of copper and investigated the wear mechanism by combining MoS_2 and $MoSe_2$. Additionally, Santos et al. [111] developed a composite that was reinforced with ZrO_2 and examined its mechanical properties, which showed enhanced strength and toughness. Li et al. created WC-reinforced materials with better mechanical properties [112]. Important aspects of matrix strengthening and power retention include the form, separation, high thermal stability, dispersion, diffusivity, and low solubility of the reinforcing components in the matrix. Hon et al. [113] explored a simple in-situ approach that is used for the manufacture of Cu matrix composites using rutile-TiO_2 nanoparticles as reinforcements. There were two approaches for the synthesis of Cu matrix composites. The first one is molecular-level mixing advancement, and the second method is vacuum hot-press sintering. Through the results and discussion, it was observed that the reinforcement with 1.72 vol% TiO_2 in the composite displayed an enormous yield strength of 290 MPa. This is 1.6 times more effective than just copper alone (110 MPa). Similarly, Somani et al. [114] used powder metallurgy to examine the production of Cu–SiC composites with varied compositions. They used a V-shaped blender to mix the powder, then used a hydraulic press to condense the powder at a pressure of 250 MPa to create a solid structure, followed by sintering at 950 °C for around 60 min. The study discovered that adding SiC to Cu as a reinforcement increases its mechanical characteristics. There are so many applications of copper-based MMCs in various industrial fields, such as electrical brushes, electrodes for automatic welding, as a composite material for the first-wall particles of nuclear reactors, and in electronic systems. Hence, a lot of studies and research are being done on these MMCs to fulfill the requirements of industrial applications. In this section, the advances in copper-based metal matrix composites have been discussed.

2.3.5 Intermetallic Composites

An intermetallic composite is a type of metallic alloy in which two or more metallic substances form an ordered solid-state compound. They can be ordered or disordered intermetallic composite materials. Long-range ordering, in which various atoms occupy specified places in the lattice, characterizes the structure of ordered intermetallic alloys [15]. It is being investigated how to significantly increase the manufacturing of advanced superalloys and other high-temperature alloys using ordered intermetallic. Because atomic diffusion and dislocation motion are more challenging in ordered alloys, they should have superior creep resistance than disordered alloys. Because they have properties that are halfway between those of metals and those of ceramics, intermetallic compounds are particularly appealing for high-temperature structural applications [115]. Even at low temperatures, some common intermetallic compounds, such as NiAl, have weak yield strengths, and in the desired temperature range, they frequently exhibit rate sensitivity. A high brittle-to-ductile transition temperature is a property of some other intermetallics that have superior strengths and sufficient creep resistance at high temperatures [116]. Reinforcements such as ceramic (Al_2O_3) or refractory metal (W, Mo) fibers, ceramic (SiC, TiB_2), or metallic (Nb) particles, as well as ceramic (SiC, TiB_2) or Nb particles, have all been used to solve the drawbacks of intermetallic composite materials. In intermetallic matrix composites, reinforcements are mostly employed to improve qualities like creep resistance, low-temperature damage tolerance, or both. There are various methods that have been used to fabricate intermetallic matrix composites. The reinforcing phase can be added to the matrix as fibers, whiskers, or other particles that can be separated from those made by in-situ reaction mechanisms, which is generally how these composites can be made. The microstructure of the reaction zone of a Ti_3Al + Nb/SiC composite was examined by Baumann et al. [117], who found that the matrix fiber reaction reduced the ductility of the Nb-rich phase near to the fiber matrix interface. The matrix area close to the interface becomes brittle as a result of the thermal expansion imbalance between the matrix and reinforcement, leading to radial cracks in the matrix that start at the contact. Ritchie et al. [118] investigation of a ductile TiNb-reinforced TiAl intermetallic matrix composite revealed that, in contrast to monolithic TiAl, which has a composite fracture toughness of just −8 Mpa, it has a composite fracture toughness of greater than 30 Mpa. Hardening was significantly weaker under cyclic load because the ductile phase was more likely to experience fatigue failure. About 140 pm diameters are used to enhance the titanium aluminide (TiAl). One of the most extensively studied intermetallic systems in SiC monofilament is reviewed by Mackay et al. [119]. Significant gains in longitudinal strength and stiffness after electrical heating have been seen, with SiC/Ti-24-Al-11 Nb composites frequently reaching 80–90% of their mixture strength. In order to strengthen N and Al with up to 20% volume percent of 12.5 pm diameter A and O_3 particles, Brennan et al. [115] used a powder metallurgy approach, and they were successful in achieving the requisite stability and strength preservation after 100 °C exposure. However, in a different study, they reinforced N and Al with A and O_3 using melt infiltration, hot extrusion, and diffusion bonding procedures, and they discovered that the strength and ductility were lower than those of an unreinforced alloy due to strong fiber bonding. Long considered an attractive material for high-temperature applications, and $MoSi_2$ is an intermetallic combination. It is commonly used in electric heating elements that work at high air temperatures. It has a suggested operating range up to 1600 °C [115] based on its great oxidizing resistance, high thermal conductivity, acute melting temperature (202 °C), good heat resistance to corrosion, and low density (6.24 g/cm^3). As a result, intermetallic composites have been the subject of much research and development to improve their properties and meet the demands of certain applications.

2.3.6 Stainless Steel-Based MMCs

Steel and bronze are now often used as the matrix and reinforcing elements in the binder jet 3D printing process for MMC production. This is as a result of the appropriate moisture and the limited solubility of one material in another [120]. In subsequent investigations, Cordero et al. [121] filled gaps and decreased stress concentrators at particle necks to increase MMC strength. They did this by using binder jet iron followed by bronze infiltration. Various stages, including sintered iron powder and bronze infiltration, were seen. The infiltrating portion showed a transverse rupture strength of 570 MPa, which was almost 4 times greater than the sintered specimen, according to the mechanical behavior of the manufactured part. A small amount of boron compounds was reportedly added to the mixture as a sintering additive element to enhance densification in subsequent investigations, according to Do et al. [122] study. The ultimate relative density of B99.7% was attained after binder-blasted stainless steel 316 L was combined with various concentrations of sintering additives (0.75 wt%) including B, BC, and BN. As expected, the sintering additives improved densification behavior and produced a smooth surface finish.

3 Properties of Metallic Composites

Materials must function according to the designer's specifications and maintain the correct qualities in the environment throughout the standard work. According to market needs, equipment must be speedier, more comfortable, and less bothersome, and machinery must be increasingly cost-effective, eco-friendly, and lightweight. Every scientifically discovered material has certain characteristics depending on how it is applied in different aspects. Similarly, it has been observed that different classes of MMCs possess different properties depending on the technological field in which they are employed. As a result, prior knowledge of the properties of synthesized MMCs is required in order to use them for various applications. MMC is made from fibrous materials that have required features such as high wear and corrosion resistance, good hardness, strength. Unprecedented demand for enhanced lifespan and weight reduction, as well as lower production costs, drives the research and implementation of innovative materials. The rising usage of metallic composites is mostly due to their superior physical, mechanical, and tribological capabilities over matrix materials [57].

3.1 Physical Properties

According to experts, the most influential physical property of composites is density. The assessment of the comparative "heaviness" of materials with a constant volume is known as density. Matrix and reinforcement in the composites are measured in weight fractions (w) or volume fractions (v) in the composites; therefore, their capabilities were affected by these changes. The inclusion of these reinforcements in the basic alloys increased the density of the composites. Furthermore, if the density of the reinforcements is lower than that of the matrix materials, the density of the composite is substantially reduced. A lot of reinforcement materials are being used to vary the characteristics of the MMCs, such as Al_2O_3, SiC, B_4C, TiC. Manoharan et al. [123] used separate casting and extrusion procedures to create the composite of AA 1050 as the base material and SiC as the reinforcement and test the density of these composites. The density of composite materials was said to rise as reinforcing content increased. The mixtures rule was used to calculate theoretical densities for the composites, which were then compared to experimental densities. Composites made by the extrusion process offered better results. The composite contains the maximum porosity, 1.2%, and includes 8 wt% SiC. Similarly, the physical properties of magnesium alloys are influenced by their density, which makes them ideal to be used as a matrix in composite materials. Composite materials constructed of magnesium alloys reinforced with silicon carbide (SiC) dispersion particles exhibit 30–40% higher tensile strengths than unreinforced magnesium alloys. Another study by Reddy et al. [124] revealed the influence of reinforcing content (TiC) on the density of the composite. With the increase in reinforcement material, the authors saw a significant improvement in density. When the reinforcement content was increased from 10 to 15%, the density increased rapidly compared to 5–10% reinforcement. Further, it can also be seen in the Mg-based MMCs that the physical properties of magnesium alloys and their unique mix of mechanical, density, and Young's modulus characteristics make them particularly advantageous for use in various applications. Composite materials based on magnesium alloys reinforced with silicon carbide (SiC) dispersion particles have 30–40% superior tensile properties than unreinforced magnesium alloys while having a relatively low density of 2.0–2.1 g/cm^3. There are now viable applications for composite materials based on magnesium alloys reinforced with ceramic particles; one such material is Melram, which is manufactured as composite pipes and is anticipated to be used in aircraft and automobile technologies. Akhlagi et al. [125] studied a matrix of aluminum (commercial-grade Al, 99% purity) and copper (97% purity and 40 mesh sizes) reinforcements and silicon carbide. They created a composite using the compocasting process and used various weight percentages of copper powder (0, 1, 2, 3, 4, and 5 wt%). Magnesium (99% pure ingots) was also used in tiny amounts (fixed weight percentage 4 wt%) to increase wettability between the metal matrix and reinforcing particles. As a result, they reported the density of the samples using the Archimedes principle. The mixture's theoretical density may be estimated using a sample weighed in an electronic balance to an accuracy of 0.1 mg (density = %Al 2.71 + 4%Mg 1.7 + %Cu 8.93 + %SiC 3.21). Furthermore, the elemental magnesium turnings were studied in this work. The base material was 99.9% pure, while the reinforcing phase was made up of elemental 2.5% titanium particles with sizes ranging from 19,610 mm. With two separate volume percentages of time, the disintegrating melt deposition procedure was used to create monolithic and reinforced magnesium. The composite was created by superheating magnesium turnings with reinforcement particles to 750 °C in a graphite crucible in an Ar gas atmosphere. The density of polished Mg and Mg/Ti samples recovered from extruded rods was determined using the Archimedes principle. Distilled water was used as the immersion fluid. It has been reported that Mg/5.6Ti was collected in wt%-to-vol% ratio (5.6:2.2), and the density is determined to be (1.815 + 0.004 g/cm^3). As a result, when pure magnesium is reinforced with titanium, its dimensional stability improves. This indicates that the physical qualities improve [126, 127]. Again, copper was employed as the primary matrix in pure and spherical form, with Ti/SiCp reinforcing particles added, and the powder metallurgy technique was used to increase wear, hardness, and density. All of the additives chosen have a high hardness and a lower density than the matrix material. It was determined that the density of composite materials may decrease as the rate of reinforcement increases. At 1050 °C, the composite materials with 2% reinforcement had the greatest relative density values of 93.13% [128]. Hence, it can be concluded that density is the most promising function when considering physical properties.

3.2 *Mechanical Properties*

Many previous studies have looked at the effect of reinforcement on mechanical parameters such as hardness, tensile, compressive, flexural, and wear resistance. Hardness is the resistance of the material to deformation, indentation, or penetration under abrasion, drilling, impact, or scratching when these mechanical qualities are improved as needed. Brinell, Rockwell, and Vickers hardness tests are examples of hardness testing. Hardness and tensile strength are statistically associated with materials such as steel. Tensile strength refers to an object's ability to endure a drawing force that causes it to expand and lengthen in the opposite direction. Its most common measurement unit is force per cross-sectional area. Tensile strength qualities are considered valuable in fields such as materials science, mechanical engineering, and structural engineering. In comparison to traditional materials, MMC has higher tensile and fatigue strengths. MMCs are used to strengthen materials like Al_2O_3, SiC, B_4C, and others, increasing their elastic modulus, hardness, and wear resistance remarkably [129]. In particular, metallic composites with very low aspect ratios of particles had good toughness. Using various hardness testing methods, it was discovered that hardness accuracy is good. Microhardness testing, which would be a simple and easy procedure with direct observation, was performed to evaluate the interface bonding strength between the matrix and reinforcement. Lloyd et al. [130] studied the role of ceramic particles in increasing the bulk hardness of Al-MMCs and Mg-MMCs. Further, Uvaraja et al. [131] conducted tests on the hardness of Al 7075 composites with B_4C held constant at 3 wt% and SiC weight percent varying between 0 and 15. With SiC at 15% weight percentage, the maximum composite hardness was attained. According to the study's findings, composites are harder than their base materials and their hardness rises as the proportion of reinforcement in the composite increases. Furthermore, Suresha et al. [132] hypothesized that the hardness of a composite material made of SiC and graphite particles increased until a level of 2.5% of each was achieved, after which it started to decline as reinforcing concentration increased. We found that the porosity levels in the composite sample rose due to the strong influence of the reinforcing particles. As the porosity of metallic composites grew, their hardness dropped. Reddy et al. [124] looked at how the SiC and TiC contents affected the hardness of hybrid composites. The observation shows that the composites with reinforcement concentrations of 2.5 wt% SiC and 7.5 wt% TiC have higher hardness values. In addition, Krishnamoorthi et al. [129] reviewed that when Al is reinforced with different ceramic materials and produced by casting and powder metallurgy, it is observed that the mechanical and physical characteristics of matrix composites vary depending on factors like reinforcement percentages, particle size and shape, and thermal processing, and their behavior is modified to yield significant goodness. It was also claimed that when there is an increase in hardness following particular reinforcement, the density matrix drops significantly, resulting in increased elastic modulus and tensile strength above the base alloys. Again, the fundamental matrix was pure and spherical copper with Ti/SiC reinforcing particles added, and the powder metallurgy method was employed to increase wear and hardness. As a result of the testing, it was determined that samples sintered at 1050 °C had the highest hardness value. Further investigation revealed that the hardness of all test samples increased significantly in composites augmented with 6% powder particles before declining again. The strong cohesive force between the main matrix and the reinforcing particles influences the reported hardness values. In general, it has been discovered that the hardness rises as the reinforcement particle ratio increases [128]. An important aspect of the MMC's overall perspective on characteristics is wear resistance. The host metal's intrinsic wear resistance is increased by the presence of strong reinforcements. Weight loss and wear rate are commonly used to assess material wear. It has a greater impact on determining the wear property of the composite material since the precise wear rate, applied load, and wear distance are all taken into account [128]. In a broad sense, composites with particles with a low aspect ratio were found to have high hardness. Because of their lower density, high elastic modulus, and outstanding cutting and molding properties, Al–Si alloys are generally utilized in automotive accessories and aircraft components for the fabrication of various sections, including exhaust valves, bearing components, and rotor blades. And their reduced friction factor and increased toughness, on the other hand, could extend the serviceability of these alloys throughout machinery operation at the extremities. As an outcome, this alloy must have excellent resistance properties and a low friction coefficient under load variations, as well as many other exceptional characteristics [133]. Nowadays, B_4C particles are suited for usage as wear and corrosion-resistant materials due to their exceptional qualities on a large scale, that is, lowest density, melting temperatures, elastic modulus, and corrosion resistance. The incorporation of SiC-graphite reinforcement in copper (Cu) enhances the wear rate of the copper as compared to that of pure cast copper. When 10% weight and volume of graphite-SiC reinforcement is applied to Cu, the wear rate is 50% lower than that of purely molded copper, and a lubricating coating forms on the edge. This material is protected against attrition and plastic deformation by a tiny SiC particle [134]. Radhika et al. [135] have synthesized Al-based LM13/AlN reinforced metal matrix composites with good efficacy. They emphasized that the wear rate and wear distance are greatly influenced by the force applied during the wear test. When inspected, it is clear that the sample wear rate (SWR) values fall as sintering temperatures generally rise. The lowest SWR values were found to be 1.9590×10^6 mm^3/Nm for 950 °C, 1.7248×10^6 mm^3/Nm for 1000 °C, and 1.5545×10^6 mm^3/Nm at 1050 °C, respectively. The technique through which temperatures rise during the wear test is quite complicated. This process is influenced by a wide range of variables, including applied stress, shear rate, and heat conduction. The temperature variations that occur during the wear test, dependent on the force applied and the strengthening ratio, may be observed in Fig. 4 [128]. It also demonstrates that for all test samples, temperatures rise in lockstep with increases in applied loads.

3.3 Thermal Properties

Temperature and humidity are the key factors during the synthesis of composite materials, so it is essential to study them. There are various techniques used to determine the thermal characteristics of synthesized materials, such as differential scanning calorimetry (DSC), oxidation induction time (OIT), thermogravimetric analysis (TGA), and dynamic mechanical analysis (DMA). The effect of temperature can have a deleterious influence on some raw materials used during the manufacturing of composite materials [136, 137]. Many studies on composite materials have recently been conducted, such as the production of nanocomposites using montmorillonite and a compatibilizer. In differential scanning calorimetry (DSC) and thermogravimetric analysis (TGA), the samples were heated to temperatures between 20 and 600 °C at a rate of 20 °C/min in either the nitrogen or the atmosphere. The samples had a mass of roughly 10 mg. Thermogravimetric and DSC signals were recorded simultaneously. Correlated DSC and TGA measurements made it possible to distinguish between exothermic effects caused by polypropylene thermo-oxidation and exothermic effects caused by a material's decreased heat capacity and related to the emission of gas degradation products. It is now clear that a nanocomposite's increased resistance to degradation, as measured by the TGA technique, is influenced by the

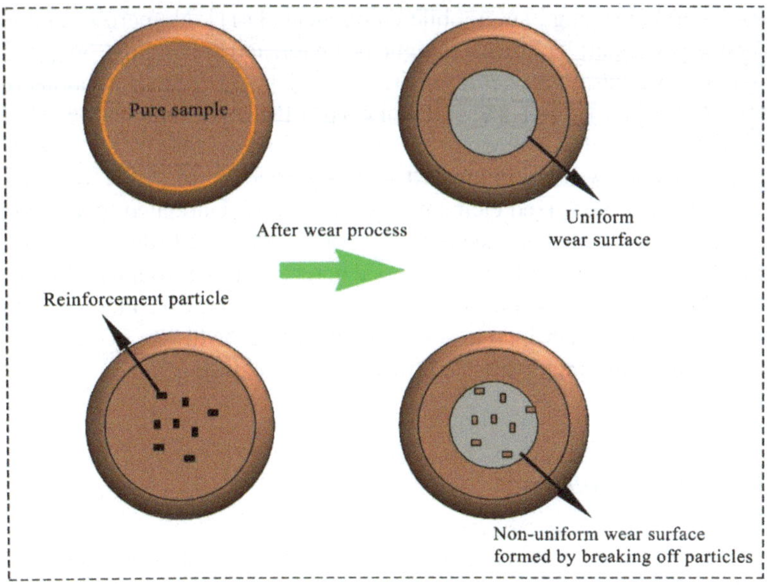

Fig. 4 Illustration showing the creation of a uniform surface following wear. Reproduced with permission from [128]

actual decreased oxygen transport and perceived increase in resistance associated with the depressed release of degradation gas products [138]. The rate of thermal expansion of ceramic materials (fibers, whiskers, or particles) is often lower than that of most metallic matrices. When the composite is exposed to a temperature change (planned or spontaneous), thermal stresses are produced in both components because of the clear restriction provided by the interfacial adhesion [15, 139]. The use of a single-crystalline copper matrix made up of large diameter tungsten fibers served as a demonstration of the relevance of thermal stresses in MMCs. They employed a displaced etch-pitting technique to identify misalignments in a single-crystal copper matrix. True, thermodynamic mismatch is challenging to avoid in any composite. Similar to this, the circulation of the reinforcing phase, the amount of matrix, and the amount of reinforcement can all be employed to change a composite's overall thermal expansion characteristics [140].

3.4 *Fatigue*

Fatigue is a dynamic loading phenomenon that causes a material's or product's mechanical properties to degrade, resulting in failure. Several high-volume composite material functionalities, such as vehicle parts, are subjected to cyclic load. The relevance of matrix microstructure on MMC fatigue behavior is exemplified by the case mentioned below, e.g., automobile components [141]. Numerous studies have examined the particulate Ti MMCs' fatigue behavior, and the findings are promising. Increased TiB concentration, according to Saito et al. [142], led to a higher increase in fatigue strength. For a Ti–6Al–4V + 10 vol% TiB MMC that could endure more than 107 cycles, it was possible to reach greater than 107 cycles at 600 MPa. Although the Ti–6Al–4V standard was not compared, this was superior to the Ti–6Al–4V alloy manufactured using the mixed element powder method. Unrelated to TiB particles, the material's pores served as the source of crack initiation; if the crack came into contact with a particle, it first became trapped before cutting through it. This demonstrates the TiB's strong interfacial connection with the matrix. Increased whisker length was linked to further increases in fatigue crack growth rate, which was found to encourage tortuous crack paths and fracture-tip shielding.

3.5 Electrical Properties

The mechanical characteristics of a hybrid composite made of sisal and Kenaf fibers improved with increase in tensile strength, flexural modulus, and impact strength. Due to the remarkable electrical properties of metallic composites, the electronic and technological era has been growing rapidly [14]. And, the need for rechargeable battery packs in electronic items like digital cameras, laptops, and electric vehicles (EVs) has skyrocketed in the recent years. The worldwide market for lithium-ion batteries (LIBs) is expected to reach $32 billion in 2020, indicating a burgeoning technological future [143–146]. Lithium iron phosphate ($LiFePO_4$) batteries are progressively replacing the traditional Pb (lead) acid batteries and are being used as electric car batteries. The resynthesis of $LiFePO_4$ from used batteries has become more cost-effective and accessible due to the expensive costs of raw resources as well as the time-consuming processing method. The $LiFePO_4$/reduced graphene oxide composite produced by the technique exhibits improved electrochemical behavior, comprising smooth CV curves, minimal electrochemical impedance, large capacity, flat voltage plateaus, and significant columbic efficiency, as well as consistent cycle outputs at 0.2C and 1C and outstanding rate capacity [14, 143]. Further, when talking about Cu-based MMCs, it should be noted that even a small amount of alloying elements can significantly reduce copper's electrical conductivity. For example, adding 0.3 percent Zn to copper reduces its electrical conductivity to 85 percent IACS, 1.25% Al to 70% IACS, and 0.1% P to 50% IACS, while keeping in mind that pure copper has an electrical conductivity of 58 MSm-1, or 100% IACS. It was shown that the mechanical characteristics of traditional alloying techniques are significantly correlated with reduced electrical conductivity [147].

4 Processing of Materials

Metal matrix composite materials are designed by merging two or more materials together in such a way to achieve specified qualities. The final material will have the best of both the metallic matrix and the reinforcing components, offering more flexibility in design [148]. Although the same composition and quantity of the constituents are involved, one can change the characteristic profile of materials by changing their manufacturing techniques, processing, and finishing as well as reinforcing components [2]. A different type of deposition technique or an in-situ method of adding a reinforcing phase may be used for designing metal matrix composites [15]. The methodologies used to manufacture MMCs can be categorized based on their treatment, whether the metal matrix is in liquid or solid form. Furthermore, both of these paths are classified in distinct ways, as seen in the diagram below. In this section, various manufacturing techniques like casting methods, plasma spray decomposition, powder metallurgy are briefly discussed. Some of the manufacturing techniques are shown in Fig. 5.

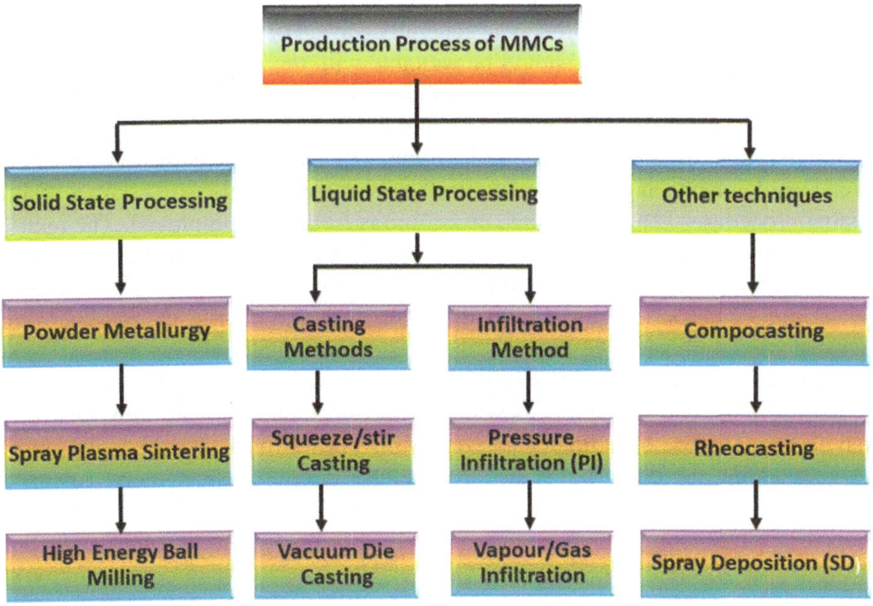

Fig. 5 Schematic representation of production of MMCs by various techniques

4.1 Liquid-State Processing

Most of the time, liquid-state processing allows for the creation of complicated geometries with strong interfacial bonds and effective dispersion. The effective bonding and uniform distribution of reinforcement are what give the material its superior mechanical properties. Stir casting, squeeze casting, and other methods of liquid-state processing are the most popular. Here, both casting methods that can be used to produce MMCs are discussed.

4.1.1 Production of MMCs by Casting Technique

The most popular and well-known technique for producing MMCs is casting. It can affordably manufacture increasingly complex MMCs in a variety of materials. The reinforcement is mixed with molten metal, poured into a mold cavity, and then allowed to settle to produce the necessary casting [55]. Casting of MMCs with particulate or reinforcement may look remarkably similar to casting metals on the surface. Due to their low melting viscosities, metals can be casted into net-shaped product forms [15]. The combining process is generally carried out at atmospheric pressure, and reinforcing components should be wettable with the molten metal alloy [147]. Among the most popular manufacturing methods for MMCs are squeeze casting and pressure casting. In this process, the molten state of the material solidifies under very high pressure after just a slow mold filling, which results in a fine-grained structure. The manufacturing of composite materials is conceivable using both pressure casting procedures [2]. A two-stage procedure is frequently employed. The melt is forced into the form at low pressure in the first step, then solidified at high pressure in the second stage. This protects the product from being damaged by overly rapid penetration. Because the duration of the infiltration and, hence, the reaction time are very short with squeeze casting, relatively reactive materials can be used. Now there are two types of squeeze casting: direct and indirect. Direct squeeze casting is used to make composite pieces with a basic form, and they are inexpensive or easy to use. The use of indirect squeeze casting allows for the production of more complex composite parts, but it also necessitates the purchase of more expensive casting dies [147]. The advantage of this method is the ability to build complexly shaped structural components with partial reinforcement to strengthen regions that are subjected to high stresses during service. The more detailed demonstration is shown in Fig. 6 [149], along with the depiction of the casting setup. In this casting method, reinforcement materials have been pictorially represented, along with their properties and applicative areas. Furthermore, a very fine representation of additives, wetting agents, and challenges should be followed during fabrication.

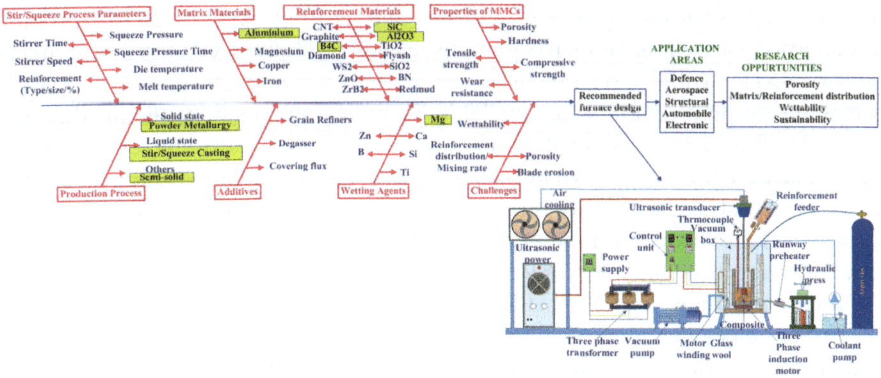

Fig. 6 Framework for producing better-quality AMMCs by a stir casting process. Reproduced with permission from [149]

4.2 *Solid-State Processing*

The process of connecting the matrix material and reinforcement during the solid-state processing of MMCs involves a greater temperature and pressure, which causes mutual diffusion between the materials. Here, the most traditional process of fabrication, namely powder metallurgy, is briefly discussed. However, there are a lot of other processes that follow the route of solid-state mechanisms [55].

4.2.1 Powder Metallurgy

Powder metallurgy technique is focused on the classic mixing of matrix powders with reinforcing components (dispersion powders, platelets, and ceramics), accompanied by cold processing and sintering, and then plastic machining occurs (forging, extrusion). Whenever cold plastic processing is utilized, a green component is initially sintered, and when hot plastic working occurs, then cold pressing is applied [147]. Powder metallurgy technology also allows for matrix alloy composition and microstructural improvements that are only attainable with quickly solidified powders. Basically, low temperatures are used in powder metallurgy processing, which theoretically allows for greater control of interface kinematics [150]. The powder metallurgy manufacturing method also enables the use of matrix alloy composition and microstructural improvements that are only feasible with quickly solidified powders. Although composites have greater mechanical properties than alloys, flaws in composites such as porosity, poor wetting, and interfacial energies can be minimized using powder metallurgy, resulting in improved mechanical performance. It is less expensive and more widely available when compared to other ways. Powder metallurgy MMCs have better strength and toughness. Using this technique, the reinforcing distribution is homogeneous and uniform. As a result, the production of powder-based metal matrix composites has revolutionized both research and industry [151]. For the fabrication of ferrous and non-ferrous metal powders, a variety of manufacturing procedures are available, including atomization, chemical reduction, and milling. Each method of production has its own set of advantages. Following the compaction process, the synthesized powders are shaped into the proper size and shape. The billets are heated in a furnace for a certain time and temperature to remove porosity and produce high strength. Most of the pieces may be used right away after sintering, but others may require further processing to relieve tension [152].

Powder production (P/M): In the P/M process, the raw material is powder, which can be pure elements, elemental blends, or pre-alloyed powders. The powders are made using a variety of techniques. Atomization is the most frequent method for producing powders due to its particular properties. Powders of stainless steel, nickel alloys, and titanium can be made via this technique. Similarly, chemical reduction is an effective process for producing different types of powders such as iron, copper, tungsten, and molybdenum. Also, electrolysis may be used to create iron and silver powders [150].

Blending: It is the process of combining powders with the same basic content but different particle sizes and forms. It is done to obtain a uniform distribution of particle sizes and to minimize porosity [152].

Compaction: Utilizing proper punch and die to generate green compacts using mechanical or hydraulic presses, powder mixtures are typically compacted. Powder mixes are cold crushed at an appropriate pressure using a uniaxial press [152].

Sintering: The procedure of binding the particles collectively by heating them in mesh belt furnaces, walking beam furnaces, and batch furnaces is among the few that are used for sintering. According to the numerous studies, components with a high surface and desirable characteristics are produced at a maximum sintering temperature. A controlled atmosphere in a furnace is referred to as "sintering." Further, a pictorial representation of powder metallurgy is shown in Fig. 7 [153].

Using mechanical mills along with a hot extrusion process, the mechanical properties of AlCuMg-B_4C nanocomposites were evaluated as a function of milling time and percentage of reinforcement material. The mechanical properties of the

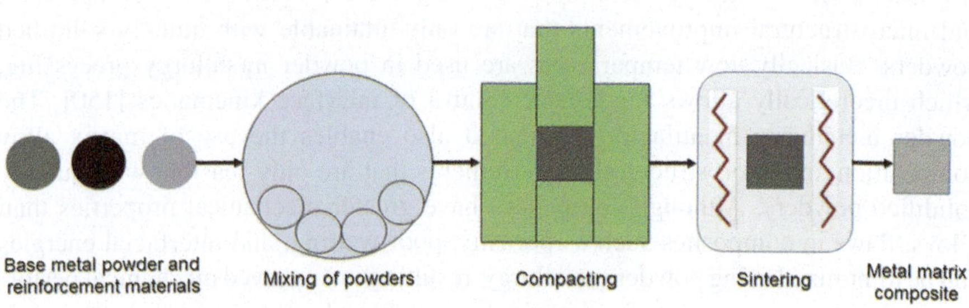

Fig. 7 Pictorial representation of the different stages involved in the powder metallurgy process. Reproduced with permission from [153]

created composites were assessed after 25 h of milling the composite powder, which contained aluminum alloy and boron carbide powders. The composite powder was then hot pressed in a vacuum at 560 °C and 300 MPa. The amount of reinforcement and milling time increase as the hot-pressed density of the composite decreases. The composite with 15% B_4C reinforcements milled for 25 h has a lower density than the alloy that is not reinforced. Tensile tests on all of the manufactured composites reveal that the 10% weight B_4C content composite material that was milled for six hours had the greatest tensile, measuring 332 MPa [154]. Furthermore, using a 316 L/polycarbosilane (PCS) combination, steel matrix composites were created using a novel semi-powder metallurgy technique that comprised SPS and solution-assisted wet mixing. By first coating the initial 316 L powder with PCS, then SPS, PCS, and 316 L powders were obtained. Tensile test results show that increasing PCS percentage enhances yield strength (YS) and ultimate tensile strength when compared to pure 316 L. The network structure, reinforced microstructures, and solid solution effects, as well as the high compatibility between M7C3 and the austenitic matrix, all contribute to the composites' increased tensile strength [155]. Hence, it was observed that the powder metallurgy route is the traditional as well as conventional approach to fabricating various MMCs.

5 Conclusion

The composite materials have gained enormous recognition in the domains of research and manufacturing as they possess useful characteristics and meet necessities that cannot be attained by the different constituting materials acting alone. This prominence can be attributed to the fact that composite materials have become more durable and lightweight. Up to this point, researchers have made attempts to combine a wide range of component materials in order to construct composites, and a number of experiments have been carried out to explore the characteristics of these composites and how they change their phase. Furthermore, in order to obtain the maximum possible material efficiency, researchers have adjusted the composition percentage as well as other key criteria that go into the construction of a better composite. Researchers have devised a large number of innovative manufacturing procedures in order to circumvent the inherent inefficiencies of traditional production methods, which include delays in time and an increased level of complexity. The vast majority of them were successful in enhancing the qualities of the materials so that they might be used for possible applications in important industries, including aerospace, automotive, chemical, and sporting goods. But in the world, we live in now, biocomposites made from natural, biodegradable materials are taking over the market. This is due to the fact that these biocomposites are showing qualities that were previously unknown to be environmentally benign. Apart from the significant research initiatives that have been undertaken in the field of composite fabrication, there is a vast amount of room for improvement in the field of developing advanced manufacturing techniques in conjunction with automation that enables an increase in the productivity and efficiency of materials. It is worthwhile to do more research into the hybridization of natural biodegradable materials with synthetic elements as components of the composite structure. This combination will eventually help in increasing the material's strength and stiffness while also being eco-friendly.

References

1. Bahl S (2020) Fiber reinforced metal matrix composites—a review. Mater Today Proc 39:317–323. https://doi.org/10.1016/j.matpr.2020.07.423
2. Kainer KU (2006) Basics of metal matrix composites. Metal Matrix Compos Cust Mater Autom Aerosp Eng 1–54. https://doi.org/10.1002/3527608117.ch1
3. Abramovich H (2017) Introduction to composite materials. In: Stability and vibrations of thin-walled composite structures, pp 1–47. https://doi.org/10.1016/B978-0-08-100410-4.00001-6
4. Dadkhah M, Mosallanejad MH, Iuliano L, Saboori A (2021) A comprehensive overview on the latest progress in the additive manufacturing of metal matrix composites: potential, challenges, and feasible solutions. Acta Metall Sin (English Lett) 34:1173–1200. https://doi.org/10.1007/s40195-021-01249-7
5. Yakout M, Elbestawi MA (2017) Additive manufacturing of composite materials: an overview. In: 6th international conference on virtual machining process technology (VMPT), Montréal, pp 1–8
6. Moheimani SK, Dadkhah M, Mosallanejad MH, Saboori A (2021) Fabrication and characterization of the modified ev31-based metal matrix nanocomposites. Metals (Basel) 11:1–9. https://doi.org/10.3390/met11010125
7. Wang B, Ruan T, Chen Y, Jin F, Peng L, Zhou Y, Wang D, Dou S (2020) Graphene-based composites for electrochemical energy storage. Energy Storage Mater 24:22–51. https://doi.org/10.1016/j.ensm.2019.08.004
8. Elanchezhian C, Vijaya Ramnath B, Ramakrishnan G, Sripada Raghavendra KN, Muralidharan M, Kishore V (2018) Review on metal matrix composites for marine applications. Mater Today Proc Elsevier 1211–1218. https://doi.org/10.1016/j.matpr.2017.11.203
9. Alaneme KK, Okotete EA, Fajemisin AV, Bodunrin MO (2019) Applicability of metallic reinforcements for mechanical performance enhancement in metal matrix composites: a review. Arab J Basic Appl Sci 26:311–330. https://doi.org/10.1080/25765299.2019.1628689
10. Miracle DB (2005) Metal matrix composites—from science to technological significance. Compos Sci Technol 65:2526–2540. https://doi.org/10.1016/j.compscitech.2005.05.027
11. Clyne TW, Withers PJ (1993) An introduction to metal matrix composites. https://doi.org/10.1017/cbo9780511623080
12. Kennedy AR, Wyatt SM (2000) The effect of processing on the mechanical properties and interfacial strength of aluminium/TiC MMCs. Compos Sci Technol 60:307–314. https://doi.org/10.1016/S0266-3538(99)00125-6
13. Miracle DB, Donaldson SL, Henry SD, Scott WW, Brown Kelly Ferjutz Edward Kubel JJ, Heather Lampman Elizabeth Marquard Beverly Musgrove Mary Jane Riddlebaugh Juli Williamson J (2001) ASM handbook, vol 21. Composites prepared under the direction of the ASM International Handbook Committee
14. Rajak DK, Pagar DD, Kumar R, Pruncu CI (2019) Recent progress of reinforcement materials: a comprehensive overview of composite materials. J Mater Res Technol 8:6354–6374. https://doi.org/10.1016/j.jmrt.2019.09.068
15. Chawla KK (2012) Composite materials. Springer, New York, NY. https://doi.org/10.1007/978-0-387-74365-3

16. Dai D, Fan M (2013) Wood fibres as reinforcements in natural fibre composites: structure, properties, processing and applications. Nat Fibre Compos Mater Process Appl 3–65. https://doi.org/10.1533/9780857099228.1.3
17. Pardo A, Merino MC, Merino S, Viejo F, Carboneras M, Arrabal R (2005) Influence of reinforcement proportion and matrix composition on pitting corrosion behaviour of cast aluminium matrix composites (A3xx.x/SiCp). Corros Sci 47:1750–1764. https://doi.org/10.1016/J.CORSCI.2004.08.010
18. Wicks SS, de Villoria RG, Wardle BL (2010) Interlaminar and intralaminar reinforcement of composite laminates with aligned carbon nanotubes. Compos Sci Technol 70:20–28. https://doi.org/10.1016/J.COMPSCITECH.2009.09.001
19. Videira-Quintela D, Martin O, Montalvo G (2021) Recent advances in polymer-metallic composites for food packaging applications. Trends Food Sci Technol 109:230–244. https://doi.org/10.1016/j.tifs.2021.01.020
20. Realini CE, Marcos B (2014) Active and intelligent packaging systems for a modern society. Meat Sci 98:404–419. https://doi.org/10.1016/j.meatsci.2014.06.031
21. Eisenbarth E, Meyle J, Nachtigall W, Breme J (1996) Influence of the surface structure of titanium materials on the adhesion of fibroblasts. Biomaterials 17:1399–1403. https://doi.org/10.1016/0142-9612(96)87281-4
22. Cheng M, Chen W, Weerasooriya T (2005) Mechanical properties of Kevlar® KM2 single fiber. J Eng Mater Technol Trans ASME 127:197–203. https://doi.org/10.1115/1.1857937
23. Guo N, Leu MC (2013) Additive manufacturing: technology, applications and research needs. Front Mech Eng 8:215–243. https://doi.org/10.1007/s11465-013-0248-8
24. Zhang C (2014) Understanding the wear and tribological properties of ceramic matrix composites. In: Advances in ceramic matrix composites. Elsevier Ltd., pp 312–339. https://doi.org/10.1533/9780857098825.2.312
25. Donald IW, McMillan PW (1976) Review ceramic-matrix composites. J Mater Sci 11:949–972
26. Kopeliovich D (2014) Advances in the manufacture of ceramic matrix composites using infiltration techniques. Adv Ceram Matrix Compos 79–108. https://doi.org/10.1533/9780857098825.1.79
27. Richerson DW, Lee WE (2018) Modern ceramic engineering. CRC Press. https://doi.org/10.1201/9780429488245
28. Naslain RR (2001) Ceramic matrix composites: matrices and processing. In: Encycl Mater Sci Technol Elsevier 1060–1066. https://doi.org/10.1016/b0-08-043152-6/00196-0
29. Schmidt S, Beyer S, Knabe H, Immich H, Meistring R, Gessler A (2004) Advanced ceramic matrix composite materials for current and future propulsion technology applications. Acta Astronaut 55:409–420. https://doi.org/10.1016/j.actaastro.2004.05.052
30. An Q, Chen J, Ming W, Chen M (2021) Machining of SiC ceramic matrix composites: a review. Chinese J Aeronaut 34:540–567. https://doi.org/10.1016/j.cja.2020.08.001
31. Benzait Z, Trabzon L (2018) A review of recent research on materials used in polymer–matrix composites for body armor application. J Compos Mater 52:3241–3263. https://doi.org/10.1177/0021998318764002
32. Advani SG, Hsiao KT (2012) Manufacturing techniques for polymer matrix composites (PMCs). https://doi.org/10.1533/9780857096258
33. Kessler MR (2012) Polymer matrix composites: a perspective for a special issue of polymer reviews. Polym Rev 52:229–233. https://doi.org/10.1080/15583724.2012.708004
34. Ray BC (2006) Temperature effect during humid ageing on interfaces of glass and carbon fibers reinforced epoxy composites. J Colloid Interface Sci 298:111–117. https://doi.org/10.1016/j.jcis.2005.12.023
35. Kaelble DH, Dynes PJ, Maus L (2006) Hydrothermal aging of composite materials part 1: interfacial aspects. 8:121–144. https://doi.org/10.1080/00218467608075078
36. Marom G, Broutman LJ (1981) Moisture penetration into composites under external stress. Polym Compos 2:132–136. https://doi.org/10.1002/PC.750020310
37. Chohan JS, Boparai KS, Singh R, Hashmi MSJ (2020) Manufacturing techniques and applications of polymer matrix composites: a brief review. Adv Mater Process Technol. https://doi.org/10.1080/2374068X.2020.1835012

38. Van De Velde K, Kiekens P (2001) Thermoplastic pultrusion of natural fibre reinforced composites. Compos Struct 54:355–360. https://doi.org/10.1016/S0263-8223(01)00110-6
39. Cheng QF, Wang JP, Wen JJ, Liu CH, Jiang KL, Li QQ, Fan SS (2010) Carbon nanotube/epoxy composites fabricated by resin transfer molding. Carbon N Y 48:260–266. https://doi.org/10.1016/j.carbon.2009.09.014
40. Masood SH, Song WQ (2004) Development of new metal/polymer materials for rapid tooling using Fused deposition modelling. Mater Des 25:587–594. https://doi.org/10.1016/j.matdes.2004.02.009
41. Allaoui A, Bai S, Cheng HM, Bai JB (2002) Mechanical and electrical properties of a MWNT/epoxy composite. Compos Sci Technol 62:1993–1998. https://doi.org/10.1016/S0266-3538(02)00129-X
42. Schwartz RT, Schwartz HS (1967) Characteristics of boron fibers and boron-fiber-reinforced plastic composites. AIAA J 5:289–295. https://doi.org/10.2514/3.3954
43. Idicula M, Malhotra SK, Joseph K, Thomas S (2005) Dynamic mechanical analysis of randomly oriented intimately mixed short banana/sisal hybrid fibre reinforced polyester composites. Compos Sci Technol 65:1077–1087. https://doi.org/10.1016/j.compscitech.2004.10.023
44. Ku H, Wang H, Pattarachaiyakoop N, Trada M (2011) A review on the tensile properties of natural fiber reinforced polymer composites. Compos Part B Eng 42:856–873. https://doi.org/10.1016/j.compositesb.2011.01.010
45. Ralph B, Yuen HC, Lee WB (1997) The processing of metal matrix composites—an overview. J Mater Process Technol 63:339–353. https://doi.org/10.1016/S0924-0136(96)02645-3
46. Bhadeshia H, Honeycombe R (2017) Steels microstructure and properties. Elsevier Publications. ISBN: 9780081002728
47. Bahl S (2020) Fiber reinforced metal matrix composites—a review. Mater Today Proc Elsevier 317–323. https://doi.org/10.1016/j.matpr.2020.07.423
48. Poovazhagan L, Kalaichelvan K, Rajadurai A, Senthilvelan V (2013) Characterization of hybrid silicon carbide and boron carbide nanoparticles-reinforced aluminum alloy composites. https://doi.org/10.1016/j.proeng.2013.09.143
49. Xavior MA, Kumar JPA (2017) Machinability of hybrid metal matrix composite—a review. https://doi.org/10.1016/j.proeng.2017.01.264
50. Lu D, Jiang Y, Zhou R (2013) Wear performance of nano-Al2O3 particles and CNTs reinforced magnesium matrix composites by friction stir processing. Wear 305:286–290. https://doi.org/10.1016/j.wear.2012.11.079
51. Dong S, Zhou J, Hui D, Wang Y, Zhang S (2015) Size dependent strengthening mechanisms in carbon nanotube reinforced metal matrix composites. Compos Part A Appl Sci Manuf 68:356–364. https://doi.org/10.1016/j.compositesa.2014.10.018
52. Mortensen A, Llorca J (2010) Metal matrix composites. Annu Rev Mater Res 40:243–270. https://doi.org/10.1146/annurev-matsci-070909-104511
53. Yu WH, Sing SL, Chua CK, Kuo CN, Tian XL (2019) Particle-reinforced metal matrix nanocomposites fabricated by selective laser melting: a state of the art review. Prog Mater Sci 104:330–379. https://doi.org/10.1016/j.pmatsci.2019.04.006
54. Barrera EV, Sims J, Provenzano V, Milliken J, Holtz RL (1994) Processing of fullerene-reinforced composites. J Mater Res 9:2662–2669. https://doi.org/10.1557/JMR.1994.2662
55. Samal P, Vundavilli PR, Meher A, Mahapatra MM (2020) Recent progress in aluminum metal matrix composites: a review on processing, mechanical and wear properties. J Manuf Process 59:131–152. https://doi.org/10.1016/j.jmapro.2020.09.010
56. Gowrishankar TP, Manjunatha LH, Sangmesh B (2020) Mechanical and wear behaviour of Al6061 reinforced with graphite and TiC Hybrid MMC's. Mater Res Innov 24:179–185. https://doi.org/10.1080/14328917.2019.1628497
57. Veličković S, Garić S, Stojanović B, Vencl A (2016) Tribological properties of aluminium matrix nanocomposites. Appl Eng Lett 1:72–79
58. Orhadahwe TA, Ajide OO, Adeleke AA, Ikubanni PP (2020) A review on primary synthesis and secondary treatment of aluminium matrix composites. Arab J Basic Appl Sci 27:389–405. https://doi.org/10.1080/25765299.2020.1830529

59. ANON (1968) Metal matrix composites. Met Matrix Compos Ind, pp 9–38. https://doi.org/10.1007/978-0-387-74365-3_6
60. Dursun T, Soutis C (2014) Recent developments in advanced aircraft aluminium alloys. Mater Des 56:862–871. https://doi.org/10.1016/j.matdes.2013.12.002
61. Rohatgi P (1991) Cast aluminum-matrix composites for automotive applications. JOM 43:10–15. https://doi.org/10.1007/BF03220538
62. Wanhill RJH (2017) Fatigue requirements for aircraft structures. Springer, Singapore, pp 331–352. https://doi.org/10.1007/978-981-10-2143-5_16
63. Stojanovic B, Babic M, Mitrovic S, Vencl A, Miloradovic N, Pantic M (2013) Tribological characteristics of aluminium hybrid composites reinforced with silicon carbide and graphite. A review. J Balk Tribol Assoc 19:83–96
64. Vasudevan AK, Doherty RD (eds) (2012) Aluminum alloys-contemporary research and applications: contemporary research and applications. Elsevier
65. Ramnath BV, Elanchezhian C, Annamalai RM, Aravind S, Atreya TS, Vignesh V, Subramanian C (2014) Aluminium metal matrix composites—a review. Rev Adv Mater Sci 55–60
66. Cavaliere P, Jahantigh F, Shabani A, Sadeghi B (2018) Influence of SiO_2 nanoparticles on the microstructure and mechanical properties of Al matrix nanocomposites fabricated by spark plasma sintering. Compos Part B Eng 146:60–68. https://doi.org/10.1016/j.compositesb.2018.03.045
67. Daimler SE, Ekambaram S, Murugan N (2015) Synthesis and characterization of aluminium alloy AA6061-alumina metal matrix composite optimization of flux cored arc welding process parameter using genetic and memetic algorithms view project. Int J Curr Eng Technol 5
68. Gómez L, Busquets-Mataix D, Amigó V, Salvador MD (2009) Analysis of boron carbide aluminum matrix composites. J Compos Mater 43:987–995. https://doi.org/10.1177/0021998308097731
69. Hashim J, Looney L, Hashmi MSJ (1999) Metal matrix composites: production by the stir casting method. J Mater Process Technol 92–93:1–7. https://doi.org/10.1016/S0924-0136(99)00118-1
70. Taha MA (2001) Industrialization of cast aluminum matrix composites (AMCCs). Mater Manuf Process 16:619–641. https://doi.org/10.1081/AMP-100108625
71. Rahimi B, Khosravi H, Haddad-Sabzevar M (2015) Microstructural characteristics and mechanical properties of Al-2024 alloy processed via a rheocasting route. Int J Miner Metall Mater 22:59–67. https://doi.org/10.1007/s12613-015-1044-8
72. Yuvaraj N, Pradeep Kumar M (2015) Multiresponse optimization of abrasive water jet cutting process parameters using TOPSIS approach. 30:882–889. https://doi.org/10.1080/10426914.2014.994763
73. Ahmadi E, Ranjkesh M, Mansoori E, Fattahi M, Mojallal RY, Amirkhanlou S (2017) Microstructure and mechanical properties of Al/ZrC/TiC hybrid nanocomposite filler metals of tungsten inert gas welding fabricated by accumulative roll bonding. J Manuf Process 26:173–177. https://doi.org/10.1016/j.jmapro.2017.02.012
74. Lindroos VK, Talvitie MJ (1995) Recent advances in metal matrix composites. J Mater Process Technol 53:273–284. https://doi.org/10.1016/0924-0136(95)01985-N
75. Kumar D, Phanden RK, Thakur L (2020) A review on environment friendly and lightweight magnesium-based metal matrix composites and alloys. Mater Today Proc 38:359–364. https://doi.org/10.1016/j.matpr.2020.07.424
76. Dehnavi V, Binns WJ, Noël JJ, Shoesmith DW, Luan BL (2018) Growth behaviour of low-energy plasma electrolytic oxidation coatings on a magnesium alloy. J Magnes Alloy 6:229–237. https://doi.org/10.1016/j.jma.2018.05.008
77. Hassan SF, Gupta M (2002) Development of a novel magnesium-copper based composite with improved mechanical properties. Mater Res Bull 37:377–389. https://doi.org/10.1016/S0025-5408(01)00772-3
78. Dash D, Samanta S, Rai RN (2018) Study on fabrication of magnesium based metal matrix composites and its improvement in mechanical and tribological properties—a review. https://doi.org/10.1088/1757-899X/377/1/012133

79. Lim CYH, Lim SC, Gupta M (2003) Wear behaviour of SiCp-reinforced magnesium matrix composites. Wear 255:629–637. https://doi.org/10.1016/S0043-1648(03)00121-2
80. Jayalakshmi S, Kailas SV, Seshan S (2002) Tensile behaviour of squeeze cast AM100 magnesium alloy and its Al2O3 fibre reinforced composites. Compos Part A Appl Sci Manuf 33:1135–1140. https://doi.org/10.1016/S1359-835X(02)00049-0
81. Jiang QC, Wang HY, Ma BX, Wang Y, Zhao F (2005) Fabrication of B4C participate reinforced magnesium matrix composite by powder metallurgy. J Alloys Compd 386:177–181. https://doi.org/10.1016/j.jallcom.2004.06.015
82. Poddar P, Srivastava VC, De PK, Sahoo KL (2007) Processing and mechanical properties of SiC reinforced cast magnesium matrix composites by stir casting process. Mater Sci Eng A 460–461:357–364. https://doi.org/10.1016/j.msea.2007.01.052
83. Tang HG, Ma XF, Zhao W, Cai SG, Zhao B, Qiao ZH (2007) The mechanical properties of magnesium matrix composites reinforced with 10 wt.% W14Al86 alloy particles. J Alloys Compd 437:285–288. https://doi.org/10.1016/j.jallcom.2006.07.103
84. Dudina DV, Georgarakis K, Li Y, Aljerf M, LeMoulec A, Yavari AR, Inoue A (2009) A magnesium alloy matrix composite reinforced with metallic glass. Compos Sci Technol 69:2734–2736. https://doi.org/10.1016/j.compscitech.2009.08.001
85. Aghion E, Bronfin B (2000) Magnesium alloys development towards the 21st century. Mater Sci Forum 350:19–28. https://doi.org/10.4028/WWW.SCIENTIFIC.NET/MSF.350-351.19
86. Joost WJ, Krajewski PE (2017) Towards magnesium alloys for high-volume automotive applications. Scr Mater 128:107–112. https://doi.org/10.1016/j.scriptamat.2016.07.035
87. Kulekci MK (2008) Magnesium and its alloys applications in automotive industry. Int J Adv Manuf Technol 39:851–865. https://doi.org/10.1007/s00170-007-1279-2
88. Dutta S, Gupta S, Roy M (2020) Recent developments in magnesium metal-matrix composites for biomedical applications: a review. ACS Biomater Sci Eng 6:4748–4773. https://doi.org/10.1021/acsbiomaterials.0c00678
89. Chen Y, Xu Z, Smith C, Sankar J (2014) Recent advances on the development of magnesium alloys for biodegradable implants. Acta Biomater 10:4561–4573. https://doi.org/10.1016/j.actbio.2014.07.005
90. Saha P, Roy M, Datta MK, Lee B, Kumta PN (2015) Effects of grain refinement on the biocorrosion and in vitro bioactivity of magnesium. Mater Sci Eng C 57:294–303. https://doi.org/10.1016/j.msec.2015.07.033
91. Brar HS, Wong J, Manuel MV (2012) Investigation of the mechanical and degradation properties of Mg-Sr and Mg-Zn-Sr alloys for use as potential biodegradable implant materials. J Mech Behav Biomed Mater 7:87–95. https://doi.org/10.1016/j.jmbbm.2011.07.018
92. Zhang S, Zhang X, Zhao C, Li J, Song Y, Xie C, Tao H, Zhang Y, He Y, Jiang Y, Bian Y (2010) Research on an Mg-Zn alloy as a degradable biomaterial. Acta Biomater 6:626–640. https://doi.org/10.1016/j.actbio.2009.06.028
93. Zhang E, He W, Du H, Yang K (2008) Microstructure, mechanical properties and corrosion properties of Mg-Zn-Y alloys with low Zn content. Mater Sci Eng A 488:102–111. https://doi.org/10.1016/j.msea.2007.10.056
94. Lei T, Ouyang C, Tang W, Li LF, Zhou LS (2010) Enhanced corrosion protection of MgO coatings on magnesium alloy deposited by an anodic electrodeposition process. Corros Sci 52:3504–3508. https://doi.org/10.1016/j.corsci.2010.06.028
95. Ali M, Hussein MA, Al-Aqeeli N (2019) Magnesium-based composites and alloys for medical applications: a review of mechanical and corrosion properties. J Alloys Compd 792:1162–1190. https://doi.org/10.1016/j.jallcom.2019.04.080
96. Yu JX, Li TH (2011) Distinct biological effects of different nanoparticles commonly used in cosmetics and medicine coatings. Cell Biosci 1:1–9. https://doi.org/10.1186/2045-3701-1-19
97. Kwon SH, Jun YK, Hong SH, Kim HE (2003) Synthesis and dissolution behavior of β-TCP and HA/β-TCP composite powders. J Eur Ceram Soc 23:1039–1045. https://doi.org/10.1016/S0955-2219(02)00263-7
98. Yu X, Tang X, Gohil SV, Laurencin CT (2015) Biomaterials for bone regenerative engineering. Adv Healthc Mater 4:1268–1285. https://doi.org/10.1002/adhm.201400760

99. Hayat MD, Singh H, He Z, Cao P (2019) Titanium metal matrix composites: an overview. Compos Part A Appl Sci Manuf 121:418–438. https://doi.org/10.1016/j.compositesa.2019.04.005
100. Liu C, Huang LJ, Geng L, Jiao Y, Tang A (2015) In Situ synthesis of (TiC + Ti_3SiC_2 + Ti_5Si_3)/Ti_6Al_4V composites with tailored two-scale architecture. Adv Eng Mater 17:933–941. https://doi.org/10.1002/adem.201400585
101. Kim IY, Choi BJ, Kim YJ, Lee YZ (2011) Friction and wear behavior of titanium matrix (TiB+TiC) composites. Wear 271:1962–1965. https://doi.org/10.1016/j.wear.2010.12.072
102. An Q, Huang LJ, Bao Y, Zhang R, Jiang S, Geng L, Xiao M (2018) Dry sliding wear characteristics of in-situ TiBw/Ti6Al4V composites with different network parameters. Tribol Int 121:252–259. https://doi.org/10.1016/j.triboint.2018.01.053
103. Stanley H, Dear M, Doel TJA, Bowen P (2018) Fatigue crack growth resistance of titanium metal matrix composites. Miner Met Mater Ser 87–102. https://doi.org/10.1007/978-3-319-72853-7_7
104. Weiner S, Wagner HD (1998) The material bone: structure-mechanical function relations. Annu Rev Mater Sci 28:271–298. https://doi.org/10.1146/annurev.matsci.28.1.271
105. Tenney DR, Davis JGJ, Pipes RB, Johnston N (2009) Composite materials development: lessons learned and future challenges. Structures.Dhu.Edu.Cn. 58
106. Singh MK, Gautam RK (2017) Synthesis of copper metal matrix hybrid composites using stir casting technique and its mechanical, optical and electrical behaviours. Trans Indian Inst Met 70:2415–2428. https://doi.org/10.1007/s12666-017-1103-0
107. Kumar Singh M, Kumar Gautam R, Prakash R, Ji G (2018) Mechanical and corrosion behaviors of developed copper-based metal matrix composites. IOP Conf Ser Mater Sci Eng 330:012021. https://doi.org/10.1088/1757-899X/330/1/012021
108. Heringhaus F, Raabe D (1996) Recent advances in the manufacturing of copper-base composites. J Mater Process Technol 59:367–372. https://doi.org/10.1016/0924-0136(95)02179-5
109. Chen B, Yang J, Zhang Q, Huang H, Li H, Tang H, Li C (2015) Tribological properties of copper-based composites with copper coated NbSe2 and CNT. Mater Des 75:24–31. https://doi.org/10.1016/j.matdes.2015.03.012
110. Kovalchenko AM, Fushchich OI, Danyluk S (2012) The tribological properties and mechanism of wear of Cu-based sintered powder materials containing molybdenum disulfide and molybdenum diselenite under unlubricated sliding against copper. Wear 290–291:106–123. https://doi.org/10.1016/j.wear.2012.05.001
111. Santos RLP, Silva FS, Nascimento RM, Motta FV, Souza JCM, Henriques B (2016) On the mechanical properties and microstructure of zirconia-reinforced feldspar-based porcelain. Ceram Int 42:14214–14221. https://doi.org/10.1016/j.ceramint.2016.05.195
112. Li Y, Zhu Z, He Y, Chen H, Jiang C, Han D, Li J (2016) WC particulate reinforced joint by ultrasonic-associated brazing of WC-Co/35CrMo. J Mater Process Technol 238:15–21. https://doi.org/10.1016/j.jmatprotec.2016.06.037
113. Han T, Li J, Zhao N, Shi C, Liu E, He F, Ma L, Li Q, He C (2017) In-situ fabrication of nano-sized TiO2 reinforced Cu matrix composites with well-balanced mechanical properties and electrical conductivity. Powder Technol 321:66–73. https://doi.org/10.1016/j.powtec.2017.08.019
114. Somani N, Sharma N, Sharma A, Gautam YK, Khatri P, Solomon JAA (2018) Fabrication of Cu-SiC Composites using powder metallurgy technique. Mater Today Proc 5:28136–28141. https://doi.org/10.1016/j.matpr.2018.10.055
115. Ward-Close CM, Minor R, Doorbar PJ (1996) Intermetallic-matrix composites—a review. Intermetallics 4:217–229. https://doi.org/10.1016/0966-9795(95)00037-2
116. Kumar KS, Bao G (1994) Intermetallic-matrix composites: an overview. Compos Sci Technol 52:127–150. https://doi.org/10.1016/0266-3538(94)90200-3
117. Baumann SF, Brindley PK, Smith SD (1990) Reaction zone microstructure in a Ti_3Al + Nb/SiC composite. Metall Trans A 21:1559–1569. https://doi.org/10.1007/BF02672571

118. Rao KTV, Soboyejo WO, Ritchie RO (1992) Ductile-phase toughening and fatigue-crack growth in Nb-reinforced molybdenum disilicide intermetallic composites. Metall Trans A 23:2249–2257. https://doi.org/10.1007/BF02646018
119. MacKay RA, Brindley PK, Froes FH (1991) Continuous fiber-reinforced titanium aluminide composites. JOM 43:23–29. https://doi.org/10.1007/BF03220564
120. Mostafaei A, Heidarzadeh A, Brabazon D (2021) Production of metal matrix composites via additive manufacturing. Encycl Mater Compos 605–614. https://doi.org/10.1016/b978-0-12-803581-8.11884-3
121. Cordero ZC, Siddel DH, Peter WH, Elliott AM (2017) Strengthening of ferrous binder jet 3D printed components through bronze infiltration. Addit Manuf 15:87–92. https://doi.org/10.1016/j.addma.2017.03.011
122. Do T, Kwon P, Shin CS (2017) Process development toward full-density stainless steel parts with binder jetting printing. Int J Mach Tools Manuf 121:50–60. https://doi.org/10.1016/j.ijmachtools.2017.04.006
123. Manoharan M, Gupta M (1999) Effect of silicon carbide volume fraction on the work hardening behaviour of thermomechanically processed aluminium-based metal-matrix composites. Compos Part B Eng 30:107–112. https://doi.org/10.1016/S1359-8368(98)00041-9
124. Reddy PV, Kumar GS, Krishnudu DM, Rao HR (2020) Mechanical and wear performances of aluminium-based metal matrix composites: a review. J Bio-Tribo-Corros 6:1–16. https://doi.org/10.1007/s40735-020-00379-2
125. Hassan AM, Alrashdan A, Hayajneh MT, Mayyas AT (2009) Prediction of density, porosity and hardness in aluminum-copper-based composite materials using artificial neural network. J Mater Process Technol 209:894–899. https://doi.org/10.1016/j.jmatprotec.2008.02.066
126. Hassan SF, Gupta M (2002) Development of ductile magnesium composite materials using titanium as reinforcement. J Alloys Compd 345:246–251. https://doi.org/10.1016/S0925-8388(02)00413-9
127. Tham LM, Gupta M, Cheng L (1999) Influence of processing parameters during disintegrated melt deposition processing on near net shape synthesis of aluminium based metal matrix composites. Mater Sci Technol 15:1139–1146. https://doi.org/10.1179/026708399101505185
128. Şap S, Uzun M, Usca ÜA, Pimenov DY, Giasin K, Wojciechowski S (2021) Investigation on microstructure, mechanical, and tribological performance of Cu base hybrid composite materials. J Mater Res Technol 15:6990–7003. https://doi.org/10.1016/J.JMRT.2021.11.114
129. Subramaniam B, Purusothaman VR, Karuppusamy SM, Ganesh SH, Markandan RK (2020) Review on properties of aluminium metal matrix composites. J Mech Energy Eng 4:57–66. https://doi.org/10.30464/jmee.2020.4.1.57
130. Lloyd DJ (1994) Particle reinforced aluminium and magnesium matrix composites. Int Mater Rev 39:1–23. https://doi.org/10.1179/imr.1994.39.1.1
131. Uvaraja VC, Natarajan N (2012) Optimization of friction and wear behaviour in hybrid metal matrix composites using Taguchi technique. J Miner Mater Char Eng 11:757–768. https://doi.org/10.4236/jmmce.2012.118063
132. Suresha S, Sridhara BK (2012) Friction characteristics of aluminium silicon carbide graphite hybrid composites. Mater Des 34:576–583. https://doi.org/10.1016/j.matdes.2011.05.010
133. Polat S, Sun Y, Çevik E, Colijn H, Turan ME (2019) Investigation of wear and corrosion behavior of graphene nanoplatelet-coated B_4C reinforced Al–Si matrix semi-ceramic hybrid composites. J Compos Mater 53:3549–3565. https://doi.org/10.1177/0021998319842297
134. Jamwal A, Prakash P, Kumar D, Singh N, Sadasivuni KK, Harshit K, Gupta S, Gupta P (2019) Microstructure, wear and corrosion characteristics of Cu matrix reinforced SiC–graphite hybrid composites. J Compos Mater 53:2545–2553. https://doi.org/10.1177/0021998319832961
135. Radhika N, Raghu R (2017) Investigation on mechanical properties and analysis of dry sliding wear behavior of Al LM13/AlN metal matrix composite based on Taguchi's technique. J Tribol 139. https://doi.org/10.1115/1.4035155
136. Jeske H, Schirp A, Cornelius F (2012) Development of a thermogravimetric analysis (TGA) method for quantitative analysis of wood flour and polypropylene in wood plastic composites (WPC). Thermochim Acta 543:165–171. https://doi.org/10.1016/j.tca.2012.05.016

137. Chiang TC, Hamdan S, Osman MS (2016) Urea formaldehyde composites reinforced with sago fibres analysis by FTIR, TGA, and DSC. Adv Mater Sci Eng. https://doi.org/10.1155/2016/5954636
138. Golebiewski J, Galeski A (2007) Thermal stability of nanoclay polypropylene composites by simultaneous DSC and TGA. Compos Sci Technol 67:3442–3447. https://doi.org/10.1016/j.compscitech.2007.03.007
139. Chawla KK, Metzger M (1972) Initial dislocation distributions in tungsten fibre-copper composites. J Mater Sci 7:34–39. https://doi.org/10.1007/BF00549547
140. Chawla KK (2012) Micromechanics of composites. Compos Mater Springer, New York, NY, 2012, pp 337–385. https://doi.org/10.1007/978-0-387-74365-3_10
141. Chawla KK (2012) Monotonic strength and fracture. Compos Mater 421–449. https://doi.org/10.1007/978-0-387-74365-3_12
142. Godfrey TMT, Goodwin PS, Ward-Close CM (2000) Titanium particulate metal matrix composites: reinforcement, production methods, and mechanical properties. Adv Eng Mater 2:85–91
143. Song W, Liu J, You L, Wang S, Zhou Q, Gao Y, Yin R, Xu W, Guo Z (2019) Re-synthesis of nano-structured LiFePO4/graphene composite derived from spent lithium-ion battery for booming electric vehicle application. J Power Sources 419:192–202. https://doi.org/10.1016/j.jpowsour.2019.02.065
144. Lu J, Chen Z, Ma Z, Pan F, Curtiss LA, Amine K (2016) The role of nanotechnology in the development of battery materials for electric vehicles. Nat Nanotechnol 11:1031–1038. https://doi.org/10.1038/nnano.2016.207
145. Larcher D, Tarascon JM (2015) Towards greener and more sustainable batteries for electrical energy storage. Nat Chem 7:19–29. https://doi.org/10.1038/nchem.2085
146. Cano ZP, Banham D, Ye S, Hintennach A, Lu J, Fowler M, Chen Z (2018) Batteries and fuel cells for emerging electric vehicle markets. Nat Energy 3:279–289. https://doi.org/10.1038/s41560-018-0108-1
147. Kaczmar JW, Pietrzak K, Wlosiński W (2000) Production and application of metal matrix composite materials. J Mater Process Technol 106:58–67. https://doi.org/10.1016/S0924-0136(00)00639-7
148. Behera MP, Dougherty T, Singamneni S (2019) Conventional and additive manufacturing with metal matrix composites: a perspective. Proc Manuf 30:159–166. https://doi.org/10.1016/j.promfg.2019.02.023
149. Ramanathan A, Krishnan PK, Muraliraja R (2019) A review on the production of metal matrix composites through stir casting—furnace design, properties, challenges, and research opportunities. J Manuf Process 42:213–245. https://doi.org/10.1016/j.jmapro.2019.04.017
150. Harrigan WC (1998) Commercial processing of metal matrix composites. Mater Sci Eng A 244:75–79. https://doi.org/10.1016/s0921-5093(97)00828-9
151. Manohar G, Dey A, Pandey KM, Maity SR (2018) Fabrication of metal matrix composites by powder metallurgy: a review. In: AIP Conf Proc, AIP Publishing LLC AIP Publishing, p 020041. https://doi.org/10.1063/1.5032003
152. Angelo PC, Subramanian R (2008) Powder metallurgy: science, technology and applications. PHI Learning Pvt. Ltd.
153. Sankhla AM, Patel KM, Makhesana MA, Giasin K, Pimenov DY, Wojciechowski S, Khanna N (2022) Effect of mixing method and particle size on hardness and compressive strength of aluminium based metal matrix composite prepared through powder metallurgy route. J Mater Res Technol 18:282–292. https://doi.org/10.1016/j.jmrt.2022.02.094
154. Ozkaya S, Canakci A (2016) Effect of the B_4C content and the milling time on the synthesis, consolidation and mechanical properties of AlCuMg-B_4C nanocomposites synthesized by mechanical milling. Powder Technol 297:8–16. https://doi.org/10.1016/j.powtec.2016.04.004
155. Guan D, He X, Zhang R, Qu X (2017) Microstructure and tensile properties of in situ polymer-derived particles reinforced steel matrix composites produced by powder metallurgy method. Mater Sci Eng A 705:231–238. https://doi.org/10.1016/j.msea.2017.07.084

Methods for the Development of High-Performance Metallic Nanocomposites

Vishnu Chauhan, Martina Saran, Jyoti Yadav, and Rajesh Kumar

Abstract Synthesis, properties and characterization of metallic nanocomposites offer lots of attention in technological and environmental research fields for their several applications. Following the featured properties of the metallic nanocomposites based on their synthesis and characterization techniques, their several applications are exhibited in biomedical, optics, data and energy storage devices, catalysis and optoelectronic devices. The current significant metallic nanocomposites cover the desirability of diverse applications. The study and investigations of metallic nanocomposites depend on their size, shape, surface chemistry that affects the particle performance for desired applications. The synthesis of metallic nanocomposites is preferred by chemical method as compared to the physical method due to their selective shape, size, composition and control over the growth, and these methods include sonochemical, reduction, microemulsion, chemical and electrochemical reduction, metal vaporization, solvothermal and microwave-assisted synthesis techniques. Moreover, the properties of metal nanocomposites such as surface plasmon resonance, catalysis, optical, magnetic, electrical properties have been described. Characterization techniques play the vital role in assessing and declaring a sample as featured nanoparticles for their desired applications without using the sensitive and precise microscopy. Several characterization techniques, such as scanning electron microscopy (SEM), X-ray diffraction (XRD), transmission electron microscopy (TEM), X-ray photoelectron spectroscopy (XPS), Raman and Fourier transform infrared spectroscopy (FTIR), are used to determine their size, morphology, structure,

V. Chauhan
Materials Research Department, GSI Helmholtz Centre for Heavy Ion Research, Planckstr. 1, 64291 Darmstadt, Germany

M. Saran
Department of Physics & Astrophysics, University of Delhi, Delhi 110007, India

M. Saran · J. Yadav
Materials Sciences Group, Inter University Accelerator Centre, New Delhi 110067, India

R. Kumar (✉)
University School of Basic and Applied Sciences, Guru Gobind Singh Indraprastha University, New Delhi 110078, India
e-mail: kumarrrpi@gmail.com

S. K. Tiwari et al. (eds.), *Nanoparticles Reinforced Metal Nanocomposites*,
https://doi.org/10.1007/978-981-19-9729-7_4

composition, porosity, band gap, porosity, functional group and their thermal stability. So, it is very much evident to carefully choose the synthesis method and characterization techniques to properly define the properties of grown metallic nanocomposites sample for their applications in diverse research areas.

Keywords Metal nanocomposites · Synthesis · Properties · Characterizations · Applications

1 Introduction

In recent decade, metallic nanocomposites have attained much attention for researchers due to their micro to nanolevel transition that resulted in change in their physical and chemical properties for various implications. The changes include the small size, quantum confinement and increase in surface-to-volume ratio. Earlier, these nanocomposites were explored for size-dependent physical and chemical properties but nowadays these nanocomposites are investigated for commercial applications [1]. The feature of surface to volume ratio of metallic nanocomposites allows them to interact with other particles that make the diffusion faster and feasible at low temperatures. Metallic nanocomposites have been the subject of research because of their uniform size and sharp size distribution in nanoscale. Metallic nanocomposites depict unique properties with suitable functional group and their synthesis allows them to bind with drugs, ligands and antibodies. There are some unique characteristics of metallic nanocomposites such as surface plasmon resonance (SPR) and optical properties. Metallic nanocomposites are made up of metal or alloy matrix in which other nanosized supplementing materials are embedded. The noble metallic nanocomposites of Ti, Al, Ag and Au have gained clear attention due to their applications in catalysis, automotive industry, anti-microbial agents, anticancer and photography [2].

Based on the starting materials or precursors metallic nanocomposites can be synthesized in two ways basically top-down or bottom-up approach. These techniques involve ball milling and mechanochemical, laser ablation, ion sputtering, physical vapor deposition method (PVD), pulsed laser deposition (PLD), chemical vapor deposition (CVD), sol–gel, chemical reduction, hydrothermal and solvothermal, electrochemical deposition, microwave-assisted and sonochemical reduction [3–5]. There are various characterization techniques that are used for the analysis of metallic nanocomposites such as absorbance and infrared spectroscopy, transmission electron microscope, scanning electron microscopy, atomic force microscopy, X-ray diffraction, Fourier transform infrared spectroscopy, extended X-ray absorption fine structure and X-ray photoelectron spectroscopy. Based on the optical properties, metallic nanocomposites have been used for imaging sensor, photocatalysis, biomedicine, rechargeable battery, electrode, shield, sensor, laser display and solar cell. Moreover, metallic nanocomposites have been used for thermal, electrical, mechanical and magnetic functions. They can be used as catalysis, fuel cell catalysts, optical detector, electrical conductive pastes, battery materials and, for medical treatment, can improve the macro- and microhardness, abrasion, scratch-resistant, to prevent environment pollution arising from coal and burning gasoline and sunscreen lotion. Metallic nanocomposites have been used for therapeutic applications such as anti-infective agents, anti-angiogenic, tumor therapy, multiple myeloma, leukemia, rheumatoid arthritis, photothermal therapy, radiotherapy and drug delivery [6, 7]. The focus toward the investigation of nanoparticles, nanocomposites, nanostructures and nanomaterial is conspicuous because of their significant use in catalysis, polymer synthesis, sensing and optoelectronic devices [8–13]. Molecularly imprinted silver nanocomposites are used for explosive tagging sensing [14]. Gold dendrimer nanocomposites (DNC) are hybrid nanoparticles that are formed by the dispersion and immobilization of guest atoms or small clusters in dendritic polymer matrices. These nanocomposites are used for imaging and drug delivery by utilizing bioactive guests or the incorporation of radioactive isotopes, like Au-198. Silver nanocomposites are used for antimicrobial and inflammatory, antifungal, antibacterial and as antioxidant purposes [15, 16]. Moreover, silver nanocomposites have been successfully used for biomedical applications in pure or alloyed form [17, 18]. Yada et al. reported the mechanical milling: a top-down technique for the synthesis of nanomaterials and nanocomposites [19]. Sellinger et al. reported the continuous self-assembly of organic–inorganic nanocomposite coatings for automotive finishes, hard coats and optical hosts. The organic and inorganic interfaces are linked covalently, and the continuous nature of dip-coating technique allows rapid synthesis of optically transparent coatings [20]. The tailoring of optical, electronic and electrical properties of metallic nanocomposites is challenge by controlling their size and shape. Synthesis, properties, characterization and applications of metallic nanocomposites are an active area of research. There are various methods (physical, chemical and biological) available for the synthesis of metal nanoparticles with some certain advantages and disadvantages. Firstly, the present chapter depicts the detailed synthesis information of metallic nanocomposites using top-up and bottom-down approaches. Second part presents the different properties of metallic nanocomposites including structural, optical and chemical with their current prospects and future scope in different areas. Furthermore, the last section identifies the characterization techniques in systematic way for the possible applications in various research areas.

2 General Methods for Synthesis of Metallic Nanocomposites

Metallic nanocomposites can be synthesized basically by top-down and bottom-up approaches depending upon the starting materials or precursors. The foremost dissimilarity in both methods is the initiated materials of nanocomposites synthesis. Bulk material is employed as the material for starting a process in top-down methods and

Fig. 1 Schematic of various synthesis methods of metal oxide-Au NCs for biological applications. Reproduced with permission [21]

particle size is diminished (coarse into fine powders) to nanosizes via different physical/chemical or mechanical processes, likewise atom/molecules may be the material in bottom-up methods to begin with. We try to discuss various methods in this chapter for the synthesis of metallic nanocomposites. Physical and chemical methods, namely electrochemical changes/chemical reduction and photochemical reduction mostly used for the purpose of synthesis and regulation of metallic nanocomposites. The choice of synthesizing method is very critical because during synthesis, they show (i) kinetics of interaction with reducing agent, (ii) adsorption process of stabilizing agent and (iii) sundry experimental techniques generate strong influence on its morphology, i.e., change in structure/size and stability/physicochemical characteristics. Figure 1 shows the schematic of various synthesis methods of metal oxide-gold nanocomposites for biological applications [21]. In top-down methods, we generally get rough edges/deformed shapes of the nanoparticles obtained due to friction and milling processes. However, in bottom-up approach we get very smooth edges, regular and uniform shapes and sizes of the prepared nanoparticle samples. Figure 2 shows the depiction of bottom-up and top-down methods for metal nanoparticles and the approaches, and methods of NPs synthesis are summarized in Fig. 3 [7]. Table 1 shows the various synthesis processes carried out for preparing metallic nanocomposites.

2.1 *Top-Down Approach*

Here in this approach, firstly bulk or micromaterial is synthesized and next step to broken down these samples into small nanosizes. This method is basically depending on the size reduction of bulk or precursors by different physical and chemical processes [22]. It accommodates techniques primarily ball milling and mechanochemical which comes under mechanical milling. Similarly, thermal and laser ablation is also helpful for nanosample preparation. These methods are easy to

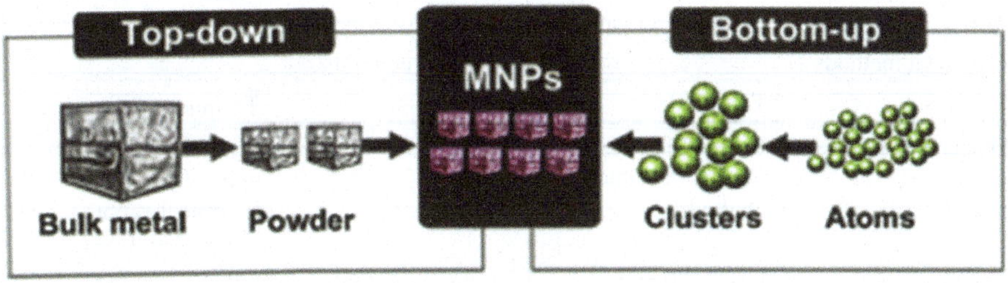

Fig. 2 Illustration of bottom-up and top-down methods for metal nanoparticles. Reproduced with permission [7]

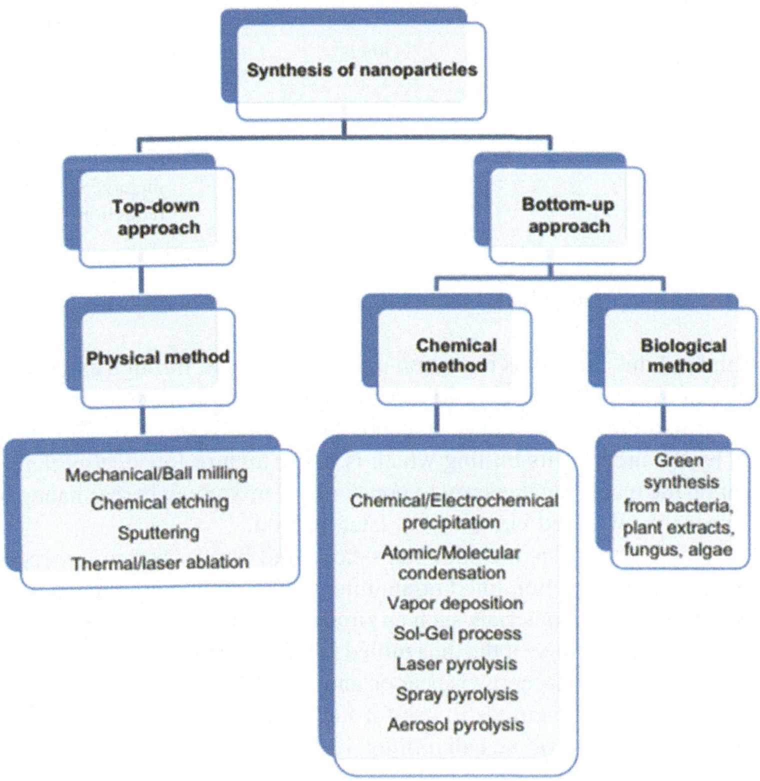

Fig. 3 Illustration of synthesis of metallic NPs using bottom-up and top-down methods. Reproduced with permission [7]

introduce but it is not the best method for regular and uniform shape with significantly small size particles. The crucial drawback in this method is change in surface chemistry/physical–chemical properties of metallic nanocomposites [23].

Table 1 Various synthesis processes carried out for preparing metallic nanocomposites

Top-down methods		Bottom-up methods	
Type	Technique	Type	Technique
Mechanical milling	Ball milling & mechanochemical method	Solid state	Physical vapor deposition (PVD) Chemical vapor deposition (CVD)
Laser technique	Laser ablation	Liquid state synthesis	Sol–gel method chemical reduction & hydrothermal method solvothermal method
Sputtering	Ion sputtering	Gas phase	Laser pyrolysis Flame pyrolysis
		Others	Electrodeposition process Microwave-assisted reduction Co-precipitation method Sonochemical reduction

2.1.1 Ball Milling and Mechanochemical

The principle of this method is reduction in the particle size through extensive high-energy zircon or steel ball milling for several hours. John Benjamin originated this technique of particle size to reduce in nanoscales in the year 1970. This method is classified by the order of its milling which is either high or low energy that can be dependent on the mechanical energy to powder bulk mixture, whereas nanoparticles are most probably obtained via using the later method.

The method of synthesis includes steps explained briefly, bulk/microcrystalline powder is added into a Teflon lined or alumina vessel along with many heavy balls mostly made up of hard materials such as zircon or steel which do not contaminate the sample. The particle size of the thus milled powder is depending upon the amount of time at high mechanical energy is bear on and was milled with high-speed rotating balls. There are various methods to use for high-energy milling, namely some are planetary ball milling, vibrating ball milling, low-energy tumbling milling and high-energy ball milling [24]. The stated methods work on the basic principle of heavy and free moving balls, which must cause friction with the bulk/micropowder material in surrounding or beneath them [19]. Nanocomposites of Fe and Nd-Fe-B alloy with size of less than 20 nm have been produced with the help of this ball milling with high-energy method also using surfactant and organic carrier liquid. Rod-shaped nanoparticles were obtained by milling [25].

In mechanochemical synthesis, samples are obtained via frequent deformation, welding or fractures in the mixture of reactants happens. We all know that high-temperature reactions are must for segregation of reactants into the final products, whereas in this method low temperature or room temperature without any external source of heating or cooling is required [26]. The reaction takes place during the process which is given by: $S_xC + {}_yR = {}_xS + R_yC$, where S_xC and ${}_yR$ are reactant, S_x is product and R_yC is by-product. This is simple yet efficient method of nanoparticle preparation [27]. Iron and copper nanoparticles (Fe–Cu) were prepared through mechanochemical method and employed the iron (III) chloride ($FeCl_3$) and copper (II) chloride ($CuCl_2$) with sodium (Na) during ball milling process. The crystallite and particle size (mean) of thus synthesized and cleaned nanoparticles after 3.5 days of milling was found to be around 9 nm and 50 nm; the size of the crystallite remains still at around 14 nm, while the size of the particle increases as measured of the sample [28].

2.1.2 Laser Ablation

In this technique, radiation of laser is fall upon on the bulk or microsample to reduce the particle size of a bulk material into nanosized level. Material is kept in between a thin layer and exposed to the beam of pulsed laser irradiation. Preferably, Nd:YAG laser and its harmonic at the wavelength of 106 μm and also Ti:Sapphire laser and copper (Cu) vapor laser are incorporated [29]. The relative amounts of atoms are ablated and, emerged particles are associated with the laser pulse time duration & energy [30]. This technique involves many variable criteria which influences ablation efficiency and characteristics of thus synthesized metal nanoparticle formed depends on duration of time, wavelength (λ), fluency of laser and surrounding of liquid solution medium in the presence or in the absence surfactant [31]. Altowyan et al. reported the effect of liquid media and laser energy on the synthesis of Ag NPs and their nanocomposites with Au nanoparticles using laser ablation technique for optoelectronic applications [32]. In the normal atmospheric conditions, the pulsed excimer laser radiation was used at wavelength of around 248 nm to prepare ~2 pm of silver (Ag), gold (Au) and permalloy of (Ni81%: Fe19%) by ablation method. Also, the mean particle radius was observed to be in the nanorange (5 to 50 nm) [33]. There is another technique of femtosecond laser radiation mostly used in a gold target in pure deionized or demineralized (DI) water to obtain colloidal gold sol (Au) nanoparticles. The procedure which involves heat free femtosecond ablation exhibits a very small 3–10 nm monodispersed gold colloids [34]. Silver (Ag) nanosized particles can be produced by laser of Nd:YAG at $\lambda = 1064$ nm by ablating the silver target dipped in different concentrations of salt (NaCl) solution along with in pure water (DI). Predominantly, the size of the silver (Ag) obtained in 5 mM NaCl solution ranged from ~5 to 50 nm in particle size (mean size of ~26 nm) [35]. Ablation of targets (silver (Ag) and gold (Au)) in water by a copper (Cu) vapor laser produces its sol formed after drying through evaporation are disk nanoshaped, with diameter (~20–60 nm) and thickness of few nanometers [36]. This method is used for synthesizing the stable NPs with the use of any chemicals or surfactants. For this purpose, this technique is used to synthesize the nanostructures, including core shells, nanocomposites, alloys and other complex structures. Mwafy et al. reported the functionalized cellulose nanofiber (CNF) with multi-walled carbon nanotubes and SnO_2 forming the hybrid nanocomposite structure [37]. Laser-assisted technique was used for the synthesis of Ag/CdO nanocomposite thin film, and their structural and optical studies have been investigated. The spectroscopic techniques were used to inform optoelectronic characteristics changes observed after embedding and post-thermal annealing effect [38].

2.1.3 Ion Sputtering

Ion sputtering is the vaporization of a target solid material via sputtering with the help of beam of different ions of inert gases. Recently, this method is popular and was used for the metals accommodating magnetron sputtering of the targets. Here in this technique the mass nanostructures films can be assembled on silicon (Si) substrate via collimated beam of the nanoparticles, at usually pressure of ~1 mTorr [39]. Deposition in sputtering method is carried via evacuated vacuum chamber, and a gas is introduced while sputtering and working pressure around ~0.05 to ~0.1 mbar are kept. The ionization of gas is done using kind of very high voltage to the cathode, and free electrons generated will then take spiral path through magnetic system and collides with sputtering gas (Ar) atoms; perpetual process shows glow discharge in plasma to ignite. Positive charge of gas ions attracts the target, impinges repeatedly and tends to the surface of the target material with energy above the binding energy (BE) and then an atom ejected [40]. The metal oxide-gold (Au) nanocomposites involve reduction of gold with photochemical from Au(III) to Au(o). The metal oxide-gold (Au) nanocomposites involve the reduction of gold with photochemical from Au(III) to Au(o). Also, the synthesis conditions can alter the size distributions during sample synthesis [41]. Tafur et al. investigated the optical properties of Ag NPs embedded in dielectric films grown by DC and RF sputtering. The nanocomposites are constituted by the intercalation of dielectric layers such as Teflon, SiO_2 and TiO_2, and the dipolar and quadrupolar plasmon resonances were observed and inferred in terms of the Mie theory [42]. This technique synthesized controllably of metal alloy nano-sized particles, formation of silver copper (Ag–Cu) alloy in nanoform on a substrate of silica with average particle size ~17–32 nm analyzed using TEM [43]. Pt/Au alloy nanoparticles at room temperature synthesized carrying sputter deposition from working magnetron sources on the liquid PEG. Prepared nanoparticles were found to be alloyed with the face-centered cubic structure [44]. Cobalt nanocomposites can be synthesized at room temperature on SiO_2/Si substrates by direct current magnetron sputtering and found that few nm to tens of nm in size can be obtained by applying positive substrate biases [45]. Magnetron sputtering commonly used techniques for the preparation of metallic nanoparticles. The formation of Silver (Ag) nanocomposites films depends on the time during accumulation and smoothness of the substrate with RMS value of roughness (~2 nm) for their morphological/topographical and optical properties [46].

2.2 *Bottom-Up Methods*

This is based on producing nanosized particles from small tiny molecules atoms/molecules or smaller particles after put them together. Here, nanosized structures constitute the nanoparticles initially processed and collect to produce final nanoparticles.

2.2.1 Physical Vapor Deposition Method (PVD)

Here, material is placed on the surface by either way as thin film or as nanosized particles. Physical vapor deposition such as pulsed vapor deposition is mostly accompanied for thin films' synthesis. In the pulsed laser deposition, laser ablation is accompanied on the target that introduces the synthesis of plasma ablated type and therefore accumulated on a substrate to produce a thin film [47]. The properties of pulsed laser-deposited nanocomposite NiO:Au thin films have been investigated for gas sensing applications. These nanocomposites thin films were analyzed as electrochemical hydrogen sensors and observed that the addition of the Au nanoparticles leads to enhance the sensor sensitivity considerably [48]. Synthesis of silver thin film of said thickness (~300 nm) on a quartz substrate, from using a pristine silver plate as a target material is also found in the literature prepared at room temperature [49]. The polyaniline/gold (PANI/Au) and polyaniline/platinum (PANI/Pt) core–shell nanocomposites were usually fabricated through mini-emulsion polymerization of aniline monomer in the presence of chloroauric acid and chloroplatinic acid, respectively [50]. Rhodium (Rh) thin film on polish stainless steel substrates was observed to be in Rh(111) plane as shown in Fig. 4, consisting dense column of nanometric crystallites. The measured thicknesses of Rh1 = 3, Rh2 = 4 and Rh3 = 5 target–substrate distance (cm) are coming out to be ~86 nm, ~82 nm and ~54 nm, respectively. The thickness of Rh1 is maximum, deposited at target–substrate distance (Dts) = 3 cm [51].

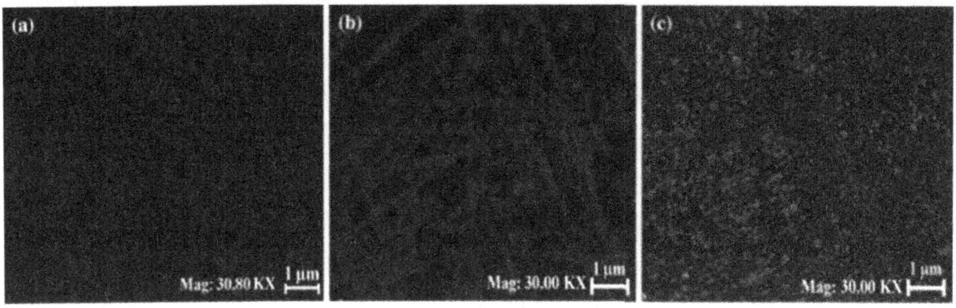

Fig. 4 SEM images of the **a** Rh1, **b** Rh2 and **c** Rh3 films, respectively. Reproduced with permission [51]

2.2.2 Chemical Vapor Deposition (CVD)

It consists of a solid element which will deposit on a surface heated enough through chemical reactions which require activation energy for initiation above 900 °C of different phases either vapor or gas phase. Experimental chamber made up of gas supply inlet–outlet system, assembly and an exhaustion facility. Solid target material is given to the system in the form of highly volatile molecules and acts initiators and then number of reactions between the fragments, precursor and substrate to synthesize nanosized thin film [52]. Non-catalytic CVD to produce Cu–Sb alloy nanostructures by using the copper foils used as both the substrates and source, Cu–Sb alloy nanosized wires and nanoparticles with various shapes/sizes and phases were prepared by fluctuating the roughness, the temperature along flow rates [53]. Piszczek et al. reported the synthesis of Ag-metal oxide nanocomposites coatings on large substrates using CVD techniques in a single synthesis process and explained the antimicrobial activity, bio-integration and toxicity [54].

2.2.3 Sol-Gel Technique

This technique has been majorly used for the preparation of metallic nanocomposites. Colloidal particles are large in size than the normal molecule or nanoparticle, and mostly on mixing with liquid colloids, they have a very bulky appearance but the nanosized molecules forms and looks clear. It occurs in stages such as hydrolysis, condensation, growth of particles and agglomeration of particles. Figure 5 shows the sol–gel synthesis of Ti/Co/Ni nanocomposites [55]. Co–Pt alloy nanoparticles with variation of concentrations of CoPt and copper were prepared at several temperatures. The pristine cobalt platinum (CoPt) and the CoPt–Cu nanosized particles (varying copper content and temperatures) are prepared. L10 cobalt platinum (CoPt) and (FCC CoPt) exist in pristine cobalt platinum nanoparticles, whereas a L10 phase formed in the CoPt–Cu alloy with 50 at.% copper doping at annealing temperature is 600 °C. Copper reduces the ordering temperature and induces the magnetization [56]. Gold (III) ions are reduced into nanoparticles through amino silica with particle size of 18–20 nm in diameter [57]. Electrically conducting thin films of thickness around ~2 nm have been synthesized by growing ultrafine particles of iron (Fe) and copper (Cu), respectively, and from using precursor sol can be modified by varying heat treatments [58].

2.2.4 Chemical Reduction

In this well-established method, different reducing agents or stabilizing agents are accompanied in the medium through surfactants using ionic salts [59]. Stability in the dispersion was observed by the absorbance analysis [60]. The size/shape of nanocomposites via this method depends on the reducing agent, precursor of usually

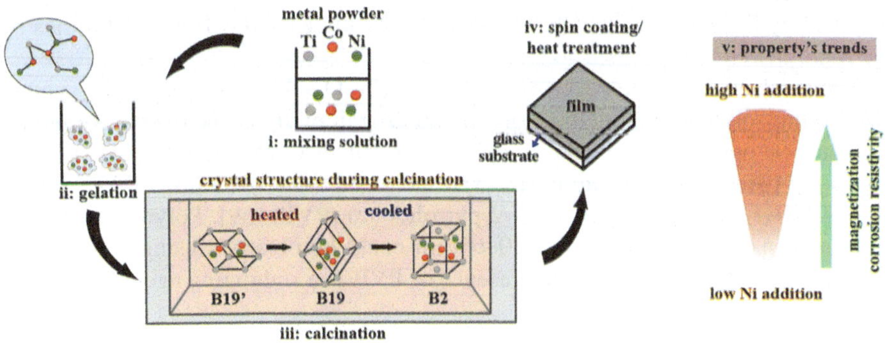

Fig. 5 Schematic of sol–gel synthesis of Ti/Co/Ni NCs by sol–gel methods. Reproduced with permission [55]

metal, choice of solvents, concentration, temperature and time involved in reaction [61, 62]. Jeyaraj et al. reported on synthesis, characterization and biomedical application of platinum nanocomposites using chemical reduction method. These NCs have been focused on therapeutic purposes, and still the investigation has to be done for a comprehensive picture of biocompatibility versus toxicity of Pt NCs [63]. Copper (Cu) nanosized particles were prepared by chemical reduction of copper sulfate ($CuSO_4$) and sodium borohydride ($NaBH_4$) in the presence of H_2O without inert gas, TEM and UV–vis spectrometry used for the determination of size and optical properties of the synthesized nanosamples (~10 nm), respectively [64]. Silver/poly(vinylalcohol) nanocomposites have been synthesized using chemical reduction method of silver salt by using two different reducing agents and TEM measurements showed their particle size less than 10 nm and composite showed higher degradation temperature than the PVA alone that have been analyzed using TGA technique [65].

2.2.5 Hydrothermal and Solvothermal

A sealed vessel (autoclave) is used in this method, where solvents can be heated till temperature above boiling points (B.P.) in vacuum ovens by increase in their autogenously produced pressures. Reaction is generally known as solvothermal method, or, in water as solvent, it is referred as hydrothermal method. The supercritical point of H_2O is above around ~374 °C at 218 atm. and exhibits characteristics of liquid and gas. Platinum alloy nanoparticles showed prominent applications as electrocatalysts in oxygen reduction reaction (ORR) fuel cell cathodes, on the use of N,N dimethylformamide (DMF) as solvent and reductant in synthesis of platinum (Pt) alloy nanosized particles. Estimation of average nanoparticle from TEM micrographs is between ~11 and 13 nm sized [66]. Antimony sulfide graphene oxide (Sb_2S_3-GO) nanocomposite has been synthesized using solvothermal technique for electrochemical excellent detection of dopamine, and results showed the sensitivity obtained by the Sb_2S_3/GO/GCE was 15 times by the RuS_2/GCE and 100 times by the MnS/GO/GCE [67].

The synthesis of facile Fe_3O_4-loaded silver (Fe_3O_4–Ag) nanocomposite synthesized using solvothermal method and in situ growth of silver nanowires for potential catalytic optics, medicine and catalysis performance [68]. Spherical silver (Ag) nanosized particles of ~10 nm in size were prepared using reduction of Ag^+ ions in aq. $Ag(NH_3)_2NO_3$ by poly(*N*-vinyl-2-pyrrolidone) (PVP) [69]. Polymer-stabilized ruthenium nanosized particles of 20–65 nm could be obtained using solutions of ruthenium chloride in E.G. in the presence of PVP as a stabilizing agent [70].

2.3 Other Popular Techniques of Nanocomposite Synthesis

2.3.1 Electrochemical Deposition

Nanostructures are produced by electrodeposition or template synthesis methods [71]. These are mechanically strong and uniform throughout. For the production of porous semiconductor structures, electrochemical anode dissolving processes are preferred [72]. The metal nanosample was prepared using baths containing metal salt, which could be acidic or basic in nature, and a 3-terminal potentiostat with negligible voltage applied for a suitable time. The shape, size and morphology can be controlled by altering various parameters like deposition time, voltage and current, which includes the preparation of nanocomposites, nanowires and nanorods [73, 74]. The ZnO nanocomposites are fabricated on PMMA powders via atomic layer deposition, where PMMA powders are synthesized using the sol–gel method [75]. The effects of the gold–silicon interface elimination are good for mono-shaped copper nanostructure particles prepared by potentiostatic deposition on an ultrathin polypyrrole (PPY) film on a Si (100) substrate sputtered coated with a thin gold film or gold film electrode (GFE) with a thickness of 80 nm [76]. A label-free immunosensor has been developed using the electrodeposition technique of highly conductive Au NPs and reduced graphene oxide (RGO) NCs for the detection of CA125 for endometriosis ELISA analysis. This immunosensor can be used for future prospective clinical diagnostic applications [77]. Cui et al. reported the electrochemical deposition of high-quality Pd, Pt, Au and Ag sub–micron tubes composed of nanoparticles in a polar aprotic solvent [78].

2.3.2 Microwave-Assisted Reduction

The microwave irradiation-induced reduction method is new to the synthesis of noble metal nanoparticles via the green route. Production is caused by the heating rather than the energy quantum of a microwave. When heating is produced by the connection of an electric field with the dipole moment of molecules, solvents with a high dipole moment are likely, while polar molecules like H_2O and ionic liquids are among the best to choose. In contrast to heating (thermolysis), the formation achieves regular nucleation and a short growth period [79]. The technique was used for the synthesis of stable and monodisperse Ag nanosized particles by using microgels as a template for in situ reduction of silver nitrate with glucose as a reducing agent. TEM images show that the spherical particles had a diameter of around 8.5 nm and were found to be stable for more than 8 months [80]. Ag nanostructures were produced via microwave-assisted synthesis using green chemistry methods. Nanospheres with a mean diameter of 7.36 nm were successfully synthesized in 15 min by reducing Ag^+ ions (from $AgNO_3$) [23]. The polyol reduction technique was applied for preparation of core–shell gold and palladium nanocomposites; the thickness of the palladium shell was obtained 3 nm, and the gold core diameter was found to be ~ 9 nm [81]. The reduction of nickel chloride and hydrazine hydrates using PVP in EG under microwave irradiation produced nickel nanosized particles with uniform, monomorphic, self-assembled flower-like microstructures. TEM and X-ray diffraction confirm the synthesized nickel nanoflower is the self-organized, assembled structure of about 6.3 nm size [82]. Passivated gold nanoparticles may be synthesized in the presence of 1-dodecanethiol, with particle sizes found to be 1.8 nm through HR-TEM and STEM. Copper-silver nanosized particles (CuAg) were created by reacting aq. copper (II) nitrate hemihydrate ($Cu(NO_3)_2.25H_2O$), silver nitrate ($AgNO_3$) and sodium acrylate ($C_3H_3NaO_2$) at room temperature, yielding 21.5 nm spherical particles [83]. Sun et al. reported the microwave-assisted synthesis of graphene nanocomposites for recent advances in lithium-ion batteries with high capacity, long cycle life and impressive high-rate capability. This technique is quite significant for the synthesis of metal nanocomposites with controlled size and shape for energy storage applications due to its various advantages, such as quick, inexpensive, uniform and energy-efficient advantages [84].

2.3.3 Sonochemical Reduction

High-intensity ultrasound shows potent technique for the nanocomposites production. Implementation of strong ultrasound radiation around ~20 kHz to ~10 MHz creates the molecules to go through a reaction chemically. The physical known process responsible is the acoustic cavitation caused in the solvent or liquid by the ultrasonic irradiation [85]. The main and foremost event in sonochemical reduction is the bubble that occurs in the liquid solvent through formation, growth and collapse. Polymer polymer–metal nanocomposites have been synthesized and resulted in excellent catalytic performance. Moreover, the catalytic activity of polymer–Au and polymer–Pd NCs was investigated for aerobic alcohol oxidation and the Suzuki–Miyaura cross-coupling reaction. This technique can be expended for the utilization of ultrasound for the preparation of polymer–metal NCs and encourages the catalytic applications of NCs [86]. Neppolian et al. successfully synthesized mono- and bimetallic Au–Ag reduced graphene oxide NCs using sonochemical reduction method and the improved catalytic activity. Effectively, the less amount of Au with Ag and GO reduced the 4-nitorphenol. The catalysts loading on GO was found to be enhanced relatively low levels of added chemicals by applying dual frequency ultrasonicator [87]. Also, the research on gold-coated magnetic nanocomposites is subjected extensively due to their significant physical and chemical properties including biocompatibility, functionality and stability that endue their great applications in various fields. NIR-absorptive bifunctional gold shells are used for the targeted delivery and MRI diagnosis [88].

3 Properties of Metallic Nanocomposites

The size, shape and various physical and chemical properties of metallic nanocomposites are greatly influenced by the parameters of an experimental condition and the kinetics of stabilizing agents. The physical and chemical properties of metallic nanocomposites, which differentiate them from bulk metals, find applications in various industries. These days, metallic nanocomposites have attracted attention in research because of their fascinating properties, like mechanical strengths, high surface areas, magnetic properties and many more. The properties of catalysts are important in applications of metallic nanocomposites [89]. The various properties of metallic nanocomposites have been addressed here.

3.1 Optical Properties

The morphology of MNPs impacts the oscillations of collective electrons on the surface which alters the optical excitation of surface plasmon resonances (SPR). The generated SPR results in giant enhancement of nonlinear optical response of MNPs. The size of MNPs plays important role in modifications of the optical properties and hence in various optical applications. The specific industrial applications are associated with the particular size of the MNPs. There are many ways by which the surface and structure of nanoparticles could be modified. The ion implantation technique can be used for almost any type of structure of metal nanoparticles [90, 91]. The optical properties of MNPs find specific applications in different wavelength regimes. For instance, the strong nonlinear optical properties of MNPs allied in near-IR region find attention in telecommunications applications [92]. The opacity of metal NCs can be ascribed to the formation of 3D network of NPs within the polymer matrix. In practice, it hinders the transmission of light through the material. The optical properties of metal oxide (Fe_2O_3, TiO_2 and $NiFe_2O_4$) dispersed high-density polyethylene (HDPE) nanocomposites have been evaluated. The (Fe_2O_3/TiO_2/$NiFe_2O_4$)/HDPE composite was found to exhibit high light absorption in UV–vis range and TiO_2/HDPE nanocomposite exhibited the lowest optical energy band gap (48.88%) [93]. The irradiation of nanoparticles has been proven a good approach for nonlinearity in optical phenomenon. The energy is transferred to electrons shell by using photographs in nanoparticles between intraband and interband transitions. When the electrons absorb the photon energy, then due to the e-e interaction and e-phonon scattering, a Fermi–Dirac state is formed. Thomas et al. investigated the size-dependent surface plasmon resonance in silver silica NCs obtained by the sol–gel technique. The phenomenon of surface plasmon resonance (SPR) bands occurs, and their evolution in the size range 5–10 nm is investigated. The decrease in NPs size causes the broadening of plasmon absorption, and redshift occurs that was modified Mie scattering theory [94].

3.2 *Magnetic Properties*

The research reports on metallic nanocomposites showed the innovative and novel techniques for synthesis of nanoparticles. The shape and the size of nanoparticles are greatly dependent on the following method of preparation. A huge number of reports have collected their nature of production, characterization techniques and their exclusive applications in the field of biomedical and nanotechnology. The vast field of research in metallic nanocomposites stated that the magnetic properties are highly influenced with the synthesis parameters. The high surface-to-volume ratio associated with nanoparticles showed that magnets of nanoscale particles are differentiated from their bulk. It is quite interesting that the factors which alter the magnetic properties of NPs not only improve but also adversely affect the same. Owing to correlation between their shape and size of metallic nanocomposites, a bunch of reports have been collected in their possible applications [95, 96]. The recent study showed the importance of embedded MNPs in multi-phase permanent magnet which strongly enhanced the coercivity via exchange coupling of hard phase and low phase [97, 98]. Zeng et al. reported that the enhanced magnetic properties FePt of high coercivity coupled with metallic nanocomposites. Also, the core shell of nanoparticles synthesized with Fe-Fe_3O_4 shell of lower coercivity greatly modified its magnetic properties by variation in the thickness of the core shell [99]. The reported hysteresis curve of the core shell FePt and Co-Fe_3O_4 showed the active coupling between core and the shell. There is found to be the inverse correlation between coercivity and the volume ratio of the core shell. The higher magnetization of Fe made it a better candidate for the potential applications in magnet industry. But with the greater chance of being oxidized, Fe has some limitations. In reported research, Qiang et al. studied the magnetization (Ms) value of the Fe core shell MNPs coated with Fe oxide. It is concluded that the diameter range of core shell between 2 and 200 nm resulted in Ms value of ~200 emu/g [100]. The interesting bonding of metallic nanocomposites with core shell seeks attention in the field of high magnetization by enhancement in the relaxation time of excited photons of the core shell [101]. Notably, the Fe core shell nanoparticles with oxides of different phases possess both types of features [102].

The various magnetic properties like the magnetization, coercivity, energy barrier are dependent on the size of the core shell metallic nanocomposites and find applications in hyperthermia [103]. The inherent characteristics of the metallic nanocomposites are reflected by the anisotropy constant (K). The specific absorption rate

(SAR) which is dependent on the K value is crucial parameters for the development of an exchange-coupled magnet. Hence, firstly the composition has become the challenge for tunable magnetic properties. In the MRI applications and according to the compatible conditions for human beings, an SAR of 1 kW/g is essential at 100 kHz and 20 mT. The carbide and Fe MNPs showed a very value of SAR, i.e., 415 W/g at 96 kHz and 20 mT for 13.6 nm diameter of nanoparticles as per the report by Meffre et al. [104]. The exchange coupling within the core shell is responsible or the enhancement of SAR value and drastic increment from 1000 to 4000 W/g as compared to single MNPs from with SAR value from 100 to 450 W/g is reported by Lee et al. [105]. Similarly, in published report by Noh et. al. the SAR value for the coupled shell-core $CoFe_2O_4$-coated $Zn_{0.4}Fe_{2.6}O_4$ MNPs is 10,600 W/g while for the single core it is only 4060 W/g [106].

A group of experiments on core shell and hollow shell of Fe-based nanoparticles are studied by Ong et. al. on the basis of the spin interaction. They have described that the Fe-Fe_3O_4 core–shell MNPs have high exchange magnetic field of 1190 Oe while the hollow-shell MNPs showed quite low value of 133 Oe [107]. Due to the surface spin defects in core shell nanoparticles, the exchange bonds prevent from breaking while it is not possible in hollow nanoparticles. But in contrast to the above-mentioned statement, Khurshid et al. stated an sevenfold enhancement of exchange bias in hollow magnetite nanoparticles itself. With hollow shell, its value is ~96 mT in 18.7 nm while for core shell it is only ~17 mT observed in solid γ-Fe_2O_3 [108]. The surface anisotropy is attributed to the spin defects on the surface of the core shell nanoparticles, and this leads to the greater blocking temperatures for hollow nanoparticles which has been studied for $NiFe_2O_4$ MNPs in solid $Ni_{33}Fe_{67}$ core/$NiFe_2O_4$ shell and $NiFe_2O_4$ shell only [109]. In a nutshell, the surface properties tune the magnetic properties via disordering in surface spin. The coercivity is associated with the diameter of the core shell; for example, the 3.5 nm FePt nanoparticles showed coercivity value of 5.5 kOe, but the 4 nm Fe showed 200 Oe. It is quite interesting that both FePt and Fe nanoparticles are ferromagnetic in nature at 10 K [110].

3.3 Electrical Properties

The novelty in the method of synthesis makes the metallic nanocomposites important for their use in potential applications in electronic devices. The exclusive parameters like the shape and size result in modifications in their electronic structure and hence are able to tune the electronic properties. The modified electrical properties play important role in the catalytic, sensing, optical, and biosensing applications [111]. The electrical properties of silver-polytetrafluoroethylene (PTFE) nanocomposites prepared by magnetron co-sputtering have been investigated. The dielectric constant of NCs relies on the metal filling factor, and it increases with achieving the percolation threshold and resistivity decrease from 10^7 to 10^{-3} Ω cm over a narrow range of metal content [112]. The polyethylene (PE)/aluminum (Al) NCs were prepared using solution compounding method. The dielectric permittivity and DC conductivity dielectric breakdown strength were also studied [113]. Jiang et al. reported the enhanced electrical properties carbon nanotubes—magnetite NCs using solvothermal technique. The addition of carbon nanotubes in composite increase the electrical conductivity by 32% from 1.9 to 2.5 Scm^{-1} [114]. The improvement in electrical properties of Al_2O_3-Ni nanocomposites (synthesized using mechanical alloying) has been explained. The electrical resistivity was found to be decreased with increasing the Ni concentration and sintering temperature. It is described that the pores exhibited the nonconductive and charge carriers confront lower pore numbers at high temperature thus resulting in decrease of electrical resistivity [115]. The reduction in size of the nanoparticles greatly increases the conductance. The temperature of the solution containing nanoparticles as well as temperature are two crucial parameters responsible for increase in electrical conductivity (as reported in alumina nanofluid with Al nanoparticles) [116]. The changes in electrical properties are attributed to the polarization effect by nanoparticles on the surface [117].

There are many reports with different concentration and different fluids which affects the electronic configuration and thus the electrical conductivity. The ions in the water have the property of the absorber surfactant, and water and mixture of sodium dodecyl sulfate are studied with different concentrations by [118]. Guo et al. investigated the electrical properties of silica NCs with multiwall carbon nanotubes. It was found that the addition of multiwall carbon nanotubes transformed the insulating silica into conductive composites. The RT conductivities of NCs were compared with pure silica over the 14 order. Significantly, the electrical conductivity linearly increased with increasing temperature from 5 to 300 K [119]

3.4 Surface Plasmon Resonance

The interaction of the metallic nanocomposites when photon energy is incident on the surface resulted in a resonance effect with generation of plasmons on the surface is known to be surface plasmon resonance (SPR). The electron cloud in the shell of the metallic nanocomposites gets excited on absorption of incident photon. Since it is the matter of e-e interaction, the whole phenomenon is dependent on the morphology and dimensions of the metallic nanocomposites. Now these days, the size of metallic nanocomposites is in surge in plasmonic applications due to the associated resonance effect. The junction between the conductor and insulators achieved the excitations of the excited electrons and the collection of these excitations on the surface are called surface plasmons [120]. The van der Waals forces interplay in generation of SPR effect observed at different dimensions of the nanoparticles. The SPR effect is mostly reported in metal NCs due to which they are in demand of industrial applications. The presence of SPR effect in NCs makes them better candidate for tunable optical applications. The strong scattering on account of the SPR affects the metallic nanocomposites and finds the applications in biological applications as dark field imaging system [121]. While the other use is in the laser therapy of cancer because of the absorption of SPR [122].

4 Unique Properties of Metallic Nanocomposites

4.1 Surface Atom and Quantum Dots as NCs

The arrangement of electrons in the core shell structure of NCs gives rise to the concept of magic number. The term surface atoms are associated with the arrangement of charge carriers within the nucleus. The surface atoms have greater tendency of being large in surface area in dispersed clusters of atoms. The surface atoms take part in the chemical reaction, and hence the higher surface area leads to the efficiency of the chemical reaction and products as well. It is very important to study the dispersed and agglomerated atoms if we setting a chemical reaction. [123]. The term tiny particles are not of the same dimension; they are further characterized like the particles in nanometer range are called as quantum dots. The electronic structure plays crucial role in various electronic devices. The dimensions of the nanoparticles alter the electronic structure with possible transitions of electrons between the band gap. The highest energy occupied molecular orbital (HOMO) and lowest energy unoccupied molecular orbital (LUMO) make the excitations of electrons and seek applications in the field of laser technology. Thus, the band theory in semiconductors gives the wide concept of applications of MNPs in optical industry. The scattering of photons on account of dimensions of MNPs resulted in interference effect and distribution of energy between the energy levels. The various statistical features of the e-photon interaction are greatly influenced by the frequency of the photon [124].

4.2 Metallic Nanocomposites as a Catalyst

The catalyst is the fuel for any chemical reactions, and the dimension of the metallic nanocomposites interplays with the efficiency of the catalytic functions. With optimizing the synthetic parameters, some catalytic sites are generated intentionally on the surface of the metallic nanocomposites. This type of technique is meant to use the metallic nanocomposites in the field of chemical industry [125]. The metallic nanocomposites have a regular arrangement of atoms within their lattice, and there is inverse correlation between the size of particle and surface atoms. The metallic nanocomposites worked as efficient catalyst under specific and stabilized conditions resulted prevention from agglomerations [126]. The specific benefits of metallic nanocomposites in the field of chemical reactions as catalysts are discussed here:

1. There is less evaporation of solvent in chemical reaction because the temperature of the catalyst is always lower. So, it the chemical reaction will be cost-effective in terms of the amount of solvent.
2. Since the metallic nanocomposites are transparent, they are better candidate for the photocatalyst.
3. The catalytic efficiency is dependent on the size of the particles, and this could be easily achieved in the metallic nanocomposites.
4. The catalytic function of metallic nanocomposites even works for the solid state and gaseous state as well.

5 Characterization and Sensing Applications of Metallic Nanocomposites

5.1 Structural Analysis of Metallic Nanocomposites

The structure and composition of the bulk materials as well as the surface are dependent to each other. The reduction of the size in metallic nanocomposites resulted in the comparable properties on the surface and in the bulk layer. The analysis of the atomic structure is equally important to further study the important properties of the materials. The characterization of the pristine composition gives the idea of the technique-dependent study not the materials and bulk. The size and the structural arrangement of the metallic nanocomposites are very sensitive to the structure and thus alter the properties which structure- and size-dependent. Figure 6 shows the XRD patterns for Ag–ZnO nanocomposites at different concentrations of Ag [127]. Vara et al. reported the transition metal oxide nanoparticles on thermal decomposition of MONs (CuZnO, CoZnO and NiZnO), and structural properties were described using XRD. The obtained results indicated the uniform size distribution with some unspecified reason larger than grain size determined using XRD [128]. The noble metal was used to with $LaMnO_3$ nanocomposites using the green bio-reduction method, and these (Ag, Au, Pd, Pt-$LaMnO_3$) nanocomposites offered the excellent activity for H_2 production. The XRD results indicated the existence of noble metals crystallites of synthesized NCs. The XRD patterns exhibited the additional reflections in Au-$LnMnO_3$ NCs of FCC structure [129]. The particles of Ni, Co and Cu nanoparticles in a polymer matrix synthesized using an aerosol single-drop reactor technique and structural properties have been investigated using XRD study [130]. The aerosols spray pyrolysis technique has been used for synthesis of metal NCs, and their structural properties have been investigated [131].

Moreover, the structural properties and images of synthesized nanosized metal materials at a spatial resolution equivalent to the level of atomic size dimensions can be imagined using TEM. It manifests size in the nm scale, degree of aggregation/agglomeration, dispersion and the heterogeneity/homogeneity of metal nanoparticles. Bright and dark field images were produced regarding metal nanosized particles as it incorporated high-energetic electrons to dispense detailed knowledge about

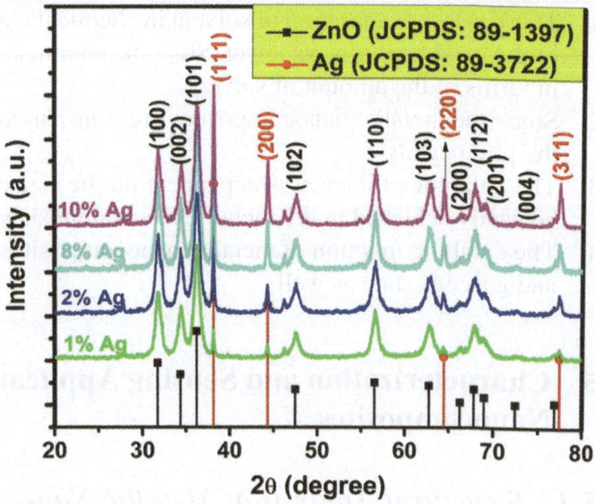

Fig. 6 XRD patterns for Ag–ZnO NCs **a** 1%, **b** 2%, **c** 8%, **d** 10% Ag. Reproduced with permission [127]

morphology of sample, composition of material and crystallographic planes information. Figure 7 shows the HR-TEM images of Au NPs and (ZnO–Au NCs) [132]. TEM image and SAED of nanostructured Al $_{2024}$ alloy under ECAP process have been reported and the preferred orientation {110} [001] was formed in Al_{2024} alloy [133]. Torrisi et al. investigated the structural properties of metal-polymer nanocomposites synthesized using (Co)evaporation/(Co)sputtering technique. TEM analysis resulted in determination of average area, surface density and surface evolution for gold particles of different thickness synthesized on polystyrene surface [134]. Laser-synthesized gold/oxide nanocomposites (Au/TiO_2 and Au/FexOy) have been analyzed using TEM with low- and high-magnification images [135]. The mixed metal oxides $Cu_{0.5}Sr_{0.5}$ and $La_{0.2}$ doped $Cu_{0.4}Sr_{0.4}$ NCs were investigated for phase purity and structural analysis and SEM analyzed particle size of NCs ranging from (43–72) nm [136].

The images can be studied for the metal nanoparticles to study the size distribution, which usually depends on the reduction (method of synthesis) and the elemental composition of the product [137]. Structural as well as analytical characterizations of metal nanoparticles are very pivotal. Parameters like particle size/grain size, lattice, morphologic, crystallography, composition, phase and distribution can be obtained by this technique.

5.2 *Fourier Transform Infrared Spectroscopy (FTIR)*

FTIR is used to determine change in the characteristic patterns of absorption band that clearly indicates a change in the metallic nanocomposites. It is useful in determining unknown elements, contaminants, additives, decomposition and oxidation. Every

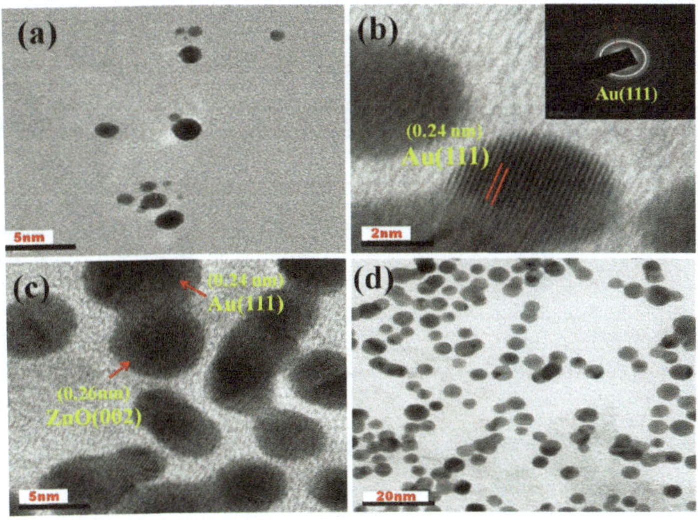

Fig. 7 HR-TEM images of **a, b** Au NPs (inset: SAED pattern) and **c, d** ZnO-Au NCs Reproduced with permission [132]

molecule of prepared metal nanoparticles has a distinctive fingerprint, implying invaluable method for chemical identification [138]. FTIR measurements have been used to identify functional groups (FG) in GO, TiO_2, TiO_2/GO-8 and TiO_2/rGO NCs and identified the number of different oxygen-containing on GO [139]. Sacourbaravi et al. reported the synthesis of Ag NPs/Zn-MOF NCs and reported their application as antibacterial agents. In Zn-MOF, bands attributed vibrations of carboxylate groups and stretching vibration [140]. Also, a scheme of different roles of NPs in IR as analyte and tools is shown in Fig. 8 [7].

5.3 *Raman Spectroscopy*

Raman spectroscopy is used to identify the phases and phase transitions of various metallic nanocomposites and other nanostructured materials. High Raman enhancement is observed when the molecules are close to highly polarizable objects, as metallic nanoparticles, due to the intense localized electric fields generated by localized surface plasmon resonance for resonant incident energies. Galaburda reported the Raman amplification in carbon-stabilized Cu/C, Co/C and Ni/C NCs exhibited G- ($\sim$1590 cm^{-1}) and D-bands ($\sim$1350 cm^{-1}) bands. Cu/C NCs demonstrated the most disordered nanographite evidenced by G- and D-bands. The most prominent peaks that appeared Co/C at 2.5 mW can be ascribed to E_g(475 cm^{-1}), F_{2g}(195, 516 and 612 cm^{-1}) and A_{1g} (681 cm^{-1}) bands of Co_3O_4 crystalline phase. Figure 9 shows the Raman spectra of Co/C, Ni/C and Cu/C nanocomposites [141]. Miedema et al.

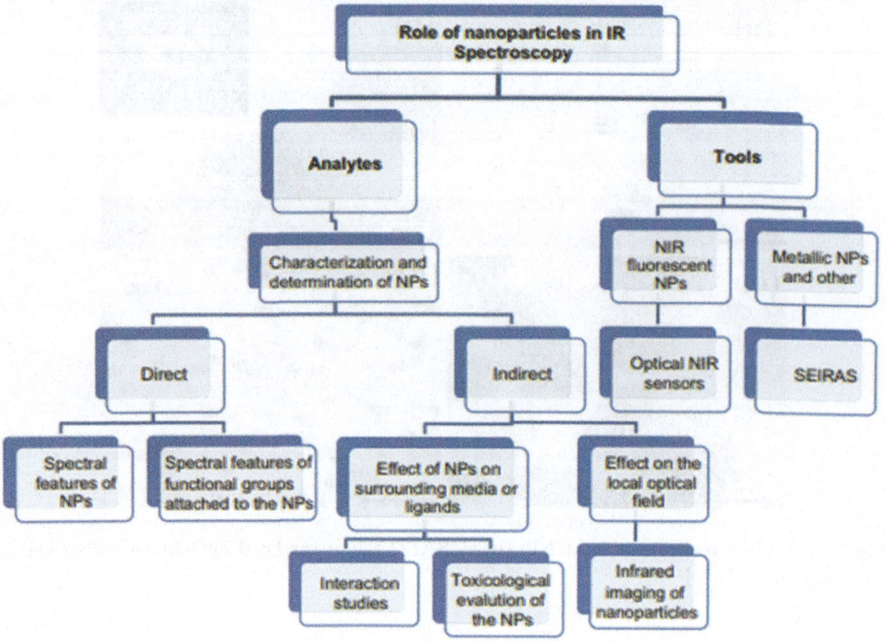

Fig. 8 Schematic diagram of different roles of infrared in the characterization of nanoparticles, and NPs as a tool for the determination of compounds and other NPs. Reproduced with permission [7]

reported the Raman spectra to study the H_2 sorption properties of Li borohydride NCs. Raman resulted that the spectra of $LiBH_4$–C are similar to bulk spectra $LiBH_4$ powder [142]. Raman spectra of GO, RGO, CeO_2 and CeO_2–RGO NCs exhibited the peaks of GO at 1360 cm^{-1} and 1599 cm^{-1}correspond to the D and G bands and exhibited a ID/IG ratio of ~0.98 with in plane crystallite size of ~19.6 nm, signify that the defects are less existing in comparison with RGO [143].

5.4 X-ray Photoelectron Spectroscopy (XPS)

XPS is used for characterization of metallic nanocomposites because of its ability for detection of thin layer and impurity. XPS data is easy to combine with modeling and computational tool to determine the information about metal nanoparticles and thickness of thin films. XPS technique has been used to determine the chemical composition and binding energy of rGO–SnO_2. XPS confirmed the presence of elements present in nanocomposite and no trace of impurities have been observed. The binding states of Sn (Sn 3s, 3p, 3d, 4s, 4p and 4d) and O 1s ascribed to SnO_2 and the addition peak of C 1s also have been observed. The binding energy of SnO_2 on rGO shifted

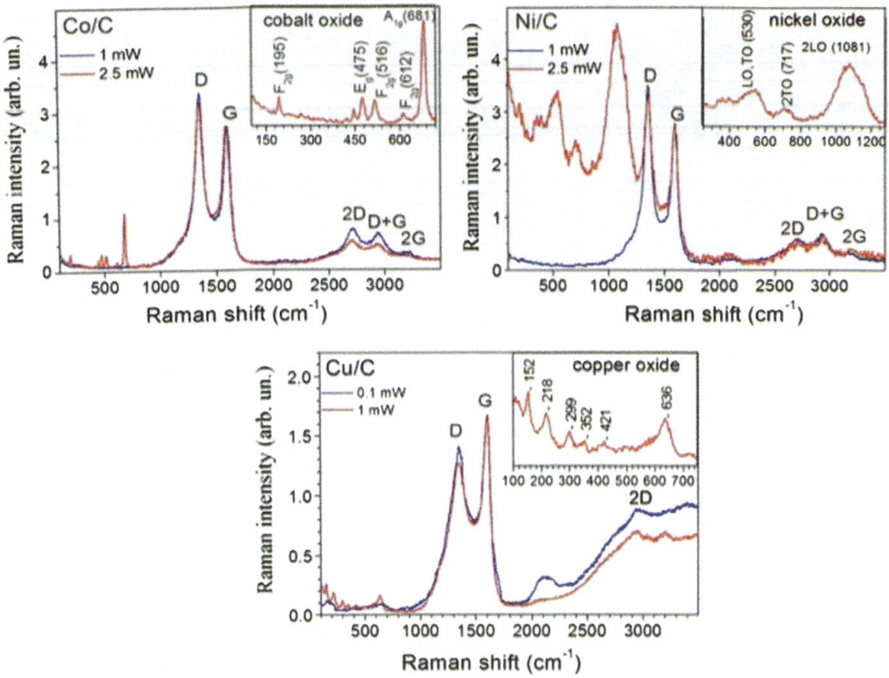

Fig. 9 Raman spectra of Co/C, Ni/C and Cu/C NCs measured at laser wavelength $\lambda_{exc} = 488$ nm and varied excitation power. Spectra are normalized to the intensity of the G-band. Reproduced with permission [141]

toward higher binding energy and peaks at 487.3 and 495.8 eV ascribed to Sn $3d_{5/2}$ and Sn $3d_{3/2}$ [144]. Singh et al. reported the synthesis of Bi_2O_3-MnO_2 NCs electrode for the application of wide-potential window high-performance supercapacitor, and XPS technique has been used to investigate the elemental composition and valence states of the ions present in NCs. The deconvolution of spin–orbit peaks signified the co-presence of Mn^{4+} and Mn^{3+} valence state at different binding energy states [145]. Fe_3O_4 NPs have been synthesized on rGO for biosensor applications, and XPS technique was used to determine the energy states (Fe 2P) and valance states Fe^{2+} and Fe^{3+} of NCs [146]. Figure 10 depicts the SEM images and XPS spectra of Cu-rGO NCs. Moreover, XPS measurements were carried out to investigate the mixture of the GO and Cu and Cu-rGO NCs electrode. These NCs were synthesized successfully for efficient electrochemical reduction of CO_2 [147].

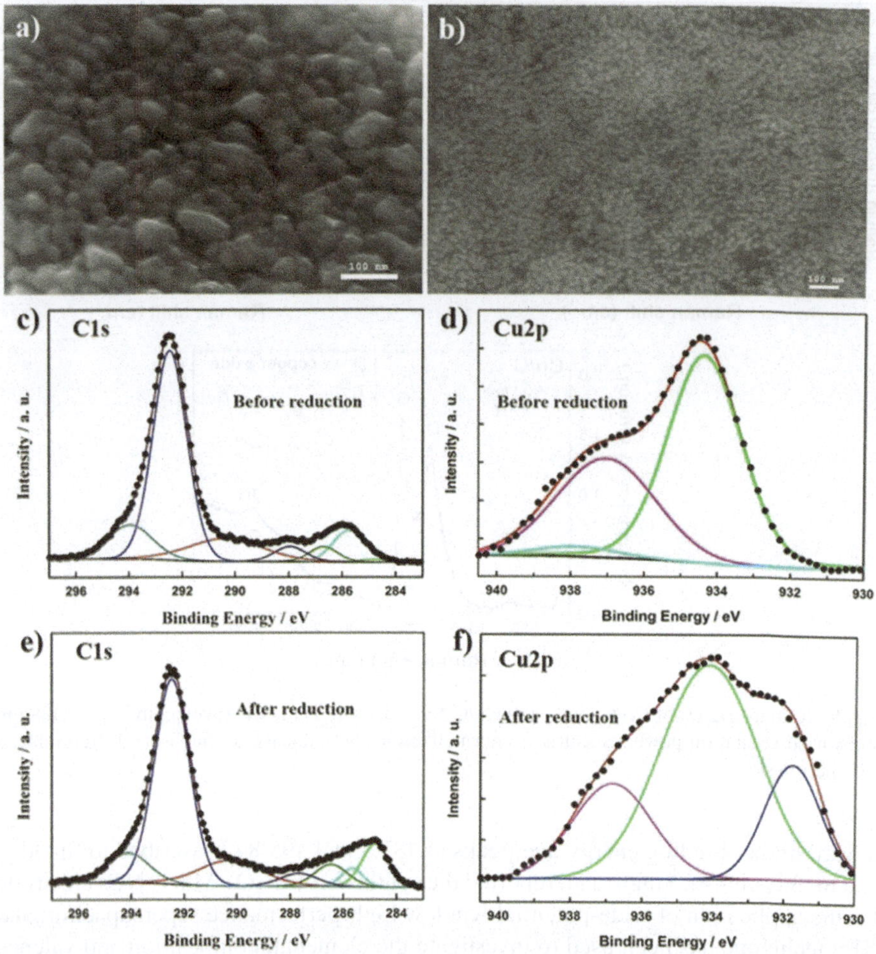

Fig. 10 SEM images of the formed Cu NPs **a** and the Cu-rGO NC **b** on a Cu substrate. High-resolution XPS spectra of the C1s region (**c** and **e**) and the Cu2p region (**d** and **f**) of the $CuSO_4$-GO thin film before the electrochemical treatment and the formed Cu-rGO NCs. Reproduced with permission [147]

6 Sensors Developed Using Metallic Nanocomposites

The unique properties of the nanoparticles are associated with their small dimension. The fantastic phenomenon observed in metallic nanocomposites differentiate from their bulk. This differentiation finds applications in variety of sensing characterizations. Here we have discussed few novel metallic nanocomposites having applications in the field of sensors technology. Figure 11 depicts the various type of metal NCs-based sensors [150].

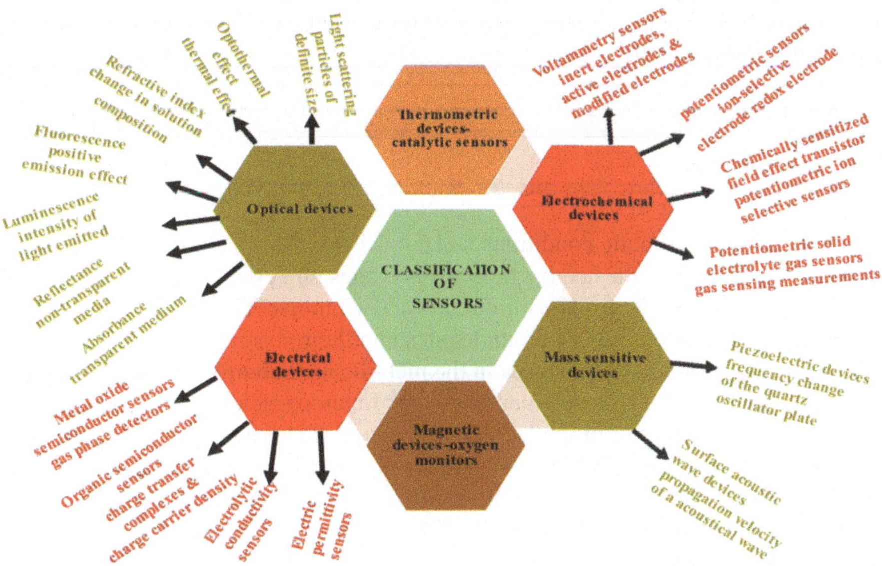

Fig. 11 Classification of Sensors. Reproduced with permission [150]

6.1 *Electrochemical Sensors*

Metal nanoparticles have engrossed researchers' consideration in electrochemical sensing applications for their unique electrocatalytic properties. Because of the possibility of the alteration in the dimensions of the nanoparticles, the electrocatalytic properties could be easily tuned. The high sensitivity of metallic nanocomposites leads the applications in sensing devices. The surface-dominated properties play important role in electrochemical sensors. However, metals in bulk form could not be used as good catalysts. The size is very crucial for metallic nanocomposites to use it as electrochemical sensors, for example the size in range 1–10 nm showed higher response to electrochemical process in different species. The extensive sought of NCs with electroanalytical systems to produce the silent features such as tunable microprobes, rugged instrumentation, portable, multiple analysis and miniaturization. Reddy et al. reported the MO_x—MNCs modified electrochemical (amperometric, voltammetric and impedimetric) sensors for application of toxic chemicals. The NCs consist of improved active surface area, excellent adsorption ability and electrocatalytic activity. The electrochemical sensor engenders subfemtomolar detection limits as well as thrives in excellent selectivity [148]. Heavy metal-based electrochemical sensor has been fabricated using layer-by-layer deposition of Ti_3C_2Tx/MWNTs NCs for detection of Cu and Zn ions that present in human biofluids. The sensor showed the excellent detection performance for Cu and Zn as 0.1 and 1.5 ppb, respectively. The concentration of Cu and Zn was determined in biofluid, in urine (Cu: 10–500, Zn: 200–600 ppb) and sweat (Cu: 300–1500; and sweat Zn: 500–1500 ppb). Sensor exhibited the advantage of ultra-repeatability and improved stability [149].

6.2 Biosensors

The changes in the climatic conditions make it is important to continuously monitor the health chart. Hence, the need of biosensors comes into effect to check the amount of essential nutrients like proteins, minerals and vitamins. The biosensors work on the principle of mobilization of biomolecules and their absorption in the body. The high specific area-related advantages in the metallic nanocomposites provide access to high surface energy. This peculiar property of nanoparticles makes it possible to absorb a greater number of biomolecules as the large area is available for absorption. The schematic diagram of biosensor is represented in Fig. 12 [150]. Selectively, the gold-based NCs biosensors are synthesized in nanorod structure that identifies E. coli. and detected the E. coli in meat, milk and in egg-shell with Au-based NCs biosensor. With immobilization of proteins and passing through the bonds between the interface, the metallic nanocomposites used as effective biosensors [150, 151]. The diverse applications of biosensor are depicted in Fig. 13 [150]. The Au-based NCs are of great importance in the field of biosensors because of their unique properties their biocompatibility and easy to synthesis with desired modifications [152]. There is another parameter which is important for the effective biosensors, i.e., the non-functionalization of nanoparticles. The biomolecules like ssDNA are also absorbed by the Au NPs and hence could be used as sensor for DNA detection as reported by Li et al. [153]. There are many industries where biosensor which are based on MNCs is used like food industry [154]. The MNCs biosensor is also used or the treatment of cancer due to the bond's formation over the cell early-stage cancer is detectable. The sensing of monoclinic antibodies is studied by Pal et al. [155]. The biosensors are based on rGo/Ag NCs functionalized textiles using advanced e-beam irradiation technique for the detection of adrenaline (epinephrine). The sensor was investigated by square wave, linear sweep and differential pulse voltammetry. The fabricated flexible textile biosensors consist of high sensitivity and strong catalytic performance for epinephrine detection. Under standard conditions (0.5–100.0 μL), low limit of detection (LOD) was found to be 9.73 nM (S/N = 3) for the detection of cotton and polyester biosensors. Figure 14 depicts the cyclic voltammetric profiles at 100 mV/s for the nanocomposite cotton and polyester fabrics. However, the fabricated electrode stands for encouraging analytical sensor for the investigation of pharmaceutical and biological samples [156].

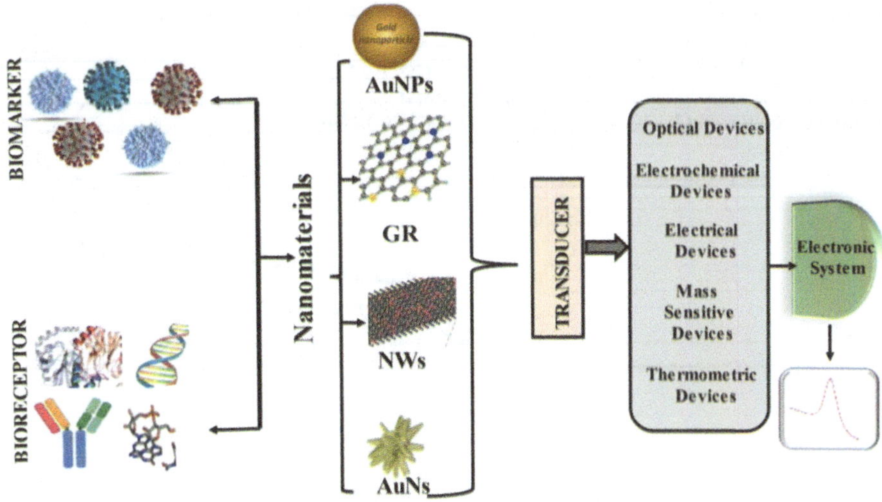

Fig. 12 Schematic representation of biosensor. Reproduced with permission [150]

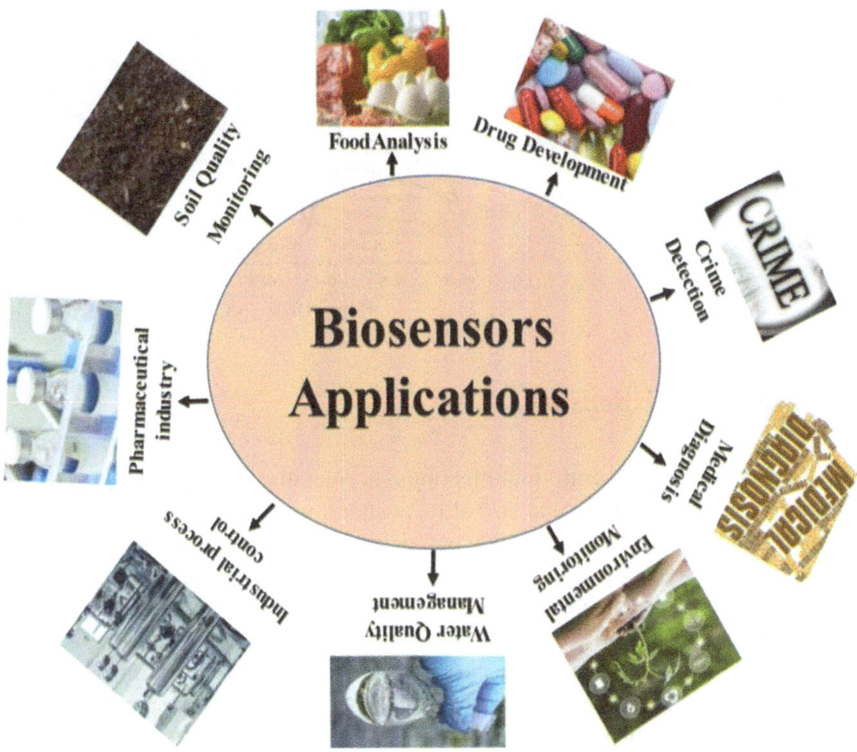

Fig. 13 Applications of biosensors. Reproduced with permission [150]

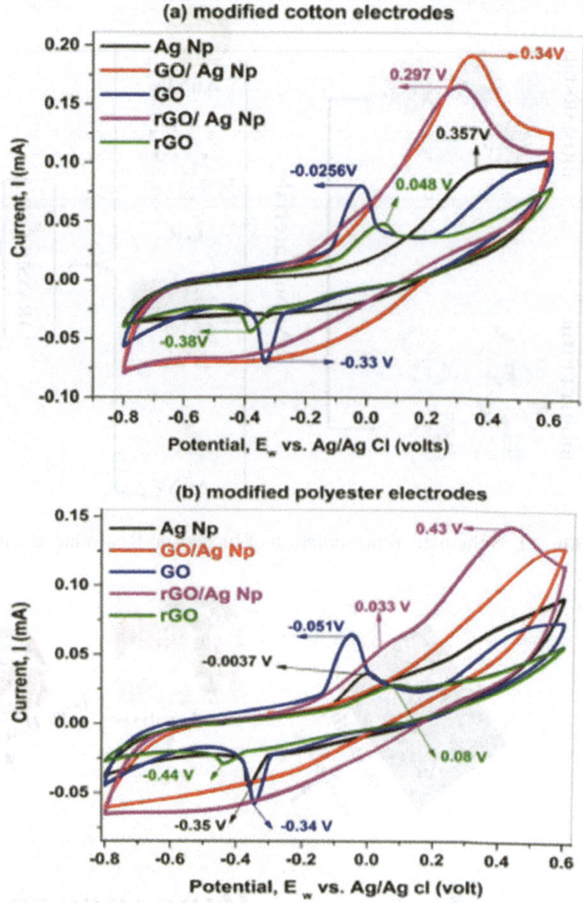

Fig. 14 CV curves in an aqueous solution 0.1 M Na_2SO_4 + 0.5 M KCl at 100 mV/s for modified electrodes **a** cotton and **b** polyester fabrics. Reproduced with permission [156]

6.3 *Photovoltaic Sensors*

The tunable band gap with modifications in metallic nanocomposites by the customization of their shape and size attracts attention in photovoltaic applications. A bunch of reports showed the efficiency of metallic nanocomposites in the solar cells. The light harvesting is another important field where the unique characteristics like generation of surface plasmons seek matter of attention. The permittivity of the absorbing materials and the thickness plays crucial role in deciding the applications. The thick layer of the substrate reduces the scattering, and a greater number of photos are absorbed. According to the survey, an array of metallic nanocomposites which are embedded in the materials could be able to produce permittivity up to $3 + 0.1i$. However, the effective absorbing medium can reach up to the value of $12.51 + 1.91i$. The applied field is associated with the possible transition within the band gap to generate the surface plasmons and absorbs the photons. This is how the field value is directly related to the area and the number of metallic nanocomposites absorbed. Hence, there is high photon–lattice interaction when there is absorption of metallic nanocomposites [157].

6.4 pH Sensors

The easy-to-tune size and the morphology of metallic nanocomposites by using unique method of synthesis opens the door to many sensing applications. This access is associated with the luminescent spot generated on the surface and also with different properties. pH sensor plays a vital role in considering the physiological/pathologic mechanism, and it is known as the abnormal pH signifies many diseases including cancer, stroke and Alzheimer's. Dual-emitting fluorescent metal–organic framework (MoF) NCs have been fabricated a broad-range pH sensor for the application of fluorescence imaging. MoF has been fabricated by summarizing the DBI-PEG-NH2 functionalized Fe3O4 into the Zr-MOFs and reacted with rhodamine B isothiocyanates (RBITC). Further, the sensor could see pH change in living cells and in water cells [158]. The sensing to pH is dependent on the equilibrium of protonation–deprotonation. The central core region of the Zn and the solution changes the fluorescence intensity of PCN-225 which is able to detect the pH value. It is observed that the range $7 < pH < 10$ is the most sensitive range for fluorescence intensity responses [159]. The properties of the metallic nanocomposites such as the large surface area and the high density allow them to remain in the solution for the longer time. The remained metallic nanocomposites are helpful in the removal of heavy metals from the water which are harmful to our health by measuring the pH value.

6.5 Photocatalytic Sensors

Nowadays, the ability to tune the band gap is very effective to semiconducting industry. The metallic nanocomposites are known to be the better candidate for the application of photocatalyst as the size is easy to adjustable. The use of all transitions of solar spectra finds application in various chemical industries. The small size of the nanoparticles provides large surface area and allows access to all-possible transitions. With the metallic nanocomposites are exposed to the light then the absorption of light excites the electrons to the higher level. The energy is released and used in the chemical reactions, and thus the metallic nanocomposites show the quality competent photocatalyst [160]. The Ag-based NCs play important role in photodegradation of dyes when exposed to UV light as reported by Tamuly et al. [161]. The large specific surface-to-volume ratio allows the higher area to degrade the dyes like methyl orange and blue which are harmful to the water bodies. Thus, the metallic nanocomposites act as barrier for the harmful effects of dyes and find applications in the dyes industries.

In recent days, the metallic nanocomposites are cluster with the semiconductors as they are affecting the photocatalytic action. The starch on leaves of the plants is also synthetically synthesized and using this the leaves absorbed the sunlight and comes out the product of photosynthesis. Similarly, the water is split to protons and O_2 using the photocatalytic activity of the embed metallic nanocomposites. However, very high amount of photon energy is required since the photocatalytic action is quite endothermic. The water splitting and the higher area are very interesting in the production of hydrogen gas. Therefore, the network of the metallic nanocomposites plays important and crucial role in semiconducting industry. The metallic nanocomposites are a good candidate which is good supplement to the action of semiconductors. However, the activity of the photocatalyst is dependent on the various electrical properties [162–164].

7 Conclusion

Metallic nanocomposites play a major role in nanotechnology and nanoscience. This chapter concludes the description of the synthesis, properties and characterizations of metal nanoparticles. In summary, a substantial amount of work has been devoted to the growth of innovative methodologies for metal nanoparticles. Different techniques have been used to analyze and characterize the synthesized NPs, including size, shape, the surface of the NP, size distribution and confirmation, to have enough information to obtain their applications in various aspects. This chapter provides access to the metal nanoparticles of a large variety of metals, each showing different properties. The characteristics and properties of metal nanoparticles are influenced by their synthesis method. The main techniques that can be used to investigate the characteristics of NPs have been described and summarized, including optical, spectroscopic, FTIR, SEM, TEM, XRD, Raman, etc. Moreover, the applications of metal NPs, especially for sensors such as those developed using metal NPs, electrochemical sensors, biosensors, photovoltaic, PH and photocatalytic sensors, have been discussed here. A multitude of opportunities for study exists to evaluate the performance of metallic nanocomposites with tunable properties. This chapter comprises the various parallel opportunities to learn the effects of arrangements of metal nanoparticles in systems and develop multifunctional materials. This provides an in-depth look at the synthesis and characterization of metal nanoparticles. Advances in synthesis techniques and their characterization will extend the limits of metal nanoparticles and open the door to new possible multifunctional applications. The applications associated with metallic nanocomposites began with a tradition of empirical work that evolved into a methodology for materials to attain improved performance. Moreover, metallic nanocomposites have shown their importance in the medical, biomedical and pharmaceutical applications of nanotechnology.

Acknowledgements One of the author's Dr. Rajesh Kumar acknowledges the Faculty Research Grant Scheme (FRGS) Project-2021 provided by Guru Gobind Singh Indraprastha University, New

Delhi, India, and Inter University Accelerator Centre (IUAC), New Delhi, India, for project (Ref: IUAC/XIII.3A/68308/2020). One of the authors (Vishnu Chauhan) is indebted to Indo-German Science & Technology Centre (IGSTC), DST Government of India for providing Post-Doctoral Industrial Fellowship (PDIF).

References

1. Akram M, Farooqi QH et al (2016) synthesis and characterization of some important metal nanoparticles and their applications. Sci Int 28(4):4049–4405
2. Venkatesh N, Harish Kumar K (2018) Biomed J Sci Tech Res 4:3765–3775
3. Chen W, Cai W, Zhang L, Wang G, Zhang L (2001) Sonochemical processes and formation of gold nanoparticles within pores of mesoporous silica. J Colloid Interface Sci 238(2):291–295
4. Geethalakshmi R, Sarada DV (2012) Gold and silver nanoparticles from Trianthema decandra: synthesis, characterization, and antimicrobial properties. Int J Nanomed 7:5375–5384
5. Vijayakumar M, Priya K, Nancy FT, Noorlidah A, Ahmed AB (2013) Biosynthesis, characterization and anti-bacterial effect of plant-mediated silver nanoparticles using Artemisia nilagirica. Ind Crops Prod 41:235–240
6. Aqoo AA, Ahmad H et al (2020) Recent advances in metal decorated nanomaterials and their various biological applications: a review. Front Chem 8:341
7. Shnoudeh AJ, Hamad I, Abdo RW, Qadumii L, Jaber AY, Surchi HS, Alkelany SZ (2019) Chapter 15—synthesis, characterization, and applications of metal nanoparticles. In: Advances in pharmaceutical product development and research, biomaterials and bionanotechnology, pp 527–612
8. Narayanan R, El-Sayed MA (2004) Shape-dependent catalytic activity of platinum nanoparticles in colloidal solution. Nano Lett 4(7):1343–1348
9. Moura D, Souza MT, Liverani L, Rella G, Luz GM, Mano JF, Boccaccini AR (2017) Development of a bioactive glass-polymer composite for wound healing applications. Mater Sci Eng C 76:224–232
10. Banerjee K, Das S, Choudhury P, Ghosh S, Baral R, Choudhuri SK (2017) A novel approach of synthesizing and evaluating the anticancer potential of silver oxidenanoparticles in vitro. Chemotherapy 62(5):279–289
11. Romero PG (2001) Hybrid organic–inorganic materials—in search of synergic activity. Adv Mater 13(3):163–174
12. Shaikh SF, Mane RS, Min BK, Hwang YJ, Joo OS (2016) D-sorbitol-induced phase control of TiO_2 nanoparticles and its application for dye-sensitized solar cells. Sci Rep 6:2010–2013
13. Gracias DH, Tien J, Breen TL, Hsu C, Whitesides GM (2000) Forming electrical networks in three dimensions by self-assembly. Science 289(5482):1170–1172
14. Aznar-Gadea E, Rodríguez-Canto PJ, Martínez-Pastor JP, Lopatynskyi A, Chegel V, Abargues R (2021) Molecularly imprinted silver nanocomposites for explosive taggant sensing. ACS Appl Polym Mater 3(6):2960–2970. https://doi.org/10.1021/acsapm.1c00116
15. Bielinska A, Eichman JD, Lee I, Baker JR Jr, Balogh L (2002) Imaging Au–PAMAM gold-dendrimer nanocomposites in cells. J Nanopart Res 4:395–403
16. Haque S, Julappagari M, Patra CR (2021) Potential application of silver nanocomposites for antimicrobial activity. In: Nayak AK, Hasnain MS (eds) Biomedical composites. materials horizons: from nature to nanomaterials. Springer, Singapore. https://doi.org/10.1007/978-981-33-4753-3_5
17. Chung IM, Park I, Seung-Hyun K, Thiruvengadam M, Rajakumar G (2016) Plant mediated synthesis of silver nanoparticles: their characteristic properties and therapeutic applications. Nanoscale Res Lett 11(1):40

18. Asharani PV, Low Kah Mun G, Hande MP, Valiyaveettil S (2008) Cytotoxicity and genotoxicity of silver nanoparticles in human cells. ACS Nano 3(2)
19. Yadav TP, Yadav RM, Singh DP (2012) Mechanical milling: a top down approach for the synthesis of nanomaterials and nanocomposites. Nanosci Nanotechnol 2(3):22–48
20. Sellinger A, Weiss P, Nguyen A et al (1998) Continuous self-assembly of organic–inorganic nanocomposite coatings that mimic nacre. Nature 394:256–260. https://doi.org/10.1038/28354
21. Bharti K, Sadhu KK (2022) Syntheses of metal oxide-gold nanocomposites for biological applications. Results Chem 4:100288
22. Meyers MA, Mishra A, Benson DJ (2006) Mechanical properties of nanocrystalline materials. Prog Mater Sci 51(4):427–556
23. Nadagouda MN, Speth TF, Varma RS (2011) Microwave-assisted green synthesis of silver nanostructures. Acc Chem Res 44(7):469–478
24. Rajput N (2015) Methods of preparation of nanoparticles: a review. Int J Adv Eng Technol 7(6):1806
25. Cha HG, Kim YH, Kim CW, Kwon HW, Kang YS (2007) Characterization and magnetic behavior of Fe and Nd-Fe-B nanoparticles by surfactant-capped high-energy ball mill. J Phys Chem 111(3):1219–1222
26. Tsuzuki T, McCormick PG (2004) Mechanochemical synthesis of nanoparticles. J Mater Sci 39(16):5143–5146
27. Murty BS, Ranganathan S (1998) Novel materials synthesis by mechanical alloying/milling. Int Mater Rev 43(3):101–141
28. Todaka Y, McCormick P, Gerard T, Koichi U, Umemoto M (2002) Synthesis of Fe-Cu nanoparticles by mechanochemical processing using a ball mill. Mater Trans 43(4):667–673
29. Simakin AV, Voronov VV, Kirichenko NA, Shafeev GA (2004) Nanoparticles produced by laser ablation of solids in liquid environment. Appl Phys Mater Sci Process 79(4):1127–1132
30. Kruis FE, Fissan H, Peled A (1998) Synthesis of nanoparticles in the gas phase for electronic, optical and magnetic applications - a review. J Aerosol Sci 29(5):511–535
31. El-Nour KM, Eftaiha AA, Al-Warthan A, Ammar RA (2010) Synthesis and applications of silver nanoparticles. Arab J Chem 3(3):135–140
32. Altowyan AS, Mostafa AM, Ahmed HA (2021) Effect of liquid media and laser energy on the preparation of Ag nanoparticles and their nanocomposites with Au nanoparticles via laser ablation for optoelectronic applications. Optik 241:167217
33. Becker MF, Brock JR, Cai H, Henneke DE, Keto JW, Lee J, Nichols WT, Glicksman HD (1998) Metal nanoparticles generated by laser ablation. Nanostruct Mater 10(5):863.https://doi.org/10.1016/s0965-9773(98)00121-4
34. Kabashin AV, Meunier M (2003) Synthesis of colloidal nanoparticles during femtosecond laser ablation of gold in water. J Appl Phys 94:7941–7943
35. Bae CH, Nam SH, Park SM (2002) Formation of silver nanoparticles by laser ablation of a silver target in NaCl solution. Appl Surf Sci 197:628–634
36. Simakin AV, Voronov VV, Shafeev GA, Brayner R, Verduraz FB (2001) Nanodisks of Au and Ag produced by laser ablation in liquid environment. Chem Phys Lett 348(3–4):182–186
37. Mwafy A, Mostafa AM (2020) Tailored MWCNTs/SnO_2 decorated cellulose nanofiber adsorbent for the removal of Cu(II) from waste water. Radiat Phys Chem 177:109172
38. Mostafa AM, Mwafy EA (2020) Laser-assisted for preparation Ag/CdO nanocomposite thin film: structural and optical study. Opt Mater 107:110124
39. Swihart MT (2003) Vapor-phase synthesis of nanoparticles. Curr Opin Colloid Interface Sci 8(1):127–133
40. Vanecht E (2012) Gold nanoparticles in ionic liquids prepared by sputter deposition. ISBN: 978-90-8649-508-5
41. Bharti K, Sadhu KK (2022) Synthesis of metal-oxide gold nanocomposites for biological applications. Results Chem 4:100288
42. Tafur G et al (2019) Optical properties of silver nanoparticles embedded in dielectric films produced by dc and rf magnetron sputtering. J Phys: Conf Ser 1173:1742–6596

43. Magudapathy P, Srivastava SK, Gangopadhyay P, Amirthapandian S, Saravanan K, Das A, Panigrahi BK (2017) Alloying of metal nanoparticles by ion-beam induced sputtering. Chem Phys Lett 667:38–44
44. Deng L, Nguyen MT, Shi J et al (2020) Highly correlated size and composition of Pt/Au alloy nanoparticles via magnetron sputtering onto liquid. Langmuir 36:3004–3015
45. Chung B-X, Liu C-P (2004) Synthesis of cobalt nanoparticles by DC magnetron sputtering and the effects of electron bombardment. Mater Lett 58(9):1440
46. Kratochvíl J, Kuzminova A, Kylián O, Ondřej B (2015) Comparison of magnetron sputtering and gas aggregation nanoparticle source used for fabrication of silver nanoparticle films. Surf Coat Technol 275:296–302
47. Gondoni P, Ghidelli M, Di Fonzo F, Bassi AL, Casari CS (2013) Fabrication of nano-engineered transparent conducting oxides by pulsed laser deposition. J Vis Exp 72
48. Fasaki I et al (2012) Properties of pulsed laser deposited nanocomposite NiO: Au thin films for gas sensing applications. Appl Phys A 107:899–904
49. Mohammed AZ, Menazea AA, Mostafa MA et al (2020) Ultra-thin silver nanoparticles film prepared via pulsed laser deposition:synthesis, characterization, and its catalytic activity on reduction of 4-nitrophenol. Surfaces and Interfaces 19:100438
50. Elhalawany N, Elmeley H (2015) Synthesis, characterization and electrical properties of highly conductive polyaniline/gold and/or platinum nanocomposites. Synthetic Metals 205:145–152
51. Mostako ATT, Khare A (2012) Effect of target–substrate distance onto the nanostructured rhodium thin films via PLD technique. Appl Nanosci 2(3):189–193
52. Pedersen H, Elliott SD (2014) Studying chemical vapor deposition processes with theoretical chemistry. Theor Chem Acc 133(5):1476
53. Chen J, Yin Z (2011) Controlled CVD growth of Cu–Sb alloy nanostructures. Nanotechnology 22(32):325602
54. Piszczek P, Radtke A (2017) Silver nanoparticles fabricated using chemical vapor deposition and atomic layer deposition techniques: properties, applications and perspectives: review. https://www.intechopen.com/chapters/57724. https://doi.org/10.5772/intechopen.71571
55. Rodchanarowan A, Chiyasak P, Kalnaowakul P, Krajaisri P, Puranasiri R, Jaturapronperm S, Thanathattakum B (2022) Effect of Ni addition on magnetism and corrosion resistivity of Ti/Co/Ni nanocomposites by sol–gel methods. J Magn Magn Mater 555:169338
56. Wang Y, Yaxin Z (2014) L10 CoPt–Cu nanoparticles for high-density magnetic recording by sol–gel technique. J Alloys Compounds 582:511–514
57. Choi YJ, Chiu CK (2010) Spontaneous deposition of gold nanoparticle nanocomposite on polymer surfaces through sol–gel chemistry. Nanotechnology 22(4):045601
58. Chatterjee A, Chakravorty D (1992) Electrical conductivity of sol-gel derived metal nanoparticles. J Mater Sci 27(15):4115–4119
59. Guzman MG, Dille J, Godet S (2009) Synthesis of silver nanoparticles by chemical reduction method and their antibacterial activity. Int J Chem Biomol Eng. 2(3):104–111
60. Chattopadhyay DP, Patel BH (2014) Nano metal particles: synthesis, characterization and application to textiles. In: Ahmed W, Ali N (eds) Manufacturing nanostructures. One Central Press (OCP), UK 184–215
61. Khan Z, Al-Thabaiti SA, Obaid AY, Al-Youbi AO (2011) Preparation and characterization of silver nanoparticles by chemical reduction method.Colloids Surf B 82:513–517
62. Wang H, Qiao X, Chen J, Ding S (2005) Preparation of silver nanoparticles by chemical reduction method. Colloids Surf A 256:11–115
63. Jeyaraj M, Gurunathan S, Qasim M, Kang MH, Kim JH (2019) A comprehensive review on the synthesis, characterization, and biomedical application of platinum nanoparticles. Nanomaterials (Basel) 9(12):1719. https://doi.org/10.3390/nano9121719
64. Dang TMD et al (2011) Synthesis and optical properties of copper nanoparticles prepared by a chemical reduction method. Adv Nat Sci: Nanosci Nanotechnol 2:015009
65. Khanna PK, Singh N, Charan S, Subbarao VVVS, Gokhale R, Mulik UP (2005) Synthesis and characterization of Ag/PVA nanocomposite by chemical reduction method. Mater Chem Phys 93:117–121

66. Carpenter MK, Moylan ET, Kukreja R et al (2012) Solvothermal synthesis of platinum alloy nanoparticles for oxygen reduction electrocatalysis. J Am Chem Soc 134(20):8535–8542
67. Gao L et al (2020) Solvothermal synthesis of Sb2S3-graphene oxide nanocomposite for electrochemical detection of dopamine. J Electrochem Soc 167:1945–7111
68. Zhan F et al (2019) Facile solvothermal preparation of Fe_3O_4–Ag nanocomposite with excellent catalytic performance. RSC Adv 9:878–883
69. Juan Z, Yao X et al (2007) Controlled growth of silver nanoparticles in a hydrothermal process. China Particuology 5(3):206–212
70. Nandanwar SU, Chakraborty M, Mukhopadhyay S, Shenoy KT (2011) Stability of ruthenium nanoparticles synthesized by solvothermal method. Cryst Res Technol 46(4):393–399
71. Dikusar AI, Globa PG, Belevskii SS, Sidelnikova SP (2009) On limiting rate of dimensional electrodeposition at meso-and nanomaterial manufacturing by template synthesis. Surf Eng Appl Electrochem 45(3):171–179
72. Hasnidawani JN, Azlina HN, Norita H, Bonnia NN, Ratim S, Ali ES (2016) Synthesis of ZnO nanostructures using sol-gel method. Procedia Chem 19:211–216
73. Rai P (2018) Plasmonic noble metal@metal oxide core–shell nanoparticles for dye-sensitized Solar cell applications. Sustain Energy Fuels. https://doi.org/10.1039/C8SE00336J
74. Bayal N, Jeevanandam P (2012) Synthesis of metal aluminate nanoparticles by sol–gel method and studies on their reactivity. J Alloys Compd 516:0–32
75. Mauro AD, Farrugia C (2020) Synthesis of ZnO/PMMA nanocomposite by low-temperature atomic layer deposition for possible photocatalysis applications. Mater Sci Semicond Process (118):105214
76. Zhou XJ, Harmer AJ (2004) Parametric study on electrochemical deposition of copper nanoparticles on an ultrathin polypyrrole film deposited on a gold film electrode. Langmuir 20(12):5109–5113
77. Sangili A (2020) Label-free electrochemical immunosensor based on one-step electrochemical deposition of AuNP-RGO nanocomposites for detection of endometriosis marker CA 125. ACS Appl Bio Mater 3(11):7620–7630
78. Cui CH, Li HH, Yu SH (2010) A general approach to electrochemical deposition of high quality free-standing noble metal (Pd, Pt, Au, Ag) sub-micron tubes composed of nanoparticles in polar aprotic solvent. Chem Commun 46:940–942
79. Chandra M, Das PK (2009) Green routes to noble metal nanoparticle synthesis. Int J Green Nanotechnol Phys Chem 1(1):10–25
80. Khan A, El-Toni AM (2011) Microwave-assisted synthesis of silver nanoparticles using poly-N-isopropylacrylamide/acrylic acid microgel particles. Colloids Surf A 377(1–3):356–360
81. Harpeness R, Gedanken A (2004) Microwave synthesis of core–shell gold/palladium bimetallic nanoparticles. Langmuir 20(8):3431–3434
82. Xu W, Liew KY (2008) Microwave-assisted synthesis of nickel nanoparticles. Mater Lett 62:2571–2573
83. Njoki PN, Rhoades AE, Barnes JI (2019) Microwave-assisted synthesis of anisotropic copper–silver nanoparticles. Mater Chem Phys 241:122348
84. Sun W, Li H, Wang Y (2015) Microwave-assisted synthesis of graphene nanocomposites: recent developments on lithium-ion batteries. Rep Electrochem 5:1–19. https://doi.org/10.2147/RIE.S65118
85. Gedanken A (2004) Using sonochemistry for the fabrication of nanomaterials. Ultrasonics Sonochem 11(2):47–55. https://doi.org/10.1016/j.ultsonch.2004.01.037
86. Wan J et al (2021) sonochemical preparation of polymer–metal nanocomposites with catalytic and plasmonic properties. Nanoscale Adv 3:3306–3315
87. Neppolian B, Wang C, Ashok Kumar M (2014) Sonochemically synthesized mono and bimetallic Au–Ag reduced graphene oxide based nanocomposites with enhanced catalytic activity. Ultrasonics Sonochem 21:1948–1953
88. Jiang H et al (2013) Preparation and biomedical applications of gold-coated magnetic nanocomposites. J Nanosci Nanotechnol 13(9):1617–1625
89. Venkatesh N (2018) Metallic nanoparticle: a review. Biomed J Sci Tech Res 4:3765–3775

90. Mazzoldi P, Arnold GW, Battaglin G, Gonella F, Haglund RF (1996) metal nanocluster formation by ion implantation in silicate glasses: nonlinear optical applications. J Nonlinear Opt Phys Mater 05:285–330
91. Xiang X, Zu XT, Zhu S, Wang LM (2004) Optical properties of metallic nanoparticles in Ni-ion-implanted α-Al_2O_3 single crystals. Appl Phys Lett 84:52–54
92. Stepanov AL (2016) Nonlinear optical properties of metal nanoparticles in silicate glass. In: Glass nanocomposites. Elsevier, In book: Glass nanocomposites, pp 165–179
93. Rahman MT et al (2019) Evaluation of thermal, mechanical, electrical and optical properties of metal-oxide dispersed HDPE nanocomposites. Mater Res Express 6:085092
94. Thomas S et al (2008) Size-dependent surface plasmon resonance in silver silica nanocomposites. Nanotechnology 19:075710
95. Colombo M, Romero SC, Casula MF, Gutiérrez L, Morales MP, Böhm IB, Heverhagen JT, Prosperi D, Parak WJ (2012) Biological applications of magnetic nanoparticles. Chem Soc Rev 41:4306
96. Singamaneni S, Bliznyuk VN, Binek C, Tsymbal EY (2011) Magnetic nanoparticles: recent advances in synthesis, self-assembly and applications. J Mater Chem 21:16819
97. Skomski R, Coey JMD (1993) Giant energy product in nanostructured two-phase magnets. Phys Rev B 48:15812–15816
98. Fullerton EE, Jiang J, Bader S (1999) Hard/soft magnetic heterostructures: model exchange-spring magnets. J Magn Magn Mater 200:392–404
99. Zeng H, Sun S, Li J, Wang ZL, Liu JP (2004) Tailoring magnetic properties of core/shell nanoparticles. Appl Phys Lett 85:792–794
100. Qiang Y, Antony J, Sharma A, Nutting J, Sikes D, Meyer D (2006) Iron/iron oxide core-shell nanoclusters for biomedical applications. J Nanopart Res 8:489–496
101. Maity D, Zoppellaro G, Sedenkova V, Tucek J, Safarova K, Polakova K, Tomankova K, Diwoky C, Stollberger R, Machala L, Zboril R (2012) Surface design of core–shell superparamagnetic iron oxide nanoparticles drives record relaxivity values in functional MRI contrast agents. Chem Commun 48:11398
102. Theil Kuhn L, Bojesen A, Timmermann L, Fauth K, Goering E, Johnson E, Meedom Nielsen M, Mørup S (2004) Core-shell iron–iron oxide nanoparticles: magnetic properties and interactions. J Magn Magn Mater 272:1485–1486
103. McKeehan LW (1950) Physical theory of ferromagnetic domains. Phys Rev 79:745–745
104. Meffre A, Mehdaoui B, Kelsen V, Fazzini PF, Carrey J, Lachaize S, Respaud M, Chaudret B (2012) A simple chemical route toward monodisperse iron carbide nanoparticles displaying tunable magnetic and unprecedented hyperthermia properties. Nano Lett 12:4722–4728
105. Lee J-H, Jang J, Choi J, Moon SH, Noh S, Kim J, Kim J-G, Kim I-S, Park KI, Cheon J (2011) Exchange-coupled magnetic nanoparticles for efficient heat induction. Nat Nanotechnol 6:418–422
106. Noh S, Na W, Jang J, Lee J-H, Lee EJ, Moon SH, Lim Y, Shin J-S, Cheon J (2012) Nanoscale magnetism control via surface and exchange anisotropy for optimized ferrimagnetic hysteresis. Nano Lett 12:3716–3721
107. Ong QK, Lin X-M, Wei A (2011) Role of frozen spins in the exchange anisotropy of core−shell Fe@Fe_3O_4 nanoparticles. J Phys Chem C 115:2665–2672
108. Khurshid H, Li W, Phan M-H, Mukherjee P, Hadjipanayis GC, Srikanth H (2012) Surface spin disorder and exchange-bias in hollow maghemite nanoparticles. Appl Phys Lett 101:022403
109. Jaffari GH, Ceylan A, Ni C, Shah SI (2010) Enhancement of surface spin disorder in hollow $NiFe_2O_4$ nanoparticles. J Appl Phys 107:013910
110. Jeyadevan B, Urakawa K et al (2003) Direct synthesis of fct-fept nanoparticles by chemical route. Jpn J Appl Phys 42:350–352
111. Barnes WL, Dereux A, Ebbesen TW (2003) Surface plasmon subwavelength optics. Nature 424:824–830
112. Schürmann U, Takele H, Zaporojtchenko V, Faupel F (2006) Optical and electrical properties of polymer metal nanocomposites prepared by magnetron co-sputtering. Thin Solid Films 515:801–804

113. Huang XY, Jiang PK (2007) Electrical properties of polyethylene/aluminum nanocomposites. J Appl Phys 102:124103
114. Jiang L, Gao L (2003) Carbon nanotubes–magnetite nanocomposites from solvothermal processes: formation, characterization, and enhanced electrical properties. Chem Mater 15(14):2848–2853
115. Taha MA, Nassar AH, Zawrah MF (2017) Improvement of wetability, sinterability, mechanical and electrical properties of Al_2O_3-Ni nanocomposites prepared by mechanical alloying. Ceram Int 43:3576–3582
116. Khdher AM, Sidik NAC, Hamzah WAW, Mamat R (2016) An experimental determination of thermal conductivity and electrical conductivity of bio glycol based Al_2O_3 nanofluids and development of new correlation. Int Commun Heat Mass Transfer 73:75–83
117. Abdolbaqi MK, Azmi WH, Mamat R, Sharma KV, Najafi G (2016) Experimental investigation of thermal conductivity and electrical conductivity of BioGlycol–water mixture based Al_2O_3 nanofluid. Appl Therm Eng 102:932–941
118. Shoghl SN, Jamali J, Moraveji MK (2016) Electrical conductivity, viscosity, and density of different nanofluids: an experimental study. Exp Thermal Fluid Sci 74:339–346
119. Guo S et al (2007) Electrical properties of silica-based nanocomposites with multiwall carbon nanotubes. J Am Ceram Soc. https://doi.org/10.1111/j.1551-2916.2007.01636.x
120. Jain PK, Lee K-S, El-Sayed IH, El-Sayed MA (2006) Calculated absorption and scattering properties of gold nanoparticles of different size, shape, and composition: applications in biological imaging and biomedicine. J Phys Chem B 110:7238
121. Baba A, Imazu K, Yoshida A et al (2014) Surface plasmon resonance properties of silver nanoparticle 2D sheets on metal gratings. Springer Plus 3(284):1–10
122. Ozbay E (2006) Plasmonics: merging photonics and electronics at nanoscale dimensions. Science 331:189–193
123. Frenkel AI, Hills CW, Nuzzo RG (2001) A view from the inside: complexity in the atomic scale ordering of supported metal nanoparticles. J Phys Chem B 105:12689–12703
124. Ridolfo A, Di Stefano O, Fina N, Saija R, Savasta S (2010) Quantum plasmonics with quantum dot-metal nanoparticle molecules: influence of the fano effect on photon statistics. Phys Rev Lett 105:263601
125. Narayan N, Meiyazhagan A, Vajtai R (2019) Metal nanoparticles as green catalysts. Materials 12:3602
126. Cuenya BR (2010) Synthesis and catalytic properties of metal nanoparticles: Size, shape, support, composition, and oxidation state effects. Thin Solid Films 518:3127–3150
127. Alharthi FA, Alghamdi AA, Al-Zaqri N et al (2020) Facile one-pot green synthesis of Ag–ZnO Nanocomposites using potato peel and their Ag concentration dependent photocatalytic properties. Sci Rep 10:20229. https://doi.org/10.1038/s41598-020-77426-y
128. Vara JA, Dave PN, Chaturvedi C (2019) The catalytic activity of transition metal oxide nanoparticles on thermal decomposition of ammonium perchlorate. Defence Technol 15:629–635
129. Jawhari A (2022) Noble metals deposited $LaMnO_3$ nanocomposites for photocatalytic H_2 production. Nanomaterials 12(17):2985
130. Yang Y et al (2018) Growth of sub-5 nm metal nanoclusters in polymer melt aerosol droplets. ACS Pub 34:585–594
131. Long NN, Vu LV, Kiem CD et al (2009) Synthesis and optical properties of colloidal gold nanoparticles. J Phys: Conf Ser 187:012026
132. Arockia Jency D, Parimaladevi R, Sathe GV, Umadevi M (2018) Detect, remove: a new paradigm in sensing and removal of PCBs from reservoir soil via SERS-Active ZnO triggered gold nanocomposites. Appl Surf Sci 449:638–646
133. Eskandari MJ, Gostariani R, Asadabad MA (2020) Transmission electron microscopy of nanomaterials. Electron Crystallogr IntechOpen. https://doi.org/10.5772/intechopen.92212
134. Torrisi V (2015) Metal-polymer nanocomposites: (Co-)evaporation/(Co)sputtering approaches and electrical properties. Coatings 5(3):378424. https://doi.org/10.3390/coatings5030378

135. Lin F et al (2010) Laser synthesis of gold/oxide nanocomposites. J Mater Chem 20:1103–1106
136. Shaheen K, Shah Z, Gulab H, Hanif HB, Faisal S, Suo H (2020) Metal oxide nanocomposites as anode and cathode for low temperature solid oxide fuel cell. Solid State Sci 102:106162
137. Ascencio JA, Liu HB, Pal U, Medina A, Wang ZL (2006) Transmission electronmicroscopy and theoretical analysis of AuCu nanoparticles: atomic distribution and dynamic behavior. Microsc Res Tech 69(7):522–530
138. Titus D et al (2019) Synthesis, characterization and applications of nanoparticles, chapter-12. Elsevier, Nanoparticle characterization techniques, pp 303–319
139. Zhang H et al (2018) Synthesis and characterization of TiO_2/graphene oxide nanocomposites for photoreduction of heavy metal ions in reverse osmosis concentrate. RSC Adv 8:34241
140. Sacourbaravi R, Ansari-Asl Z, Kooti M, Nobakht V, Darabpour E (2020) Fabrication of Ag NPs/Zn-MOF nanocomposites and their application as antibacterial agents. J Inorg Organomet Polym Mater 30. https://doi.org/10.1007/s10904-020-01601-x
141. Galaburda MV (2019) Mechanochemical synthesis of carbon-stabilized Cu/C, Co/C and Ni/C nanocomposites with prolonged resistance to oxidation. Sci Rep 9(1)
142. Miedema PS et al (2014) In situ X-ray Raman spectroscopy study of the hydrogen sorption properties of lithium borohydride nanocomposites. Phys Chem Chem Phys 16:22651–22658
143. Dezfuli A (2015) A high performance supercapacitor based on a ceria/graphene nanocomposite synthesized by a facile sonochemical method. RSC Adv 5:46050–46058
144. Thiyagarajan K, Sivakumar K (2017) Oxygen vacancy-induced room temperature ferromagnetism in graphene–SnO_2 nanocomposites. J Mater Sci 52. https://doi.org/10.1007/s10853-017-1016-7
145. Singh S et al (2019) Synthesis of Bi_2O_3-MnO_2 nanocomposite electrode for wide-potential window high performance supercapacitor. Energies 12:3320. https://doi.org/10.3390/en12173320
146. Zhu S, Guo J, Dong J, Cui Z, Tao L, Zhu C, Zhang Di, Ma J (2013) Sonochemical fabrication of Fe_3O_4 nanoparticles on reduced graphene oxide for biosensors. Ultrason Sonochem 20:872–880
147. Hossain MN, Wen J, Chen A (2017) Unique copper and reduced graphene oxide nanocomposite toward the efficient electrochemical reduction of carbon dioxide. Sci Rep 7:3184. https://doi.org/10.1038/s41598-017-03601-3
148. Koteshwara Reddy K, Yugender Goud K, Satyanarayana M, Kummari S, Sunil Kumar V, Bandal H, Jayaramudu T, Pyarasani RD, Kim H, Amalraj J, Vengatajalabathy Gobi K (2021) Chapter four—Metal oxide-metal nanocomposite-modified electrochemical sensors for toxic chemicals. In: Metal oxides, metal oxides in nanocomposite-based electrochemical sensors for toxic chemicals. Elsevier, pp 79–137
149. Hui X et al (2020) High-performance flexible electrochemical heavy metal sensor based on layer-by-layer assembly of Ti_3C2Tx/MWNTs nanocomposites for noninvasive detection of copper and zinc ions in human biofluids. ACS Appl Mater Interfaces 12:48928–48937
150. Arunadevi N (2022) Metal nanocomposites for advanced futuristic biosensing applications. Mater Lett 309:131320
151. Gole A, Vyas S, Phadtare S, Lachke A, Sastry M (2002) Studies on the formation of bioconjugates of endoglucanase with colloidal gold. Colloids Surf B 25:129–138
152. Tripathy N, Kim DH (2018) Metal oxide modified ZnO nanomaterials for biosensor applications. Nano Convergence 5:27. https://doi.org/10.1186/s40580-018-0159-9
153. Li H, Rothberg LJ (2004) Label-free colorimetric detection of specific sequences in genomic DNA amplified by the polymerase chain reaction. J Am Chem Soc 126:10958–10961
154. Zhang H, Zhang H, Aldalbahi A, Zuo X, Fan C, Mi X (2017) Fluorescent biosensors enabled by graphene and graphene oxide. Biosens Bioelectron 89:96–106
155. Meng L, Gan N, Li T, Cao Y, Hu F, Zheng L, Three-Dimensional A (2011) Magnetic and electroactive nanoprobe for amperometric determination of tumor biomarkers. Int J Mol Sci 12:362–375
156. Yousif NM, Attia RM, Balboul MR (2022) Adrenaline biosensors based on rGo/Ag nanocomposites functionalized textiles using advanced electron beam irradiation technique. J Organomet Chem 972:122392

157. Watanabe R, Miyano K (2011) Metal nanoparticles in a photovoltaic cell: effect of metallic loss. AIP Adv 1:042154
158. Chen H et al (2018) Dual-emitting fluorescent metal-organic framework nanocomposites as a broad-range pH sensor for fluorescence imaging. Anal Chem 90(11):7056–7063
159. Ma D, Li B, Zhou X, Zhou Q, Liu K, Zeng G, Li G, Shi Z, Feng S (2013) A dual functional MOF as a luminescent sensor for quantitatively detecting the concentration of nitrobenzene and temperature. Chem Commun 49:8964
160. Sarina S, Waclawik ER, Zhu H (2013) Photocatalysis on supported gold and silver nanoparticles under ultraviolet and visible light irradiation. Green Chem 15:1814. https://doi.org/10.1039/c3gc40450a
161. Tamuly C, Hazarika M, Bordoloi M, Das MR (2013) Photocatalytic activity of Ag nanoparticles synthesized by using *Piper pedicellatum* C. DC fruits. Mater Lett 102–103:1–4. https://doi.org/10.1016/j.matlet.2013.03.090
162. Nasrallah H, Douma F, Hamoud HI, El-Roz M (2021) Metal nanoparticles in photocatalysis: advances and challenges. In: Nanostructured photocatalysts. Elsevier, pp 119–143. https://doi.org/10.1016/B978-0-12-823007-7.00019-5
163. Attia Y, Samer M (2017) Metal clusters: new era of hydrogen production. Renew Sustain Energy Rev 79:878–892. https://doi.org/10.1016/j.rser.2017.05.113
164. Hisatomi T, Takanabe K, Domen K (2015) Photocatalytic water-splitting reaction from catalytic and kinetic perspectives. Catal Lett 145:95–108. https://doi.org/10.1007/s10562-014-1397-z

Metallic Nanoparticles: Status and Prospect

Umer Mehmood, Sadia Yasmeen, Rabia Nazar, and Santosh K. Tiwari

Abstract Metallic nanoparticles (NPs) have fascinated scientists all over the world for over a century. They grab this attention because of their huge potential in nanotechnology. Nanoparticles are a very special class of materials, with sizes ranging from 1 to 100 nm. The uniqueness of these nanoparticles is due to the fact that their properties—physical, chemical, optical, and mechanical—are sensitive to their size, shape, and structure. They can be grouped into fullerenes, metal nanoparticles, ceramics, and polymeric-based nanoparticles. The most important application and industrial demand project is the use of these specialized materials in various fields like coatings, nanolubrication, and catalysis. In this chapter, a brief introduction to metal nanoparticles followed by their synthesis methods is presented. The main objective of this chapter is to discuss the principles underlying the mechanical properties of these metal nanoparticles, e.g., hardness, creep, adhesion, and friction. Pursuant to these, several of the main applications, i.e., lubricant additives, nanoparticles in nanomanufacturing, and nanoparticle-reinforced composite coatings because of their special mechanical properties are discussed.

1 Introduction

Nanotechnology is the science concerned with managing nano-sized matter that is one billionth part of a meter (10^{-9}). It encompasses the study of matter at a submicroscopic level to manipulate it into newer products, designs, and structures. A nanoparticle is in the size range that falls between the macro and atomic level, very small than everyday objects as concerned with Newton's laws of motion but larger than sizes concerned with quantum mechanics like atoms or molecules. Generally, nanoparticles vary from 1 to 100 nm on a size scale. Different characteristics of metallic

U. Mehmood · S. Yasmeen · R. Nazar (✉)
Polymer and Process Engineering (PPE) Department, University of Engineering and Technology (UET), Lahore, Pakistan
e-mail: rabia.nazar@uet.edu.pk

S. K. Tiwari
Department of Chemistry, University of Warsaw, Warsaw, Poland

S. K. Tiwari et al. (eds.), *Nanoparticles Reinforced Metal Nanocomposites*,
https://doi.org/10.1007/978-981-19-9729-7_5

nanoparticles like mechanical strength, lower melting points, specific optical properties, higher specific surface areas, magnetization, etc., are in contrast compared to the bulk metals. This feature can be important in different industrial applications. This is the reason that these metal NPs qualify to be a potential candidate to be used in different applications such as catalysis, photonics surface-enhanced Raman spectroscopy (SERS), nanoelectronics, and biological and physical sense [1–7]. The inherent properties of metal NPs can be tuned further by manipulating their size and shape which facilitates their use in a variety of medical and industrial applications [8]. In catalysis, the exposed facets, and defects of crystals (including the corners and edges) are dependent on the shape of the crystals that affect their selectivity and reactivity toward different chemical compounds [9, 10]. The catalytic performance of NPs is enhanced in the presence of high-index planes of these nanocrystals which show higher activity in fragmenting chemical bonds [11].

A lot of endeavors have been going on to develop innovative synthesis routes for controlling the shape of nanoparticles and to investigate and tune properties dependent on their shape. Particularly wet synthesis methods have been widely used to produce several morphologies of crystals like spherical, wires, cubical, plates, cages, polyhedral, spikes, meatballs, flowers, snowflakes, prisms, combs, and various unique shapes [2, 5, 10, 12–17]. Despite the availability of a wide range of synthetic shapes, it is quite challenging to produce these metal nanostructures in high yields and devise a synthesis method that can produce different controllable morphologies by adjusting the reactant compositions [2]. Solution phase synthesis method that employs surfactants or polymers as capping agents has been studied for developing effective strategies. This owes to the advantage of precise morphological control with the production of metallic nanocrystals in bulk quantities when solution phase synthesis is used. The presence of both hydrophilic and hydrophobic groups makes them amphiphilic organic compounds. Due to their strong interactions with metal particles that control nucleation, the kinetics of their growth, and hence the final morphology.

A deeper study and understanding of growth mechanism aids in the selection of chemical precursors, seed utilization and physical stimuli (e.g., pH, electromagnetic field, temperature, etc.) is a crux of designing newer synthesis routes. These variables have a controlling effect on the kinetics of growth and nucleation. The investigation of shape and dimensional variations of metallic nanoparticles in their growing stage is carried out with the help of microscopic characterization techniques like scanning electron microscope (SEM) and transmission electron microscope (TEM). These techniques enable the interpretation of samples up to a sub-nonmetric scale. Different spectroscopic techniques like energy dispersive spectroscopy (EDS), selected area electron diffraction (SAED), X-ray diffraction (XRD), and X-ray photoelectron spectroscopy (XPS) can provide basic information regarding the chemical constitution, configuration, and crystalline assembly of NPs. This chapter is focused on metal-based nanoparticles, no complicated procedure is involved in the synthesis of these metals with different shapes, and they also have a wide range of applicability. For example, silver NPs, are broadly used in drug delivery systems owing to their biocompatibility and photo-thermal characteristics, which allow for localized heating. The anti-bacterial and anti-fungal characteristics of these NPs make them suitable for application in bioengineering, health, and wastewater treatment [18]. Furthermore, Ag and Au NPs have a high affinity for binding to a variety of biological molecules due to their narrow plasmon resonance, which has sparked a number of research initiatives and technology applications [3, 5, 19, 20].

The optical feature is one of the prime attractions and characteristics in the case of nanoparticles, for instance, a gold nanoparticle 20 nm in size has a distinctive wine-red color. The Ag NPs will give off yellow color. The NPs have been utilized since remote ages for visual appeal in paintings and architecture earlier in the fourth century AD.

This unique cup is the only intact historical example of dichroic glass, an extraordinary glass type that changes color on exposure to light. It turns from a translucent glow to red when light is incident internally through it, which happens when the angle of incidence is 90° from the viewing direction. The glass contains a tiny amount of silver and gold NPs in an estimated 14:1 fraction which gives it these remarkable photosensitive qualities. These nanocrystals are the reason behind the special color display of the Lycurgus Cup. This amazing relic is on display in the British Museum [21]. All the developments related to nanoparticles have been studied and summarized by Astruc and Daniel [19]. Francisci Antonii who was a philosopher as well as a medical professional wrote a book about colloidal gold in 1618, which became the first book to be published on the said topic. The book entails that soluble gold was first discovered around the fourth or fifth century B.C. in China and Egypt. Industrial manufacturing of these with stained glass started in the seventeenth century (1676) by Kunckle, who also published a book, Chap. 7 of that book contains information about drinkable gold composed of a slightly pink and neutral solution of metallic gold which has curing power for different diseases [22]. He summarized the discussion with the deduction that aqueous gold solutions must contain gold within a limit that should not be observable to the naked human eye. The "Purple of Cassius" which is a colloid made from Au particles and tin oxide and used as a colorant for glasses, remained quite popular around the seventeenth century. Helcher published a comprehensive dissertation on colloidal gold in 1718 [23]. He wrote about the use of different thickeners to prepare the consumable gold solution. This was quite common at that time; this can also be found in the French chemical dictionary dating to 1796 [24]. Another scientist, Fulhame claimed in a book published in 1974 that gold solution was used to dye the silk [25]. The color difference was observed while preparing various drinkable colloidal gold solutions like pink gold solutions containing gold in a very fine subdivision, whereas the yellow color is observed when the finer gold particles aggregate together. The famous publication of Michael Faraday that appeared in 1857 described that deep red colloidal gold solution could be formed by reducing the aqueous chloroaurate ($AuCl_4^-$) solution with phosphorus in CS_2 environment that forms a two-phase system. Faraday further explains about optical properties of dried colloidal in the form of thin films that show a reversible change in color when mechanically compressed (change between bluish-purple and green color). Since then, metallic nanoparticles have remained under intensive research and many scientific papers were published regarding the synthesis, assembly, and alteration of the properties of the NPs by employing various solvents and substrates [26, 27].

2 Methods of Nanoparticle Synthesis

Since ancient times, two methods are known for manufacturing sub-microscopic particles. One approach is called the top-down method, in which an external force is applied to a solid substance, converting it into small particles. The bottom-up approach, which involves the synthesis of nanoparticles initiated with atoms of gas or liquids is another technique that is further classified into dry and wet grinding. The grain refining process has the advantage that it increases the surface energy by increasing the aggregation of particles. The dry grinding method grounds the solid substance by compression, shock, or friction, which is achieved through the shearing mill, hammer mill, shock shearing mill, ball mill, and tumbling mill. It becomes difficult to obtain particles smaller in size than 3 μm with the refining grain method because condensation and pulverization are taking place simultaneously whereas, wet grinding employs a wet jet mill, agitating bead mill, ball mill, vibratory ball mill, and annular gap bead mill. The wet process is recommended because it does not allow the condensation of the nanoparticles that are produced, thus highly dispersed particles can be obtained easily. Bottom-up approaches include gaseous and liquid-phase procedures. Particles produced by the gaseous phase method contain lesser organic impurities than those produced by the liquid phase method, but they are complicated because of the necessary vacuum equipment required that is expensive and less productive.

Gaseous phase chemical vapor deposition produces 1 μm ultrafine particles. In the gas phase, nanoparticles within the 10–100 nm range can be obtained by controlling the parameters of the chemical reaction. Different heat sources like chemical flame, plasma process, laser, or electronic furnace are required because the chemical reaction must be performed at high temperatures. Physical vapor deposition involves the evaporation of solid or liquid material, which forms vapors that are then rapidly cooled, producing nanoparticles. The arc discharge method can be used for achieving evaporation. The thermal decomposition procedure is very productive for obtaining metal oxide or other particles; hence, it is an extensively used industrial process for this purpose.

Liquid phase approaches have been the most common way of nanoparticle synthesis for many years, further divided into sedimentation and liquid/liquid methods. Facile fabrication of different nanostructured particles can be achieved by the chemical reduction of metal ions. In this approach, fine-tuning the morphology of NPs is possible by controlling the reaction time, temperature, and the number of dispersing agents. By chemical reduction the metal ions can be converted to their 0-oxidation state (i.e., $M^{n+} \rightarrow M^0$); the chemical reduction method does not require complicated equipment and is a low-cost procedure to produce nanoparticles in bulk.

Microwave radiation is another type of heat source in this procedure and is particularly compelling as it is a fast technique and can synthesize nanoparticles of high quality. Reduction techniques other than direct reduction by the chemical way with the addition of reducing agents, such as γ rays, UV rays, and ultrasonic waves can also be employed.

As there is no need for reducing agents in these methods, hence extraneous impurities in nanoparticles can be avoided which is an attractive feature of these methods. Other methods include spray drying, spray pyrolysis, solvothermal synthesis, and supercritical method, extensively applied for metal oxide nanoparticle fabrication. By hydrolysis, a solution of a metal alkoxide becomes a sol, which is then polycondensed to produce a gel. The description of a sol–gel process is given in many books (see e.g. [28]). The liquid phase or wet process ensures high nanoparticle dispersivity in comparison to the dry method. However, if these resulting nanoparticles are dried, the particles begin to clump together. To address this problem, re-dispersion is required which can be carried out as done for the solid-phase method. Scheme 1 summarizes various techniques; some characteristics are the same for all the methods following that synthesis procedure satisfies the conditions listed below.

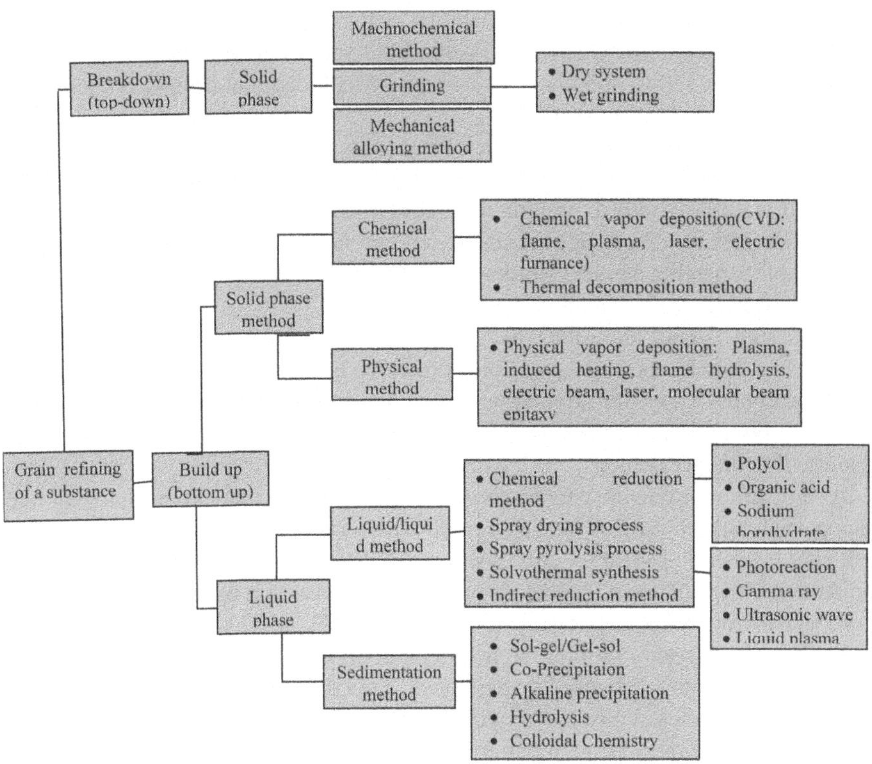

Scheme 1 Different approaches for nanoparticles synthesis

- Particle size, shape, crystal structure, size distribution, and component distribution are all under control
- Lower impurities in the nanoparticles
- Preventing agglomeration
- The physical structures and reactants are stabilized
- Greater reproduction ability
- Scale-up and mass production with lower cost
- Alteration of size, shape, and structure as needed

3 Morphology Control of Nanoparticles (Size, Shape, and Structure)

3.1 Size Control of Nanoparticles

The chemical and physical characteristics of nanoparticles are particle size [29] and shape-dependent [30]. One of the examples of the dependence of particle size on different properties of NPs is shown in Fig. 1, i.e., when the particle size of gold nanoparticles is increased, the visible light spectrum shifts to a longer wavelength side from 530 to 600 nm [31]. It can be concluded that optical properties are sensitive to the size of nanoparticles.

The size distribution of nanoparticles becomes a very essential issue in optical applications of nanoparticles. The slow development of NPs after the seeds are rapidly produced is necessary for obtaining nano-dispersed particles [32]. Reduction in the size of nanoparticles increases the surface energy facilitating aggregation. Thus, the addition of dispersing agents is important so that particulate matter can be stabilized once the desired size is obtained.

Dispersing agents have long been used in the synthesis of nanoparticles; for instance, Lea et al. reported Ag colloids protected by citrate in 1889 [33]. In case of an unusually high concentration of nanoparticles, stabilization will be decentralized because aggregation can no longer be prevented by the action of organic substrate (citrate) that has weakened.

As a result, multiple investigations of dispersing agents that aid in achieving high nanoparticle dispersivity at varying compositions are available. As per the hard and soft acid bases rule [34], Pt^{2+}, Au^{+}, Ag^{+}, and Pd^{2+}, etc. are all Lewis acids, while substrates with thiol and phosphine group are grouped as soft bases; they have shown to be satisfactory dispersing agents [35].

Brust and coworkers [36] researched organic thiol compounds as suitable dispersion agents., if 1-dodecanethiol is employed as the dispersing agent in the synthesis of gold nanoparticles, a monolayer will be formed on the surface as shown in Fig. 2. This stabilizes the dispersion of nanoparticles. Since 1965 this publication is said to have had a notable impact in the field of chemistry and is the third most cited article in the scientific journal [26].

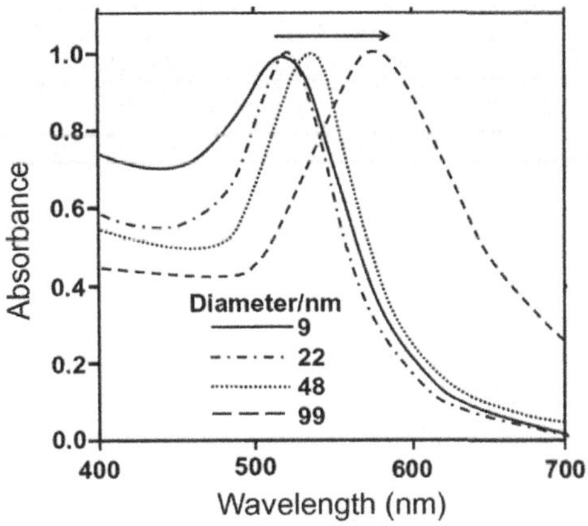

Fig. 1 Visible-light spectrum of gold NPs with different particle diameters [31]

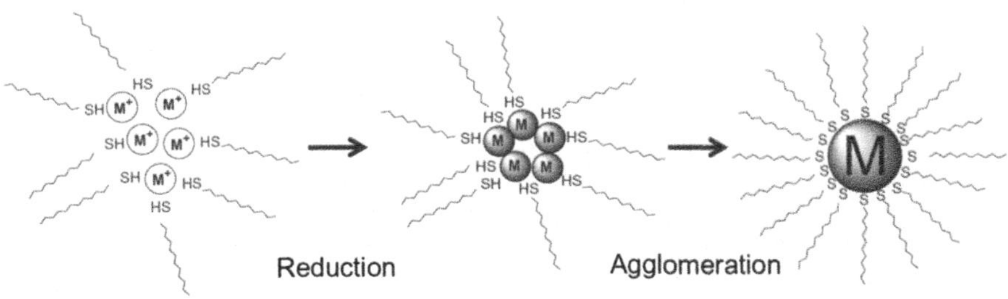

Fig. 2 Metallic NPs capped with 1-dodecanethiol [26]

Furthermore, increasing the size of the nanoparticles by modifying the alkyl chain length from 1-dodecanethiol to alkyl chains of octane, decane, and hexadecane, or by minimizing steric hindrance. 1-dodecanethiol which is a dispersion agent can thus be used to modify particle size. Studies have been done to use polymers as dispersing agents. Their protective action is determined by the surface affinity of the NPs and the polymer's molecular weight; this also made a significant impact on chemistry. The images of the state of aggregation are shown in Fig. 3. A randomly dispersed form is shown in Fig. 3a. The fractal form manifests the electronic properties (Fig. 3b). On the other hand, electronic transport properties are due to the structural orientation as shown in Fig. 3c. The closed-packed structure demonstrates the optical properties (Fig. 3d), whereas the orderly structure shown in Fig. 3e, f tells about the physical, optical, magnetism, and electronic properties.

Many methods have been established to separate desired particle sizes from the colloidal solutions of nanoparticles. These methods are (i) gel filtration column, (ii) gel electrophoresis, (iii) centrifugal separation, and (iv) separation by precipitation. Each of these screening methods is suitable, but precipitation separation is well suited to large distributions of nanoparticle colloids in solutions. Narrow-size distribution

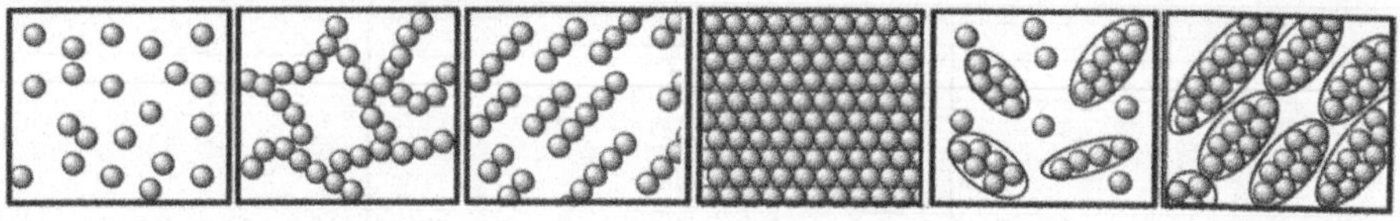

Fig. 3 Different structural orientation of NPs; **a** Random form **b** fractal form **c** alignment; **d** close-packed structure; **e** dispersed-ordered structure; **f** dense form of ordered structure [37]

solutions of colloidal nanoparticles should be screened with centrifugal separation and gel filtration. Gel electrophoresis separates nanoparticles based on the charge density difference between the nanoparticles. It is also a better way to separate particles having smaller cluster sizes. For best screening, these methods should be employed in combination instead of relying on one method alone.

However, one of the limitations of these methods is that screening of nanoparticles by the stipulated procedure will yield only a fraction of nanoparticles, and they cannot be collected in large quantities. To counter this problem, digestive ripening and melting techniques can be employed [38].

3.2 Shape Control of Nanoparticles

The shape of nanoparticles is also a very important characteristic that governs the nature of the surface plasmon resonance band. Figure 4 shows the absorption spectra of various nanorods of Au in the visible spectrum with respect to the change in aspect ratio. In this experimentation, Au nanorod-shaped particles had a diameter in the range of 5–20 nm and lengths between 20 and 150 nm. It was observed that any change in the size of the crystal face was related to the other aspects of the physical form of nanorods. As shown in Fig. 4, the wavelength of the absorption spectrum shifts to a higher wavelength with the increase in L/D ratios of nanorods. The spectroscopic features can be improved by changing the physical arrangement of the nanorods; many studies have been done to understand these characteristics.

Yu and coworkers [38] reported on the use of surfactants in the manufacture of Au nanorods. Gold nanorods were made by ultrasonically irradiating an Au anode with a template made of the cationic surfactant hexadecyltrimethylammonium bromide (CTAB). After the interaction with the CTAB micelle (above CMC concentration), Au particles will cluster together from the electrode and form the shape of a rod. CTAB acts as a dispersing agent being adsorbed selectively on the Au crystal faces [99, 100] in the growth of nanorods.

A rod-shaped metallic nanoparticle is attained with the growth of the crystal faces. Many reports are available on the use of CTAB as a dispersing agent and that led to newer research on the nanoparticles. Adsorption of dispersing agents using different techniques leads to nanoparticles of various forms and shapes. A special composition

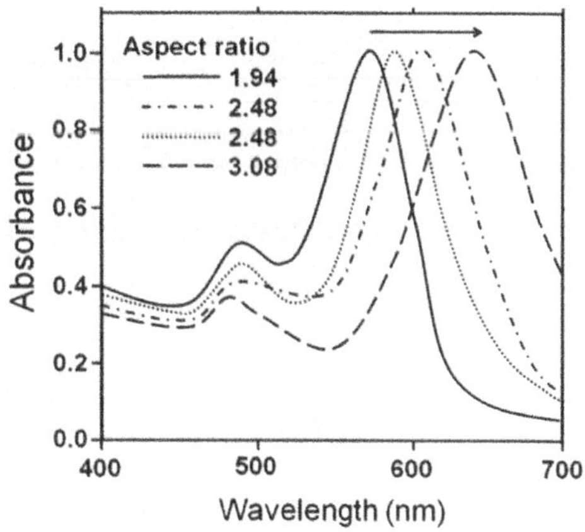

Fig. 4 Visible-light spectra of gold rod-shaped NPs with different aspect ratios [39]

of branched nanoparticles of Au was reported by Chen et al. who achieved this because of higher contents of dispersing agents CTAB.

It is concluded that there is a strong effect of molecular associations of surfactants and dispersing agents on the shape of metallic NPs.

Au nanorods prepared with a hard template like mesoporous alumina have similar physical characteristics to those prepared with soft templates like CTAB. According to an early report, Au nanoparticles of rod shape can be synthesized in fine pores of soft template mesoporous alumina [40]. Initially, the nanosized porous electrodes of alumina are produced with electrochemical deposition of metal in form of fine pores to make a firm mold. This short-axis diameter of the Au nanorod growing in those pores can be controlled by pore size. Afterward, the mold of alumina is removed when it is dissolved, and the nanoparticles produced are taken out. Nanorods with multiple layers of various metals can be fabricated with this procedure, such as Au–Ag–Au. Thus, the formation of nanoparticles with distinctive unique features is possible. Applied research on multiple-layer fabrication can produce nano size system that can function as a nanosized bar code [41].

3.3 *Structure Control of Nanoparticles*

Nanoparticles containing two or more metal elements that will have different properties from their uni-metal nanoparticles.

Categorically, such a class of metallic nanoparticles has been further divided into the following:

(a) Random alloy structures in the crystal lattice (Fig. 6a)
(b) Core–shell assembly where the central metal atom is different from the boundary or peripheral atoms (Fig. 6b)

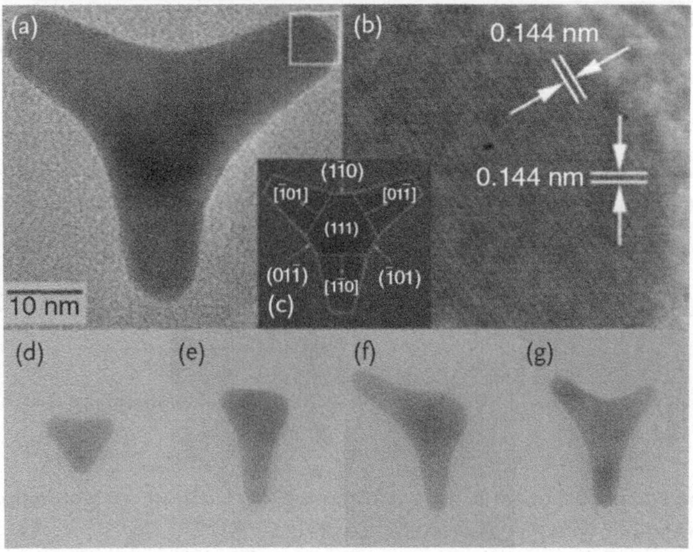

Fig. 5 **a** TEM image of a tripod-shaped NP and **b** HR-TEM image of the tripod nanocrystal; **c** the directions of pod and planes in a crystal. The second row of Fig shows the different stages of development of particles: **d** initiation of a trilateral form, **e** monopod, **f** V-shaped bipod, and **g** Y-shaped tripod [42]

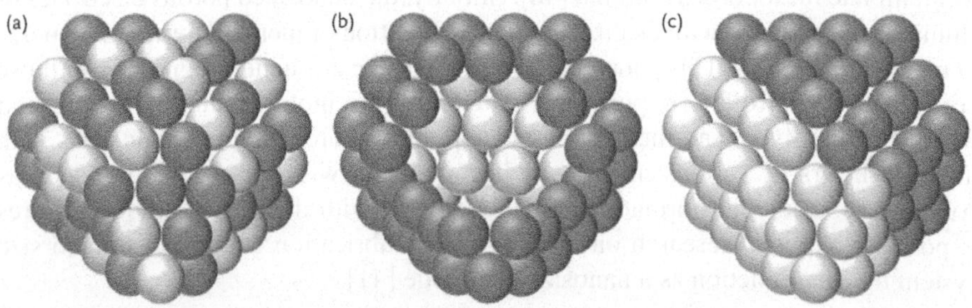

Fig. 6 Illustrative description of bimetal nanoparticles: **a** alloy form, **b** core–shell form **c** and heterojunction assembly of complex metal NPs

(c) Double hemisphere structure where two hemispheres are joined together to form a twinned assembly

Phase separation can occur as a result of the heterojunction arrangement (Fig. 6c). Complicated metal nanoparticle structures can obscure many of the new features. It is comparatively easy to produce the core–shell structure due to the easy fabrication of complex metallic nanoparticles with efficient functional control. Many reports and literature are available in this regard. These heterojunction structures influence the color of dispersions, e.g., the purplish-red of Au NPs and the yellowish appearance of Ag NPs changes to orange-red when Au and Ag core–shell structure is formed,

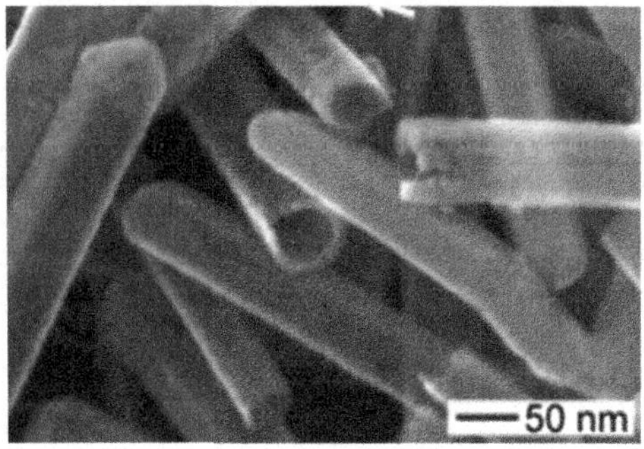

Fig. 7 SEM image of gold nanoparticles cross-section (broken by sonication). The gold nanotubes were produced by silver nanowires reaction with an aqueous $HAuCl_4$ solution [44]

instead of a single metal structure. Other properties are also important like in the case of magnetic structured matter, for instance, the core should be based on magnetite particles and the shell should be composed of the other metal so that the resulting assembly now has both magnetic and optical characteristics.

Core–shell NPs can be produced by different methods classified as simultaneous reduction reactions and sequential one-electron reduction. Consider the synthesis via simultaneous reduction reaction, in which the core is synthesized of Pt NPs and a shell made up of Pd NPs entities [43]. The oxidation potential difference of both metals should also be considered during the synthesis and one such unique method has been proposed [44], in which an oxidation–reduction reaction takes place after silver particles are added to $HAuCl_4$ solution (Eq. 1) and gold gets deposited on the Ag NPs surface forming the desired core–shell structure. With the development of this method, Ag nanotubes can be fabricated by employing Ag nanowires in pentagonal prismatic as a template (Fig. 7).

$$3Ag(s) + HAuCl_4(aq) \rightarrow Au(s) + 3AgCl(aq) + HCl(aq) \tag{1}$$

4 Physicochemical Properties of NPs

As previously said, these specialized particles are exclusive in nature and due to their special physical and chemical characteristics can be employed in a wide variety of applications. In the following section, some of their most crucial characteristics are reviewed.

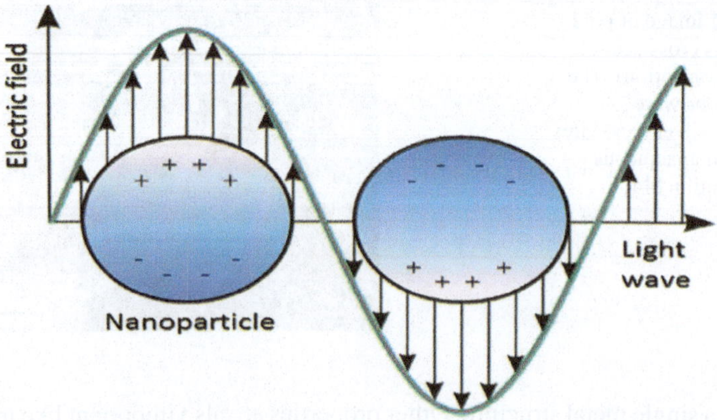

Fig. 8 Graphical demonstration of LSPR on an outer surface of the nanoparticle [47]

4.1 *Electronic and Optical Properties*

The electrical and optical characteristics of nanoparticles are related to each other such that they are interdependent. The optical characteristics of noble metals are size dependent, and the UV–vis spectrum of their nanoparticles exhibits an 'extinction band' that is not present in the spectrum of bulk material. The extinction band called localized surface plasma resonance (LSPR) is formed because the frequency of incident radiations (photons) is constant with the combined excitations of conduction electrons. This type of excitation results in selective absorption of wavelength and a high coefficient of molar excitation; the resonance ray light scattering takes place with an efficiency comparable to ten fluorophores. The spectroscopies are enhanced by better electromagnetic fields near the surface of NPs.The shape, size, spacing between nanoparticles, dielectric properties, and environment that includes a solvent, substrate, and adsorbent are all factors that affect the peak wavelength in the LSPR spectrum [45, 46]. Gold colloidal NPs caused the rusty hues found in marred glass doors and windows, whereas silver particles are typically yellowish. The surface-free electrons of these NPs can easily move throughout the nanomaterial. The free path of Ag and Au is about 50 nm which is in fact greater than their particle size. As a result, following light engagement, no dispersion is predicted from the bulk; instead, a standing resonance is observed which is responsible for the LSPR spectrum of these nanoparticles (Fig. 8).

4.2 *Magnetic Properties*

The magnetic properties of NPs favor their use in different domains like biomedicine, data storage, or magnetic fluids. They are also of interest in heterogeneous and homogenous catalysis, MRI and water purification, etc. As per the literature, the best properties are exhibited by nanoparticles when their size is between 10 and 20 nm, which is less than the critical value [48]. They are valuable in a wide range of applications because their magnetic characteristics are very dominant at such small scales [48–51]. The magnetic properties of NPs are due to their unequal electrical dispersion. These features are also reliant on synthetic protocols and methodology [52, 53].

4.3 Mechanical Properties

4.3.1 Strength

The most significant parameter to consider when designing structural components is strength. The evaluation of material strength is required in functional materials and/or microelectromechanical systems (MEMS) to increase their reliability. According to a recent study, bending or tensile tests are used to determine strength. Because of machinability, tensile tests are frequently used to determine metallic materials' strength. If grain sizes are larger than a micrometer, according to the Hall–Petch relationship, the strength will increase as grain size decreases.

Furthermore, the metallic material is ductile, and exhibits work hardening. Work hardening is limited as grain refining progresses. As a result, as grain size falls, strength diminishes (inverse Hall–Petch relationship). Tensile tests are used to compare ceramics to metallic materials. In tensile testing, a specimen having 6 mm dia and a gauge length of almost 30 mm is recommended. The radius of curvature at the gauge's shoulder must be greater than 30 mm, and a universal joint should be used for specimen holding to avoid any bends or twists. Because ceramics are more difficult to process than metallic materials, making accurately shaped tensile test samples is difficult.

4.3.2 Fracture Toughness

Strength is measured with the help of fracture toughness KIC and flaw size. As a result, fracture toughness is a crucial metric for determining strength and dependability. JIS G 0564 and ISO 12737 govern the measurement of plane-strain fracture toughness in metallic materials. The test for fracture toughness (JIS R1607) is standardized as an indentation fracture method and a single-edge-precracked-beam (SEPB) method. The SEPB method ISO 15732:2003, the surface crack in flexure method ISO 18756:2003, and the chevron notched beam method ISO/DIS24379:2003 are all ISO standards.

4.3.3 Creep/Plasticity

Design engineers must understand the mechanical reaction of nanoparticles to utilize them in structural applications. An applied force will lead to a change in shape which is called deformation. When it comes to elastic deformation, it is reversible. On the other hand, plastic deformation is permanent and cannot be reversed. Creep is the process through which a substance continues to swell over time after being stressed. As the grain size of metals approaches the nanoscale, a new type of material emerges. The grain boundaries reveal where an increasing number of atoms in the solids are discovered. Diffusional creep occurs at a temperature range between (0.5–0.4 Tm) since grain boundary diffusion increases the strain rate significantly higher than in normal microcrystalline materials [54]. Creep is a mechanical strength property of nanocrystalline materials. Superplasticity, on the other hand, refers to a polycrystalline solid's propensity to exhibit extremely significant elongations in tension at high temperatures [55]. When the grains are very tiny, i.e., even lesser than a few microns for metals and almost 1 μm for ceramics, this property is very noticeable in metals, alloys, intermetallic, and ceramics.

The unique and distinctive mechanical characteristics of NPs help research for finding new uses in various fields, including nanomanufacturing, nanofabrication, surface engineering, and tribology. The actual mechanical nature can be calculated with the help of some metrics like elastic modulus, stress, strain, hardness, and friction. As well as some other characteristics like surface coating, coagulation, and lubrication are also helpful to modify the mechanical performance of nanocrystals (see Scheme 2).

The mechanical nature of nanoparticles is substantially different from their respective micro and bulk entities. Indentation of nanoparticles or their distortion in lubricated/greased contact is dependent on stiffness contrast between NPS and contacting external surface when the contact pressure applied is very high. The behavior of nanoparticles in a contact environment can be predicted from this. The surface quality can be easily enhanced by fine-tuning the mechanical properties of NPs and controlling surface interactions. In order to provide useful results in these sectors, a detailed understanding of the fundamentals of the mechanical properties of nanoparticles is essential [56].

4.3.4 Hardness and Elastic Modulus of Nanoparticles

A correct choice of particle design to be utilized further in specific applications, the fundamentals of different mechanical characteristics of nanoparticles, e.g., hardness and elastic modulus are needed to help in assessing their functions and mechanisms. In the last few decades, the measurement of the mechanical characteristics of microparticles has been developed. The micro-indentation technique was employed to measure the hardness of microparticles with indented areas of more than 100 μm^2 and an indenter size of 20 μm^2. Shorey et al. utilized nanoindentation to test the elastic characteristics of particles around 10 years ago. [57] Instead of monitoring

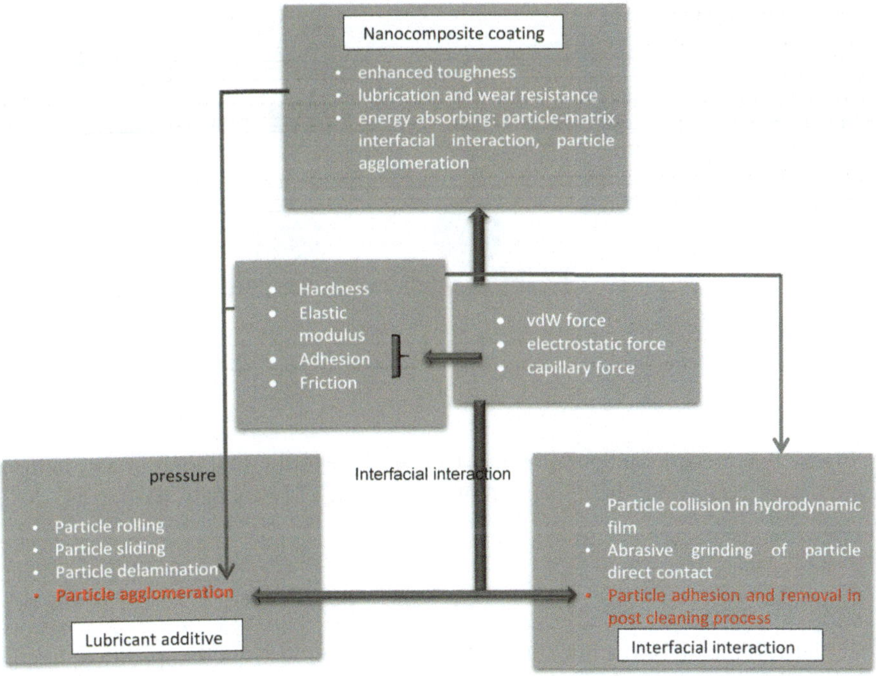

Scheme 2 Different mechanical properties and their applications [56]

individual particles, their approaches were designed to measure the film of particles. The deformation behavior of polystyrene microspheres was studied by Biggs and Spinks by using AFM on a mica surface for the first time in 1998 [58] After that discovery, rapid progress is witnessed in the development of techniques for determining the mechanical properties of nanoparticles by using AFM.

Indentation measurements are needed in order to evaluate elastic moduli for the nanoparticles, which can be done by conversion of AFM force versus displacement graphs to force versus indentation graphs. This is done because force indentation curves cannot be obtained directly. Hooke's law describes a new method in which the force was applied with the help of a cantilever to the surface's tip. In order to determine the elastic modulus of NPs, the depth of indentation was measured, for which we need to transform force vs. displacement data into force vs. indentation curve. Directly obtaining the latter is difficult. Hooke's law may be used to explain the load P applied to the tip through the cantilever (k being its spring constant).

$$P = k \cdot \delta_c \tag{2}$$

where δc is the deformation of a cantilever. The depth of indentation 'h' on the surface is given by:

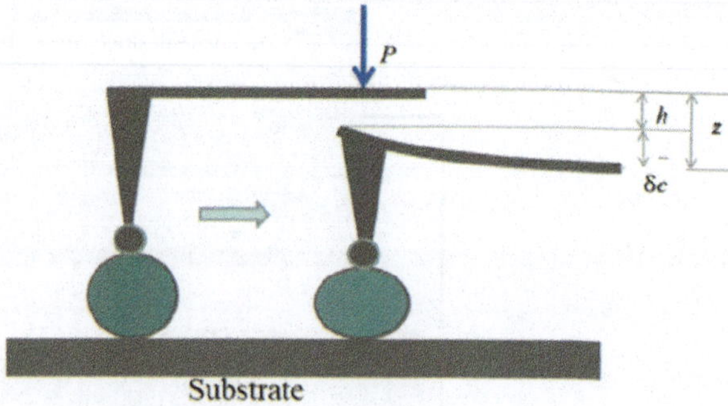

Fig. 9 Particle AFM tip system displays relative displacements and deformations. Left: the tip of AFM brushes against the particle, causing it to deform. Right: the AFM tip applies force to the particles, causing them to deform [56]

$$h = z - \delta_c \tag{3}$$

Figure 9 depicts the indentation mechanism. Due to thermal drift in a system, it is important to consider the deflection offset, δ_{c0}. Eq. (2) may be rewritten as:

$$P = k \cdot (\delta_c - \delta_{c0}) \tag{4}$$

Initially, if the tip is touching the surface of the sample at another position, height offset z_0 should be incorporated, Eq. (3) takes the form of,

$$h = (z - z_0) - (\delta_c - \delta_{c0}) \tag{5}$$

Analyzing slope of loading region on the force versus indentation curve, elastic modulus of the nanoparticles can be determined. Further information on how to calculate the elastic modulus of compressed NPs can be obtained from these references [59–61]. The elastic modulus of different nanoparticles has been determined, largely by using AFM, by compressing or bending particles. The hardness values along with the elastic modulus are given in Table 1. It is shown that these properties significantly from their bulk forms, and few of them exhibit size-dependent behavior.

The following three categories can be used to categorize typical related outcomes and underlying mechanisms.

1. There is no same mechanical behavior observed for spherical-shaped NPs till now. Hydrated ionic functional groups cause the compression moduli of polystyrene nanoparticles with a diameter of 200 nm to be smaller than their bulk form [62]. Instead, the elastic moduli of polypropylene NPs were found to be greater than that of bulk material [60]. There are different reasons reported for such behavior. The deformation of the polymer chain inside the particle can be influenced by many factors like glass transition temperature (Tg), crystalline phase, and crystallinity, resulting in a change in the elastic modulus of the particle.
2. Contrary to general assumption, crystalline metallic nanoparticles do contain dislocations inside their structure, and this is one of the contributing factors of

Table 1 Summary of the elastic moduli and hardness value of various NPs with different size ranges

Particle material		Diameter/size	Hardness/(bulk value)	Elastic modulus/(bulk value)	Indentation depth	Notes
Organic nanospheres	Polystyrene (PS) [62]	58–194 nm [62];		8.0–4.1 GPa [62];	3–6 nm [62]	Modulus increases with the decrease of particle size [62]
		180–250 nm [63]		1–2 GPa/(3–3.6 GPa) [63]	5–6 nm [63]	Vinylbenzyl (trimethyl) ammonium chloride units inside [63]
	Polypropylene (PP) [60]	200–500 nm		1.3–2.8 GPa/(1.5–2 GPa)	1.5 nm	
	Polyesters nanoscale stiffness [64]	2–3 nm		0.1–0.3 GPa		Hyperbranched, molecular weight = 3000–7000
	Polyethylenimine (PEI) [65]	15 nm		5–160 MPa	Up to 10 nm	Bigger pressure resulted in larger modulus
	Poly- (methylmetha crystal) (PMMA) [61]	350 nm		4.3 GPa/(4 GPa) [63]	up to 60 nm	6.6 GPa (200 °C°C heat treatment)
	Liquid crystal [66]	95–150 nm		0.1–0.6 GPa	10 nm	4-pentyl-4-cyanobiphenyl (5CB) (main component)
	Coreshe 11 PS/CeO_2 [59]	130–260 nm		5–15 GPa	20–30 nm	Modulus increases with particle size

(continued)

Table 1 (continued)

Particle material		Diameter/size	Hardness/(bulk value)	Elastic modulus/(bulk value)	Indentation depth	Notes
	PMMA/silica [59, 61]	450 nm [61]; 350 nm [59] 22 nm		10.3 GPa [61]; 9–11 GPa [59]	up to 80 nm	PMMA-based terpolvmer [59]
Metal nanoparticle	Gold [67]	2 nm	1.72 GPa/(Vckers hardness 216 MPA)	100 GPa/(79 G Pa)	3–5 nm	Six-fold symmetry gold nanoparticles
	Gold modified with proteins [69]	10 and 20 nm	0.12 and 0.08 GPa (*a*) 0.22 and 0.13 GPa (*b*)	1.3 GPa (*a*) 9.5 and 1.0 GPa (*b*)		Protein: (*a*) bovine serum albumin; (*b*) streptavidin pure gold particle: hardness = 0.4 GPa; modulus =5.2 GPa
	Silver [69]	13 nm	3.12 GPa/(Vickers hardness 251 MPa)	103.9 Gpa/(83 GPa)		
Silicon nanoparticle		40–140 nm [70]; 5–40 nm [71]; 40–100 nm [72]	25–34 GPa [71]; 20–50 GPa/(12 GPa) [72]	600–180 GPa [70]/(around 150 GPa)	13–36 nm [70] 3–24 nm [72]	Modulus increases with decrease of particle size [70]; simulation result [71]
Nanowire, nanotube, etc	Gold nanowire	40–250 nm		70 ± 11 Gpa	400 nm (displacement)	
	Silver nanowire [73, 74]	20–140 nm		75–160 Gpa		
	Lead nanowire [74]	30–280 nm		14–30 Gpa/(16 Gpa)		
	ZnO nanowire [75]	70, 99 nm		120, 83 Gpa/(140 GPa)		
	WS_2 nanotube [76]	20 nm		171 GPa/(150 GPa)		
	Boron nitride (BN) nanotubes [77]	0.58–2.38 nm		40.78–1.85 GPa/(30–40 or 74 GPa)		Single-walled, modulus decreases with diameter increases

(continued)

Table 1 (continued)

Particle material		Diameter/size	Hardness/(bulk value)	Elastic modulus/(bulk value)	Indentation depth	Notes
	Carbon nanotubes [78]	0.92–0.91 nm		57–9 Gpa/(36.5 GPa, bulk graphite)		
	Carbon nanotubes [79]	~9 nm		~16 GPa		Multi-walled
	Silicon nitride nanobelts [80]	20–50 nm (thickness)		570 GPa (bending modulus)/(120–330 GPa)	150 nm (displacement)	
	Cellulose nanocrystals [81]	4.2 nm (wood), 5.9 nm (cotton) [81]		24.8, 17.7 GPa [81]; 8.1 GPa (mean value) [82]		Cellulose nanocrystals are crystalline, rod-like shaped particles

change observed in the mechanical behavior of nanoparticles. The hardness and elastic modulus of gold nanoparticles were greater than that of the bulk phase, according to the experimental work done by Ramos et al [63]. Mordehai and Nix et al. [64, 65] demonstrated single-crystal gold nanoparticles on sapphire substrate deformation behavior with nanoindentation and compression testing collectively with theoretical simulation. Interestingly the strength of particles that are under indentation is increased with the lateral dimension as a result of competitive dislocation formation and drainage under the indenter [57]. The lateral dimension of the particle caused an increase in the particle strength under indentation due to the competition between the generation of dislocations beneath the indenter and their drainage from the particle. Because the nucleated dislocations resulted in a stress gradient along the slip planes, the compressive stress of the particle rose as the particle size decreased when compressed with a flat diamond punch [64]. As shown in Fig. 10, in situ TEM nanoindentation tests, revealed the existence of dislocations in metal nanoparticles during deformation, but they vanished after the unloading process. Wang et al. [66] recently revealed a new type of stacking defect in gold nanocrystals that may nucleate, move, and annihilate under mechanical stress applied with in situ TEM and MD simulation. Gerberich et al. [67] discovered comparable behavior with silicon nanoparticles, claiming that their hardness (particle diameter: 40 nm) was four times greater than that of bulk silicon. They explained that the key reasons for resisting high pressures are dislocations or line faults inside the particle. Zhang et al. [68] used atomistic modeling to establish that the super-hard silicon nanoparticles were formed by the nucleation and migration of dislocations. Changes in the lattice strain and the bond energies of nanoparticles in response to compressive stress have been hypothesized as another mechanism for the strengthening and weakening of nanoparticle mechanical characteristics [69].

3. The elastic modulus has an inverse relationship with the radial diameter of silver and leads nanowires. The scientists suggested that the modulus increase was due to the impacts of surface stress, oxidation layer, and surface roughness, or the surface tension effect [70, 71]. The bulk modulus of Ni/Ni_3Al nanowires rose as the wire perimeter size grew, while the surface energy dropped [72]. Surface influences, on the other hand, influenced only the fracture characteristics of ZnO nanowires, not their elastic behavior, due to the existence of surface cracks and flaws [73].

Therefore, it is concluded that determining the mechanical characteristics of individual nanoparticles is a complicated process with numerous variables that might influence the results.

4.3.5 Friction and Adhesion of Nanoparticles

The frictional forces and adhesive forces are crucial in many of the nanoparticle applications like nanofabrication, lubrication, micro/nanodevice design, colloidal

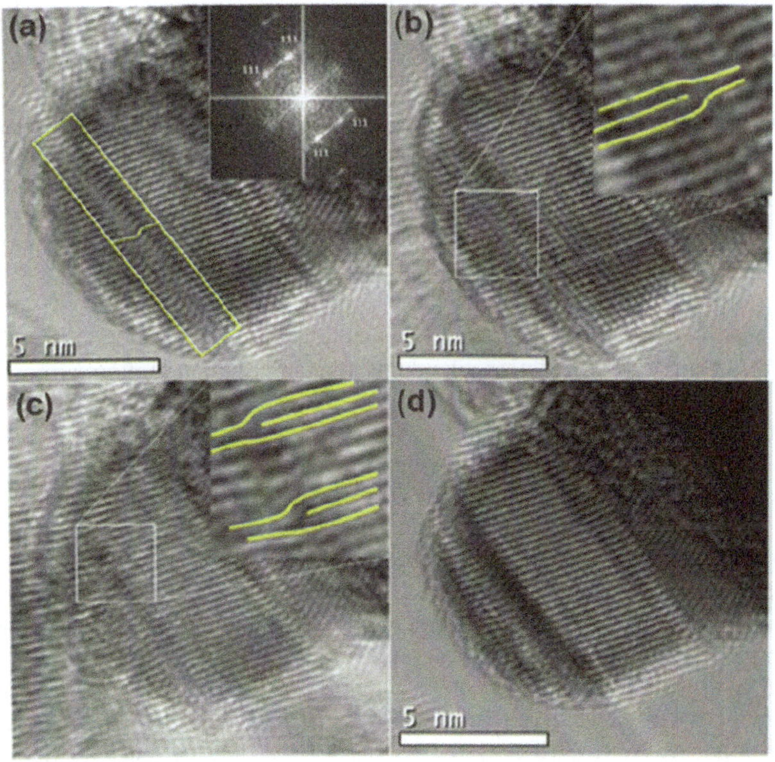

Fig. 10 HR-TEM images of a silver NP before and after compression: **a** No compression; **b** an edge dislocation is highlighted after compression at an earlier phase; **c** two more dislocations can be observed after further compression; **d** no dislocation observed after removing the compression [74]

stability, and drug delivery. Characterization of the adhesion and friction properties of nanoparticles has piqued researchers' interest in this area during the last decade [75]. AFM has already been shown to be a useful technology for the determination of friction and adhesion of a nanoparticle and a solid substrate. The AFM tip is also considered a NP, and the adhesion and friction forces can be easily determined by the deflection of the cantilever [76, 77]. The tip material and geometric shape, on the other hand, limit the application of AFM. The force between a solid substrate and NP was measured by Ducker et al. [78] by connecting the force sensor to the particle in the microscope. Because the connected particle's parameters, such as size, shape, and substance, could be controlled, inaccuracies in force measurement induced by the AFM tip's irregular form could be avoided. As a result, the colloidal probe approach is more successful in measuring micro/nanoparticle adhesion and friction [79, 80]. Nonetheless, the attachment of one NP with a diameter of fewer than 1 μm to the force sensor of AFM is extremely challenging; most references use colloid probes with diameters more than 1 μm [79]. Au NPs (20–40 nm) were placed on the tip of an AFM cantilever individually by Vakarelski et al. using a chemical technique to quantify the adhesion force between NPs and mica [81]. Ong and Sokolov [82] used epoxy glue to bind 50 nm cerium oxide NPs to the AFM tip in order to calculate the adhesion force between a flat surface and NPs. Other approaches include evaluating the tip's adhesion force to a coating of nanoparticles and fabricating a cantilever tip with a particular shape via thermal oxidation, among others [83, 84].

4.3.6 Movement of Nanoparticles

When nanoparticles are present in any particular medium, their motion is governed and influenced by several factors, like the Brownian diffusion motion, viscous flow forces, forces acting on the surfaces, and gravitational forces, which will be constituted by the buoyancy forces [85, 86]. However, the tiny size of particles prevents the use of the most regularly used imaging methods, and investigations for direct observation of nanoparticle mobility are limited. Fortunately, advances in measuring technology have made it possible to detect specific nanoparticles. Several approaches are utilized in the past to make high-resolution measurements of single nanoparticle motion. The approaches may be divided into two categories: passively tracking the movement of the particle without adding major external load and measuring particle movements under external mechanical forces.

The first method uses fluorescence technology to track particles. A fluorescence microscope is used in which fluorescent core–shell SiO_2 NPs of a particular size are employed as seed NPs [87]. The system was used to observe and study the Marangoni flow, the velocity profile in a channel flow, and nanoparticle–wall collision behaviors. As demonstrated in Fig. 11, a stagnation point is seen during the Marangoni flow in a droplet when the orientations of surface flow, surface tension gradient, and temperature gradient are altered.

The detection of nanoparticle motion is a very complex phenomenon due to many reasons, e.g., complex forces involved in it, the environment, and the medium of their synthesis. Therefore, there are qualitative studies available on the nanoparticle's movement; in the future, more accurate methods and instruments are needed for the quantitative analysis of their movement. Previous research on the mobility of a single nanoparticle was primarily qualitative. However, in the future, more accurate measuring techniques or tools are required for quantitative analysis.

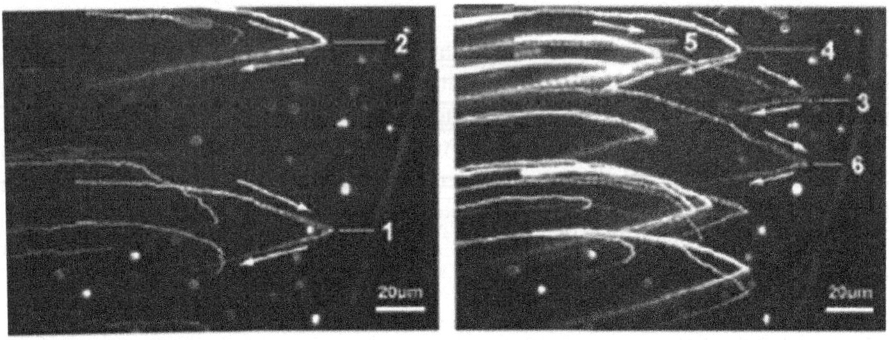

Fig. 11 Movement of particles in a droplet of water through the evaporation process [88]

5 Applications

5.1 Nanoparticles in Coatings

Nanocomposites can be designed by adding different types of NPs in a metal or polymer-based matrix to provide superior mechanical characteristics. Nanocomposites are divided into two groups, which are outlined below.

(a) Metal or composite materials are mostly used due to some of the matrix's intrinsic qualities, such as wear resistance, superior thermal and electrical properties, greater strength, and modulus. Compared to polymeric composites, metal alloy matrix composite coatings have significant benefits. Ceramic (Al_2O_3, TiO_2, etc.) is used in these coatings. Carbon-based (graphite and CNTs) and silicon-based (SiC) nanoparticles are often used [89–92].

The reasons behind using ceramic-based particles as reinforcement are given below:

- The high strength and hardness of particles
- The particles in the matrix can prevent grain boundary migration and dislocation motion [92]
- The influence of heterogeneous nucleation in metals and metal alloys [93]

On the one hand, adding graphite NPs or carbon nanotubes (CNTs) to a metal substrate might lower the pure metal coating porosity level, making it more impenetrable and compact with fewer fractures. Alternatively, the inclusion of nanoparticles might refine the size of crystals in the coatings. Another major component contributing to the strengthening effect is the chemical and structural stability of carbon nanotubes, which have better stiffness and strength than the metal matrix.

Adding inorganic or organic nanoparticles to a polymer-based matrix can change its physical characteristics, resulting in novel polymer features. Inorganic nanoparticles embedded polymer composite coatings can prevent the beginning of fracture and its propagation, fill voids, and can promote crack bridging, deflection, and bending. In general, numerous variables influence the mechanical characteristics of polymer-based nanocomposites, however the availability of larger quantity of interfacial area with respect to volume, that is result of synergy between the nanofillers and polymer chains, plays a prominent role [90, 94]. As a result, a good nanocomposite design considers the complex interaction of matrix, interface, and nanoparticles may tune the composite material system to have ideal physical characteristics.

The contact between the NPs and matrix can be improved with the increase in interactions either chemical or physical.

(b) Good mechanical qualities of nanocomposite coatings need uniform NPs dispersion in the matrix because the particles with a larger surface area have an inadequate maximum filling capacity. When the nanoparticle concentration of nanocomposite coatings hits a certain threshold, Particle agglomeration would occur at a critical value, causing mechanical characteristics to deteriorate, i.e., an inverse relationship of young's modulus decreases with the rate of wear observed.

Appropriate preparation and processing methodology are required to produce excellent particle dispersion. The interface resulting between the polymer matrix and the nanoparticle filler may be carefully changed on the molecular or atomic level to generate some fascinating shapes, such as core/shell hybrid nanoparticles by employing molecular assembly, atomic layer deposition (ALD) or any other such techniques [95].

5.2 *Nanoparticles in Catalysis*

Because at least one catalytic process is involved in approximately 90% of manufactured chemicals, catalysis plays a critical role in modern chemical industries [96]. The rapid development of catalysis science is dependent on the quick growth of nanoscience and nanotechnology [97]. Nano-catalysis in which nanoparticles are exploited as catalytically active materials, has grown rapidly in recent decades as an important field of nanoscience because it combines the advantages of fast turnover in homogeneous catalysis with recyclability in heterogeneous catalysis [98]. Because of their high surface-to-volume ratio, abundant active surface atoms, and distinct electronic structures compared to their bulk counterparts, metal nanoparticles show considerable potential in heterogeneous catalysis.

Metal NPs were known as part of heterogeneous catalysts (see, for example, the pioneering research of Sabatier [99] and Rampino and Nord [100]), but there has been a shift in interest in the creation of better-defined systems in the last three decades [101]. Surprisingly, most of the erstwhile heterogeneous catalysis community is now integrating into the nanoparticle community. As can be seen, by a large number of research publications, significant efforts are being devoted to finding better routes for the synthesis of metal nano species even at an atomic level for greater precision, [102–104] as well as the study of their properties.

The physical and chemical characteristics of nano-sized metal species are influenced by their specific matter state and electronic parameters. Small nanoparticles are generally referred to as nanoclusters, however as elucidated by the molecular orbitals and bigger NPs characterized by their energy band structures, there is a continuous range from molecules to solid state. The number of atoms of metal, the nature of the stabilizing ligands, and the dispersion of these distinct kinds varied. The molecular assembly of polymer and metal with ligands, whose X-ray crystal structure is defined can be referred to as clusters and nanoclusters. On the other hand mixtures of polydisperse nanoclusters as defined by a histogram that is shown by TEM analysis, are referred to as nanoparticles [105]. Nanosized metal particles have unique features that make them appealing materials for a variety of applications, including optoelectronics or optronics, sensor technology, bio-medicine, catalysis, energy conversion, and energy storage to name a few [106–108]. Because of their high surface-to-volume ratio, metal nanoparticles are an especially attractive catalytic species. Metal nanoparticles in the size range of even below 1 nm are called sub nanoparticles, this ratio is promising since the number of surface atoms might be more than 90%, which results in many active potential sites.

5.3 *Nanoparticles in Lubrication*

Properties of nanoparticles related with aspects of friction, wear and lubrication of interacting surfaces, termed as tribological properties of nanoparticle lubricated systems are strongly influenced by its mechanical properties. Mechanical properties affect the tribological properties of the nanoparticles as lubricant additives, which vary with the different materials. Tribofilms, creation of third body layer, sliding, and rolling, as briefly explained in the following sections, are the key causes for the improved lubricating behavior after introducing nanoparticles [109]. Nanoparticles that are present in the lubricated contact regions will have very low friction and wear; this is highly reliant on specific variables, such as the shape, size, and concentration of nanoparticles in the lubricant [110]. Sliding mode of NPs can also minimize friction and wear. When a particle is not perfectly spherical and has poor adhesion to the tribopair surfaces, sliding friction occurs. Another element that could cause sliding friction during shearing is particle aggregation in the contact area. The nanoparticles operate as a spacer in this scenario, preventing direct contact between the two surfaces two shearing surfaces' irregularities.

The nanoparticles operate as a spacer in this scenario, preventing direct contact between the two surfaces and two shearing surfaces' irregularities [111, 112].

5.4 Nanoparticles in Nanomanufacturing

CMP is an essential flattening aid in the nanomanufacturing of ICs. To create a flat surface on a wafer small topographic characteristics should be physically ground and eliminated chemically. The abrasiveness of NPs plays a critical role in the slurry, allowing for regulated material removal without sacrificing planarity. The surface quality and rate of removal depend on the mechanical interface between the wafer surface and NPs. There are two models to understand the mechanical behavior of the material removal process, the hydrodynamic model and the solid contact model [113, 114].

(1) In the first model, a thin liquid film separates the wafer from the polishing pad; material removal is predominantly caused by abrasive nanoparticles colliding with the wafer. Various factors like the effects of incidence speed, particle size, and angles on nanoparticle-wafer surface collisions have been studied. If the angle of incidence and the speed of impact of NPs were increased, the surface of the wafer may be damaged. For example, there were several dents and scratches on the surface with 45° incidence angle and 50 ms^{-1} speed, mixed and substantial deformation on the surface layer was recorded with the help of TEM. The action of nanoparticles colliding onto a silicon or on surface of silica has been studied through Molecular Dynamic (MD) simulations, proving that with increasing incident angle, the damage might be minimized. Furthermore, the incidence angle, not the particle size, influences the threshold velocity for pile-up development on the surface of silicone.

(2) Analyzing with the solid contact model, it can be seen that the surface of a wafer is in direct contact with a segment of the polishing pad. The particles implanted move against the substrate surface. Those particles which are absorbed in the slurry which is present between the wafer and the pad are called free particles. An experimental setup (fluorescence based) was devised to observe and study the motion of discreet particles present between the solid surface and the polishing pad. The results obtained showed that a few of these particles become trapped and get rotated along with the polishing pad, and the rest of the particles get transmitted along with the flow of the slurry, moving freely. The number of fixed particles to free particles ratio is very important in the material removal mode.

6 Challenges and Future Prospects

Numerous theoretical and experimental literature research on nanotechnology and nanomaterials have been published recently. Effective manipulation and modification of materials at nanoscale for various applications will have a great role in shaping future technologies. Development and effective application of nanomaterials, however, brings in some new challenges. One of the major challenges is search for economically viable synthesis route for nanomaterials. Large-scale production of high-quality nanomaterials is hindered by requirement of advanced equipment and controlled environment. This issue becomes more evident and challenging in case of 2D nanomaterials synthesis. The majority of low-cost techniques adopted for large scale production, however, result in low-quality materials with defects. Controlled production of nanomaterials is still a challenge at present. For example, in synthesizing carbon nanotubes, attainment of chiral selectivity, conductivity, and precise controlled diameters becomes critical challenge [115]. The only method to obtain the theoretically predicted properties stated in the literature is to be able to reach structurally pure nanomaterials. To create novel synthesis techniques that overcome the drawbacks of existing ones, more concentrated efforts are needed. The aggregation of particles at the nanoscale level is a fundamental problem that gravely impairs performance in pertinent disciplines. When nanomaterials interact with one another, the majority begin to aggregate together. Agglomeration can be caused by physical entanglement, electrostatic interactions, or high surface energy. Graphene agglomeration is also triggered by the basal planes of graphene sheets because of p-p interactions and van der Waals forces. CNTs can also aggregate due to van der Waals interactions and form bundles, which makes it challenging to align or distribute them properly in polymer matrices [116]. Some basic properties of graphene for example its high surface area, become effected as a result of severe aggregation. High-throughput electrode materials or composite materials cannot be used practically because of these obstacles [117]. With ever increasing demand of nanomaterials in industry owing to their applications, there is growing need for higher rates of nanoscale material synthesis. Furthermore, the research in nanotechnology has a very broad scope; with the exploration of new nanomaterials with intriguing properties, new fields will be unveiled in the future. One of the keys about nanomaterials that cannot be disregarded is their toxicity, which is still not well recognized, and is a substantial issue to regard in light of their domestic, environmental, and industrial use. Uncertainty exists over the potential role of nanoparticle-based compounds in cellular toxicity [118]. The scientific community needs to work to close the information gap between the rapid development of nanomaterials and possible in vivo toxicity. For the safe design and commercialization of nanotechnology, a thorough understanding of how nanoparticles interact with organisms, tissues, and proteins is essential.

Nanotechnology breakthroughs are connected with the future of modern technology. The development of nanomaterial-based engineering techniques is making the goal of producing clean energy a reality. Nanomaterials have shown propitious results, leaving us more optimistic for future. They have produced novel types of solar and hydrogen fuel cells, served as effective catalysts for water splitting, and demonstrated good hydrogen storage capabilities. Nanomedicine holds a bright future for nanomaterials. Therapeutic compounds can be delivered via nanocarriers.

7 Conclusions

This book chapter introduced the different methods to synthesize metal nanoparticles, their physicochemical properties, and their applications in various fields. However, the major focus here was to discuss the different factors affecting the mechanical properties of these metal nanoparticles and their applications related to their mechanical properties. Research studies show that the mechanical nature of nanoparticles is substantially different from their respective micro and bulk entities. The surface quality of MNPs can be easily enhanced by fine-tuning their mechanical properties. A lot of data is available stating the electronic, magnetic, and optical properties of metal nanoparticles. But, the area related to the mechanical properties and their industrial production still needs to be explored. Some studies on the improvement of mechanical properties and their subsequent use in industrial sectors have achieved substantial results. However, in order to provide useful results in industrial sectors, a detailed understanding of the fundamentals of the mechanical properties of nanoparticles is essential. The unique properties of nanomaterials allow for a wide range of potential uses and significant value in the future. Therefore, a continuous investigation of metal nanoparticles and modification methods to improve their properties is needed.

References

1. Meng XK, Tang SC, Vongehr S (2010) A review on diverse silver nanostructures. J Mater Sci Technol 26:487–522
2. Wiley B, Sun Y, Mayers B, Xia Y (2005) Shape-controlled synthesis of metal nanostructures: the case of silver. Chem A Eur J 11(2):454–463
3. Jain PK, Huang X, El-Sayed IH, El-Sayed MA (2008) Noble metals on the nanoscale: optical and photothermal properties and some applications in imaging, sensing, biology, and medicine. Undefined 41(12):1578–1586
4. Gold nanostructures: engineering their plasmonic properties for biomedical applications. Chem Soc Rev (RSC Publishing)
5. Wustholz KL, Brosseau CL, Casadio F, Van RP, Deckert-Gaudig T, Bailo E, Deckert V, Guerrini L, Izquierdo-Lorenzo I, Garcia-Ramos JV, Domingo C, Sanchez-Cortes S, Galloway CM, Ru Le EC, Etchegoin PG, Chem P, Kranich A, Naumann H, Molina-Heredia FP, Justin Moore H, Randall Lee T, Lecomte S, Rosa de la MA, Hildebrandt P, Murgida DH, Boyack R, Ru Le EC, Buchanan S, Tognalli NG, Scodeller P, Flexer V, Szamocki P, Ricci A, Tagliazucchi M, Calvo EJ, Fainstein A, Michael Morton S, Ewusi-Annan E, Jensen L, Zuo P, Albrecht T, Barker PD, Murgida DH, Sardo M, Ruano C, Luis Castro J, López-Tocón I, Soto J, Ribeiro-Claro P, Carlos Otero J, Park H, Lee S, Chen L, Kyu Lee E, Young Shin S, Han Lee Y, S Wook Son S, Hwan Oh C, Myong Song J, Ho Kang S, Choo J, Shuang Shen X, Zhong Wang G, Hong X, Zhu W, J Bell SE, McCourt MR, Maher RC, Zhang T, Cohen LF, Gallop JC, Liu FM, Green M, Cortés E, Vela ME, Salvarezza RC, Jabeen S, Dines TJ, Withnall R, Leharne SA, Kamal Hossain M, Gary Huang G, Kaneko T, Ozaki Y, Claudio Santos Costa J, Augusto Ando R, Carlos Sant Ana A, Marcia Rossi L, Sérgio Santos P, Laudelina Arruda Temperini M, Corio P, Chem Chem P, Jehn C, Küstner B, Adam P, Marx A, Ströbel P, Schmuck C, Schlücker S, Ando RA, W Pieczonka NP, Santos PS (2009) Nanospheres of silver nanoparticles: agglomeration,

surface morphology control and application as SERS substrates. Phys Chem Chem Phys 11(34):7450–7454
6. Bao Y, Hu Z, Li Z, Zhu X, Fang Z (2015) Plasmonics: magnetic plasmonic Fano resonance at optical frequency (Small 18/2015). Small 11(18):2102–2102
7. Huang X, El-Sayed IH, Qian W, El-Sayed MA (2006) Cancer cell imaging and photothermal therapy in the near-infrared region by using gold nanorods. J Am Chem Soc 128(6):2115–2120
8. Sau TK, Rogach AL, Jäckel F, Klar TA, Feldmann J (2010) Properties and applications of colloidal nonspherical noble metal nanoparticles. Adv Mater 22(16):1805–1825
9. Xia Y, Xiong Y, Lim B, Skrabalak SE (2009) Shape-controlled synthesis of metal nanocrystals: simple chemistry meets complex physics? Angew Chem Int Ed 48(1):60–103
10. Liao H-G, Jiang Y-X, Zhou Z-Y, Chen S-P, Sun S-G, Liao HG, Jiang YX, Zhou ZY, Chen SP, Sun SG (2008) Shape-controlled synthesis of gold nanoparticles in deep eutectic solvents for studies of structure-functionality relationships in electrocatalysis. Angew Chem Int Ed 47(47):9100–9103
11. Barbosa S, Agrawal A, Rodríguez-Lorenzo L, Pastoriza-Santos I, Alvarez-Puebla RA, Kornowski A, Weller H, Liz-Marzán LM (2010) Tuning size and sensing properties in colloidal gold nanostars. Langmuir 26(18):14943–14950
12. Xionghui C, Aixia Z (2010) Preparation of microsized silver crystals with different morphologies by a wet-chemical method. Rare Met 29(4):407–412
13. Hong L, Li Q, Lin H, Li Y (2009) Synthesis of flower-like silver nanoarchitectures at room temperature. Mater Res Bull 44(6):1201–1204
14. Sau TK, Murphy CJ (2004) Room temperature, high-yield synthesis of multiple shapes of gold nanoparticles in aqueous solution. J Am Chem Soc 126(28):8648–8649
15. Zhao N, Wei Y, Sun N, Chen Q, Bai J, Zhou L, Qin Y, Li M, Qi L (2008) Controlled synthesis of gold nanobelts and nanocombs in aqueous mixed surfactant solutions. Langmuir 24(3):991–998
16. Xia Y, Li W, Cobley CM, Chen J, Xia X, Zhang Q, Yang M, Cho EC, Brown PK (2011) Gold nanocages: from synthesis to theranostic applications. Acc Chem Res 44(10):914–924
17. Cauda V, Schlossbauer A, Kecht J, Zürner A, Bein T (2009) Multiple core-shell functionalized colloidal mesoporous silica nanoparticles. J Am Chem Soc 131(32):11361–11370
18. Morones JR, Elechiguerra JL, Camacho A, Holt K, Kouri JB, Ramírez JT, Yacaman MJ (2005) The bactericidal effect of silver nanoparticles. Nanotechnology 16(10):2346–2353
19. Daniel M-C, Astruc D (2004) Gold nanoparticles: assembly, supramolecular chemistry, quantum-size-related properties, and applications toward biology. catalysis, and nanotechnology
20. Jeong GH, Lee YW, Kim M, Han SW (2009) High-yield synthesis of multi-branched gold nanoparticles and their surface-enhanced Raman scattering properties. J Colloid Interface Sci 329(1):97–102
21. Roman Empire | British Museum
22. Kunckel J (1679) Ars Vitraria Experimentalis, Oder Vollkommene Glasmacher-Kunst/ Lehrende [...] Die allerkurtz-bündigsten Manieren/das reineste Chrystall-Glas; alle gefärbte oder tingirte Gläser; künstliche Edelstein oder Flüsse; Amausen/ oder Schmeltze; Doubleten; Spiegeln/das Tropff-Glas; die schönste Ultramarin, Lacc- und andere nützliche Mahler-Farben; Jngleichen wie die Saltze zu den allerreinesten Chrystallinen Gut/nach der besten Weise an allen Orten Deutschlands mit geringer Müh und Unkosten copieus u, Selbstverlag, Frankfurt (Main); Leipzig; Jena
23. Aurum Potabile Oder Gold-tinctur: Dessen Praeparation Daß Sie Sicher, Samt Des Goldes Vortrefflichkeit Und Analogie Mit Unserm Coerper, Würckung Und ... Als Praeservative, ..., Geanwtwortet Wird: Helcher, Hans Heinrich: 9781173872878: Amazon.com: Books
24. Joseph P (1769) Dictionnaire de chymie, contenant la theorie, la pratique de cette science, son application a la physique, a l'histoire, 1st edn from Nick Bikoff, Bookseller (SKU: 16734)
25. Fulhame (1794) An essay on combustion, with a view to a new art of dying and painting. Wherein the phlogistic and antiphlogistic hypotheses are proven erroneous, Printed for the author, by J. Cooper, London

26. Satoshi Horikoshi NS Introduction to nanoparticles
27. Jin R, Cao Y, Mirkin CA, Kelly KL, Schatz GC, Zheng JG (2001) Photoinduced conversion of silver nanospheres to nanoprisms. Science (New York, N.Y.) 294(5548):1901–1903
28. Brinker CJ, Scherer GW (2013) Sol-Gel science: the physics and chemistry of sol-gel processing. Sol-Gel Sci: Phys Chem Sol-Gel Process 1–908
29. Henglein A (2002) Small-particle research: physicochemical properties of extremely small colloidal metal and semiconductor particles. Chem Rev 89(8):1861–1873
30. Burda C, Chen X, Narayanan R, El-Sayed MA (2005) Chemistry and properties of nanocrystals of different shapes. Chem Rev 105(4):1025–1102
31. Liz-Marzán LM (2005) Tailoring surface plasmons through the morphology and assembly of metal nanoparticles. Langmuir 22(1):32–41
32. Sugimoto T (2000) Fine particles : synthesis, characterization, and mechanisms of growth. Marcel Dekker, Basel, New York
33. Lea CM (1889) Allotropic forms of silver. AmJS 37(222):476–491
34. Pearson RG (2002) Hard and soft acids and bases. J Am Chem Soc 85(22):3533–3539
35. Prasad BLV, Stoeva SI, Sorensen CM, Klabunde KJ (2003) Digestive-ripening agents for gold nanoparticles: alternatives to thiols. Chem Mater 15(4):935–942
36. Brust M, Walker M, Bethell D, Schiffrin DJ, Whyman R (1994) Synthesis of thiol-derivatised gold nanoparticles in a two-phase Liquid-Liquid system. J Chem Soc, Chem Commun 7:801–802
37. Yamaguchi Y, Matsubara Y, Ochi T, Wakamiya T, Yoshida ZI (2008) How the π conjugation length affects the fluorescence emission efficiency. J Am Chem Soc 130(42):13867–13869
38. Stoeva S, Klabunde KJ, Sorensen CM, Dragieva I (2002) Gram-scale synthesis of monodisperse gold colloids by the solvated metal atom dispersion method and digestive ripening and their organization into two- and three-dimensional structures. J Am Chem Soc 124(10):2305–2311
39. Yu YY, Chang SS, Lee CL, Wang CRC (1997) Gold nanorods: electrochemical synthesis and optical properties. J Phys Chem B 101(34):6661–6664
40. van der Zande BMI, Bohmer MR, Fokkink LG, Schonenberger C (1997) Aqueous gold sols of rod-shaped particles. J Phys Chem B 101(6):852–854
41. Nicewarner Peñ SR, Raina S, Goodrich GP, Fedoroff NV, Keating CD (2002) Hybridization and enzymatic extension of Au nanoparticle-bound oligonucleotides
42. Chen S, Wang ZL, Ballato J, Foulger SH, Carroll DL (2003) Monopod, bipod, tripod, and tetrapod gold nanocrystals. J Am Chem Soc 125(52):16186–16187
43. Toshima N, Yonezawa T, Kushihashi K (1993) Polymer-protected palladium–platinum bimetallic clusters: preparation, catalytic properties and structural considerations. J Chem Soc, Faraday Trans 89(14):2537–2543
44. Sun Y, Mayers B, Herricks T, Xia Y (2003) Polyol synthesis of uniform silver nanowires: a plausible growth mechanism and the supporting evidence. Nano Lett 3(7):955–960
45. Liu Y, Zhu J, Weng G, Li J, Zhao J (2020) Gold nanotubes: synthesis, properties and biomedical applications. Microchim Acta 187(11):612
46. Eustis S, El-Sayed MA (2006) Why gold nanoparticles are more precious than pretty gold: noble metal surface plasmon resonance and its enhancement of the radiative and nonradiative properties of nanocrystals of different shapes. Chem Soc Rev 35(3):209–217
47. Khan I, Saeed K, Khan I (2019) Nanoparticles: Properties, applications and toxicities. Arab J Chem 12(7):908–931
48. Reiss G, Hütten A (2005) Magnetic nanoparticles: applications beyond data storage. Nat Mater 4(10):725–726
49. Faivre D, Bennet M (2016) Magnetic nanoparticles line up. Nature 535(7611):235–236
50. Priyadarshana G, Kottegoda N, Senaratne A, De Alwis A, Karunaratne V (2015) Synthesis of magnetite nanoparticles by top-down approach from a high purity ore. J Nanomater 2015
51. Zhu Y, Goodridge AG, Stapleton SR (1994) Zinc, vanadate and selenate inhibit the tri-iodothyronine-induced expression of fatty acid synthase and malic enzyme in chick-embryo hepatocytes in culture. Biochem J 303(1):213–216

52. Qi M, Zhang K, Li S, Wu J, Pham-Huy C, Diao X, Xiao D, He H (2016) Superparamagnetic Fe_3O_4 nanoparticles: synthesis by a solvothermal process and functionalization for a magnetic targeted curcumin delivery system. New J Chem 40(5):4480–4491
53. Wu W, He Q, Jiang C (2008) Magnetic iron oxide nanoparticles: synthesis and surface functionalization strategies. Nanoscale Res Lett 3(11):397–415
54. Fukui T, Baba T, Katamoto T, Suda A, Nogi K (2018) Evaluation methods for properties of nanostructured body. In: Nanoparticle technology handbook, pp 301–363
55. Nieh TG, Wadsworth J, Sherby OD Superplasticity in metals and ceramics
56. Guo D, Xie G, Luo J (2013) Mechanical properties of nanoparticles: basics and applications. J Phys D Appl Phys 47(1):013001
57. Shorey AB, Johnson KM, Kwong KM, Jacobs SD (2000) Nanoindentation hardness of particles used in magnetorheological finishing (MRF) Appl Opt 39(28):5194–5204
58. Biggs S, Spinks G (2012). Atomic force microscopy investigation of the adhesion between a single polymer sphere and a flat surface. 12(5):461–478. https://doi.org/10.1163/156856198X00164
59. Chen Y, Mu W, Lu J (2012) Young's modulus of PS/CeO_2 composite with core/shell structure microspheres measured using atomic force microscopy. J Nanopart Res 14(2):1–9
60. Paik P, Kar KK, Deva D, Sharma A (2007) Measurement of mechanical properties of polymer nanospheres by atomic force microscopy: effects of particle size. Micro Nano Lett 2(3):72–77
61. Armini S, Vakarelski IU, Whelan CM, Maex K, Higashitani K (2007) Nanoscale indentation of polymer and composite polymer−silica core−shell submicrometer particles by atomic force microscopy. Langmuir 23(4):2007–2014
62. Tan S, Sherman RL, Ford WT (2004) Nanoscale compression of polymer microspheres by atomic force microscopy. Langmuir 20(17):7015–7020
63. Ramos M, Ortiz-Jordan L, Hurtado-Macias A, Flores S, Elizalde-Galindo JT, Rocha C, Torres B, Zarei-Chaleshtori M, Chianelli RR (2013) Hardness and elastic modulus on six-fold symmetry gold nanoparticles. Materials 6(1):198–205
64. Mordehai D, Lee SW, Backes B, Srolovitz DJ, Nix WD, Rabkin E (2011) Size effect in compression of single-crystal gold microparticles. Acta Mater 59(13):5202–5215
65. Mordehai D, Kazakevich M, Srolovitz DJ, Rabkin E (2011) Nanoindentation size effect in single-crystal nanoparticles and thin films: a comparative experimental and simulation study. Acta Mater 59(6):2309–2321
66. Wei Wang J, Narayanan S, Yu Huang J, Zhang Z, Zhu T, Mao SX (2013) Atomic-scale dynamic process of deformation-induced stacking fault tetrahedra in gold nanocrystals. Nat Commun 4(1):1–8
67. Gerberich WW, Mook WM, Perrey CR, Carter CB, Baskes MI, Mukherjee R, Gidwani A, Heberlein J, McMurry PH, Girshick SL (2003) Superhard silicon nanospheres. J Mech Phys Solids 51(6):979–992
68. Zhang N, Deng Q, Hong Y, Xiong L, Li S, Strasberg M, Yin W, Zou Y, Taylor CR, Sawyer G, Chen Y (2011) Deformation mechanisms in silicon nanoparticles. J Appl Phys 109(6):063534
69. Ouyang G, Zhu WG, Sun CQ, Zhu ZM, Liao SZ (2010) Atomistic origin of lattice strain on stiffness of nanoparticles. Phys Chem Chem Phys 12(7):1543–1549
70. Cuenot S, Frétigny C, Demoustier-Champagne S, Nysten B (2004) Surface tension effect on the mechanical properties of nanomaterials measured by atomic force microscopy. Phys Rev B 69(16):165410
71. Jing GY, Duan HL, Sun XM, Zhang ZS, Xu J, Li YD, Wang JX, Yu DP (2006) Surface effects on elastic properties of silver nanowires: contact atomic-force microscopy. Phys Rev B Condens Matter Mater Phys 73(23):235409
72. Yang X, Xiao S, Hu W (2013) Atomistic simulation for the size effect on the mechanical properties of Ni/Ni_3Al nanowire. J Appl Phys 114(9):094303
73. Jing G, Zhang X, Yu D (2010) Effect of surface morphology on the mechanical properties of ZnO nanowires. Appl Phys A 100(2):473–478
74. Carlton CE, Ferreira PJ (2012) In situ TEM nanoindentation of nanoparticles. Micron 43(11):1134–1139

75. Luan B, Robbins MO (2005) The breakdown of continuum models for mechanical contacts. Nature 435(7044):929–932
76. Larson I, Grieser F, Drummond CJ, Chan DYC (2002) Direct force measurements between titanium dioxide surfaces. J Am Chem Soc 115(25):11885–11890
77. Mate CM, McClelland GM, Erlandsson R, Chiang S (1987) Atomic-scale friction of a tungsten tip on a graphite surface. Phys Rev Lett 59(17):1942
78. Ducker WA, Senden TJ, Pashley RM (1991) Direct measurement of colloidal forces using an atomic force microscope. Nature 353(6341):239–241
79. Kappl M, Butt H-J The colloidal probe technique and its application to adhesion force measurements
80. Butt HJ, Cappella B, Kappl M (2005) Force measurements with the atomic force microscope: technique, interpretation and applications. Surf Sci Rep 59(1–6):1–152
81. Vakarelski IU, Higashitani K (2006) Single-nanoparticle-terminated tips for scanning probe microscopy. Langmuir 22(7):2931–2934
82. Ong QK, Sokolov I (2007) Attachment of nanoparticles to the AFM tips for direct measurements of interaction between a single nanoparticle and surfaces. J Colloid Interface Sci 310(2):385–390
83. Salameh S, Schneider J, Laube J, Alessandrini A, Facci P, Seo JW, Ciacchi LC, Mädler L (2012) Adhesion mechanisms of the contact interface of TiO_2 nanoparticles in films and aggregates. Langmuir 28(31):11457–11464
84. Sokolov I, Ong QK, Shodiev H, Chechik N, James D, Oliver M (2006) AFM study of forces between silica, silicon nitride and polyurethane pads. J Colloid Interface Sci 300(2):475–481
85. Medintz IL, Uyeda HT, Goldman ER, Mattoussi H (2005) Quantum dot bioconjugates for imaging, labelling and sensing. Nat Mater 4(6):435–446
86. Xu J, Luo JB, Lu XC, Wang LL, Pan GS, Wen SZ (2005) Atomic scale deformation in the solid surface induced by nanoparticle impacts. Nanotechnology 16(6):859
87. Xu X, Luo J, Yan J (2008) A PIV system for two-phase flow with nanoparticles. Int J Surf Sci Eng 2(1–2):168–175
88. Xu X, Luo J (2007) Marangoni flow in an evaporating water droplet. Appl Phys Lett 91(12):124102
89. Kang YC, Chan SLI (2004) Tensile properties of nanometric Al_2O_3 particulate-reinforced aluminum matrix composites. Mater Chem Phys 85(2–3):438–443
90. Bakshi SR, Lahiri D, Agarwal A (2013). Carbon nanotube reinforced metal matrix composites. A Rev 55(1):41-64 https://doi.org/10.1179/095066009X12572530170543
91. Hanemann T, Szabó DV (2010) Polymer-nanoparticle composites: from synthesis to modern applications. Materials 3:3468–3517
92. Shao W, Nabb D, Renevier N, Sherrington I, Luo JK (2012) Mechanical and corrosion resistance properties of TiO_2 nanoparticles reinforced Ni coating by electrodeposition. IOP Conf Ser Mater Sci Eng 40(1):012043
93. Wang ZH, Wang XD, Zhao YX, Du WB (2010) SiC nanoparticles reinforced magnesium matrix composites fabricated by ultrasonic method. Trans Nonferrous Metals Soc China 20(SUPPL. 3):s1029–s1032
94. Wang H, Lu R, Huang T, Ma Y, Cong P, Li T (2011) Effect of grafted polytetrafluoroethylene nanoparticles on the mechanical and tribological performances of phenol resin. Mater Sci Eng, A 528(22–23):6878–6886
95. Wang J, Shi TJ, Jiang XC (2009) Synthesis and characterization of core-shell ZrO_2/PAAEM/PS nanoparticles. Nanoscale Res Lett 4(3):240–246
96. Thomas JM (2018) The enduring relevance and academic fascination of catalysis. Nat Catal 1(1):2–5
97. Grunes J, Zhu J, Somorjai GA (2003) Catalysis and nanoscience. Chem Commun 3(18):2257–2260
98. Wang D, Astruc D (2017) The recent development of efficient Earth-abundant transition-metal nanocatalysts. Chem Soc Rev 46(3):816–854

99. Sabatier P (1913) La Catalyse en chimie organique, par Paul Sabatier ..., C. Béranger, Paris, Liége
100. Rampino LD, Nord FF (2002) Preparation of palladium and platinum synthetic high polymer catalysts and the relationship between particle size and rate of hydrogenation. J Am Chem Soc 63(10):2745–2749
101. Polshettiwar V (2013) Nanomaterials in catalysis. In: Serp P, Philippot K (eds) Angewandte Chemie Int Edn 52(43):11199–11199
102. Jin R, Zeng C, Zhou M, Chen Y (2016) Atomically precise colloidal metal nanoclusters and nanoparticles: fundamentals and opportunities. Chem Rev 116(18):10346–10413
103. Jin R, Pei Y, Tsukuda T (2019) Controlling nanoparticles with atomic precision. Acc Chem Res 52(1):1
104. Kang X, Zhu M (2019) Transformation of atomically precise nanoclusters by ligand-exchange. Chem Mater 31(24):9939–9969
105. Astruc D (2020) Introduction: nanoparticles in catalysis. Chem Rev 120(2):461–463
106. Talapin DV, Shevchenko EV (2016) Introduction: nanoparticle chemistry. Chem Rev 116(18):10343–10345
107. Roucoux A, Philippot K (2021) New trends in the design of metal nanoparticles and derived nanomaterials for catalysis
108. Gilroy KD, Ruditskiy A, Peng HC, Qin D, Xia Y (2016) Bimetallic nanocrystals: syntheses, properties, and applications. Chem Rev 116(18):10414–10472
109. Wang X-B, Liu W-M (2013) Nanoparticle-based lubricant additives. Encycl Tribol 2369–2376
110. Rapoport L, Leshchinsky V, Lapsker I, Volovik Y, Nepomnyashchy O, Lvovsky M, Popovitz-Biro R, Feldman Y, Tenne R (2003) Tribological properties of WS2 nanoparticles under mixed lubrication. Wear 255(7–12):785–793
111. Chinas-Castillo F, Spikes HA (2003) Mechanism of action of colloidal solid dispersions. J Tribol 125(3):552–557
112. Zhou J, Wu Z, Zhang Z, Liu W, Xue Q (2000) Tribological behavior and lubricating mechanism of Cu nanoparticles in oil. Tribol Lett 8(4):213–218
113. Seok J, Sukam CP, Kim AT, Tichy JA, Cale TS (2003) Multiscale material removal modeling of chemical mechanical polishing. Wear 254(3–4):307–320
114. Luo J, Dornfeld DA (2001) Material removal mechanism in chemical mechanical polishing: theory and modelling. IEEE Trans Semicond Manuf 14(2):112–133
115. Yang F, Wang X, Zhang D, Yang J, Luo D, Xu Z, Wei J, Wang J-Q, Xu Z, Peng F, Li X, Li R, Li Y, Li M, Bai X, Ding F, Li Y (2014) Chirality-specific growth of single-walled carbon nanotubes on solid alloy catalysts. Nature 510(7506):522–524
116. Sahoo NG, Rana S, Cho JW, Li L, Chan SH (2010) Polymer nanocomposites based on functionalized carbon nanotubes. Prog Polym Sci 35(7):837–867
117. He Y, Chen W, Li X, Zhang Z, Fu J, Zhao C, Xie E (2013) Freestanding three-dimensional graphene/MnO_2 composite networks as ultralight and flexible supercapacitor electrodes. ACS Nano 7(1):174–182
118. Metabolic effects of TiO_2 nanoparticles, a common component of sunscreens and cosmetics, on human keratinocytes—PubMed

Synthesis, Properties and Characterization of Metal Nanoparticles

K. Thummavichai, Y. Chen, N. N. Wang, Y. Q. Zhu, and O. Ola

Abstract Precious metals including Au, Pd, Pt, Ag and Pd and other metals such as Fe, Co, Mg and Ni as solid powder, dispersion in solution and deposition as thin films have attained wide interest in the last decades. They have induced intense research interest in nanotechnology due to their exciting properties including good conductivity, magnetic recording, localized surface plasmon resonance, antibacterial and catalytic effects [1, 2]. This chapter will introduce several main synthesis and characterization methods of metallic nanoparticles (NPs). The unique feature, key parameters and especially advantages and disadvantages of top-down (i.e. physical vapour deposition, bill milling and lithography) and bottom-up (e.g. chemical vapour deposition, sol–gel, hydrothermal/solvothermal, etc.) methodologies are discussed to trigger advances in nanotechnology advancement. Alternative green synthesis approaches are also included in this chapter. Furthermore, the basic characterization techniques for metallic NPs are pointed out for improving synthesis strategies, deciphering the topography evolution and comprehending the potential applications. Finally, emphasis has been placed on some main properties of metallic NPs for the potential of a wide range of applications.

Keywords Metallic nanoparticles · Physicochemical characterization · Nanoparticle synthesis · Properties of metallic nanoparticles

---

K. Thummavichai
Department of Mathematics, Physics and Electrical Engineering, Northumbria University, Newcastle Upon Tyne, UK

Y. Chen · Y. Q. Zhu
College of Engineering, Mathematics and Physical Sciences, University of Exeter, Exeter, UK

N. N. Wang
Guangxi Institute Fullerene Technology (GIFT), Laboratory of New Processing Technology for Nonferrous Metals and Materials, Ministry of Education, School of Resources, Environment and Materials, Guangxi University, Nanning, China

O. Ola (✉)
Advanced Materials Research Group, Faculty of Engineering, The University of Nottingham, University Park, Nottingham, UK
e-mail: Oluwafunmilola.Ola1@nottingham.ac.uk

S. K. Tiwari et al. (eds.), *Nanoparticles Reinforced Metal Nanocomposites*,
https://doi.org/10.1007/978-981-19-9729-7_6

1 Introduction

Nanoparticles (NPs) have attracted research interests for over a century and are strongly believed to be the key to future technology due to their large proportion of high-energy nanosized surface atoms compared to bulk materials, which gives rise to outstanding chemical, magnetic, physical, mechanical, catalytic and optical properties. NPs have emerged as an amazing class of materials which is defined within the dimensional range of ~1–100 nm. Depending on the overall shape, these materials can be zero-, one-, two- or three-dimensional. For decades, researchers have been paying more and more attention to different metal NPs such as Au, Ag, Pt, Pd, Fe, Co, Mg and Ni, focussing on their synthesis method, characterisations and applications in energy, magnetic imaging, drug delivery, information technology and optoelectronics. Numerous synthesis methods are being established to either enhance the properties or reduce production costs. Furthermore, synthetic techniques and fabrication tools have been continuously studied to permit the production of reproducible nanostructures. It is worth noting that various parameters should also be monitored when NPs are stored due to the changes with time in various environments. Compared to bulk materials, it is sometimes incomplete and inherently difficult to properly analyse the nanomaterials due to the small size and low quantity (i.e. laboratory-scale production). Moreover, the arrangement of particles or atoms in nanostructures brings forth unusual, sometimes exotic forms (e.g. core–shell NPs, fullerenes, nanostructured metals, dendrites, etc.). The NPs characterization is understudied and therefore demands reproducing and validating both theoretical and experimental findings for better scientific understanding, which will benefit the development of new technologies and address important issues, e.g. product lifetimes, etc. Hence, it is important to update new development in the synthesis and characterization of nanomaterials, especially the analysis of their structures, which is believed essential for further progress in nanotechnology. Here in, we explore the nature and causes of a few main NPs synthesis methods and their analytical encounters.

2 Method in Metallic Nanoparticles Synthesis

A wide range of methods have been introduced to produce metal nanoparticles which are categorized into two main types as top-down and bottom-up approaches depending on the starting materials. Hence, bulk materials are usually used as starting materials for the top-down approach while atoms or molecules are typically used for the bottom-up process (Fig. 1). Ball milling, physical vapour deposition (PVD) and lithography are typical top-down methods, while laser pyrolysis, chemical vapour deposition, hydrothermal, sol–gel and electrodeposition are classic methods for the bottom-up approach. However, these all-synthesis techniques are toxic, resulting in environmental contamination; thus, biological methods as an alternative green

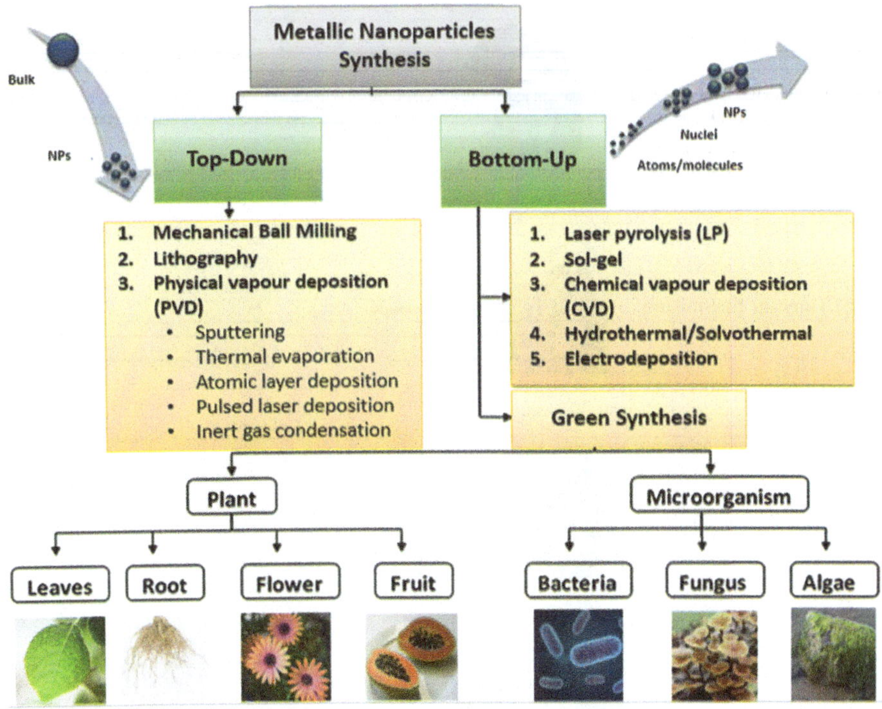

Fig. 1 Summary of metallic nanoparticle synthesis process from both top-down and bottom-up approaches

synthesis were found to be less toxic and environmental friendly. Different preparation methods of metallic NPs create various single or multiple metal (alloy) with different sizes, shapes and structures because of the variation of the parameters during the synthesis process (e.g. stabilizing agent for metallic NPs during the adsorption process, etc.).

2.1 *Top-Down Methods*

2.1.1 Mechanical Ball Milling

In this method, bulk material which is usually in the micro-dimensions is grounded down to the nanoscale by applying strong mechanical shear forces. Four types of attrition devices are generally used, namely vibration mills, planetary ball mill, tumbler ball milling and attrition mill (as shown in Fig. 2).

Among all top-down methods, ball milling has been widely used for the synthesis of various alloy nanoparticles and composites such as Al, Co, Mg, Ti, Cu and Fe

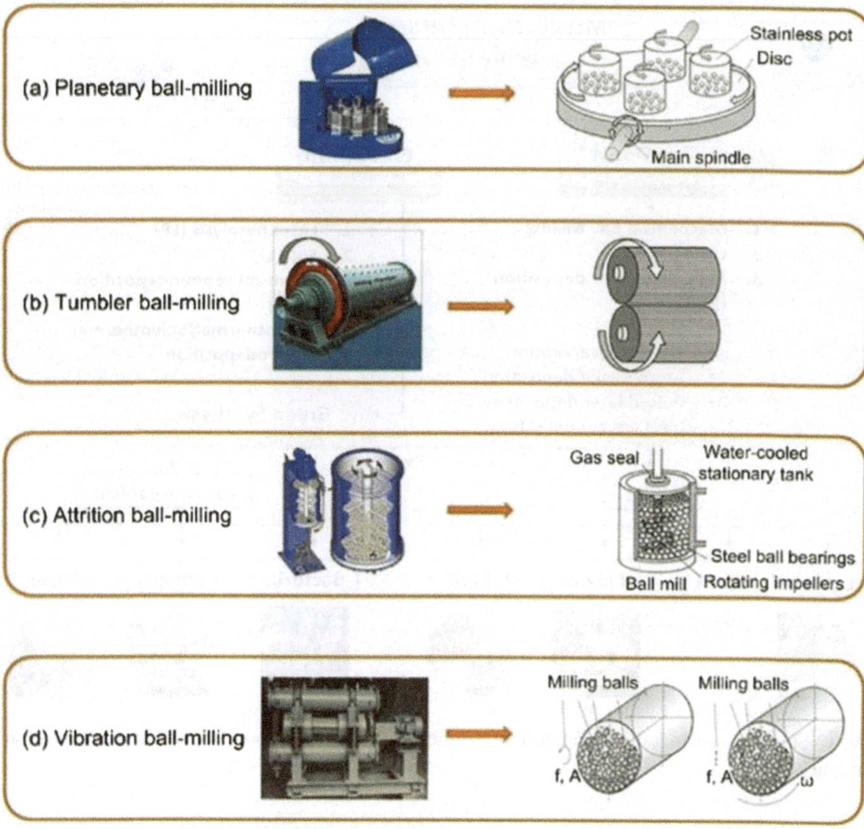

Fig. 2 Different types of ball milling and its working principles: **a** Planetary ball milling, **b** Tumbler ball milling, **c** Attrition ball milling and **d** Vibration ball milling (where f is vibration frequency, A is vibration amplitude and ω is angular velocity) [3]

[4–8]. By using this technique, the nanostructures are fabricated by mechanical attrition, where kinetic energy from grinding balls is used to reduce the size of material [9]. Milling conditions including the size of the milling vial, type of mills (low- or high-energy), milling speed, shape of milling media (balls or rods), milling atmosphere (e.g. inert gas or hydrogen), ball-to-powder weight ratio, milling time, milling environment (dry or wet milling) and milling temperature have a direct effect on properties, stoichiometry, particle size distribution and degree of disorder or amorphization of the final products. A more comprehensive control and monitoring of the milling conditions are suggested to improve the quality of the products.

The raw material's temperature influences the defect concentration and diffusivity inside it, which also influences the phase transformations during milling [10]. Meanwhile, processing time is also a significant factor for this technique. Based on the literature, an increase in time of milling process improves the microhardness of as-prepared materials [11]. Generally, the rate of the internal structure refinement (e.g. crystallite size, particle size, lamellar spacing, etc.) is roughly logarithmic with processing time and thus the size of the starting materials is relatively unimportant. The lamellar spacing usually becomes small, and the grain size is refined to nanoscale within a few minutes to an hour (Fig. 1); therefore, the milled powders will exhibit increased lattice strain while decreased grain size, hence the resulting milled powders will be in a highly energetic condition [3].

Advantages: Ball milling is well known as a simple, low-production-cost and sustainable technique, and it also possesses the capability to achieve very high yields. The nanoparticle's size ranging from 2 to 20 nm can be achieved through the different speeds of the rotation of the balls. This technique is one of the most reliable, easy-operation and reproducible process because of the speed and energy controls; moreover, it is suitable in both dry and wet conditions for a wide range of materials [4, 7, 12].

Disadvantages: Key serious issue with the milling of fine powders is the possibility of significant contamination from the milling atmosphere or media [4, 7, 9]. Iron contamination can be a problem for steel balls and containers. It has been reported that 10 atomic percentage of the iron contamination in some refractory metallic powders have been found by extended milling times in a high-energy shaker mill. On the other hand, if milling is carried out in open atmosphere, contamination with nitrogen or oxygen can occur. However, optimized milling speed and milling time may effectively help to minimize the contamination issue.

2.1.2 Nanolithography

Nanolithography is one of the most accurate and classic methods for the synthesis of nanoparticle pattern, and it provides high-resolution structures over large areas (>1 cm^2) with good control of all the dimensions (length, width, height) and other features such as roughness, edge shape and inter-diffusion of the as-prepared materials [13]. This technique is widely used in integrated circuits manufacturing and nanoelectromechanical systems [14]. Lithography can be categorized into 2 main types including masked and maskless lithographies. Masked Lithography utilizes the mask or template to transfer pattern over the large area. The form of mask lithography includes soft lithography, nanoimprint lithography, [15] X-ray lithography and photolithography. On the other hand, maskless lithography technique (including electron beam lithography, Scanning Probe Lithography, etc.) fabricates subjective patterns by a series writing without the use of the mask. In general, the masked lithography has better controllability compared to maskless lithography. Masked lithography is commonly used to prepare nanomaterials that are highly dependent on the specific shape and size of the selected template. The principle of a few different types of lithography is presented in Fig. 3.

Additionally, mask can be classified into 2 types, soft mark and hard mask. Surfactant molecules (e.g. CTAB, TEOS, etc.) are usually used as a soft mask, while polymeric materials (e.g. PDMS PMMA, PS, PFPE and other elastomeric material) are usually used as hard mask [17]. In general, the lithography technique is a three-step

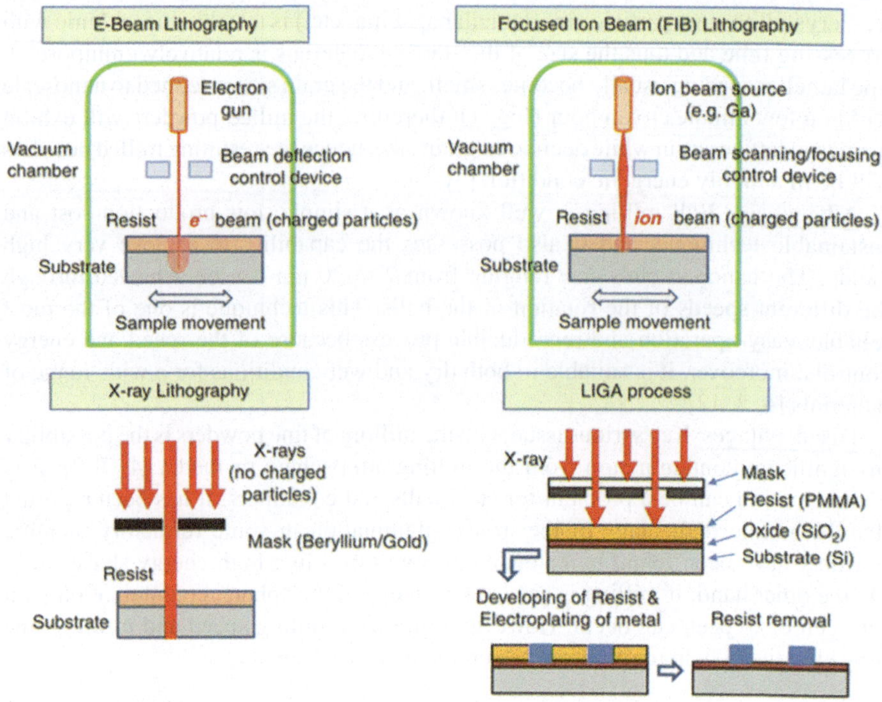

Fig. 3 Different type of mask and maskless lithography technique [16]

process: (1) preparation of a colloidal crystal mask made of nanospheres; (2) deposition of the desired product through the mask; and (3) etching of the mask from the layer.

Sub-10 nm Mo gratings have been successfully prepared by using photolithography [18]. However, the photolithography demands the use of a stepper, which is too expensive to be afforded by a standard research laboratory. Dot arrays can either be achieved using four-beam interference lithography or achromatic Talbot lithography (i.e. high-resolution photolithography) [19, 20]. E-beam lithography can produce features as small as 20 nm of Au NPs, [21] but is very expensive and time-consuming. Hence, it would be reasonable to use photolithography for 5 min while using e-beam lithography for approximately another 5 h to complete a lithographic process. Pros and cons of each lithography techniques are outlined in Table 1.

2.1.3 Physical Vapour Deposition (PVD)

PVD involves either high-temperature evaporation or ions bombardment of solid form materials in vacuum atmosphere; meanwhile, the reactive gas is also introduced into the system to form a compound with metal vapour and then deposited onto the

Table 1 Summery of advantage and disadvantage of lithography technique [22–25]

Type of lithography technique	Advantage	Disadvantage	Applications
Photolithography	• Well controlled features and high throughput compared to other lithography • High-resolution in the range of millimetres to micrometres • Low-cost, simple and time-efficient method (i.e. it can create patterns over an entire wafer in a single step)	• High up-front cost, e.g. requirement of flat substrate and a mask optical scissors technique for photolithography applications, requirement of clean room facilities, etc • Limitation of diffraction at wavelength of 193 nm	Fabricating integrated circuits such as computer chips and other semiconductor related devices
X-ray Lithography (an advanced version of photolithography)	• Shorter wavelengths (0.1–10 nm) can be used to solve diffraction limits of photolithography Smaller features can be patterned compared to original photolithography	• Deformation during the process • Vibrations during the process • Time-consuming process • X-ray masks are very expensive	Fabricating integrated circuits such as computer chips and other semiconductor related devices
E-beam lithography	• Capability to provide a small feature compared to those X-ray- and photolithography (be able to custom patterns with sub-10 nm resolution) • Fast turn-around time • Small wavelength that diffraction no longer defines the lithographic resolution	• Scaling error or pattern under-sizing • Not ideal process for industrial processing due to high cost of an operation • More complex system than photolithography • Low production yield per hour upon the different type of electron beam resists	Cryo-electric, optoelectronic, optical devices, etc
Nanoimprint lithography	• High cost-effective synthesis technique which is essential for industrial sector • Offer a high-resolution structure as fabricated pattern is only limited by the template only • Can be used to fabricate nanopatterns at a large scale in a short time (15 wafers/hours per imprint station) • High replication accuracy and high preparation efficiency • Simply operation process	• Require an improvement of the throughput of current approaches to meet the market demand for commercial production • The physical nanoimprint mask must be remanufactured by using inflexible mask (hence similar mark that used in e-beam lithography or reactive ion etching • The adhesion between resist and mask is a significant mould problem, it could cause the demand during the procedure of mould release	Solar cell, Light emitting, plasmonic devices, memory devices, flat panel display, etc

(continued)

Table 1 (continued)

Type of lithography technique	Advantage	Disadvantage	Applications
Scanning probe lithography	• Nanopatterns can be created without optical apparatus (i.e. direct writing in nature of the process) • It is capable of highest resolutions less than 10 nm of the pattern, as well as shaping in one-step 3D process • The methods operate under controlled atmospheric conditions, which reduces the tool overhead and costs • The scanning probe capable of detecting surface features down to atomic resolution	• The method is slow inherently slower than, e.g. photolithography or nanoimprint lithography (Super slow 80 nm/s for atomic scale) • The intrinsically low throughput of single probe systems (however, the multiple probes system can be applied to improve this problem) • Limited to specialized device	Protein patterns, the placement of ferritin nanostructures, electronic devices using quantum effects, etc

substrate as a highly adherent thin film or NPs [26]. PVD can be separated into different types such as *electron beam evaporation deposition (EBD), Ion plating (IP), thermal evaporation deposition (TED), pulsed laser deposition (PLD), atomic layer deposition (ALD), cathode arc deposition (CAD), dynamic ion mixing (DIM), electrophoretic deposition and sputtering technique.* The type of evaporation source and deposition substrate is the essential parameter that affect the deposition quality and efficiency. The evolution of size, sharpness and phase of the product also depends on the gas pressure and deposition temperature in the deposition chamber [27]. Typically, this process takes place at the temperature in the range of 100–600 °C and is widely used for fabrication of inorganic thin films (thickness less than 5 μm). PVD coating can be used in a broad range of applications in automotive, aerospace, optical, medical, thin films (e.g. food packaging, Window tint, etc.) and textile industry.

Among all PVD techniques, **PLD** is carried out at low substrate temperatures with a stoichiometry of the target retained in the evaporated films; hence, it is often used for the thin film electrolytes deposition [28]. Two advantages of PLD are the simplicity in process design and the multi-choice of the target forms (e.g. sintered pellet, powder, single crystal, etc.). For the **EBD** technique, both the conductor and insulator can be used as a target. Meanwhile, EBD of organic materials is limited by the decomposition of the molecules. In general, a vacuum environment of 10^{-2} – 10^{-4} Pa, with deposition rates of at least 25 μm/min, is typical condition required for EBD [29]. Moreover, components are commonly preheated in vacuum at temperatures between 800 and 1100 °C, and rotation process is required during the deposition period. EBD is a more efficient choice for evaporating materials with high melting point than other PVD heating and evaporate techniques. However, it is limited for coating the inner surface of complex geometries with the thickness of thin film below 5 nm. The filament degradation in the electron gun could be a cause of non-uniform evaporating rate. The deposition rate of EBD can be obtained up to 100 μm/min depending on the materials and set up conditions [30].

TED technique involves heating a solid material (located in evaporation source, e.g. boat, coil and basket at the bottom of the chamber) to its melting point [31]. The evaporated metal constitutes vapour steam inside a vacuum chamber and then travels across the chamber to the deposition target (located at the top of the chamber), sticking to it as a coating or film. In general, TED of metals with low to moderate deposition rates at 10–100 nm/min provides a simple and reliable manner to produce nanoparticles with a wide range of size; nevertheless, this method is not applicable to substances with high cohesive energies that require high temperature for the evaporation [32]. Material holders are normally made from W, Ta, Mo or ceramic materials that capable of bearing high temperatures. Many materials can be deposited using this method, including Al, Ag, Ni, Cr, Mg, etc. The selection of the PVD technique depends on the material type and possible application. If we are looking for growing thin metallic film, EBD might be the preferred choice compared to the TED. None of PLD technique is better in all aspects. If we are looking for a good step coverage, faster deposition rate, and efficiently, EBD can be preferred, while the sputtering gives you good control on the thickness and density of the films.

Sputtering is good for the deposition of high melting point materials like refractory metals and ceramics, which are tough to be transformed to nanomaterials by evaporation technique. It has been reported that the deposition rates of sputtering technique are much lower than the EBD route (i.e. up to 50 μm/h for a magnetron system or about 10–20 μm/h for a diode system) [33]. Additionally, sputtering offers greater stoichiometric control of the thin film compared to the TED techniques. Thin films from sputtering generally have a higher density than those thin films obtained from the evaporation process while the evaporated films are likely to cause less contamination than those sputtering films because of the lower purity of the sputtering target. **Inert gas condensation (IGC)** is a PVD method with extra function of using the inert gas to reduce the mean value of the free path of the species. IGC technique involves two steps: evaporation of target materials in nanoscale and rapid condensation of the evaporated material. Several different techniques have been employed to evaporate inorganic or metallic materials into a vaporized form, e.g. Joule-heated refractory crucibles, laser/plasma heating or electron beam evaporation devices, etc. Typically, inert gas including He, Ne, Ar, helium, neon, argon, Kr, Xe and Ra with pressures >3 mPa is required during the process [34]. The size and shape of the as-prepared NPs can be managed by different factors including the temperature and pressure of the chamber, evaporation rate and molecular weight of the inert gas that injected into the chamber. A high pressure in the growth region could obtain large particle size due to the consecutive agglomeration and less sintering can be used to synthesize high-purity metallic/bimetallic nanoparticles. It also offers a high surface cleanliness and well-defined grain size with a narrow size distribution of the as-prepared NPs compared to other PVD method. The cost of operation for ultra-high vacuum-based deposition systems of IGC is remarkably high. IGC involves an extremely slow process and suffers from other limitations, e.g. temperature ranges, the source-precursor incompatibility, dissimilar evaporation rates in an alloy, etc.

Advantage: PVD technique can be utilised to deposit a virtually wide range of inorganic materials including metal alloys, ceramics, glass and polymer as well as some of the organic materials. Moreover, it is more environmentally and user friendly compared with other top-down process. The process demands the high vacuum condition to minimize an unwanted reaction within the free space, which helps to shape the film composition easily and causes less contamination on any substrate surface [35]. The purity of the deposited film depends on the vacuum and the quality of the source material. Coating thin film on materials via this technique will offer higher hardness value, more corrosive resistance, good impact strength and excellent abrasion.

Disadvantage: Difficult to coat undercuts and similar surface feature, high capital cost but low deposition rate. Some PVD processes are typically required to be carried out under high temperature and high vacuum, which cause a high-energy consumption and demands extra attention for operation. Water cooling systems are also needed to dissipate the large heat loads (Table 2).

Table 2 An example of materials that synthesized via different PVD technique

Techniques	Materials	Percussors	Process parameter	Achievement	Application	References
Sputtering deposition	Gd-WO_3 NPs 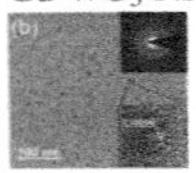	Metallic W target (99.99%) Gd metal slice	RF power 200 W, Ar/O_2 flow rate ratio 2:3, deposition at room temperature with 0.46 Pa working pressure	139–170 nm of Gd-WO_3 thin film depending on different amount of Gd dopant	Electrochromic glass	[36]
	Au-GO NPs 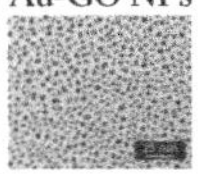	Metallic Au target	Based pressure of 10^{-8} mTorr with Ar pressure of 2.5 mTorr	Au nanoparticle diameter of 1–6 nm on the GO thin film	–	[37]
	Mg NPs 	Mg target (99.95%)	A base pressure of $\sim 1 \times 10^{-8}$ mbar, O_2 pressure of $\sim 10^{-9}$ mbar, an inert krypton atmosphere (pressure of $\sim$0.25 mbar)	Mg nanoparticle with diameter of 10–50 nm	Hydrogen storage	[2]
Pulsed laser deposition	Ag NPs 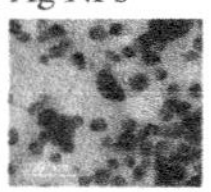	Plates of Ag of 99.99%	Laser pulses at 532 nm with pulse duration of 14 ns, a repetition rate of 20 kHz and an average output power of 6.0 W, electric field of 15 V was applied between the electrodes	Ag nanoparticle with diameter of 20 nm	Development of the synthesis technique	[38]
	Au NPs 	Au target	Laser pulses of energy 0.5 mJ, repetition rate was increased to 3 and 10 Hz, the translation speed at 0.25 mm s^{-1}. The ablation depth per shot was $\approx$100 nm, and the ablated mass was 23 ng. shot-to-shot spacing was 250 μm	Thickness of NPs ranging from 0.4–28 nm upon the changing of the target-substrate separation and the shot-to-shot spacing	Development of the synthesis technique	[39]
Atomic layer Deposition	Pt-Graphene 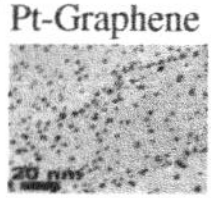	Pt (MeCp)Me_3, graphene nanoplatelets	About 20 min/cycle. Deposit for 3 cycles at 100 °C. Synthetic air (20 wt % O_2) was used as the oxygen source, Pt (MeCp)Me_3 exposure time was 4 min, N_2 was used as the purging step with step of 5	Ag nanoparticle with diameter below 5 nm	–	[27]

(continued)

Table 2 (continued)

Techniques	Materials	Percussors	Process parameter	Achievement	Application	References
	Ru@CNTs 	Bis(ethylcyclopentadienyl)ruthenium(II)	O_3 was used as counter reactant. N_2 was used as carrier with a flow rate of 20 sccm. Reaction chamber temperature of 290 °C. 150 cycles of Ru were deposited to acquire the nanoparticles	3.4 ± 0.4 nm	Catalytic activity in hydrolytic dehydrogenation of methylamine borane (MeAB; $CH_3NH_2BH_3$) and ammonia-borane (AB; NH_3BH_3)	[40]
	Lu@TiO_2 NPs 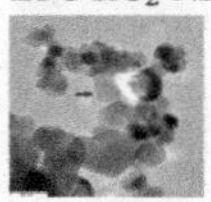	Lu $(TMHD)_3$ and Lu $(HMDS)_3$	The required dosing time for Lu $(TMHD)_3$ and Lu $(HMDS)_3$ were 23 and 10 min, respectively. Dosing times were 1.62 (for O_3) and 2 (for NH_3) min. During the purge, an additional N_2 flow of 0.1 l/min (0.30 × 10^{-2} m/s) was added. Percussor temperature was 210 °C and deposited temperature was 230 °C	No mention clearly about the size of particle	Radio nuclide generator	[41]
Thermal evaporation	Mg NPs 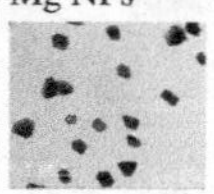 Mg-Zn Nanotriangular 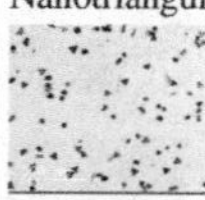	Mg and Zn metal powders	Base pressure of 10^{-5}–10^{-6} mbar, temperatures ranging from 600–800 °C at different holding durations ranging from 1 to 15 min	140–400 nm for Mg nanoparticle, ~45 nm in average. Mg-Zn is hexagonal phase with triangular, platelet like nano structures	Development of the synthesis technique	[1]
	Ni NPs 	Ni powder	Vacuum at 2 × 10^{-4} mbarr Deposition rate: 0.18 nm/Sec at 1.19 Amps to 0.5 nm/Sec at 25 Amps	10 nm of the thin film thickness, size of Ni nanoparticle after heat treatment of the thin film at 700 °C	The annealing effect of thin film	[42]

(continued)

Table 2 (continued)

Techniques	Materials	Percussors	Process parameter	Achievement	Application	References
	In NPs 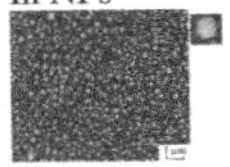	In powder (99.999% purity)	High-purity Ar atmosphere at a pressure of 6×10^3 Pa at different evaporation time of 10, 18 and 30 s	Nanoparticles with dimensions of 10–100 nm and crystallites with dimensions up to 500 nm	Development of the synthesis technique	[31]
	Ag NPs 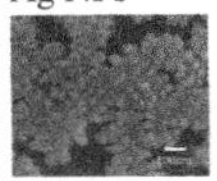	Ag wire	Deposited at room temperature in a vacuum chamber under pressure of 2.2×10^{-4} Pa	Ag diameter generally in the range of 40–100 nm	Surface-enhanced Raman scattering (SERS)	[43]
Inert gas condensation	Ni@CoO core-shall NPs 	Ni wire and CoO metal target	Evaporation temperature of 1450 °C by using 150 V ac voltage. W boat and a CoO target was ablated with 20 Hz repetition rate and 4 J/cm^2 fluence. Pressure during preparation was 6 Torr. The evaporation rate was 5 g/h and material collection rate at the filter surface was 0.2 g/h	Approximately 4.5 nm diameter core with a shell whose thickness varies in the range of 1.5–2.5 nm	Development of the synthesis technique	[44]
	Cu NPs 	Cu target	A supersaturated vapour of Cu atoms is originated by sputtering a Cu target. The system pressure was 1×10^{-8} Torr. He and Ar are used as inert gases at pressure of 10^{-1} Torr during aggregation. Pressure at the filtering zone is10^{-4} Torr). Depositing at rates 0.001–0.5 nm/s at condensation zone of 100 mm	Cu clusters in the size range of 1–5 nm	Development of the synthesis technique	[45]

(continued)

Table 2 (continued)

Techniques	Materials	Percussors	Process parameter	Achievement	Application	References
	Fe NPs 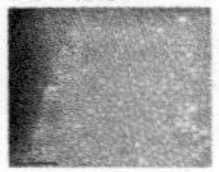	Fe target	Sputtering technique is used as ion source condensation zone length 30–150 mm. magnetron power (25–100 W). Flow of gases, Ar and He with partial pressure 1–2 × 10^{-1} Torr. Deposition time was few minutes up to 30 min	5 nm in average	Development of the synthesis technique	[46]
	Mg NPs 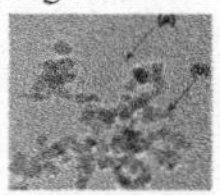	Manganese granule	The chamber was then sealed and evacuated to around 1.1 × 10^{-9} mbar. Thermal evaporation at 1533 K. 40 mbar of He is used as carried gas and liquid nitrogen cold trap was used to collect the particles	2–10 nm	Development of the synthesis technique	[47]

2.2 *Bottom-Up Methods*

2.2.1 Laser Pyrolysis (LP)

LP technique involves the use of a continuous flowing of CO_2, leading to molecular decomposition to form vapours to initiate nucleation, followed by the growth of NPs [48]. On the other hand, some of the other gases such ammonia (NH_3), sulphur hexafluoride (SF_6) and ethylene gas (C_2H_4) are also in use [49]. Next, NPs will be further transported to a filter by an inert gas depending on its amount. The main criterion is that either reactant or precursor should be able to absorb the energy that is supplied through the resonant vibrational mode of infrared CO_2 laser radiations. Inside the system, in most of the time, gaseous precursors are used in LP synthesis; however, solid or liquid precursors can be preferable in many cases, due to safety issues or the cost and availability of the precursor; hence, volatile precursor is not abundantly available for some metals.

LP provides superfast heating and quenching of particle growth on a small area, hence resulting in rapidly nucleation within few ms. It has been acknowledged that coalescence is higher at the high temperature, resulting in spherical particles, while other shapes are obtained at low temperature. In addition, the final particle size also depends on the aggregation of the initial nuclei. Gas pressure plays a significant role in determining the particle size and their distribution. Moreover, other parameters including laser intensity, the pressure in the reactor cell, the temperature of precursor were also varied in order to obtain different crystallinity and particle size [50]. In general, LP provides more possibilities for NPs with narrow size distribution of 5–200 nm.

Advantage: LP mode provides a very rapid, in-depth heating to a steady-state temperature, offering a high and continuous production of well-dispersed NPs [51]. Moreover, it is the simplest method for producing many nanostructures even at pilot plant dimensions by balancing CO_2 laser exposure and continuous flow of reactor.

Disadvantage: Elevated costs are the main constraints of this method. The requirement of a specific reagent/laser resonance with a specific installation is essential.

2.2.2 Sol–Gel

Sol–gel is a well-known wet-chemical process where a chemical solution acts as a precursor for an integrated system of discrete particles. Among bottom-up synthesis routes, sol–gel is one of the most preferred methods because of its simplicity, easy control of the particle morphology and size, as well as the ability for large volume production of metallic nanomaterial. In general, the method involves two main reactions including hydrolysis of the precursor in the acidic or base mediums (reaction 1) and polycondensation of the hydrolysed products (reactions 2 and 3) [52, 53].

$$\equiv M - OR + HOH \rightleftarrows \equiv M - OH + ROH \quad (1)$$

$$\equiv M - OH + \equiv M - OH \rightleftarrows \equiv M - O - M + HOH \quad (2)$$

$$\equiv M - OH + \equiv M - OR \rightleftarrows \equiv M - O - M + ROH \quad (3)$$

In hydrolysis reaction, an alkoxy group [OR] is replaced with a hydroxyl [OH-]. The part of hydrolysed alkoxide molecules may either react with another OH–species by removing water or react with alkoxy group to produce an alcohol molecule. The hydrolysis reaction rate can be accelerated by using catalyst such as HCI and NH_3. Other parameters such as reagent concentration, ageing and drying process are also important for the structure and properties of final materials network for sol–gel method.

Metal alkoxides (M_x $(OR)_y$) and chloride compounds are usually used as starting precursors and dissolved in a suitable solvent, which is often ethanol. After completing the hydrolysis/polycondensation reaction process, wet gel product will be dried to obtain the final material (i.e. Aerogel, Xerogel or Cryogel). Ageing process could help to cushion a warping and cracking phenomenon of the gel matrix. Generally, the drying temperature is in the range of 300–500 °C to remove residual organics. Additional calcination or sintering process (temperature up to 800 °C) can also be applied to remove the organic liquid and densify the product in the final step, if required. Normally, the sintering is used for improving the density and reducing of the pore volume and surface area of as-prepared products. On the other hand, calcination often helps in term of mechanically stable of the materials. Xerogel can be obtained from the uncontrolled drying process. It is characterized by disordered porosity in the absence of structure directing agent. Meanwhile, well-controlled drying process to achieve moderate shrinkage of the gel matrix will offer the aerogel type. Hence, drying is the key process to minimize the impact on the porous structure of gel matrix. Figure 4 shows the schematic of sol–gel process from precursor to aerogel.

Parameters such as the ratio of water to alkoxide, the nature of the R-group (e.g. inductive effects) and amount of catalysts strongly affected the reaction of sol–gel process [55]. As an example of silica preparation, the sol–gel chemical reaction normally requires either base or acid catalysts as the neutral reaction is very slow. The structure of the synthesized gel is substantially different depending on the type of additional catalyst as it can offer different in the relative rates of the hydrolysis and condensation reactions. Example of some metallic nanomaterials that synthesized via sol–gel technique is presented in Table 3.

Advantage: This method is simple, highly controllable, economical and efficient to produce high quality and high yield of nanoparticles. Moreover, it can be used for a thick coating to provide protection layer that help to against corrosion of substrate surfaces or used for thin layer coating to ensure a great bond between the top layer and substrate. This method consumes less energy compared to other techniques, especially top-down techniques. There is no need to reach the melting temperature of the precursors for low temperature reaction.

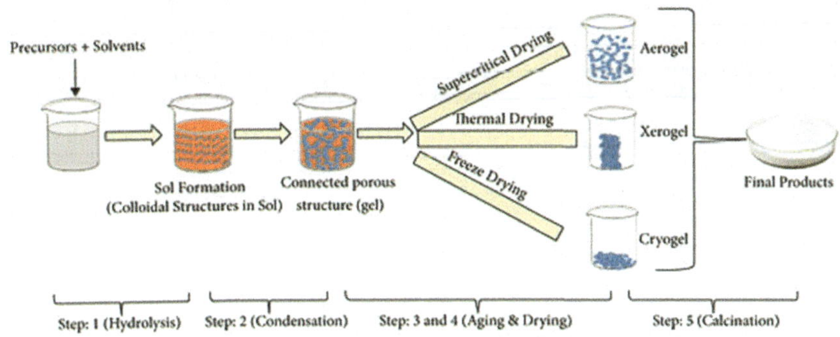

Fig. 4 Schematic of sol–gel process: from precursor to aerogel [54]

Disadvantage: The products often shrinkage and there could be crack formation during drying process, making it difficult to attain a large monolithic piece. The cost of the precursors is high. The process in general involves the use of organic solution that is toxic and harmful to environment. It is hard to avoid residual porosity and OH groups. Finally, the agglomeration issue might occur during the heating process [56].

2.2.3 Chemical Vapour Deposition (CVD)

The key difference between CVD and PVD is that the raw material/percussors for CVD is presented into the reaction chamber in the vapour phase, while the percussors are in solid form for the PVD process. The reaction of CVD process prefers to occur at the substrate rather than in the gas phase. Generally, the temperature ranges from 300 to 1200 °C (higher than PVD process) at the substrate and gas pressures are suggested to be in the range of 0.1 to 1.0 torr. Different factors can affect the deposition process and final produce quality including the precursor delivery method, carrier gas and its chemical properties, the reactor chamber pressure, deposition time and rate, substrate temperature, flow rate of precursor vapour and substrate properties [65]. CVD can be categorized into several types, such as low-pressure, atmospheric-pressure, photo-enhanced, metalorganic and thermal activated CVDs [66].

From Fig. 5, the deposition process can be summarized as three main stages (1) the volatile precursors are introduced to the reactor chamber via carrier gas; (2) the precursor vapours are adsorbed on the substrate surface and then form the intermediate products; and (3) these products are decomposed on the heated substrate, nucleated and grown as the solid layer/grains. The volatile by-products are generated and removed from the chamber by the carrier gas. Growth rate and quality of the

Table 3 Example of some materials that synthesis via sol–gel technique

Metallic nanoparticles	Precursors	Solvent	Catalysts	Polymer skeleton	Ageing temperature and time	Size	Application	References
Au	$HAuCl \cdot 4H_2O$	De-ionized water	$Na_3C_6H_5O_7 \cdot 2H_2O$	–	–	20–40 nm	None specify	[57]
Au-siloxane gel	$HAuCl \cdot 4H_2O$	Methanol and distilled water	Hydrogen tetrachloroaurate(III)	MPTMS and MTMS	Ageing process: 40 °C for 24 h	20–30 nm	None specify	[58]
Au-graphite	$HAuCl_4 \cdot 3H_2O$, $NaAuCl_4 \cdot 2H_2O$ and pure graphite powder	Distilled water	HCl, $Na_3C_6H_5O_7 \cdot 2H_2O$, $NaBH_4$	MTMOS	Room temperature or 80 C	<10–20 nm from $NaAuCl_4$ 10–20 nm from $HAuCl_4$	Amperometric sensing	[59]
Au-Cu-TiO_2	Au_2Cl_6, $Cu(NO_3)_2$, $C_{12}H_{28}O_4Ti$	Distilled water and ethanol	CH_3COOH	–	Ageing: 70 °C for 6 h Calcination: 3 h at 420 °C	3–4 nm of Au NPs	Photocatalyst	[60]
Pt- porous SiO_2	$PtCl_4$	Methanol and acetone	NH_4OH	TMOS	Synthesized at room temperature Calcination: 300 °C for 2 h	5–10 nm Pt NPs	Catalyst in heterogeneous reactions	[61]

(continued)

Table 3 (continued)

Metallic nanoparticles	Precursors	Solvent	Catalysts	Polymer skeleton	Ageing temperature and time	Size	Application	References
Ag-SiO_2	$AgNO_3$	Ethanol	HNO_3	TEOS	Ageing: 25 min at room temperature Drying: 600 °C for 3 h Calcination: 550 °C for 30 min	Average particle size about 10 nm	None specify	[62]
Ag, Cu and Ni	$CuSO_4$, $AgNO_3$, $NiCl_2$	De-ionized water and geraniol	NaOH	Starch, PEG and Gelatin	-	70–90 nm for Cu NPs 80–100 nm for Ag NPs 100–120 for Ni NPs	Antibacterial activity	[63]
Au, Ag, Pt, Pd	HAuCl4, $AgNO_3$, H_2PtCl_6 and $PdCl_2$	Milli-Q water	Sugar	–	Ageing: 2 h at about 70–75 °C	1, 3, 10 and 20 nm sizes for gold, platinum, silver and palladium, respectively	Synthesis technique	[64]

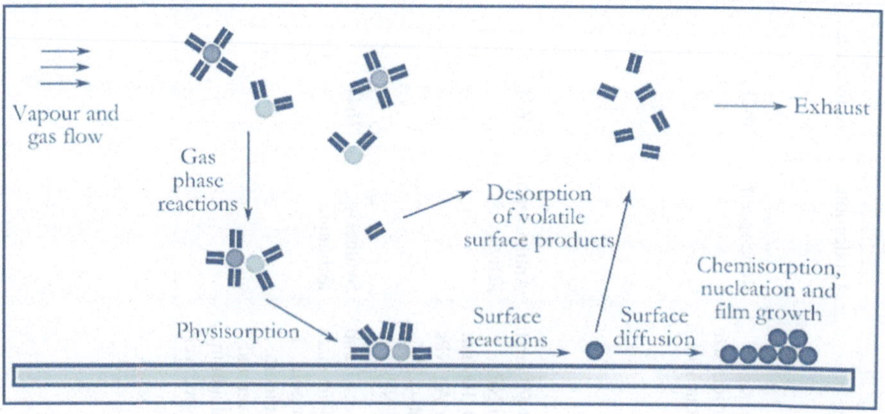

Fig. 5 Reaction in the chemical vapour deposition [67]

thin film depend on the gas pressure inside the reaction chamber and the substrate temperature.

Advantage: CVD offers high uniformity and good reproducibility and adhesion, with acceptable deposition rate [51]. It is not limited to a line-of-sight deposition which is a general characteristic of most of PVD processes. Complex three-dimensional configurations like recesses and holes can be relatively easily coated.

Disadvantage: The synthesis of nanostructures using CVD technique is limited by reason of the involvement of toxic precursors, trace impurities, processing time and prolonged reaction [51]. Possible chemical hazards due to toxic, corrosive and explosive precursor gases. The process typically happened at high temperatures depending on the evaporation temperature of the precursor. Size is limited to reaction chamber capacity.

2.2.4 Hydrothermal/Solvothermal

Hydrothermal/solvothermal synthesis is a solution reaction-based approach, and the reaction temperature could be in a wide range (typically between 100 to 1000 °C). Briefly, the mixed percussor solutions was transferred into lined Teflon and sealed in autoclave made of stainless steel, followed by heating in an oven at a specific temperature and reaction time [68]. Various metal NPs such as Ag, Ni, Cu and Ru [72, 75–78] have been successfully synthesized by using hydrothermal/solvothermal. An image of different NPs is shown in Fig. 6. Crystals with different morphologies (3D-sphere, 2D-rod or 1D-wire structures) are formed by varying the solvent type, concentration of precursor and kinetic control (e.g. reaction time and temperature, etc.).

In general, the reaction medium of hydrothermal synthesis is aqueous solution, while the chemical reaction of solvothermal synthesis happens in various organic

Fig. 6 SEM image of (**a**) Ag NPs, synthesized at 200 °C for 6 h [71], (**b**) Cu NPs, synthesized at 100 °C for 10 h, (**c**) Au NPs, synthesized at 110 °C for 12 h [72], (**d**) CuS nanostructure, synthesized at 150 °C for 5 h [73], TEM image of (**e**) Al/Zn NPs, synthesized at 200 °C for 6 h [74] and (**f**) Ru@C nanocomposites, synthesized at 200 °C for 24 h [75]

mediums. Water has a low boiling point which limits its ability to many reactions that require high-temperature condition. Moreover, water has a high polarity which unable to solubilize many organic and non-polar molecules; hence, capping agents/surfactant (e.g. oleic acid (OA), polyethylenimine (PEI), Ethylenediaminetetraacetic acid (EDTA) and cetyltrimethylammonium bromide (CTAB)) might be needed to stabilize inorganic nanomaterials [69, 70]. Typical precursors for hydrothermal are usually metal-nitrate, -chloride or -acetate compounds.

Advantage: The compositions of nanomaterials can be well simplified and the size, shape distribution and crystallinity in hydrothermal synthesis can be precisely controlled through varying parameters, e.g. temperatures and time of the reaction, the types of precursors, surfactants and solvents. The process can be used to prepare many geometries including bulking powders, single crystals, thin film and nanocrystals.

Disadvantage: Safety issue during the reaction process, high equipment cost (i.e. autoclave) and low products yielded could be the main drawbacks of this technique. Moreover, the reaction process during the reaction cannot be observed. Similar to sol–gel technique, hydrothermal/solvothermal require long time reaction period. This method requires soluble precursor, and large amount of solvent waste can be generated.

2.2.5 Green or Biological Synthesis

Typically, NPs can be achieved by using either microorganisms (such as bacteria, yeast, actinomycetes and fungi) or plant tissues (such as leaf, fruit, root, stem, peel and flower). The synthesis of NPs using a biological system can be categorized into 3 options: (1) using of the solvent medium, (2) using of an eco-friendly and environmentally benign reducing agent and (3) using of a non-toxic material as a capping agent [76]. Several plants and microorganisms have been successfully used for efficient extracellular synthesis of different metal NPs such as Co, Cu, Ag, Au, Pd and Pt as presented in Table 4. Factors including pH, reaction time, reactant concentration and reaction temperature can be adjusted to control the size, shape, yield and stability of as-prepared NPs from the biological synthesis [77].

NPs produced by bacteria offer a very good stability, well dispersion and activities against various pathogens. Fungi are able to produce metal nano- and meso-structures by reducing enzyme intra- or extra-cellularly with the biomimetic mineralization procedure [77]. Using different species of fungi as nano-biofactories for synthesis process is considered more straight forward and easy for stable production of NPs as compared to bacteria. Fungi offers several advantages over bacteria including (1) higher bioaccumulation of metabolites, (2) higher biomass and easy mode of culture, (3) higher wall binding capacity of metals and (4) higher tolerance and uptake capability of metals.[78] Yeasts according to invention are classified into the kingdom Fungi and class Saccharomyces. One main advantage of using yeast cells as NP-carriers is simple encapsulation mechanism which implies that the synthesis process does not require stabilizers compared to other NP-carrier system.

Advantage: Green synthesis eliminates the use of chemicals and offers non-toxic products and by-products. Moreover, it can also generate a large amount of highly stable NPs with a better-size distribution than those chemical and physical methods due to non-nutrient bioactive compounds (as stabilizing agents) that are used in the reaction act [79].

Disadvantage: The large-scale production protocols of this technique require further modification to make them cost-effective and comparable to other methods. Instability and aggregation of NPs, control of morphology, crystallization and size of NPs via this technique are under the development stage. Moreover, the separation and purification process of NPs is another key issue that needs further exploration.

3 Methods Used in Metal Nanoparticles Characterization

Understanding the characteristics of NPs benefits in validating the fabrication technique, succeeding the morphology evolution, improving the method protocols and realizing the potential applications of the NPs. Characterization techniques can be classified based on the concept of the techniques used, the provided information or the types of materials that required for the technique. The main features of the techniques and their key benefit, advantages and limitations are explained and pointed

Table 4 Example of NPs synthesis from various biological sources [80–84]

	Type of metallic nanoparticles	Bio-reducing agent	Size of nanoparticles	Application	References
Plants	Au 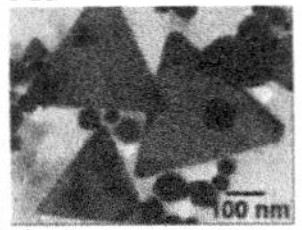 Ag 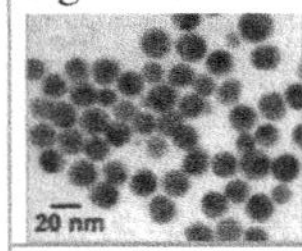	Aloe vera (Leaf)	10–30 nm	Optical coatings and cancer hyperthermia	[85]
	Ag 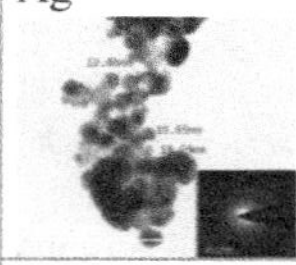	Tectona grandis (seed)	<30 nm	Antimicrobial activity	[86]
	Au 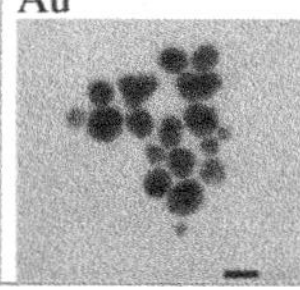	Polyscias scutellaria (Leaf)	~5–20 nm	Catalytic Activity to Reduce Methylene Blue	[87]

(continued)

Table 4 (continued)

	Type of metallic nanoparticles	Bio-reducing agent	Size of nanoparticles	Application	References
	Cu 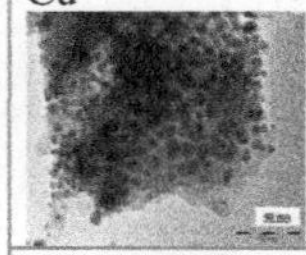	Celastrus paniculatus (leaf)	~2–10 nm	The antifungal activity	[88]
	Cu 	Fortunella margarita (Leaf)	~51–56 nm	Development of synthesis process	[89]
	Cu 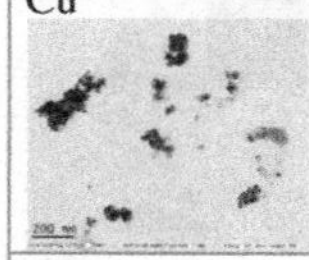	Cissus vitiginea	10–20 nm	The antioxidant and antibacterial activity against urinary tract infections pathogens	[90]
	Co 	Asparagus racemosus (root)	20 nm	Antibacterial activity	[91]

(continued)

Table 4 (continued)

	Type of metallic nanoparticles	Bio-reducing agent	Size of nanoparticles	Application	References
Bacteria	Ag 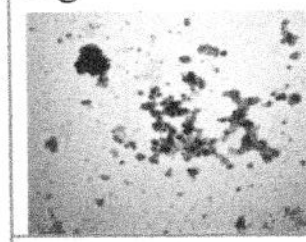	Bacillus cereus	20–40 nm	Antibacterial activities	[92]
	Au 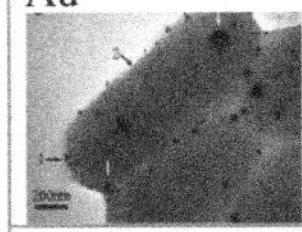	Escherichia coli DH5α	about 20 nm	electrochemistry of haemoglobin	[93]
	Pd 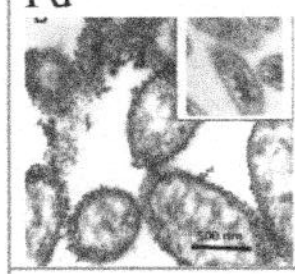	Sulfidogenic	~10 nm	Development of synthesis process	[94]
	Ni 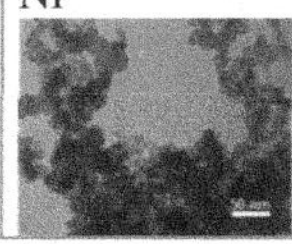	Escherichia coli and Staphylococcus aureus	70–90 nm in average	Magnetic mirror with antibacterial activities	[95]

(continued)

Table 4 (continued)

	Type of metallic nanoparticles	Bio-reducing agent	Size of nanoparticles	Application	References
Fungi	Ag 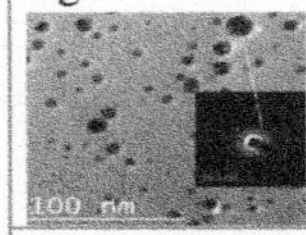	Rhizopus stolonifer	NPs size ~2.86 nm was produced at 40 °C NPs size ~25 and ~48 nm was produced at 20 and 60 °C, respectively	Development of synthesis process	[82]
	Zn 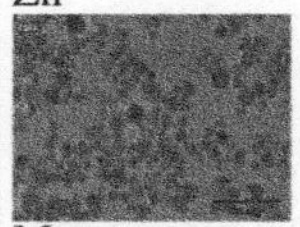 Mg 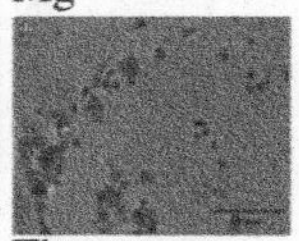 Ti 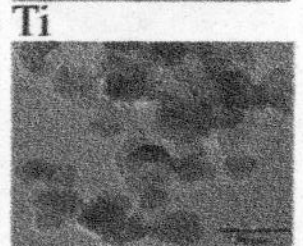	Soil-borne	Zn: 15–88 nm Mg: 10–96 nm Ti: 13–17 nm The size varies depended on various precursor compound	Development of synthesis process	[96]
	Au 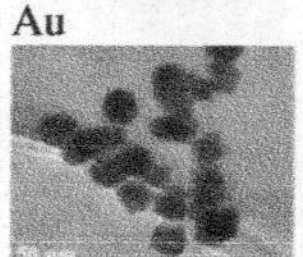	Bacillus niabensis 45	~20 nm	Antibiofilm activity	[97]

(continued)

Table 4 (continued)

	Type of metallic nanoparticles	Bio-reducing agent	Size of nanoparticles	Application	References
Yeast	Au 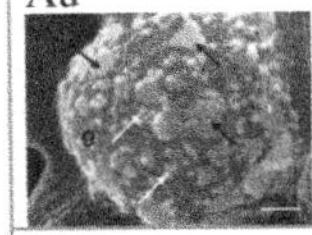	Yarrowia lipolytica NCIM 3589	Varying	Development of synthesis process	[98]
	Zr 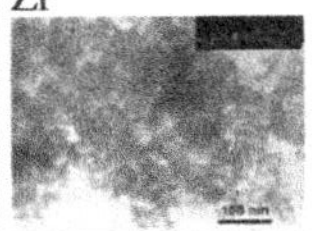	–	2.7 nm	Fuel cell technologies	[99]
Algae	Cu  Ag 	Botryococcus braunii	Ag and Cu NPs were found to be in the range of 40–100 nm and 10–70 nm, respectively	Antimicrobial Activity	[100]

(continued)

Table 4 (continued)

	Type of metallic nanoparticles	Bio-reducing agent	Size of nanoparticles	Application	References
	Ag 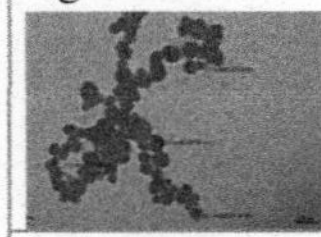	Gelidium corneum	20–40 nm	Antibiofilm activity	[101]
	Au 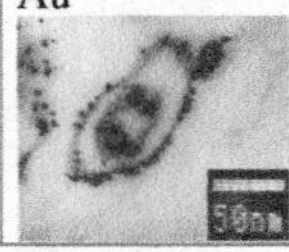	Tetraselmis cochinensis	~15 nm	Catalysis, electronics, coatings	[102]

out in this session. Table 5 summarizes the techniques that are suitable for different focus characterized parameter of NPs.

Table 5 Characterization techniques for NPs parameters [103, 104]

Entity characterized	Suitable characterization techniques
Size (structural properties)	TEM, XRD, DLS, NTA, SAXS, HRTEM, SEM, AFM, EXAFS, FMR, DCS, ICP-MS, UV–Vis, MALDI, NMR, TRPS, EPLS, magnetic susceptibility
Shape	TEM, HRTEM, AFM, EPLS, FMR, 3D-tomography
Elemental-chemical composition	XRD, XPS, ICP-MS, ICP-OES, SEM–EDX, NMR, MFM, LEIS
Crystal structure	XRD, EXAFS, HRTEM, electron diffraction, STEM
Size distribution	DCS, DLS, SAXS, NTA, ICP-MS, FMR, superparamagnetic relaxometry, DTA, TRPS, SEM
Chemical state–oxidation state	XAS, EELS, XPS, Mössbauer
Growth kinetics	SAXS, NMR, TEM, cryo-TEM, liquid TEM
Ligand binding/composition/density/arrangement/mass, surface composition	XPS, FTIR, NMR, SIMS, FMR, TGA, SANS
Surface area, specific surface area	BET, liquid NMR
Surface charge	Zeta potential, EPM
Concentration	ICP-MS, UV–Vis, RMM-MEMS, PTA, DCS, TRPS
Agglomeration state	Zeta potential, DLS, DCS, UV–Vis, SEM, Cryo-TEM, TEM
Density	DCS, RMM-MEMS
Single particle properties	Sp-ICP-MS, MFM, HRTEM, liquid TEM
3D visualization	3D-tomography, AFM, SEM
Dispersion of NP in matrices/supports	SEM, AFM, TEM
Structural defects	HRTEM, EBSD
Detection of NPs	TEM, SEM, STEM, EBSD, magnetic susceptibility
Optical properties	UV–Vis-NIR, PL, EELS-STEM
Magnetic properties	SQUID, VSM, Mössbauer, MFM, FMR, XMCD, magnetic susceptibility

3.1 X-Ray-Based Techniques

- *X-ray diffraction (XRD)* is a basic technique for NP characterization which provides the information of the crystalline structure, phase, lattice and grain size of NPs [105]. XRD is commonly used for powder or thin film samples. The composition of the NPs can be identified by comparing the position of the 2-theta peaks with the standard patterns (i.e. ICDD, or known as JCPDS) database [103]. In general, the intensity of X-ray source is about [108] times lower than that of the electron diffraction.
- *X-ray photoelectron spectroscopy (XPS)* is a highly surface-sensitive analysis method, giving information of electronic structure of NPs and charge transfer between constituent elements in alloy and/or heterostructure NP's surface [106]. Moreover, it is also possible to extract the element and phase composition of the NPs from surface analysis data. Binding energy presented an information of the element, whereas the peak intensity reflects the relative concentration of the elements [107]. Background subtraction and peak fitting methods are required for an analysis of the complex element that has several oxidation states.
- *X-ray absorption spectroscopy (XAS) includes both extended X-ray absorption fine structure (EXAFS) and X-ray absorption near edge structure (XANES, also known as NEXAFS).* XAS is involved in the measurement of X-ray absorption coefficient of NPs as function of energy (Fig. 7.) [108]. In principle, each element has a set of characteristic absorption edges (i.e. K, L, M, etc., absorption edges) according to the different binding energies of its electrons, offering XAS element selectivity. In other words, this technique offers result of the chemical composition, unoccupied electronic states and bonding information of the NPs [109]. A highly sensitive EXAFS techniques can be used to gain the information about the interatomic distances, near neighbour coordination numbers and lattice dynamics. On the other hand, XANE gives information about the oxidation states, vacant orbitals, electronic configuration and site symmetry of the absorbing atom.
- *Energy-dispersive X-ray spectroscopy (EDS) and X-ray scattering (SAXS)* are used for the elemental analysis or chemical characterization of NPs. Each chemical element has characteristic X-ray energy that emitted from the specimen. EDS is typically combined with SEM and TEM, allowing a clear identification of the composition of elements heavier than oxygen.

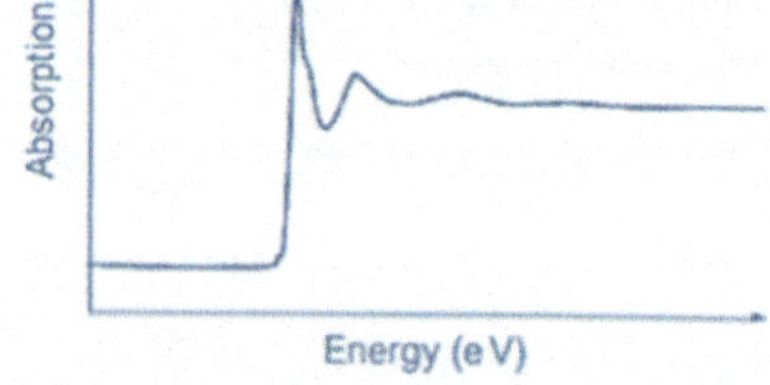

Fig. 7 XAS spectrum shows the edge XANES (within ca. 50 eV) and the edge EXAFS (>1000 eV above) of the structured absorption [110]

3.2 Fourier Transform Infrared Spectroscopy (FTIR)

FTIR is used for studying the vibration of the functional groups associated with NPs. FTIR records the absorption of electromagnetic radiation with wavelengths within the mid-infrared region (400–4000 cm^{-1}) [111]. If a molecule of NPs can absorb infrared region radiation, the dipole moment somehow changed, and the molecule becomes IR active. A spectrum not only offer the information of band position related to nature and strength of bonds, but also determine the specific functional groups, which can help to obtain the information of molecular structures and interactions of NPs. In general, powder samples will be grounded with KBr (about 5% of the weight of the samples) and pressed to form the hard pellet [112].

3.3 Transmission Electron Microscopy (TEM)

The interaction between an electron beam with uniform current density and a thin sample can be observed by using TEM as the beam is transmitted through that thin film samples to form an image. In general, the energies of the electron beam are within a range of 60–150 keV. TEM is the most common technique to analyse morphologies of NPs as it can provide not only direct images of the sample but also the most accurate estimation of the NPs homogeneity due to its powerful magnification with the potential of over 1 million times compared to other SEM techniques.

High-resolution TEM (HRTEM) is an imaging mode of TEM that uses phase-contrast imaging, using the combination of transmitted and scattered electrons to produce the image of the internal structure of NPs (i.e. the arrays of atomic level in crystalline structures of NPs). It is worth mentioning that the internal structure characterization of amorphous-based structure is not always practicable by this HRTEM technique due to the random orientation of their crystals relative to the electron beam; hence, the atom directions are not well aligned, obtaining a complex image that cannot be directly used for structural analysis. Selected area electron diffraction (SAED) or an electron backscatter diffraction (EBSD) in TEM is used for the study of the crystal structure of NPs, obtaining the reverse space of the lattice planes as it can be used to determine the d-spacing value of the crystal planes of the NPs.

3.4 Scanning Electron Microscopy (SEM)

SEM is another common technique for imaging of nanomaterial surface with a resolution down to about 1 nm [113]. SEM uses electrons where the incident beam of electrons transversely scans the sample, offering data on the composition of atoms along with the physical feature of NPs [114]. The focussed beam of electrons can be able to create a magnified image with much improved magnification of 10 to

1,000,000 times. SEM usually uses three analysis modes including (1) secondary electron (SE) mode, (2) backscattered electron (BSE) mode and (3) X-ray energy dispersive spectroscopy [115]. A classic SE mode can obtain image with up to 1 nm resolution. However, SEM is limited when used to characterize non-conductive materials and coloured images (i.e. SEM only provide black and white image), or when used to acquire the height of the specimen. Coating the samples (sputter coating) with an additional conductive material layer of ~3–10 nm (such as Pt, Ag, Pd and Au) is required, especially for non-conductive sample to remove charging effects and get better quality images of the sample. However, the thin layer of conductive materials (from sputtering) may affect the atomic percentage and elemental composition analysis of the NPs.

3.5 *Atomic Force Microscopy (AFM)*

AFM is capable of creating 3D images of surfaces with high magnification. The measurement of this technique is based on the interacting forces between the sample surface and a probe. AFM can scan under three different modes (i.e. contact, non-contact and tapping mode) depending on the degree of proximity between the probe and the sample. Parameters including tip curvature radius and elasticity and surface energy of NPs influence the final topological values. AFM does not require any surface modification or coating prior to imaging; hence, it does not require a coating of conductive layer on the surface of non-conductive materials as SEM technique. Moreover, the topological analysis of small NPs (≤6 nm) can be obtained by AFM without requirements of the special treatment. Low density materials, which present poor contrast in electron microscopy, can also be characterized. For comparison, Fig. 8 shows the image of copper/1,4-benzenedicarboxylate (N-Cu(BDC) composite that taken by AFM, TEM and SEM.

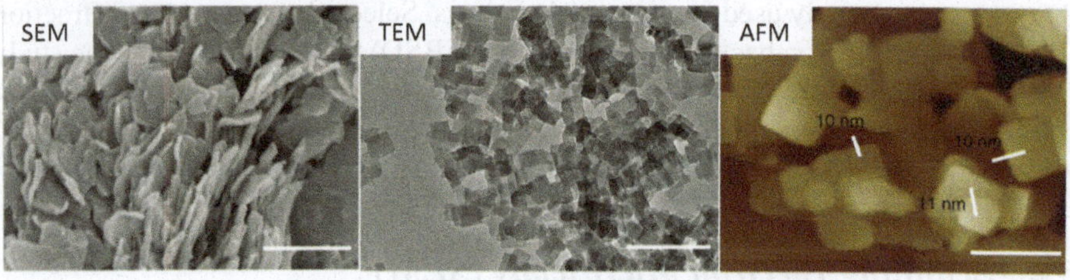

Fig. 8 An example of SEM, TEM and AFM of N-Cu(BDC) composite [104]

3.6 Dynamic Light Scattering (DLS)

DLS provides the measurement of the size of the NPs from the scattered light in a solution; hence, it is difficult to imaging dried samples and very sensitive to agglomeration and dynamic aggregation [116]. As the DLS directly characterizes particles in dispersion, dispersion conditions including the effect of salt or pH on colloidal stability an be monitored [117]. This method can be used to analyse in a great diameter ranging from nano- to micro-scale. However, it cannot directly observe the size of NPs from the DLS result as the mechanism of method is based on diffusion coefficient value, and the spherical shape of the particles is therefore assumed. Hence, the dimensions of a particle can only be defined by using an assumption of shape by Stokes–Einstein equation [115].

3.7 Zeta Potential Measurement

Zeta potential or an electrokinetic potential is a measurement of the "effective" electric charge at the slip plane between the bulk of base solution and the bound layer of diluent molecules that surrounds the NPs. Hence, this technique measures the charge stability of colloidal NPs, which is a key parameter that governs the electrokinetic behaviour of NPs in the solution [118]. In general, NPs with a zeta potential value between -10 and $+10$ mV are considered as neutral, while NPs with zeta potentials value of less than -30 mV or greater than $+30$ mV are considered strongly anionic or strongly cationic [119]. Zeta potential is dependent on pH and the conductivity of the dispersing medium; thus, it is important to accurately measure and report it. For example, if acid is added in nanofluid, pH will decrease, leading to the increase of positive charges on the particle surface. The minimum concentration required for analysis depends on the relative refractive index and particle size. Additionally, this technique some time can be used to predict the long-term stability of colloidal NPs. For example, NPs with zeta potentials larger than $\pm$ 60 mV have great stability, whereas when zeta values of NPs are between -10 mV and $+10$ mV, a rapid agglomeration can be observed unless they are sterically protected [120].

3.8 Secondary Ion Mass Spectrometry (SIMS)

SIMS is one of the key techniques for surface chemical and imaging analysis in the field of material sciences. This technique is recognized as isotopic surface analysis and the most sensitive elemental analysis technique. In the chamber, the sample is bombarded with an ion beam in vacuum atmosphere; thus, the secondary ions are sputtered from the sample and conveyed into a mass spectrometer for analysis [121]. The SIMS technique provides a unique combination of extremely high sensitivity which limits down to the ppm level for all elements from Hydrogen to Uranium and above. Moreover, it also offers high lateral resolution imaging (≥40 nm), and a very low background that allows more than 5 decades of dynamic range.

3.9 UV–VIS Spectrophotometry

It is widely known that metallic NPs possess muti-colours and, therefore, best matched for photo-related applications. Figure 9 shows characteristic colours and properties of Au NPs with the variation of sizes and shapes, which can be utilized in imaging-related applications [105]. Ultraviolet–visible (UV–Vis) and photoluminescence (PL) are the well-known techniques for the optical study of NPs materials [105]. Both techniques offer extra information about the absorption or emission capacity of the NPs and their effect on the overall excitation time of photo-excitons. Additionally, UV–Vis can quantitatively monitor the formation and provide information about the size of NPs through different responses to the electromagnetic waves, ranging from 200–700 nm [122] (Table 6).

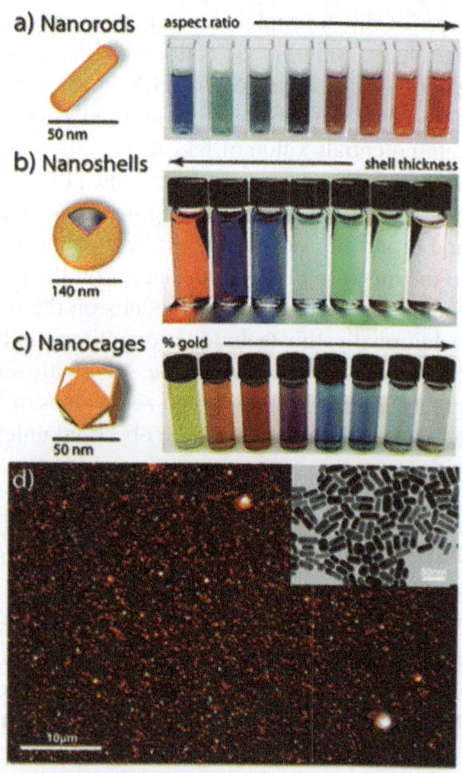

Fig. 9 Colour dependence of Au NPs on size and shape [105]

Table 6 Summary of advantage and limitation of the techniques that are used for NPs characterization in this paper [123, 124]

	Role	Advantage	Limitations	Sensitivity	References
SEM	Topographical, morphological and compositional information (i.e. combined with EDS system)	• Imaging at all directions through x–y-z (3D) rotation of sample is possible • Easy to be integrated with digital camera systems for data storage and analysis • Allows a large amount of the sample to be focussed at one time compared to TEM	• Limited to solid sample and must be operated under vacuum conditions • SEM must be in an area free of any possible electric, magnetic or vibration interference • Low resolution compared to TEM, usually up to sub-micron or a few hundreds of nanometres • Required surface stain-coating with metals for electron conducting	Down to 1 nm	[125]
TEM	Morphology (particle size and shape); crystallographic information (detection of atomic scale defects); compositional information (STEM); the phases that present (i.e. lattice spacing measurement)	• TEM offers the highest resolution compared to other morphological analysis technique • Direct imaging of crystalline lattice of sample • No metallic stain-coating is required to assign the defects inside the sample	• Required ultra-thin sample (thickness below 200 nm) • Sample might be damaged by electron beam during process • Expensive equipment installation, i.e. vacuum system, is required for atomic scale resolution • The difficulty in quantifying muti-particles or misleading images due to orientation effects • Limited to solid sample	0.1 nm	[124, 126]
AFM	Shape, texture and roughness of individual particles and their distribution for an assembly of particles	• Can be operated in vacuum/air/liquid, unlike TEM and SEM that required to operate in the vacuum chamber • High lateral magnification is not required for AFM, while lateral magnification is an important factor in TEM and SEM method • Incapable of characterizing the height or z-axis of NPs	• AFM has slower scanned times than any kind of electron microscopy • Tip shape can become convoluted with measurements • Unable to obtain qualitative or quantitative information of the sample composition (i.e. cannot provide direct chemical information)	1 nm (XY), 0.1 nm (Z)	[104]

(continued)

Table 6 (continued)

	Role	Advantage	Limitations	Sensitivity	References
DLS	Measuring the hydrodynamic size and size distribution of molecules and particles	• Fast, easy handling, non-invasive and used under a wide variety of solvent conditions • Requiring a low volume of the sample and can be easily implemented as an in-line quality control method • less labour intensive and an extensive experience is not required for routine measurement • Fast, no sample preparation and sample preservation from downstream analysis	• Measurement of size is inaccurate with heterogeneous samples • Requires adapted sample dilutions • Offering highly sensitive analysis of NPs, which might be possible to detect unwanted size fractions from unclean lab-ware, dust and aggregated sample • Cannot separate particles and agglomerations	3 nm	[124, 127, 128]
XRD	Size and crystal structure (long range order) measurement of NPs- lattice parameter, internal stress/strain (elastic), coherently scattering domain size (crystallite size)	• The most convenient, inexpensive compared to those other technique • Best method for phase analysis	• High intensity x-ray beam required (i.e. synchrotron x-ray source) • X-ray do not interact very strongly with lighter element • Complex composition of NPs and plasmon cannot be found • The shrinkage of the lattice space by XRD may be complicated, as instrumental parameters, reflection broadening due to a very small NP size and matrix effects can lead to unclear XRD results	Down to 1 nm	[129]
XPS	Qualitative and quantitative elemental composition of the surface layers, chemical state identification and density of electronic states	• High accuracy in identifying samples • Suitable for both conductive and non-conductive materials • Suitable for both inorganic and non-organic	• Very expensive technique • Slow process (30 min–8 h per sample) • H and He cannot be identified • Sample must be compatible with high vacuum environment • 10% relative error in repeated analyses	3–92 nm	[106–107, 130[]

(continued)

Table 6 (continued)

	Role	Advantage	Limitations	Sensitivity	References
XAS	Absorption energy, element valence state, charge transfer and type of bonding information	• Can focus on one element without interference from other elements present in the sample • Ability to analyse almost any type of samples including amorphous (non-crystalline) materials • Determination of the element oxidation state, data which often difficult to obtain by other spectroscopic methods	• Cannot be used to study low atomic number element • Difficult to deconvolute the bulk data when the sample is composed of a mixture of structures of the absorber element	5–6 Å	[108, 131, 132]
EDS	Evaluation of chemical composition of the particles	• Composition of NPs can be analysed via various modes, i.e. Elemental mapping, point analysis, line analysis and area analysis	• EDS is generally not a particularly sensitive technique. If the concentration of an element in the sample is too low, EDS might not be enough to adequately measure its proportion • EDS generally does not work for elements with a low atomic number, e.g. H and He	<2 nm	[104]
FTIR	Nature of bonds and functional groups measurement	• FTIR provides the fastest scan rate among all of the dispersive instruments • The FTIR instrument does not limit the amount of light reaching the detector using a slit, unlike other structure instruments (hence, more energy reaches the sample, thus less reflective losses occur)	• The changes in their spectra are not significant enough to provide clear evidence for nanocomplex formation (because of low sensitivity or peak overlapping) • Structure and size of NPs cannot be measured	20 Å–1 μm	[112, 133]

(continued)

Table 6 (continued)

	Role	Advantage	Limitations	Sensitivity	References
SIMS	Presence of surface coatings or contaminants on collections of nanoparticles, functional groups on surface of NPs	• Can provide molecular information about film and particles surfaces • Determine the presence of trace elements	• Most instruments cannot characterize individual particles, therefore requires a collection of particles • Needs appropriate sample preparation and handling to minimize information loss • Sputter rates accelerated for nanoparticles • Nanoparticles can melt or transform, sputtering can destroy the size, shape and composition of the particles • Requires vacuum	50 nm (x, y), 1 nm (z)	[134]
UV–VIS	Optical properties, concentration and shape of NPs can be measured	• Rapid analysis, reliable, easy to handle, high precision and accuracy • Available on various sample forms, e.g. colloidal solutions, suspensions, thin films and powders	• Solid particles in a heterogeneous sample scatter light more than they absorb it; hence, the data are doubtful. Restriction to turbid samples	UV–visible regions 200–800 nm	[124, 135]

4 Basic Properties of Metallic Nanomaterials

4.1 *Surface Plasmon Resonance Properties*

Among metals, noble metals including Au, Ag, Cu, Pd and Pt gain a great attention due to their exclusive optical properties on account of their surface plasmon resonance (SPR), which can be applied in a broad range of applications such as photocatalysis, biomedicine, surface-enhanced Raman spectroscopy (SERS), plasmonic devices, sensors and photothermal therapy [136, 137]. Among these plasmonic NPs, silver and gold are most used as plasmonic metals because of their chemical stability and visual colour change to the naked eye. On the other hand, Pt or Pd exhibit only broad absorption which continuously extend throughout the near UV and visible range.

Noble metal NPs exhibit a strong UV–vis absorption band, which is not possessed by bulk metals. This absorption band is provided by the collective oscillations of electrons in the conduction band that excited by light with appropriate frequency from both visible and near-infrared ranges, which is known as localized SPR [138]. The position of the plasmon band (extinction spectrum) is best measured on a conventional UV–visible spectrophotometer, observing a band with high extinction coefficients (up to 10^{11}/M.cm) [139]. The optical features of the localized SPR (e.g. peak absorption, peak extinction wavelength, scattering, linewidth) depend on the size, shape, composition of the metal NP, surface-adsorbed species, surface charge, interparticle interactions and the refractive index of the surrounding medium [140]. An example of optical value change of Au and Ag NPs that depend on their shape is present in Fig. 10.

4.2 *Magnetic Properties*

Magnetic properties of metal NPs are of great interest for a wide range of disciplines, such as magnetic fluids, catalysis, biomedicine, magnetic energy storage, information storage and spintronics [141]. Fine particle magnetism comes from size effects, which are based on the magnetic domain structure of ferromagnetic materials. When the size of single-domain particles is reduced below a critical diameter (<15 nm for the common materials), the coercivity is zero and such particles become superparamagnetic, which is caused by thermal effects. Each potential application of the magnetic nanoparticles requires different properties. Materials that hold ferromagnetism (e.g. Fe, Ni and Co) have aligned atomic magnetic moments of equal magnitude; thus, their crystalline structures allow for direct coupling interactions between the moments, which helps to enhance the flux density of material [142]. Many studies have confirmed low toxicity of AuNPs compared to other metal-based NPs [143], Au (3.9 ± 0.2 nm) has been used to decorate peI-Fe_3O_4 (Au@PEI-Fe_3O_4) NPs and their DC magnetization. The estimated effective magnetic anisotropy constant (K_{eff},

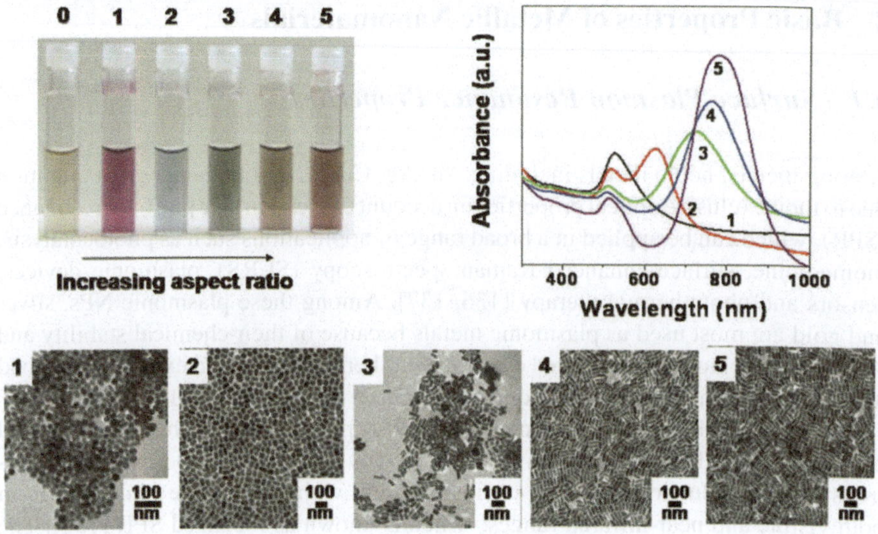

Fig. 10 A different shape of Au and Ag NPs can directly influence the change in their optical performance. Vial 0 presented the aqueous solution of 4 nm Au nanospheres and Vial 1–5 presented the higher aspect ratio gold nanorods [140]

at 5 K) was 2.0×10^4 j/m^3, nearly 50% larger than K_{eff} of bulk magnetite (1.1–1.3 $\times 10^4$ j/m^3) (Fig. 11) [144].

4.3 Mechanical Properties

The mechanical properties of NPs have gained a lot of attention over the last few years especially the tribological properties for lubricants and as reinforcements for composite coating technologies [145, 146]. In general, the mechanical properties of material involve elastic–plastic deformation, hardness, bulk modulus, Young's modulus, scratch resistance, time-dependent creep and relaxation properties, residual stresses, fracture toughness, fatigue and yield strength. Au NPs thin films have been well studied for their high hardness, creep and strain rate effects [147]. However, the obtained results are still insufficient and some are controversial [148, 149]. There is still no conclusion whether the elastic modulus of NPs is affected by the particle size and the indentation depth. Additionally, their frictional and mechanical behaviours have not been fully understood. A study has recently reported the ultra-high compressive strength of the Ni NPs of 34 GPa (D $\approx$ 210 nm), as shown in Fig. 12 [150].

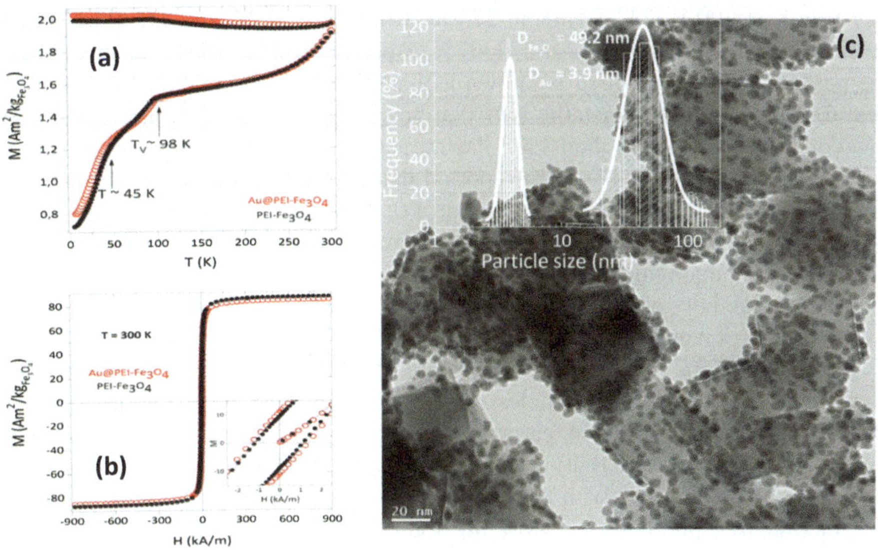

Fig. 11 **a** DC magnetization curves gained in zero-field-cooled (ZFC, lower branch) and field-cooled (HFC = 2.39 kA/m, upper branch) modes for PEI-Fe_3O_4 (filled black circles) and Au@PEI-Fe_3O_4 (open red circles) NPs. **b** M vs. H curves at T = 300 K. Inset: magnification of the low-field region of the hysteresis loops. **c** TEM of Au@PEI-Fe_3O_4 NPs, inset: the histogram of particle fitted with a lognormal distribution (solid line) [145]

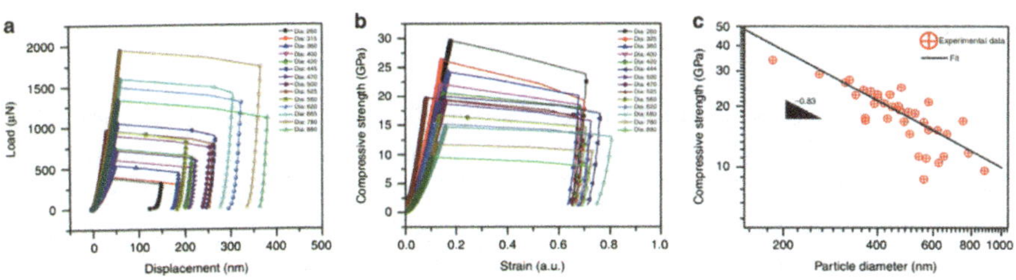

Fig. 12 **a** Load–displacement curves for different particle sizes of NPs. **b** Stress–strain curves demonstrate that the strength increases with decreasing particle diameter. **c** Compressive strength as a function of particle diameter [150]

4.4 *Thermal Properties*

Water, phase change materials, oil and ethylene glycol are typical heat transfer fluids, and they are mostly used as cooling fluid in many thermal engineering applications. In general, the thermal conductivity of a standard heat transfer fluid is lower than <1 W/mK [151]. Among them, water offers the highest value of thermal conductivity (0.613 W/mK at 303 K). This is still not comparable to the value that metal NPs can offer. Ag, Al, Cu, Fe and Au NPs can offer the high thermal conductivity of

429 W/mK at 300 K, 237 W/mK at 273 K, 401 W/mK at 273–373 K, 80.4 W/mK at 273–373 K and 318 W/mK at 273–373, respectively [152, 153]. For this reason, the heat transfer fluid, which is known as nanofluid (a combination of traditional heat transfer fluids with metallic NPs that have size less than 100 nm and at very low concentration of less than 1%), has attracted huge attention to researchers in the past decade [154]. It has been reported that thermal conductivity of water can be improved by 54% through an addition of 0.1% weight of 20 nm Ag NPs [152]. Many reports agree that size, shape and volume fraction of NPs can directly affect the thermal conductivities of typical heat fluids [154, 155]. In the case of Ag, the thermal conductivity of nanofluids increases with decreasing size and increasing the concentration of nanoparticles [156].

For given volume fraction and temperature, the thermal conductivity of nanofluid containing rectangular nanoparticles was higher than those with NPs in spherical shape. Additionally, metal nanoparticles can be added into polymers, such as epoxy, to enhance thermal conductivity and to maintain electrical insulation below percolation threshold. Compared with the electrical properties in metal–polymer composites, thermal transport does not show any dramatic increase [157]. However, similar trend for metal/heat transfer fluids, it has been reported that Ni NPs with the size of 40 nm show higher thermal conductivities of 0.37 W/mK due to its wider-spread aggregation structures in epoxy compared to those larger Ni NPs with size of 70 nm (0.31 W/mK) [158].

5 Conclusion

Bottom-up techniques are more precise than top-down techniques in controlling particle size distribution. Controlling the component concentration, reaction time and temperature can be achieved with different particle sizes and shapes, or NPs. Despite the fact that these growth conditions are extremely sensitive to process variables, limiting the variation of NP morphologies and structures, top-down methods, on the other hand, have advantages over bottom-up methods, such as simple production, the use of non-organic precursors and the high potential for large-scale production with high yields. The size distribution can be controlled via the evaporation rate, system pressure or system geometry. Although many researchers have successfully prepared various types of NPs using various techniques, the precise information of the parameter conditions of each technique that allow to optimize the designed size and shape of various types of NPs is unclear and necessitates additional research for an accurate reproduction process. Mass production could be taken into consideration for further development of the metal NPs. The long-term stability of NPs and their toxicity, especially in biotechnologies, are still unclear. An intensive study must be conducted to investigate the composition, morphology, size, shape, structure and side effects of metal NPs for diverse applications in the fields of magnetic, thermal, optical, etc.

Acknowledgements This work was supported by the Leverhulme Trust Early Career Fellowship (ECF-2021-657) and Nottingham Research Fellowship (A7X164).

Conflicts of Interest There are no conflicts to declare.

References

1. Haritha VS, Balan M, Hosson JTM, Krishnan G (2020) Nanoscale Adv 2:9
2. Krishnan G, Kooi BJ, Palasantzas G, Pivak Y, Dam B (2010) J Appl Phys 107:5
3. Liang L, Wang F, Rong M, Wang Z, Yang S, Wang J, Zhou H (2020) J Mater Sci Chem Eng 8:12
4. Munoz JE, Cervantes J, Esparza R, Rosas G (2007) J Nano Res 9:5
5. Zarrouk T, Nouari M, Salhi JE, Makich H, Salhi M, Atlati S, Salhi N (2022) Int J Adv Manuf Technol 119
6. Dercz G, Matula I, Zubko M, Liberska A (2016) Acta Physica Polonoca A 130:4.7
7. Pradeep NB, Rajath Hegde MM, Manjunath Patel GC, Giasin K, Pimenov DY, Wojciechowski S (2022) J Mater Res Technol 16
8. Hu X, Sun Z, Zhang C, Wang X, Wu K (2018) J Magnes Alloy 6:2
9. El-Eskandarany MS, Al-Hazza A, Al-Hajji LA, Ali N, Al-Duweesh AA, Banyan M, Al-Ajmi F (2021) Nanomaterials (Basel) 11:10
10. Oleszak D, Pawlyta M, Pikula T (2021) Materials (Basel) 14:24
11. Toozandehjani M, Matori KA, Ostovan F, Abdul Aziz S, Mamat MS (2017) Materials 10:11
12. Piras CC, Fernández-Prieto S, De Borggraeve WM (2019) Nanoscale Adv 1:3
13. Lopes WA, Jaeger HM, (2001) Nature 414.
14. Colson P, Henrist C, Cloots R (2013) J Nanomater 2013:948510
15. Kothari R, Beaulieu MR, Hendricks NR, Li S, Watkins JJ (2017) Chem Mater 29:9
16. Kim DE, Sung IH (2013) In: Wang QJ, Chung YW (eds) Encyclopedia of Tribology. Springer, Boston, MA
17. Lipomi DJ, Martinez RV, Cademartiri L, Whitesides GM In: Matyjaszewski K, Moller M (eds) Polymer science: a comprehensive reference. Elsevier BV, Amsterdam, NL
18. Paivanranta B, Langner A, Kirk E, David C, Ekinci Y (2011) Nanotechnology 22:37
19. Auzelyte V, Dais C, Farquet P, Gruetzmacher D, Heyderman L, Luo F, Olliges S, Padeste C, Sahoo P, Thomson T, Turchanin A, David C, Solak H (2009) J Micro/Nanolithogr 8:2
20. Vala M, Homola J (2014) Opt Express 22:15
21. Kim P, David E, Raboin L, Ribbe AE, Russell TP, Hoagland DA (2013) Microsc Microanal 19:6
22. Luo S, Hoff BH, Maier SA, De-Mello JC (2021) Adv Sci 8:24
23. Wu D, Rajput SN, Luo X (2016) Curr Nanosci 12:6
24. Leggett GJ (2011) ACS Nano 5:3
25. Fan P, Gao J, Mao H, Geng Y, Yan Y, Wang Y, Goel S, Luo X (2022) Micromachines 13:2
26. Savale PA, (2016) Arch Appl Sci Res 8
27. Grillo F, Van Bui H, Moulijn JA, Kreutzer MT, Van-Ommen JR (2017) J Phys Chem Lett 8:5
28. Kuwata N, Kawamura J, Toribami K, Hattori T, Sata N (2004) Thin-film lithium-ion battery with amorphous solid electrolyte fabricated by pulsed laser deposition. Electrochem Commun 6:4
29. Saunders SRJ, Nicholls JR (1996) In: Cahn RW, Haasen P (eds) Physical metallurgy, 4th edn.. North-Holland, Oxford, UK
30. Kerdcharoen T, Wongchoosuk C (2013) In: Jaaniso R, Tan OK (eds) Semiconductor gas sensors. Woodhead Publishing
31. Kozhemyakin GN, Kiiko SA, Bryl OE (2019) Crystallogr Rep 64:3

32. Luttge R (2011) In: Luttge R (eds) Microfabrication for industrial applications. William Andrew Publishing, Boston, US
33. Baptista A, Silva FJG, Porteiro J, Miguez JL, Pinto G (2018) Coatings 8:11
34. Suryanarayana C, Prabhu B (2007) In: Koch CC (eds) Nanostructured materials, 2nd edn. William Andrew Publishing, Norwich, NY
35. Shahidi S, Moazzenchi B, Ghoranneviss M (2015) Eur Phys J Appl Phys 71:3
36. Yin Y, Lan C, Hu S, Li C (2018) J Alloys Comp 739
37. Pandey PA, Bell GR, Rourke JP, Sanchez AM, Elkin MD, Hickey BJ, Wilson NR (2011) Small 7:22
38. Fernandez-Arias M, Zimbone M, Boutinguiza M, Del-Val J, Riveiro A, Privitera V, Grimaldi MG, Pou J (2019) Coatings 9:9
39. Donnelly T, O'Connell G, Lunney JG (2022) Nanomaterials 10:11
40. Khalily MA, Yurderi M, Haider A, Bulut A, Patil B, Zahmakiran M, Uyar T (2018) ACS Appl Mater Interfaces 10:31
41. Moret JLTM, Griffiths MBE, Frijns JEBM, Terpstra BE, Wolterbeek HT, Barry ST, Denkova AG, Ommen JRV (2020) J Vac Sci Technol A 38:2
42. Kumar G, Jagirdar Rao V (2013) Int J Nanotechnol Appl 3:1
43. Sun G, Ye G, Wang K, Lou M, Jia X, Xu F, Ye Z (2020) ACS Omega 5:13
44. Ceylan A, Rumaiz AK, Shah SI (2007) J Appl Phys 101:9
45. Gracia-Pinilla M, Martinez E, Vidaurri GS, Perez-Tijerina E (2009) Nanoscale Res Lett 5:1
46. Silva LG, Solis-Pomar F, Gutierrez-Lazos CD, Melendrez MF, Martinez E, Fundora A, Perez-Tijerina E (2014). J Nanomater 2014:643967
47. Ward MB, Brydson R, Cochrane RF (2006) J Physics: Conf Ser 26:296
48. Alexandrescu R, Morjan I, Dumitrache F, Scarisoreanu M, Soare I, Fleaca C, Birjega R, Popovici E, Gavrila L, Prodan G, Ciupina V, Filoti G, Kuncser V, Vekas L (2008) Int J Photoenergy 2008:604181
49. Bendre K, Bhat MP, Lee KH, Altalhi T, Ayad Alruqi M, Kurkuri M (2022) Mater Today Adv 13:100205
50. Spreafico C, Russo D, Degl-Innocenti R (2022) J Intell Manuf 33:2
51. Jamkhande PG, Ghule NW, Bamer AH, Kalaskar MG (2019) J Drug Deliv Sci Technol 53:101174
52. Esposito S (2019) Materials (Basel) 12:4
53. Modan EM, Plaiasu AG (2020) The annals of "Dunarea de Jos" University of Galati, Fascicle IX. Metall Mater Sci 43:1. Accessed 15 Mar 2020
54. Bokov D, Turki Jalil A, Chupradit S, Suksatan W, Javed Ansari M, Shewael IH, Valiev GH, Kianfar E (2021) Adv Mater Sci Eng 2021:5102014
55. Danks AE, Hall SR, Schnepp Z (2016) Mater Horiz 3:2
56. Mahmud NA, Habiballah AS, Affandi NSM, Osman N, Jani AMM (2018) AIP Conf Proc 2031:1
57. Jameel ZN (2017) Energy Procedia 119
58. Hamada Y, Nishi M, Shimotsuma Y, Miura K, Hirao K (2011) IOP Conf Ser: Mater Sci Eng 8:3
59. Ligabue ML, Terzi F, Zanardi C, Lusvardi G (2019) J Mater Sci 54:13
60. Gondal MA, Rashid SG, Dastageer MA, Zubair SM, Ali MA, Lienhard JH, McKinley GH, Varanasi KK (2013) IEEE Photonics J 5:3
61. Ingale SV, Wagh PB, Bandyopadhyay D, Singh IK, Tewari R, Gupta SC (2015) IOP Conf Ser: Mater Sci Eng 73:1
62. Ahlawat DS, Kumari, R, Rachna, Yadav I (2014) Int J Nanosci 13:1
63. Mohindru JJ, Garg UK (2017) Int J Theor Appl Sci 9:2
64. Panigrahi S, Kundu S, Ghosh S, Nath S, Pal T (2004) J Nanopart Res 6:4
65. Piszczek P, Radtke A. (2018) In: Seehra MS, Bristow AD (eds) Noble and precious metals—properties, nanoscale effects and applications. IntechOpen, London, UK. Accessed 20 Dec 2017

66. Xia L (2021) In: Osaka A, Narayan R (eds) Advanced ceramic materials: bioceramics. Elsevier, Osaka, JP
67. Chew CKT (2016) Chemical vapour deposition of gold nanoparticles and metal oxide composites. Dissertation, University College London
68. Ng JJ, Leong KH, Sim LC, Oh WD, Dai C, Saravanan P (2020) In: Abdeltif A, Assadi AA, Nguyen-Tri P, Nguyen TA, Rtimi S (eds) Micro and nano technol: nanomaterials for air remediation. Elsevier, Amsterdam, NL
69. Dunne PW, Munn AS, Starkey CL, Huddle TA, Lester EH (2015) Philos Trans A Math Phys Eng Sci 373:2057
70. Qiu J, Li Y, Jia Y (2021) In: Qiu J, Li Y, Jia, Y (eds) Persistent phosphors from fundamentals to applications: synthesis method. Woodhead Publishing, Oxford, UK
71. Tippayawat P, Phromviyo N, Boueroy P, Chompoosor A (2016) Peer J 4
72. Liu Y, Yang L, Shen Y (2018) J Mater Res 33:18
73. Murugan S, Grace A (2012) J Nano Res 18:1
74. Lozhkomoev AS, Kazantsev SO, Pervikov AV (2020) AIP Conf Proc 2310:1
75. Cored J, Garcia-Ortiz A, Iborra S, Climent MJ, Liu L, Chuang CH, Chan TS, Escudero C, Concepcion P, Corma A (2019) J Am Chem Soc 141:49
76. Parveen K, Banse V, Ledwani L (2016) AIP Conf Proc 1724:1
77. Boroumand Moghaddam A, Namvar F, Moniri M, Md Tahir P, Azizi S, Mohamad R (2015) Molecules 20:9
78. Roychoudhury A (2020) Indian J Pham Biol Res 8:3
79. Sun Y, Wang Q, Chen J, Liu L, Ding L, Shen M, Li J, Han B, Duan Y (2017) Theranostics 7:18
80. Ijaz I, Gilani E, Nazir A, Bukhari A (2020) Green Chem Lett Rev 13:3
81. Zhang D, Ma XI, Gu Y, Huang H, Zhang G (2020) Front Chem 8
82. AbdelRahim K, Mahmoud SY, Ali AM, Almaary KS, Mustafa AEZMA, Husseiny SM (2017) Saudi. J Biol Sci 24:1
83. Guilger-Casagrande M, Lima RD (2019) Front Bioeng Biotechnol 7:287
84. AlNadhari S, Al-Enazi NM, Alshehrei F, Ameen F (2021) Environ Res 194:110672
85. Chandran SP, Chaudhary M, Pasricha R, Ahmad A, Sastry M (2006) Biotechnol Prog 22:2
86. Rautela A, Rani J, Debnath M (2019) J Anal Sci Technol 10:1
87. Yulizar Y, Utari T, Ariyanta HA, Maulina D (2017) J Nanomater 2017:3079636
88. Mali SC, Dhaka A, Githala CK, Trivedi R (2020) Biotechnol Rep 27:e00518
89. Amjad R, Mubeen B, Ali SS, Imam SS, Alshehri S, Ghoneim MM, Alzarea SI, Rasoo R, Ullah I, Nadeem MS, Kazmi I (2021) Polymers (Basel) 13:24
90. Wu S, Rajeshkumar S, Madasamy M, Mahendran V (2020) Artif Cells Nanomed B 48:1
91. Varaprasad T, Govindh B, Venkateswara Rao B (2017) Int J ChemTech Res 10:9
92. Sunkar S, Nachiyar CV (2012) Asian Pac J Trop Biomed 2:12
93. Du L, Jiang H, Liu X, Wang E (2007) Electro Comm 9:5
94. Mikheenko IP, Bennett JA, Omajali JB, Walker M, Johnson DB, Grail BM, Wong-Pascua D, Moseley JD, Macaskie LE (2022) Appl Catal B: Environ 306:121059
95. Ahghari MR, Soltaninejad V, Maleki A (2020) Sci Rep 10:1
96. Raliya R, Tarafdar JC (2014) Int Nano Lett 4:1
97. Li Y, Li Y, Li Q, Fan X, Gao J, Luo Y (2016) J Chem 2016:2781347
98. Pimprikar PS, Joshi SS, Kumar AR, Zinjarde SS, Kulkarni SK (2009) Colloids Surf B: Biointerfaces 74:1
99. Tian X, He W, Cui J, Zhang X, Zhou W, Yan S, Sun X, Han X, Han S, Yue Y (2010) J Colloid Interface Sci 343:1
100. Arya A, Gupta K, Chundawat TS, Vaya D (2018) Bioinorg Chem Appl 2018:7879403
101. Yılmaz Ozturk B, Yenice Gursu B, Dag I (2020) Process Biochem 89
102. Senapati S, Syed A, Moeez S, Kumar A, Ahmad A (2012) Mater Lett 79
103. Mourdikoudis S, Pallares RM, Thanh NTK (2018) Nanoscale 10:27
104. Tiede K, Boxall ABA, Tear SP, Lewis J, David H, Hassellov M (2008) Food Addit Contam: Part A 25:7

105. Khan I, Saeed K, Khan I (2019) Arabian J Chem 12:7
106. Lubenchenko AV, Batrakov AA, Pavolotsky AB, Lubenchenko OI, Ivanov DA (2018) Appl Surf Sci 427
107. Baer DR, Engelhard MH (2010) J Electron Spectrosc Relat Phenom 178–179
108. Yano J, Yachandra VK (2009) Photosynth Res 102:2–3
109. Terzano R, Denecke MA, Falkenberg G, Miller B, Paterson D, Janssens K (2019) Pure Appl Chem 91:6
110. Penner-Hahn JE, X-ray absorption spectroscopy. In eLS
111. Manjumeena R (2018) In: Tiwari A (eds) Handbook of antimicrobial coatings. Elsevier
112. Faghihzadeh F, Anaya NM, Schifman LA, Oyanedel-Craver V (2016) Nanotechnol Environ Eng 1:1
113. Venkatesh N (2018) Biomed J Sci Technol 4
114. Shnoudeh AJ, Hamad I, Abdo RW, Qadumii L, Jaber AY, Surchi HS, Alkelany SZ (2019) In: Tekade RK (eds) Biomaterials and bionanotechnology. Academic Press
115. Zavasnik J, Sestan A, Shvalya V (2021) In: Milacic R, Scancar J, Goenaga-Infante H, Vidmar J (eds) Comprehensive analytical chemistry. Elsevier, Amasterdam, NL
116. Eaton P, Quaresma P, Soares C, Neves C, De-Almeida MP, Pereira E, West P (2017) Ultramicroscopy 182
117. Malm AV, Corbett JCW (2019) Sci Rep 9:1
118. Griffiths D, Hole WBP, Smith J, Malloy A, Carr B (2011) NSTI-Nanotech 1
119. Clogston JD, Patri AK (2011) In: McNeil S (eds) Characterization of nanoparticles intended for drug delivery. Humana Press, Maryland, USA
120. Nanocomposix, Zeta Potential Measurements (2022) https://nanocomposix.com/pages/zeta-potential-measurements
121. Bonnin EA, Rizzoli SO (2020) Front Behav Neurosci 14
122. Wang C, Gao X, Chen Z, Chen Y, Chen H (2017) Polymers 9:12
123. Din M, Arshad F, Hussain Z, Mukhtar M (2017) Nanoscale Res Lett 12
124. Linkov P, Artemyev M, Efimov AE, Nabiev I (2013) Nanoscale 5:19
125. Choudhary OP, Choudhary P (2017) Int J Curr Microbiol Appl Sci 6
126. Smith DJ (2015) Nanocharacterisation. The Royal Society of Chemistry
127. Chiriaco M, Bianco M, Nigro A, Primiceri E, Ferrara F, Romano A, Quattrini A, Furlan R, Arima V, Maruccio G (2018) Sensors 18:3175
128. Lim J, Yeap SP, Che HX, Low SC (2013) Nanoscale Res Lett 8:1
129. Grunder Y, Lucas C (2016) Nano Energy 29:1
130. Baer DR (2022) J Vac Sci Technol 38:3
131. San-Miguel A (2005) Acta Crystallogr Sec A 61.
132. Bak SM, Lin R, Yu X, Yang XQ (2018) NPG Asia Mater 10
133. Thermoscientific, Advantages of a Fourier Transform Infrared Spectrometer (Technical Note). https://www.thermoscientific.com/content/dam/tfs/ATG/CAD/CAD%20Documents/Application%20%26%20Technical%20Notes/Molecular%20Spectroscopy/FTIR/FTIR%20Spectrometers/TN50674-E-0215M-FT-IR-Advantages.pdf
134. Baer DR, Engelhard MH, Johnson GE, Laskin J, Lai J, Mueller K, Munusamy P, Thevuthasan S, Wang H, Washton N, Elder A, Baisch BL, Karakoti A, Kuchibhatla SVNT, Moon D (2013) J Vac Sci Technol A 31:5
135. Khandel P, Shahi KS (2016) Int J Nanomater Biostruc 6:1
136. Fratoddi I, Matassa R, Fontana L, Venditti I, Familiari G, Battocchio C, Magnano E, Nappini S, Leahu G, Belardini A, Li Voti R, Sibilia C (2017) J Phys Chem C 121:33
137. Zhu S, Zhou W (2010) J Nanomater 2010:562035
138. Sau TK, Rogach AL, Jackel F, Klar TA, Feldmann J (2010) Adv Mater 22:16
139. Murphy CJ, Gole AM, Hunyadi SE, Stone JW, Sisco PN, Alkilany A, Kinard BE, Hankins P (2008) Chem Comm 5
140. Mulvaney P (1996) Optical properties of metal clusters By U. Kreibig, M. Vollmer, Springer Series in Materials Science, Advanced Materials. Wiley, Hardcover, DM
141. Issa B, Obaidat IM, Albiss BA, Haik Y (2013) Int J Mol Sci 14:11

142. Diaz C, Valenzuela ML, Laguna-Bercero MA, Orera A, Bobadilla D, Abarca S, Pena O (2017) RSC Adv 7:44
143. Sani A, Cao C, Cui D (2021) Biochem Biophys Rep 26
144. Leon Felix L, Sanz B, Sebastian V, Torres TE, Sousa MH, Coaquira JAH, Ibarra MR, Goya GF (2019) Sci Rep 9:1
145. Luo X, Morrin A, Killard AJ, Smyth MR (2006) Electroanalysis 18:4
146. Hussain F, Hojjati M, Okamoto M, Gorga RE (2006) J Comp Mater 40:17
147. Maharaj D, Bhushan B (2014) Beilstein J Nanotechnol 5
148. Guo D, Li J, Chang L, Luo J (2013) Langmuir 29:23
149. Ritter C, Heyde M, Schwarz UD, Rademann K (2002) Langmuir 18:21
150. Sharma A, Hickman J, Gazit N, Rabkin E, Mishin Y (2018) Nat Comm 9:1
151. Mbambo MC, Khamlich S, Khamliche T, Moodley MK, Kaviyarasu K, Madiba IG, Madito MJ, Khenfouch M, Kennedy J, Henini M, Manikandan E, Maaza M (2020) Sci Rep 10:1
152. Iyahraja S, Rajadurai JS (2015) AIP Adv 5:5
153. Farid N, Sedigheh A (2017) J Adv Mater Proc 5:2
154. Liu M, Ma Y, Wu H, Wang RY (2015) ACS Nano 9:2
155. Navarrete N, Gimeno-Furio A, Mondragon R, Hernandez L, Cabedo L, Cordoncillo E, Julia JE (2017) Sci Rep 7:1
156. Warrier P, Teja A (2011) Nanoscale Res Lett 6:1
157. Guoqing Z, Yanping X, Hui W, Yu T, Guoliang T, Shantung T, Haiping W (2009) J Comp Mater 44:8
158. Li X, Park W, Chen YP, Ruan X (2013) J Heat Transfer 139:2

Advantages and Disadvantages of Metal Nanoparticles

Sanjay Kumar, Bharat Kumar, Rishabh Sehgal, M. F. Wani, Deepak Kumar, Mukund Dutt Sharma, Vivek Singh, Rakesh Sehgal, and Vijay Kumar

Abstract Nanoparticles have scientific importance because they act as a link between bulk materials and atomic or molecular structures. The physical properties of the bulk material are constant because of its size, but in the case of nanoparticles, size-dependent characteristics are observed. Particle size reduction to the nanoscale reveals unexpected and improved characteristics such as particle size distribution and shape. The large surface area of nanoparticles is attributable to their remarkable and unexpected characteristics, and this exceeds the contributions made by the small bulk of the materials. Due to these properties, nanoparticles have been used in different industries such as energy, electronics, optical, medicine, mechanical, and pharmaceutical in recent years. Different approaches are being used for the synthesis of nanoparticles, such as bottom-up and top-down approaches. Metal nanoparticles are increasingly being used in biomedicine and other fields around the world. Because of their prominent features, metal nanoparticles, nanostructures, and nanomaterial production are currently gaining the attention of researchers. Various aspects of the synthesis of metal nanoparticles and the advantages and disadvantages of metal nanoparticles are summarized in this chapter.

S. Kumar · M. F. Wani · M. D. Sharma · R. Sehgal
Department of Mechanical Engineering, National Institute of Technology Srinagar, Srinagar 190006, Jammu and Kashmir, India

B. Kumar · D. Kumar
Center for Automotive Research and Tribology, Indian Institute of Technology Delhi, Delhi 110016, New Delhi, India

R. Sehgal
Department of Electrical and Computer Engineering, University of Texas at Austin, Austin, Texas 78751, USA

V. Singh
Department of Mechanical Engineering, National Institute of Technology Hamirpur, Hamirpur 177005, Himachal Pradesh, India

V. Kumar (✉)
Department of Physics, National Institute of Technology Srinagar, Srinagar 190006, Jammu and Kashmir, India
e-mail: vj.physics@gmail.com

S. K. Tiwari et al. (eds.), *Nanoparticles Reinforced Metal Nanocomposites*,
https://doi.org/10.1007/978-981-19-9729-7_7

Keywords Metal nanoparticles · Bottom-up approach · Top-down approach · Advantages of nanoparticles · Disadvantages of nanoparticles

1 Introduction

Nanotechnology is a branch of science concerned with the fabrication of nanoparticles with average particle sizes ranging from 1–100 nm using various synthesis techniques [1]. Because of their small dimensions, nanomaterials have unique features such as excellent electrical, mechanical, and thermal characteristics, a large surface area, and good optical and magnetic properties [2, 3]. Nanomaterials are being used in various fields nowadays, such as molecular biology, physics, mechanical engineering, material science, wastewater technology [4, 5]. Various nanostructures and hybrid nanomaterials are currently being studied by many researchers for the preparation of composite polymers [6], disease diagnosis and treatment [7], sensor technology [8, 9], catalysis [10], and labeling of optoelectronic recorded media due to their unique properties [11]. Figure 1 depicts a few metal nanoparticles that are crucial in a range of applications. The first category of nanomaterials includes the metals such as titanium, platinum, silver, copper, gold, zinc, magnesium, iron, and alginate nanoparticles. Metal oxide nanoparticles, such as titanium dioxide, silver oxide, and zinc oxide, make up the second category of nanomaterials. The third type of nanoparticle is thought to include doped metal, metal oxide, and metal nanomaterials [12, 13]. Furthermore, metal–organic frameworks (MOFs) and metal sulfide nanomaterials have drawn a lot of interest because of their intriguing properties and potential uses in a number of biological fields. Nanoparticles made of Zn, Cu, AgS, CuS, FeS, MOF, and Mn-based MOF are frequently used in drug delivery and have antibacterial properties [14]. The creation and stability of metallic nanoparticles are frequently achieved by a variety of physical and chemical techniques [15–17]. Their morphology (structure and size), stability, and physicochemical properties are strongly influenced by choice of metallic nanoparticle preparation processes, such as the kinetics of metal ions' interaction with the reducing agent, the adsorption process of a stabilizing agent with metal nanoparticles, and various experimental techniques [17].

In mechanical engineering, using nanoparticles as additives in nanofluids and nanolubrication is a relatively new concept. The major benefit of employing nanolubricants is that they are generally temperature-insensitive and have fewer tribochemical interactions [18]. Another benefit of using nanoparticles in lubricating oils is that they do not get trapped in filters [19]. Another application of these metal nanoparticles in the mechanical sector is metal matrix nanocomposites (MMNCs). The nanomaterials have a significant impact on the characteristics of metal matrix composites (MMCs). It is now simple to create MMNCs with a variety of increased properties thanks to recent advancements in the synthesis of nanoparticles, such as those with a size smaller than 100 nm. It is anticipated that MMNCs will be able to overcome a

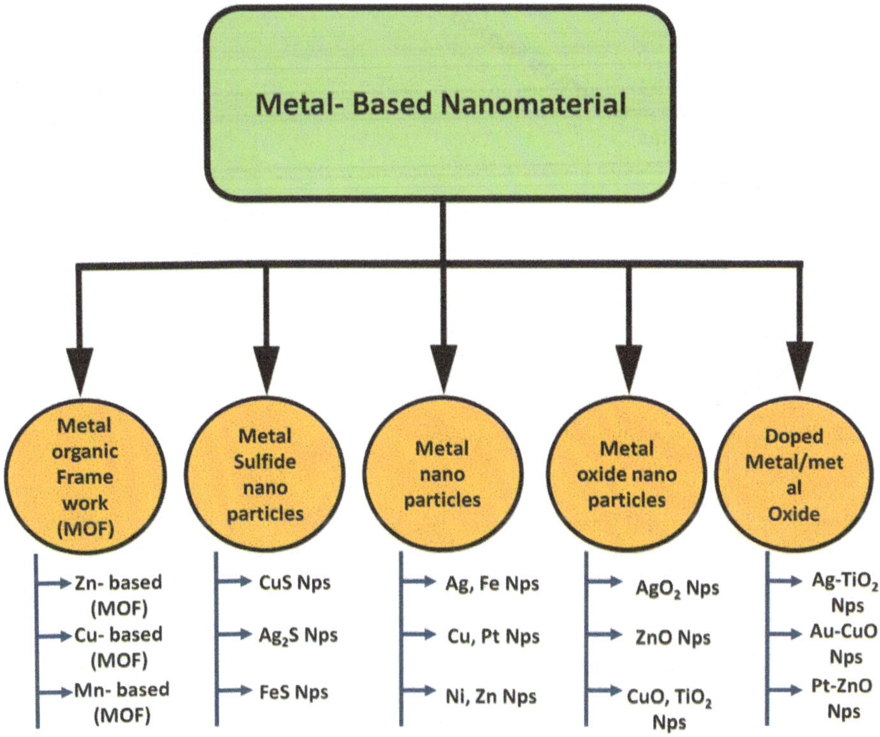

Fig. 1 Different types of metal-based nanomaterial

few MMC drawbacks, including limited ductility, machinability, and fracture toughness [20]. Because of a combination of Orowan strengthening, grain refinement, and high-temperature creep resistance, MMNCs have dramatically improved mechanical properties, especially when lightweight metals like Al or Mg are used as the matrix [21, 22]. Metal nanoparticles are progressively being used in medical, engineering, and related fields across the world. In this context, the present chapter is an assortment of the different roots of the fabrication of metallic nanoparticles, the advantages and disadvantages of these metallic nanoparticles, and their application.

2 Routes of Fabrication

According to the starting material for nanoparticle preparation, two categories of approaches are employed to synthesize metallic nanoparticles: a bottom-up approach and a top-down approach. The starting material, as depicted in Fig. 2, is the main distinction between the two procedures. Atoms or molecules are used as the starting material in the bottom-up technique, whereas bulk material is used in the top-down

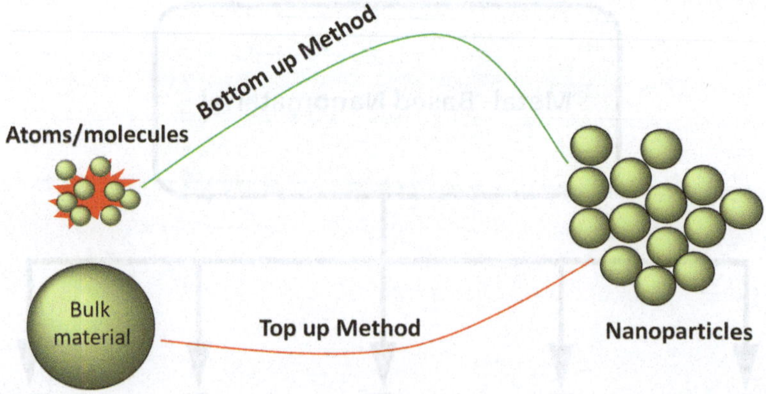

Fig. 2 Routes for the fabrication of nanomaterials

approach. Top-down approaches use a variety of physical, chemical, and mechanical methods to reduce particle size from micro- to nanoscale [23, 24].

2.1 *Top-Down Approach*

This method results in the creation of small nanoparticles from bulk material. The first step in the creation of nanoparticles is the size reduction of the starting material using a variety of physical and chemical procedures [25]. Processes, including ball milling, mechanochemical synthesis, laser ablation, and ion sputtering, are examples of top-down methods. Although the top-down method is easy to apply, it cannot produce very small and irregularly shaped particles. This method has a significant problem with changes to the surface chemistry and physicochemical properties of nanoparticles. The advantages and disadvantages of the top-down approach are summarized in Table 1.

2.1.1 Ball Milling Technique

A powerful method for creating high-strength materials with a precise microstructure and metastability is ball milling [26]. Reduced particle size is achieved through mechanical milling, also referred to as high-energy ball milling [27]. This method was developed by Johan Benjamin in 1970; it reduced particle size while simultaneously creating novel phase and surface qualities. High-energy ball milling is a common method for creating nanoparticles, and it is particularly popular for creating intermetallic nanoparticles. The ball milling method applies significant mechanical energy to the bulk powder by means of these high-speed rotating balls. The bulk powder is placed in a container with a number of heavy balls. Several types of ball

Table 1 Advantages and disadvantages of the top-down approach

Top-down approach	Advantages	Disadvantages
Ball milling technique	This method can be used to create high-purity nanoparticles with exceptional physical properties on a wide scale [29, 31]. This method gives the component several new and enhanced qualities [29]	It takes a lot of energy and a long time to grind something [32]
Mechanochemical synthesis	The use of nanoparticles in this technique is relatively straightforward and effective [35]	Long-term milling is necessary for the creation of smaller-sized particles (less than 20 nm) [35]
Laser ablation technique	A method that is both simple and effective for creating lots of nano-sized particles in suspension. Nanoparticles can develop in a liquid medium without the need for a surfactant [30]	Long-term laser ablation causes the colloidal solution to produce a lot of nanoparticles, which restrict the laser path and lower the ablation rate [36]
Ion sputtering technique	In this method, the target material's composition and the sputtered material's composition are the same [31] Impurities are created in smaller amounts than those created by chemical processes Better compositional alloy nanoparticles can be produced than those from other chemical reduction techniques [37] This method is flexible and can produce large-sized ionic nanoparticles [38]	Sputtering gases (He, Ne, Ar, Kr, and Xe) alter the surface morphology, composition, texture, and optical properties of nanocrystalline [39]

mills, such as low-energy tumble mills, attrition mills, vibrating mills, planetary mills, and high-energy ball mills, can be used to reduce particle size (Fig. 3).

2.1.2 Mechanochemical Synthesis

Mechanochemical synthesis is a process that involves repeatedly deforming, welding, and breaking a reactant mixture. At the interface of nano-sized particles, many chemical changes occur during the milling process. For a variety of reasons, including isolating the reacting phases from the product phase, high temperatures are often needed to initiate chemical reactions. In a ball mill, nanoparticles can be produced at low temperatures without the need for extra heating [28]. The reactants are distorted, separated, and welded during the milling process. High-temperature reactions can

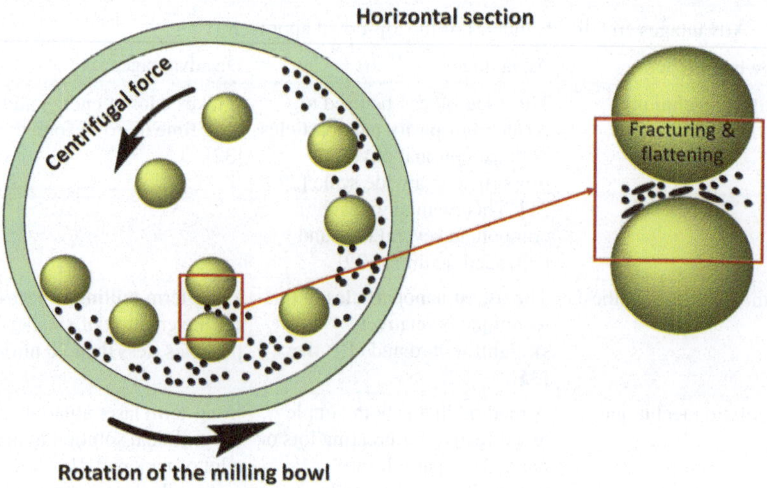

Fig. 3 Schematic diagram of ball milling technique

happen at low temperatures without the use of external heat sources, thanks to the numerous chemical reactions that occur at the interface between the substrate and the reagent's surface [29].

2.1.3 Laser Ablation Technique

The particle size is reduced to the nano-level by using the laser ablation technique and pulse laser irradiation. Ti: sapphire (titanium-doped sapphire) lasers, Nd: YAG (neodymium-doped yttrium aluminum garnet) lasers, and copper vapor lasers are the most often utilized lasers in this technology [30]. The solid material disintegrates into tiny particles known as nanoparticles when it is exposed to laser light. These particles remain in the surrounding liquid and combine to produce a colloidal solution. The laser pulse duration and energy are used to calculate the quantities of ablated atoms and particles produced [31]. The ablation effectiveness and characteristics of produced metal particles are influenced by several parameters, such as the effective surrounding liquid medium, wavelength, laser fluency, laser pulse duration, and ablation time [32].

2.1.4 Ion Sputtering Technique

In this method, the solid material is vaporized by sputtering with an ion beam from an inert atmosphere. Using magnetron sputtering of metal targets, this technique was recently employed to create nanoparticles from several different metals [33]. Utilizing sputtering gas, sputter deposition occurs in a sealed vacuum chamber at a constant operating pressure (between 0.05 and 0.1 mbar) (argon). High voltage is delivered to the target (cathode) during the sputtering process. As a result of the high voltage, a magnetic system pushes free electrons in a spiral direction, where they smash with atoms of the sputtering gas (argon), which causes gas ionization. This ongoing process produces a glow discharge (plasma), which is used to ignite the fuel. The target is shifted in the direction of the positively charged gas ions, which continue to impact it [34]. In a vacuum container, persistent collisions between gas molecules and metal atoms result in atom scattering and the creation of a diffuse cloud [34].

2.2 *Bottom-Up Approach*

Biological systems, where nature has employed chemical forces to produce nearly all of the structures required for existence, serve as an inspiration for bottom-up approaches. Nanomaterials are created using a bottom-up approach by combining atoms or tiny molecules, as well as by using forces that operate at the nanoscale for chemical and physical reactions. Bottom-up methods are very important and an alternative to the top-down method in nanofabrication as the particle sizes get smaller. Additionally, in this procedure, the final nanoparticle is created by combining the nanostructured building blocks that were initially created [40]. The advantages and disadvantages of the bottom-up approach are summarized in Table 2.

2.2.1 The Physical Vapor Deposition Technique

"Physical vapor deposition" refers to a group of vacuum deposition techniques used to create thin films and coatings. The process of a substance transitioning from a condensed to a vapor phase and then returning to a thin-film condensed phase is known as physical vapor deposition. Thermal evaporation and sputtered deposition are two methods used in the physical deposition technique to deposit material as a thin film or as nanoparticles on a surface under the control of a highly regulated vacuum [41]. A solid target is ablated with a laser in pulsed laser deposition, producing plasma of ablated molecules. The resulting film, as depicted in Fig. 4 [42], is subsequently put on a substrate. Carbon nanotubes are frequently coated with thin coatings and metal nanoparticles using this technique. PVD thin-film technology includes processes including reactive evaporation, ion plating, and electron-beam or hot-boat evaporation.

2.2.2 Chemical Vapor Deposition Technique

Using the CVD technique, solid nanomaterials of higher quality and greater performance can be created. This method of deposition forms a thin film on the surface of

Table 2 Advantages and disadvantages of bottom-up approach

Bottom-down approach	Advantages	Disadvantages
Physical vapor deposition technique	An easy method for producing thin metal films	This is very expensive method for creating the nanoparticles
Chemical vapor deposition technique	Nanostructure particle characteristics such as surface morphology and crystal structure can be altered in this technique The chemical vapor deposition coating technique has good film durability and controlled surface morphology	Chemical hazards are possible as a result of toxic, corrosive, and explosive precursor gases In this technique, multicomponent material deposition is difficult
Sol–gel approach	Easiest approach. Controlling particle size and shape is achieved by systematic monitoring of reaction parameters [49]	
Hydrothermal approach	Desired size and shape of nanoparticles can be prepared [35] Well-crystallized powder with high crystallinity can be formed [50]	Hydrothermal processes are difficult to control In this approach, there are limitations in terms of reliability and reproducibility [35]

a target material by utilizing a chemical reaction and a gaseous molecule containing atoms that are advantageous for film formation [43]. The precursor material, which is the target material, is released as an explosive molecule. A series of chemical interactions between the precursor, substrate surface, and explosive molecule result in the development of a thin-film coating. In this technique, a surface chemical process is commonly used to produce atomic layer deposition thin films [43–45]. The instrument's schematic is shown in Fig. 5. Through the control valve, the gas stream is introduced into the dynamically pumped vacuum chamber at a controlled rate. The pressure in the chamber is kept low by high-speed pumping (1–50 mbar). The precursor's molecules begin to break up and mix during the brief holding period in the heated tube to generate tiny clusters or particles. The cluster or particle beam rapidly expands near the reactor's exit in order to prevent particle growth and agglomeration. The powders can then be scraped off and collected while the particle beam condenses on a rolling platform that is chilled with liquid nitrogen [46, 47].

2.2.3 Sol–Gel Approach

The sol–gel method of making nanomaterials entails one of three steps: (a) open mixing of metal and metal oxide or nanoparticles in a pre-hydrolyzed silica sol;

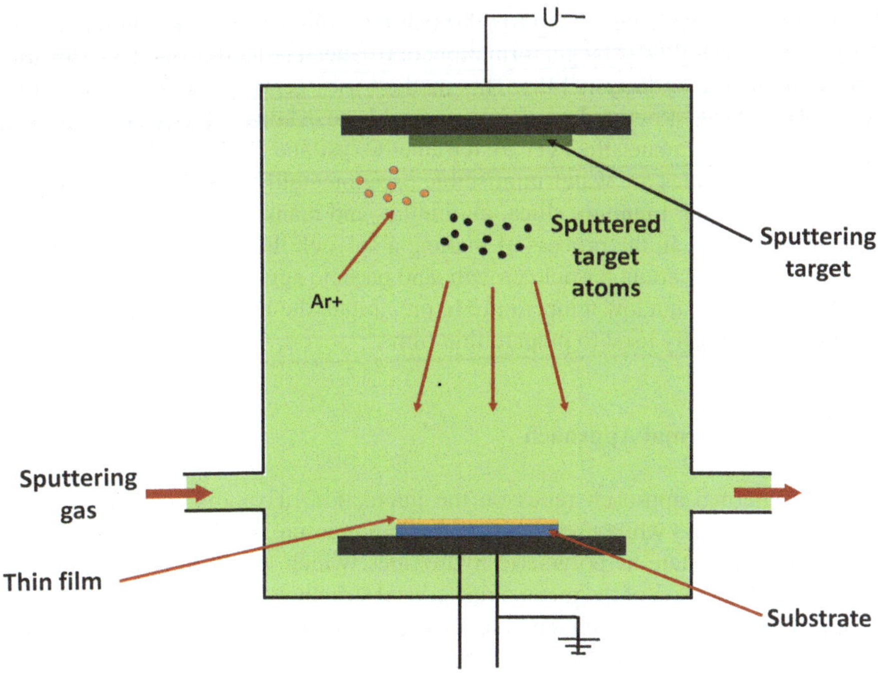

Fig. 4 Schematic image of physical vapor deposition technique

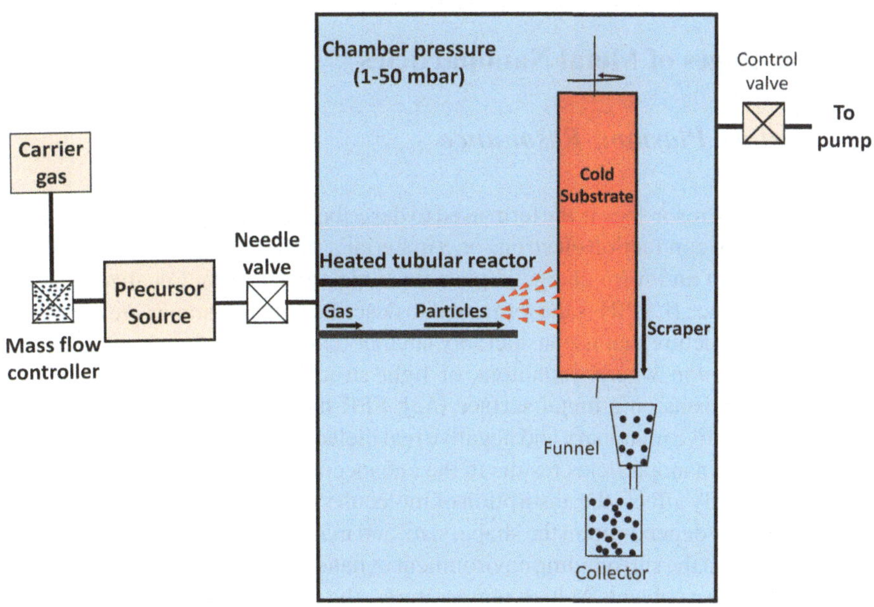

Fig. 5 Schematic diagram of chemical vapor deposition technique

(b) mixing of prefabricated metal (oxide) colloids with a sol containing the matrix-forming species, followed by gel formation; or (c) metal reduction and complexation with silane before hydrolysis [48]. This method uses gelatin and colloidal suspension (sol) to create a network in a continuous phase. The two types of silane most frequently used to create silica gel are tetramethoxysilane (TMS) and tetraethoxysilane. Metal alkoxides are water-immiscible organometallic precursors for a variety of metals, including titanium, silica, aluminum, and many others, where alcohol is a common solvent. In the process of making a sol–gel, there are four basic stages: hydrolysis, condensation, particle growth, and particle agglomeration [24]. Usually, silica sol is heated at a low temperature to precipitate the metal oxide particles, and this method is mostly used to prepare thin films.

2.2.4 Hydrothermal Approach

The hydrothermal approach relies on the interaction of vapors from an aqueous solution with a solid while under extremely high pressure and heat. In this process, the cations precipitate as polymeric hydroxides, which are further dehydrated to promote the formation of the metal oxide crystal structure. The base is introduced to the metal salt solution without complex hydroxide being produced, resulting in the development of the second metal cation, which helps control the particle formation process [35].

3 Advantages of Metal Nanoparticles

3.1 Surface Plasmon Resonance

Surface plasmon resonance is the term used to describe the resonance effect caused by the oscillation of conduction electrons on a material's surface when they interact with the incident light and form energetic plasmonic electrons (SPR). Localized surface plasmon resonance (LSPR) is the term used to describe when plasmons are contained within a nanostructure that is considerably smaller than the wavelength of light [51]. It is a phenomenon whereby photons of light struck at a specific angle activate conduction electrons on a metal surface [52]. SPR has been observed in materials with a small positive imaginary and negative real dielectric constant. The excitation of SPR by metallic nanoparticles results in the enhancement of a local electromagnetic field, which highly affects the absorption of molecules on the surface of nanoparticles [53]. SPR highly depends upon the shape, size, and material of nanoparticles [53, 54]. It also depends on the surrounding environment of nanoparticles, such as the dielectric constant of air or a solvent. At high temperatures, the tips of triangular nanoparticles become rounded and ultimately hemispherical, which reduces the electromagnetic (EM) enhancement. SPR and LSPR occur mostly in metals such as gold and silver.

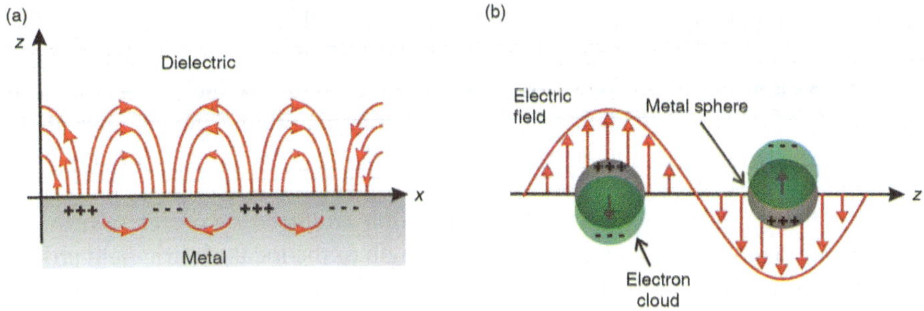

Fig. 6 Schematic diagrams of: **a** SPR and **b** LSPR. Reused with permission from [53]

LSPR is also dependent on the interaction between the surrounding nanoparticles, and this is measured by the change in the interparticle spacing of nanoparticle arrays. LSPR is shifted toward a lower wavelength when the lattice spacing reduces [53], as shown in Fig. 6. A strong electric field is produced when metal nanoparticle surfaces are present, which has sparked a lot of interest in the study of the plasmonic excitation of metal nanoparticles in recent years. The detection of chemical signals is one application for this characteristic [54].

Pereira et al. [55] reported the optical properties of a nanocomposite system consisting of plasmonic Au nanoparticles embedded in a CuO semiconductor matrix, resulting in the coupling between localized surface plasmons (LSPs) and excitons (free carrier interbond transition). The shape of the absorption band line is non-Lorentzian, which is similar to the typical Fano spectrum in the case of the overlapping of narrow band exciton transitions with nanoparticle surface plasmon resonance (SPR). Such nanostructures, beyond the improved catalytic properties, also enhance the sensitivity.

3.2 *Surface-Enhanced Raman Scattering*

Being a weak phenomenon, the Raman effect has some disadvantages. Raman scattering happens to one out of every 107 photons. Low analyte concentrations or inadequate Raman scattering both contribute to the comparatively weak Raman signal. Sometimes, the Raman signals are hidden by the molecule's intense fluorescence. Surface-enhanced Raman scattering (SERS) objective is to considerably boost the Raman signals from molecules [56]. SERS is a method that produces considerably increased scattering effectiveness when molecules are adsorbed on metal colloidal nanoparticles or rough metal surfaces [57]. In some cases, by factors of 108 or even more, molecules' inelastic light scattering is dramatically amplified when they are adsorbed onto corrugated metal surfaces, such as silver or gold nanoparticles, allowing for single-molecule SERS (NPs). A powerful imaging approach known as tip-enhanced Raman scattering (TERS), which combines SERS with the tips of

AFM or STM techniques, has also been developed. SERS varies from many other analytical techniques due to the wealth of vibrational spectroscopic information it provides. As a result, it has found use in a number of fields, including electrochemistry, catalysis, biology, medicine, art conservation, materials science, and others. Since the low intensity of Raman scattering is evidently insufficient for many practical applications, it is frequently advantageous to find methods to improve the Raman process. The significant optical field augmentation produced by correctly resonant components enables such a technique. The strength of the local electric field around the molecule, which plasmons in noble metal nanostructures can greatly increase in comparison to the intensity of incoming light, is particularly important for the early absorption phase. It is preferable to use a more complex structure, such as sandwiching the molecules between two metal particles in the space between them (referred to as hotspots), which enables intensities of up to EF 105–106 to be achieved on a regular basis [58]. Although SERS can be achieved by increasing the electric field at a single NP, this method is not as effective as using a more complex structure.

Saviello et al. recommended using Raman and SERS spectroscopy to examine various blue and black writing inks on paper surfaces. Raman and SERS spectroscopies were used to analyze black and blue writing inks that were applied to the paper in order to identify the primary dye components. SERS measurements were performed by placing a microscopic droplet of plasmonic nanomaterial on the analytical surface. When compared to Raman spectra and SERS spectra obtained at 514 nm illumination wavelength using Au nanopaste, blue and black Bic and Staedler inks treated with Ag nanopaste produced SERS spectra with significantly higher intensities. The interaction of EM, CT, and MR effects was said to be the main cause of the noticeable improvement. SERS also evaluated the deterioration of blue writing inks using simulated aging techniques that involved exposure to light and high temperatures. Interestingly, it was discovered that different types of ancient colored paper had very different colors despite having an initially comparable blue tone. The Pilot paper had very slight color differences, the Bic paper had a darker blue hue, and the Staedler-colored paper was almost completely discolored. The Staedler pen included both triarylmethane dye and phthalocyanine dye, whereas the Bic pen contained solely triarylmethane dye. This was due to the different chemical compositions of the inks. In light of the spectrum findings, a degradation pathway involving N-demethylation of triarylmethane components was proposed. However, the extreme browning in the Staedler pen showed that photoreduction processes were also producing colorless leucon forms [59].

3.3 Enhanced Rayleigh Scattering

Rayleigh scattering is the elastic scattering of radiation by particles that are much smaller than the wavelength of the radiation and is inversely proportional to the fourth power of the radiation's wavelength [60]. Backscatter reflectometry uses the refractive index variation of the optical distribution characteristics along the sensing fiber to measure temperature and strain. It has been widely used in civil engineering and aerospace engineering for structure health monitoring. Because the sensor used to measure these parameters depends on the intensity of backscattered light, one of the method's drawbacks is the low strength of the scattering signal due to the inherent refractive index variation of the optical fiber [61]. Backscatter reflectometry can benefit from the addition of nanoparticles (NP) into the optical fiber's core to enhance the scattered light signals. Blanc and Dussardier [62] reported the techniques for making NP-doped silica optical fiber and used Rayleigh scattering to measure the optical intensity loss. Molardi et al. [63] conducted additional research on the characteristics of MgO-doped optical fiber and achieved a 50 dB amplification of scattered light in the laboratory. They suggested that this kind of spatial multiplexing might be used in optical backscatter reflectometry. Beisenova et al. [64] presented the experimental results of 3D shape sensing in a spatial multiplexing device for an epidural needle using a MgO-doped optical fiber. Bulot et al. [65] showed the enhanced stability of temperature sensing at a high temperature by employing zirconia-coated gold NP-doped optical fiber in optical frequency domain reflectometry (OFDR) distributed sensing. These studies show that doping NPs into the core of optical fibers for distributed sensing is favorable and practicable.

3.4 Targeting Capabilities

Recently, efforts to enhance biological detection and imaging have been sparked by nanomaterials because of their distinctive passive, active, and physical targeting characteristics. The concentrations of local contrast agents have risen a little in tumors as a result of the enhanced permeability and retention (EPR) effects of nanoparticles in tumors [66]. The size of a nanoparticle is among its most crucial characteristics for tumor imaging. Nanoparticle size influences biodistribution, blood circulation half-life, cellular absorption, tumor penetration, and targeting [67]. Nanoparticles smaller than 10 nm are easily excreted by the kidneys since the normal renal filtration pore is 10 nm in size [68, 69]. On the other hand, macrophages, which are found in organs with the mononuclear phagocyte system (MPS), such as the lymph nodes, liver, spleen, and lung, are able to recognize nanoparticles larger than 100 nm [70]. In addition, numerous reviews discovered that nanoparticles with diameters ranging from 10 to 60 nm consistently enhanced cellular absorption [67, 71]. By precisely binding to target receptors in lesions, nanoparticle surface labeling with different ligands for target receptors might improve imaging contrast agent localization in addition to passive targeting techniques [72–74]. For the purpose of imaging prostate cancer cells, the CT density of gold nanoparticles coated with a prostate-specific membrane antigen RNA aptamer has been shown to be greater [75]. Additionally, lung tumors can be targeted by MRI using nanoscale superparamagnetic iron oxide agents that have been surface coated with a high-affinity anti-EGFR antibody [76].

3.5 Biocompatibility

Due to their similar size (nanoscale range), biocompatibility, and simplicity of interaction with receptors, proteins, and nucleic acids, metal nanoparticles have been used widely in biology and medicine [77, 78]. Additionally, it has been found that metal nanoparticles can bind medicines, peptides, nucleic acids, antibodies, and targeted agents after being properly functionalized, as illustrated in Fig. 7 [79, 80]. Additionally, numerous studies have demonstrated that metal nanoparticles are eliminated through urine and feces, demonstrating their biodegradability [81–85]. Because of this, researchers have spent a lot of time over the past few decades developing alternative therapeutic and diagnostic nanomedicine approaches using metal nanoparticles to treat a number of diseases such as cancer, diabetes, ischemia diseases, neurodegenerative diseases, and others [79, 86–88]. Numerous papers [89, 90] claim that these metals' nanoparticulate forms, such as nanoparticles, nanorods, nanospheres, may be used for in vivo medical applications due to their distinct and peculiar physicochemical and biological properties in comparison to their bulk forms.

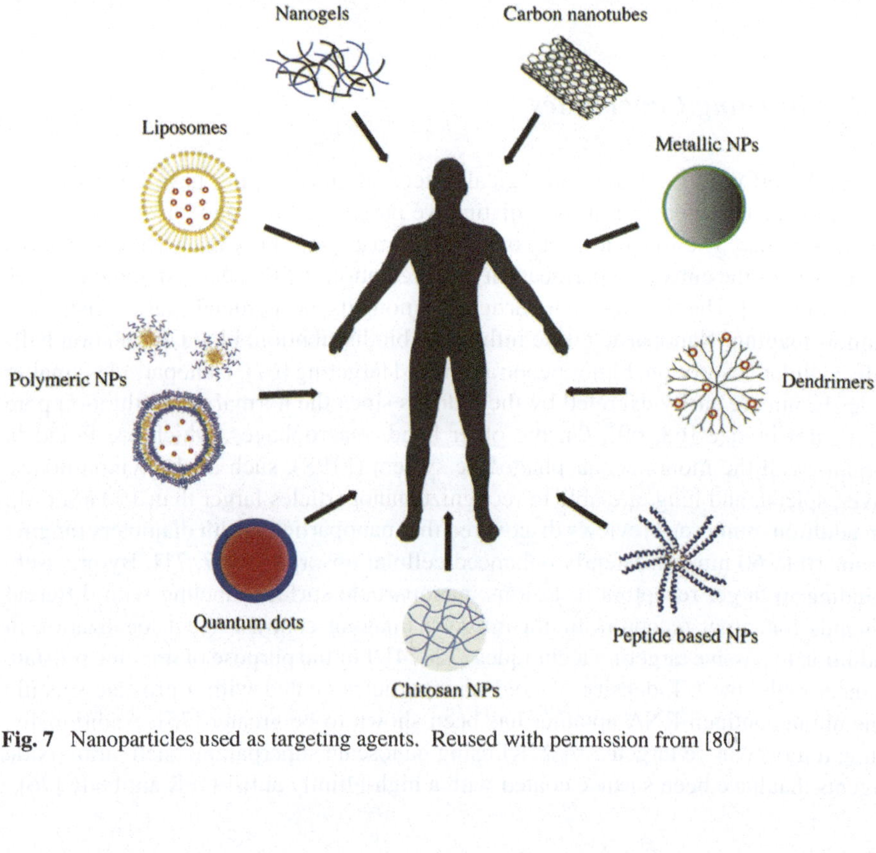

Fig. 7 Nanoparticles used as targeting agents. Reused with permission from [80]

3.6 *Anti-friction and Antiwear Effect in Nanolubrication*

Nanomaterials have the potential to be additives for lubricants that are good for the environment. Compared to organic chemicals, which are typically used as lubricant additives, they have a number of advantages. Nanomaterial additions can aid in energy conservation, improve dynamic transmission efficiency, dampen vibrations, and reduce noise by increasing lubricity, lowering friction, reducing wear, and boosting load-bearing capacity. Nanoparticles have been used as lubricant additives to improve traditional lubricants due to their small-scale effects, temperature insensitivity, and high tribological qualities [91]. Different nanoparticle additions can work in different ways. By reducing wear and friction through the colloidal effect, rolling effect, small-size effect, protective layer effect, and third-body effect, lubricants containing nano-additives can improve boundary lubrication [92]. Soft nanomaterials function as cohesive third bodies that can be positioned between two surfaces. Under stress, they exhibit extreme deformation, and they predominantly adhere to surfaces and shear in the bulk medium to adjust for changes in surface velocities. Hard particles with limited cohesion, known as granular nanomaterials, can maintain their spherical shape under stress and correct for variations in surface velocity by sliding and rolling at low shear rates [93]. The degree to which friction and wear can be decreased depends on the characteristics of nanoparticles, including size, shape, and concentration [94].

3.7 *Combined Propertied in Nanocomposites*

Because of the potential applications for nanomaterials in electronics, nonlinear optics, and magnetics, there has been a lot of interest in these fields recently. The extensive distribution of the tools needed to characterize these materials has facilitated development in this area. Because they have the potential to exhibit unusual mechanical, electrical, or chemical behavior, nanocomposites, which combine the properties of two or more distinct materials on a nanoscale, are of significant interest. Metal nanoparticles (NPs) have properties that set them apart from bulk metal, such as peculiar electrical, optical, and chemical behavior, as a result of quantum size effects. Shape, size, interparticle spacing, and the dielectric environment can all be controlled using different methods. Additionally, attempts are being made to organize NPs in two and three dimensions in order to build nano-electric devices based on these materials. The habitat for metal NPs in conjugated materials with p-conjugated backbones is particularly notable. Recently, a range of metals, conjugated polymers, and oligomer linkers have been used to create nanocomposites of conjugated materials and metal NPs [95].

4 Disadvantages of Metal Nanoparticles

4.1 Particle's Instability

Despite the frequent use of nanolubricants in friction testing, little is known regarding their stability and degree of dispersion in viscous liquids [96–99]. Colloidal nanoparticles are thermodynamically unstable in ambient circumstances because of their high Gibbs energy [100]. Primary nanoparticles interact with one another to generate aggregates that eventually settle, as shown in Fig. 8 [101]. When nanoparticles are subjected to chemical, mechanical, and thermal stressors while an engine is running, the likelihood of agglomeration increases [101].

Research on colloidal stability under challenging and ambient circumstances is essential to prevent a change in the thermal conductivity and tribological properties of the nanolubricant over time [96, 98, 99, 102]. It is obvious that strong liquid dispersion of the nanoparticles is necessary for optimal colloidal nanoparticle stability. The topic of inadequate nanoparticle dispersion in viscous liquids has been discussed in the scientific community. When Mahbubul et al. came across inconsistent assertions in the literature about the influence of chemical and physical factors on colloidal viscosity, they turned to poorly dispersed colloids to help explain their findings [97]. These poorly distributed colloids are created through a two-step synthetic process that includes synthesis and then dispersion in oil. In order to obtain high nanoparticle dispersion, the nanoparticles are first mechanically processed using ultrasonication or a planetary ball mill [103]. Prior to tribological studies, the level of dispersion

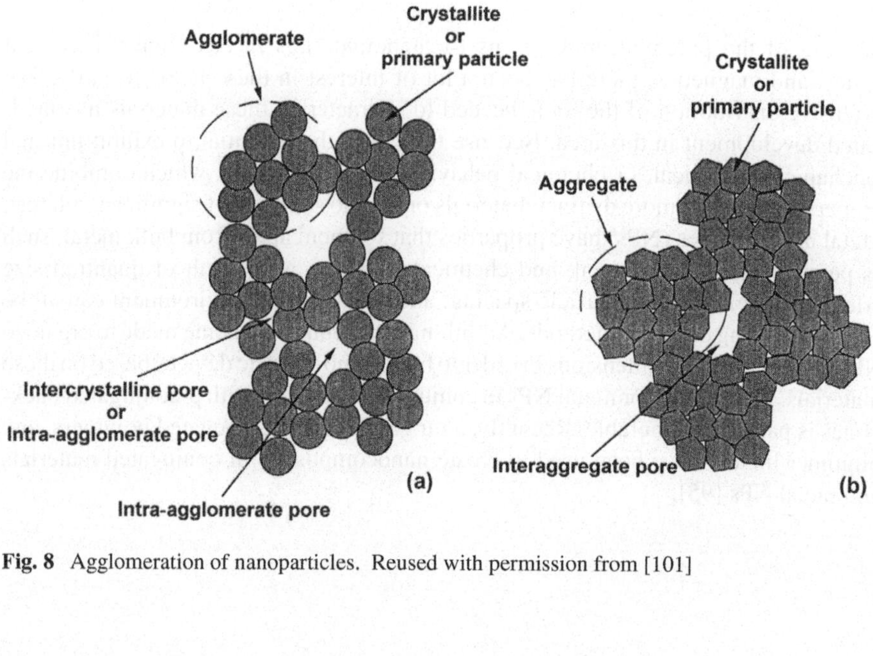

Fig. 8 Agglomeration of nanoparticles. Reused with permission from [101]

was not further studied. Most researchers choose this two-step approach due to its simplicity [97], even if it is improper for metals [103].

4.2 *Toxicity in the Human Body*

Serious concerns concerning the safety of using manufactured nanoparticles in human applications are emerging as they are used increasingly frequently in biological applications. Due to their tiny size and unique characteristics, nanoparticles (NPs) are frequently used in nanomedicine and as therapeutic nanocarriers [104, 105]. Figure 9 shows how many of their characteristics, including their size [106], morphology, surface functional groups [107], and dose-dependent qualities [108], may make them poisonous to healthy, normal human cells, tissues, and organs. Because they contain synthetic chemicals as surface functional and capping agents, chemically synthesized nanoparticles (NPs) are much more hazardous to human cells than biosynthesized nanoparticles with benign surface functional groups [109]. However, some biologically generated nanoparticles have the potential to cause harm when they come into contact with cells, degrade into smaller sizes, or aggregate [110, 111].

The ratio of surface area to volume rapidly increases as nanoparticle size falls, boosting biological and chemical reactivities [112]. For instance, when the NP size is reduced from 30 to 3 nm, the percentage of expressed surface molecules rises from 10 to 50% [113]. Contact between a nanomaterial's surface and a biological component causes cytotoxicity. As a result, depending on surface area and particle

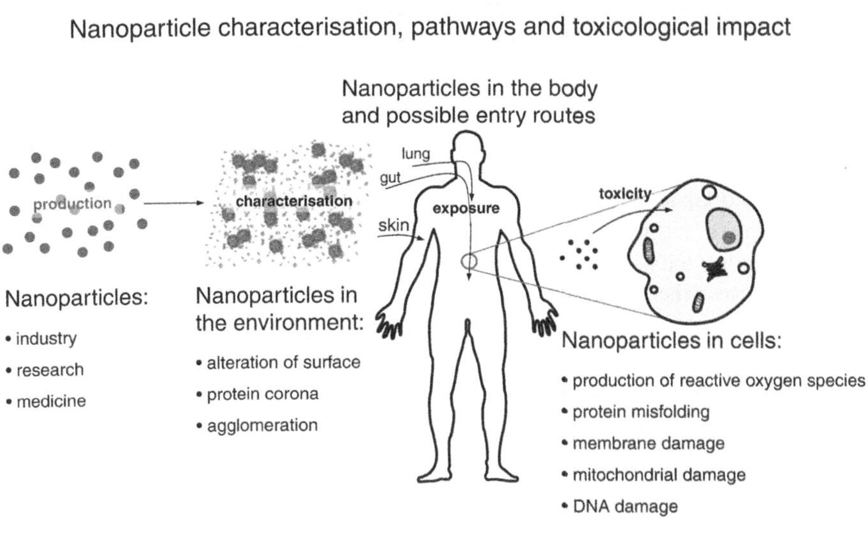

Fig. 9 Toxicity effect of nanoparticles on the human body. Reused with permission from [106]

size, nanoparticles with the same chemical composition may exhibit considerably varying levels of cytotoxicity. To put it another way, NPs pose a greater risk than larger particles with similar compositions [114]. The epithelial cells of the stomach and gill showed cellular uptake of spherical-shaped carbon nanomaterial and tubes constructed of multi-graphitic sheets but not of cube-shaped carbon nanoparticles [115]. In zebrafish (Danio rerio) embryos, silver nanoplates were said to be more harmful than silver nanospheres [116]. Cells absorb spherical nanoparticles more frequently than other types [117]. Compared to gold nanospheres, gold nanorods induce reduced autophagosome accumulation [118]. The cytotoxic effects of gold nanoparticles (gold NP) on human fetal osteoblast, osteosarcoma, and pancreatic duct cell lines were investigated by Steckiewicz et al. [119] using the MTT assay. The gold nanoparticles that resemble stars are the most hazardous for human cells [114].

4.3 *Toxicity in Plants*

Plants have a huge basic impact because of photosynthesis and the release of oxygen into the atmosphere. Plants may be more impacted by NP contamination than other living things since all of their parts, including their roots, branches, and leaves, are in close contact with the environment's matrix, which includes the air, water, and soil. The stomatal openings on leaves allow NPs from the air to enter the plant, whereas roots are preferentially able to absorb NPs from water and soil [120]. It has been established that NPs harm plants and obstruct their growth. The primary contributor to plant toxicity is ROS production, which leads to lipid peroxidation, damages DNA, and lowers the amount of soluble protein, photosynthetic pigments, and biomass in plants [288]. Although enzymatic and non-enzymatic antioxidants are a part of plants' defense mechanism against oxidative stress, they may lose their effectiveness at higher oxygen concentrations [121]. In Dunaliella tertiolecta and Chlorella vulgaris green algae, Ag NPs demonstrated enhanced levels of lipid peroxidation and reactive oxygen species (ROS) formation [122]. Additionally, Pithophora oedogonia and Chara vulgaris were shown to be adversely affected by Ag NPs in terms of growth and development [123]. DNA damage in plant cells can be used to detect genotoxicity, whereas morphological and physiological alterations can be utilized to assess the phytotoxic effects of NPs on plants. The phytotoxicity levels of NPs on plants can be determined using a variety of physiological indicators, such as germination, biomass output, leaf number, photosynthetic capacity, root and shoot lengths. The simplest method for observing how genotoxic compounds affect plants is to examine the cytology of plant roots for the detection of mitotic index, chromosomal abnormalities, etc. [124].

4.4 Contamination During Synthesis

Ionic liquids based on imidazolium are increasingly exploited as intriguing dual-function solvents and stabilizers in the production of metal nanoparticles [125–128]. For instance, 1-butyl-3-methylimidazolium tetrafluoroborate (BMIM-BF4) has been successfully used to create silver nanoparticles (Ag NPs) of varied sizes and shapes using a range of reducing agents [129–131]. Ionic liquids are thought to stabilize metal nanoparticles by coordinating to their surfaces in a manner similar to that of conventional surface ligands, but it is generally believed that these interactions are much weaker than those of conventional ligands (such as thiols, carboxylates, and amines) [127]. Ionic liquids based on imidazolium have a condensed phase that appears to be very well ordered, with supramolecular networks stacking between imidazolium rings and connected by hydrogen bonds [132]. The local hydrophilic and hydrophobic zones that are produced as a result of the prolonged ordering of ionic liquids based on imidazolium may affect the size and shape of the developing metal nanoparticles [133, 134]. It is well known that impurities in reagents and solvents can change the quality of the nanoparticles made in different syntheses from batch to batch. For instance, it has been shown that halide impurities in a surfactant, which are necessary for the production of gold nanorods, substantially limit the formation of the nanorods [135, 136]. A common solvent used in quantum dot synthesis has been demonstrated to include certain impurities that are necessary for the creation of high-quality Cd-Se nanocrystals [137]. According to Lazarus et al. [138], silver nanoparticles in various batches of commercially produced BMIM-BF4 with similar preparation showed significantly different LSPR bands in their UV–vis spectra. This revealed that potential impurities in the ionic liquid BMIM-BF4 had a negative impact on the final nanoparticle quality, which is consistent with past research that demonstrated how highly susceptible the catalytic performance of ionic liquids is to impurities. Purified BMIM-BF4 was spiked with trace amounts of 1-methylimidazole, chloride, and water, which are all known to be typical impurities in imidazolium-based ionic liquids, in order to examine the impact of impurities on the production of Ag NP. Each of these contaminants has a negative impact on the quality of Ag NP.

4.5 Challenges in Their Synthesis

For the synthesis of size-, shape-, and composition-controlled noble metal NPs, solution-based batch techniques, such as the reduction or breakdown of the salt precursor in the presence of capping and stabilizing agents, have proved extremely successful [50, 139–143]. Despite the advancement and achievement, there are still some difficulties. Diluted media are widely used in solution-based syntheses, resulting in suspensions with low nanoparticle concentrations. Scaling up and concentrating the suspension through precipitation and/or centrifugation are essential for many practical applications. Reproducibility problems could result from the scale up, and the necessary workup could change or agglomerate the nanoparticles. To get rid of excess surfactants and unreacted precursors, purification is also necessary. The synthesis of NPs can be extremely sensitive to changes in the reaction parameters and microscopic quantities of ions in the solvent or solution, which presents another problem with repeatability [144]. Another factor is the customary requirement for high temperatures in solution-phase synthesis techniques such as polyol [145, 146] and the Turkevich or Frens processes [147–150]. Compared to the current stage of development of solution-phase synthesis, the mechanochemical synthesis of noble metal NPs still faces a number of challenges. The reactivity of solid reductants differs from that of the solution phase (it remains poorly understood how different reducing agents affect the kinetics and shape of the resulting NPs). The diffusion mechanisms in the liquid and solid phases are altered. Surfactants or stabilizers should interact with the growth species differently in the mechanochemical synthesis than they do in the solution phase. The reactivity, reduction kinetics, and morphology/size of produced nuclei can all significantly vary as a result of these events, which do not take place in mechanochemistry. Before reduction or breakdown produces the growth species from the metal NP precursors, the metal might dissociate or form dimers in solution [139].

5 Conclusions

Based on the many origins of fabrication, the benefits and drawbacks of synthesis techniques, and metal nanoparticles, this chapter can be concluded. Although the top-down method produces nanoparticles using very simple methods, it consumes a lot of energy. A bottom-up strategy can produce the correct shape and size, but it is costly to make thin films using this strategy. Nanoparticles' surface plasmon resonance feature is particularly helpful in improving metals' surface sensitivity and in molecular detection methods. Metal nanoparticles are highly advantageous to utilize in the optical industry due to their improved Rayleigh scattering property, such as in the doping of nanoparticles in the core of the optical fiber. They are helpful in the realm of biological detection and imaging, such as the identification of cancers, due to their strong targeting capabilities. They are frequently utilized in medication delivery, medicine, and the treatment of cancer because of their good biocompatibility. Nanoparticles are excellent for nanolubrication because of their rolling impact. Due to their great tensile strength and toughness, metal nanoparticles are frequently employed in nanocomposites. It is highly challenging to synthesize nanoparticles that are precisely sized and devoid of contaminants. One of the biggest issues facing the lubrication business is the aggregation of nanoparticles in lubricants. Chemically produced nanoparticles have the potential to be extremely harmful to plants and humans and can be toxic in nature.

References

1. Jamkhande PG, Ghule NW, Bamer AH, Kalaskar MG (2019) Metal nanoparticles synthesis: an overview on methods of preparation, advantages and disadvantages, and applications. J Drug Deliv Sci Technol 53:101174. https://doi.org/10.1016/j.jddst.2019.101174
2. Khoshnevisan K, Maleki H, Honarvarfard E, Baharifar H, Gholami M, Faridbod F, Larijani B, Faridi Majidi R, Khorramizadeh MR (2019) Nanomaterial based electrochemical sensing of the biomarker serotonin: a comprehensive review. Microchim Acta 186:1–21
3. Yaqoob AA, Parveen T, Umar K, Mohamad Ibrahim MN (2020) Role of nanomaterials in the treatment of wastewater: a review. Water 12:495
4. Heiligtag FJ, Niederberger M (2013) The fascinating world of nanoparticle research. Mater Today 16:262–271
5. De M, Ghosh PS, Rotello VM (2008) Applications of nanoparticles in biology. Adv Mater 20:4225–4241
6. Moura D, Souza MT, Liverani L, Rella G, Luz GM, Mano JF, Boccaccini AR (2017) Development of a bioactive glass-polymer composite for wound healing applications. Mater Sci Eng C 76:224–232
7. Banerjee K, Das S, Choudhury P, Ghosh S, Baral R, Choudhuri SK (2017) A novel approach of synthesizing and evaluating the anticancer potential of silver oxide nanoparticles in vitro. Chemotherapy 62:279–289
8. Gomez-Romero P (2001) Hybrid organic–inorganic materials—in search of synergic activity. Adv Mater 13:163–174
9. Shaikh SF, Mane RS, Min BK, Hwang YJ, Joo O (2016) D-sorbitol-induced phase control of TiO_2 nanoparticles and its application for dye-sensitized solar cells. Sci Rep 6:1–10
10. Narayanan R, El-Sayed MA (2004) Shape-dependent catalytic activity of platinum nanoparticles in colloidal solution. Nano Lett 4:1343–1348
11. Gracias DH, Tien J, Breen TL, Hsu C, Whitesides GM (2009) Forming electrical networks in three dimensions by self-assembly. Science (80):1170–1172
12. Dar AA, Umar K, Mir NA, Haque MM, Muneer M, Boxall C (2011) Photocatalysed degradation of a herbicide derivative, Dinoterb, in aqueous suspension. Res Chem Intermed 37:567–578
13. Umar K, Haque MM, Mir NA, Muneer M, Farooqi IH (2013) Titanium dioxide-mediated photocatalysed mineralization of two selected organic pollutants in aqueous suspensions. J Adv Oxid Technol 16:252–260
14. Yaqoob AA, Ahmad H, Parveen T, Ahmad A, Oves M, Ismail IMI, Qari HA, Umar K, Mohamad Ibrahim MN (2020) Recent advances in metal decorated nanomaterials and their various biological applications. Rev Front Chem 8:1–23. https://doi.org/10.3389/fchem.2020.00341
15. Chen W, Cai W, Zhang L, Wang G, Zhang L (2001) Sonochemical processes and formation of gold nanoparticles within pores of mesoporous silica. J Colloid Interface Sci 238:291–295
16. Geethalakshmi R, Sarada DVL (2012) Gold and silver nanoparticles from Trianthema decandra: synthesis, characterization, and antimicrobial properties. Int J Nanomed 7:5375
17. Vijayakumar M, Priya K, Nancy FT, Noorlidah A, Ahmed ABA (2013) Biosynthesis, characterisation and antibacterial effect of plant-mediated silver nanoparticles using Artemisia nilagirica. Ind Crops Prod 41:235–240
18. Rane AV, Kanny K, Abitha VK, Thomas S (2018) Methods for synthesis of nanoparticles and fabrication of nanocomposites. In: Synthesis of inorganic nanomaterials. Elsevier, pp 121–139
19. Koo JH (2019) Polymer nanocomposites: processing, characterization, and applications. McGraw-Hill Education
20. Moghadam AD, Schultz BF, Ferguson JB, Omrani E, Rohatgi PK, Gupta N (2014) Functional metal matrix composites: self-lubricating, self-healing, and nanocomposites-an outlook. Jom 66:872–881. https://doi.org/10.1007/s11837-014-0948-5

21. Ezatpour HR, Sajjadi SA, Sabzevar MH, Huang Y (2014) Investigation of microstructure and mechanical properties of Al6061-nanocomposite fabricated by stir casting. Mater Des 55:921–928
22. Zhang Z, Chen DL (2006) Consideration of Orowan strengthening effect in particulate-reinforced metal matrix nanocomposites: a model for predicting their yield strength. Scr Mater 54:1321–1326
23. Pacioni NL, Borsarelli CD, Rey V, Veglia AV (2015) Synthetic routes for the preparation of silver nanoparticles. In: Silver nanoparticle application. Springer, pp 13–46
24. Rajput N (2015) Methods of preparation of nanoparticles-a review. Int J Adv Eng Technol 7:1806
25. Meyers MA, Mishra A, Benson DJ (2006) Mechanical properties of nanocrystalline materials. Prog Mater Sci 51:427–556
26. Benjamin JS (1992) Fundamentals of mechanical alloying. In: Material Science Forum. Trans Tech Publication, pp 1–18
27. Chung KH, Rodriguez R, Lavernia EJ, Lee J (2002) Grain growth behavior of cryomilled INCONEL 625 powder during isothermal heat treatment. Metall Mater Trans A 33:125–134
28. Tsuzuki T, McCormick PG (2004) Mechanochemical synthesis of nanoparticles. J Mater Sci 39:5143–5146
29. Seyedi M, Haratian S, Khaki JV (2015) Mechanochemical synthesis of Fe_2O_3 nanoparticles. Procedia Mater Sci 11:309–313
30. Simakin AV, Voronov VV, Kirichenko NA, Shafeev GA (2004) Nanoparticles produced by laser ablation of solids in liquid environment. Appl Phys A 79:1127–1132
31. Kruis FE, Fissan H, Peled A (1998) Synthesis of nanoparticles in the gas phase for electronic, optical and magnetic applications—a review. J Aerosol Sci 29:511–535
32. Abou El-Nour KMM, Eftaiha A, Al-Warthan A, Ammar RAA (2010) Synthesis and applications of silver nanoparticles. Arab J Chem 3:135–140
33. Swihart MT (2003) Vapor-phase synthesis of nanoparticles. Curr Opin Colloid Interface Sci 8:127–133
34. Vanecht E (2012) Gold nanoparticles in ionic liquids prepared by sputter deposition
35. Tavakoli A, Sohrabi M, Kargari A (2007) A review of methods for synthesis of nanostructured metals with emphasis on iron compounds. Chem Pap 61:151–170
36. Ghorbani HR (2014) A review of methods for synthesis of Al nanoparticles. Orient J Chem 30:1941–1949
37. Nie M, Sun K, Meng DD (2009) Formation of metal nanoparticles by short-distance sputter deposition in a reactive ion etching chamber. J Appl Phys 106:54314
38. Johnson GE, Moser T, Engelhard M, Browning ND, Laskin J (2016) Fabrication of electrocatalytic Ta nanoparticles by reactive sputtering and ion soft landing. J Chem Phys 145:174701
39. Chandra R, Chawla AK, Ayyub P (2006) Optical and structural properties of sputter-deposited nanocrystalline Cu2O films: Effect of sputtering gas. J Nanosci Nanotechnol 6:1119–1123
40. Mukherjee P, Ahmad A, Mandal D, Senapati S, Sainkar SR, Khan MI, Parishcha R, Ajaykumar PV, Alam M, Kumar R (2001) Fungus-mediated synthesis of silver nanoparticles and their immobilization in the mycelial matrix: a novel biological approach to nanoparticle synthesis. Nano Lett 1:515–519
41. Pandey PA, Bell GR, Rourke JP, Sanchez AM, Elkin MD, Hickey BJ, Wilson NR (2011) Physical vapor deposition of metal nanoparticles on chemically modified graphene: observations on metal–graphene interactions. Small 7:3202–3210
42. Gondoni P, Ghidelli M, Di Fonzo F, Bassi AL, Casari CS (2013) Fabrication of nano-engineered transparent conducting oxides by pulsed laser deposition. JoVE e50297
43. Pedersen H, Elliott SD (2014) Studying chemical vapor deposition processes with theoretical chemistry. Theor Chem Acc 133:1–10
44. George SM (2010) Atomic layer deposition: an overview. Chem Rev 110:111–131
45. Miikkulainen V, Leskelä M, Ritala M, Puurunen RL (2013) Crystallinity of inorganic films grown by atomic layer deposition: overview and general trends. J Appl Phys 113:2

46. Chang W, Skandan G, Hahn H, Danforth SC, Kear BH (1994) Chemical vapor condensation of nanostructured ceramic powders. Nanostruct Mater 4:345–351
47. Chang W, Skandan G, Danforth SC, Rose M, Balogh AG, Hahn H, Kear B (1995) Nanostructured ceramics synthesized by chemical vapor condensation. Nanostruct Mater 6:321–324
48. Cushing BL, Kolesnichenko VL, O'connor CJ (2004) Recent advances in the liquid-phase syntheses of inorganic nanoparticles. Chem Rev 104:3893–3946
49. Hasnidawani JN, Azlina HN, Norita H, Bonnia NN, Ratim S, Ali ES (2016) Synthesis of ZnO nanostructures using sol-gel method. Procedia Chem 19:211–216
50. Burda C, Chen X, Narayanan R, El-Sayed MA (2005) Chemistry and properties of nanocrystals of different shapes. Chem Rev 105:1025–1102. https://doi.org/10.1021/cr030063a
51. Jana J, Ganguly M, Pal T (2016) Enlightening surface plasmon resonance effect of metal nanoparticles for practical spectroscopic application. RSC Adv 6:86174–86211
52. Zhu X, Gao T (2019) Chapter 10—Spectrometry. In: G.B.T.-N.-I.B. for P.A. with C.A. Li (Ed.). Elsevier, pp 237–264. https://doi.org/10.1016/B978-0-12-815053-5.00010-6
53. Kosuda K, Bingham J, Wustholz K, Van Duyne R (2010) Nanostructures and surfaceenhanced Raman spectroscopy. Handb Nanoscale Opt Electron 309
54. Bonatti L, Gil G, Giovannini T, Corni S, Cappelli C (2020) Plasmonic resonances of metal nanoparticles: atomistic vs. Continuum approaches. Front Chem 8:340
55. Pereira RMS, Borges J, Smirnov GV, Vaz F, Vasilevskiy MI (2018) Surface plasmon resonance in a metallic nanoparticle embedded in a semiconductor matrix: exciton-plasmon coupling. ACS Photonics 6:204–210
56. Otto AJ (1991) Surface-enhanced Raman scattering of adsorbates. J Raman Spectrosc 22:743–752
57. Fleischmann M, Hendra PJ, McQuillan AJ (1974) Raman spectra of pyridine adsorbed at a silver electrode. Chem Phys Lett Chem Phys Lett 26:163
58. Langer J, Jimenez de Aberasturi D, Aizpurua J, Alvarez-Puebla RA, Auguié B, Baumberg JJ, Bazan GC, Bell SEJ, Boisen A, Brolo AG (2019) Present and future of surface-enhanced Raman scattering. ACS Nano 14:28–117
59. Saviello D, Trabace M, Alyami A, Mirabile A, Baglioni P, Giorgi R, Iacopino D (2019) Raman Spectroscopy and Surface Enhanced Raman Scattering (SERS) for the analysis of blue and black writing inks: identification of dye content and degradation processes. Front Chem 727
60. Strutt JW (1871) XV on the light from the sky, its polarization and colour. Lond, Edinb, Dublin Philos Mag J Sci 41:107–120
61. Wang X, Benedictus R, Groves RM (2021) Optimization of light scattering enhancement by gold nanoparticles in fused silica optical fiber. Opt Express 29:19450–19464
62. Blanc W, Dussardier B (2016) Formation and applications of nanoparticles in silica optical fibers. J Opt 45:247–254
63. Molardi C, Korganbayev S, Blanc W, Tosi D (2018) Characterization of a nanoparticles-doped optical fiber by the use of optical backscatter reflectometry. In: Advanced sensor systems and application VIII. International Society for Optics and Photonics, p 1082121
64. Beisenova A, Issatayeva A, Iordachita I, Blanc W, Molardi C, Tosi D (2019) Distributed fiber optics 3D shape sensing by means of high scattering NP-doped fibers simultaneous spatial multiplexing. Opt Express 27:22074–22087
65. Bulot P, Cristini O, Bouet M, Demol A, Bigot L, Bouwmans G, Plus S, Habert R, Laffont G, Douay M (2018) OFDR distributed temperature sensing at 800 °C on a fiber with enhanced Rayleigh scattering profile by doping. In: Bragg Gratings, Photosensit. Poling Glas. Waveguides Mater., Optical Society of America, pp BM3A-2
66. Oh I, Min HS, Li L, Tran TH, Lee Y, Kwon IC, Choi K, Kim K, Huh KM (2013) Cancer cell-specific photoactivity of pheophorbide a–glycol chitosan nanoparticles for photodynamic therapy in tumor-bearing mice. Biomaterials 34:6454–6463
67. Hoshyar N, Gray S, Han H, Bao G (2016) The effect of nanoparticle size on in vivo pharmacokinetics and cellular interaction. Nanomedicine 11:673–692

68. Scott RP, Quaggin SE (2015) The cell biology of renal filtration. J Cell Biol 209:199–210
69. Longmire M, Choyke PL, Kobayashi H (2008) Clearance properties of nano-sized particles and molecules as imaging agents: considerations and caveats
70. Zhou Y, Dai Z (2018) New strategies in the design of nanomedicines to oppose uptake by the mononuclear phagocyte system and enhance cancer therapeutic efficacy. Chem Asian J 13:3333–3340
71. Huang Y, He S, Cao W, Cai K, Liang X-J (2012) Biomedical nanomaterials for imaging-guided cancer therapy. Nanoscale 4:6135–6149
72. Chen LQ, Xiao SJ, Hu PP, Peng L, Ma J, Luo LF, Li YF, Huang CZ (2012) Aptamer-mediated nanoparticle-based protein labeling platform for intracellular imaging and tracking endocytosis dynamics. Anal Chem 84:3099–3110
73. Das M, Duan W, Sahoo SK (2015) Multifunctional nanoparticle–EpCAM aptamer bioconjugates: a paradigm for targeted drug delivery and imaging in cancer therapy. Nanomed Nanotechnol Biol Med 11:379–389
74. Ke R, Yang W, Xia X, Xu Y, Li Q (2010) Tandem conjugation of enzyme and antibody on silica nanoparticle for enzyme immunoassay. Anal Biochem 406:8–13
75. Kim D, Jeong YY, Jon S (2010) A drug-loaded aptamer—gold nanoparticle bioconjugate for combined CT imaging and therapy of prostate cancer. ACS Nano 4:3689–3696
76. Wang Z, Qiao R, Tang N, Lu Z, Wang H, Zhang Z, Xue X, Huang Z, Zhang S, Zhang G (2017) Active targeting theranostic iron oxide nanoparticles for MRI and magnetic resonance-guided focused ultrasound ablation of lung cancer. Biomaterials 127:25–35
77. Lee KJ, Nallathamby PD, Browning LM, Osgood CJ, Xu X-HN (2007) In vivo imaging of transport and biocompatibility of single silver nanoparticles in early development of zebrafish embryos. ACS Nano 1:133–143
78. Winter JO (2006) Nanoparticles and nanowires for cellular engineering. Tissue, Cell Organ Eng 9:388–460
79. Patra CR, Bhattacharya R, Wang E, Katarya A, Lau JS, Dutta S, Muders M, Wang S, Buhrow SA, Safgren SL (2008) Targeted delivery of gemcitabine to pancreatic adenocarcinoma using cetuximab as a targeting agent. Cancer Res 68:1970–1978
80. Lehner R, Wang X, Marsch S, Hunziker P (2013) Intelligent nanomaterials for medicine: carrier platforms and targeting strategies in the context of clinical application. Nanomed: Nanotechnol, Biol Med 9(6):742–757
81. Liu J, Yu M, Ning X, Zhou C, Yang S, Zheng J (2013) PEGylation and zwitterionization: pros and cons in the renal clearance and tumor targeting of near-IR-emitting gold nanoparticles. Angew Chemie 125:12804–12808
82. Liu J, Yu M, Zhou C, Yang S, Ning X, Zheng J (2013) Passive tumor targeting of renal-clearable luminescent gold nanoparticles: long tumor retention and fast normal tissue clearance. J Am Chem Soc 135:4978–4981
83. Alric C, Miladi I, Kryza D, Taleb J, Lux F, Bazzi R, Billotey C, Janier M, Perriat P, Roux S (2013) The biodistribution of gold nanoparticles designed for renal clearance. Nanoscale 5:5930–5939
84. Yang S, Sun S, Zhou C, Hao G, Liu J, Ramezani S, Yu M, Sun X, Zheng J (2015) Renal clearance and degradation of glutathione-coated copper nanoparticles. Bioconjug Chem 26:511–519
85. Zhou C, Long M, Qin Y, Sun X, Zheng J (2011) Luminescent gold nanoparticles with efficient renal clearance. Angew Chemie Int Ed 50:3168–3172
86. Boisselier E, Astruc D (2009) Gold nanoparticles in nanomedicine: preparations, imaging, diagnostics, therapies and toxicity. Chem Soc Rev 38:1759–1782
87. Patra CR, Bhattacharya R, Mukhopadhyay D, Mukherjee P (2010) Fabrication of gold nanoparticles for targeted therapy in pancreatic cancer. Adv Drug Deliv Rev 62:346–361
88. Klein J (2007) Probing the interactions of proteins and nanoparticles. Proc Natl Acad Sci 104:2029–2030
89. Mody VV, Siwale R, Singh A, Mody HR (2010) Introduction to metallic nanoparticles. J Pharm Bioallied Sci 2:282

90. Barui AK, Kotcherlakota R, Patra CR (2018) Medicinal applications of metal nanoparticles. In: Metal nanoparticles synthesis and applications in pharmaceutical sciences. Wiley, pp 101–153
91. Wu YY, Tsui WC, Liu TC (2007) Experimental analysis of tribological properties of lubricating oils with nanoparticle additives. Wear 262:819–825
92. Li X, Cao Z, Zhang Z, Dang H (2006) Surface-modification in situ of nano-SiO_2 and its structure and tribological properties. Appl Surf Sci 252:7856–7861
93. Deepika (2020) Nanotechnology implications for high performance lubricants. SN Appl Sci 2:1128. https://doi.org/10.1007/s42452-020-2916-8
94. Battez AH, González R, Viesca JL, Fernández JE, Fernández JMD, Machado A, Chou R, Riba J (2008) CuO, ZrO_2 and ZnO nanoparticles as antiwear additive in oil lubricants. Wear 265:422–428
95. Sih BC, Wolf MO (2005) Metal nanoparticle—conjugated polymer nanocomposites. Chem Commun 3375–3384
96. Bakunin VN, Suslov AY, Kuzmina GN, Parenago OP, Topchiev AV (2004) Synthesis and application of inorganic nanoparticles as lubricant components–a review. J Nanoparticle Res 6:273–284
97. Yu W, France DM, Timofeeva EV, Singh D, Routbort JL (2012) Comparative review of turbulent heat transfer of nanofluids. Int J Heat Mass Transf 55:5380–5396
98. Mekhilef S, Saidur R, Kamalisarvestani M (2012) Effect of dust, humidity and air velocity on efficiency of photovoltaic cells. Renew Sustain Energy Rev 16:2920–2925
99. Hwang Y, Park HS, Lee JK, Jung WH (2006) Thermal conductivity and lubrication characteristics of nanofluids. Curr Appl Phys 6:e67–e71
100. Lu Z, Xiao J, Wang Y, Meng M (2015) In situ synthesis of silver nanoparticles uniformly distributed on polydopamine-coated silk fibers for antibacterial application. J Colloid Interface Sci 452:8–14
101. Trunec M, Maca K (2014) Advanced ceramic processes. In: Advanced ceramics for dentistry. Butterworth-Heinemann, pp 123–150
102. Chen S, Liu W (2006) Oleic acid capped PbS nanoparticles: synthesis, characterization and tribological properties. Mater Chem Phys 98:183–189
103. Manna I (2009) Synthesis, characterization and application of nanofluid—an overview. J Ind Inst Sci 89:21–33
104. Farah FH, Farah FH (2019) Nanocarriers as delivery systems for therapeutics agents. Int J Pharm Sci Res 10:3487–3507
105. Zhu X, Vo C, Taylor M, Smith BR (2019) Non-spherical micro-and nanoparticles in nanomedicine. Mater Horiz 6:1094–1121
106. Elsaesser A, Vyvyan Howard C (2012) Toxicology of nanoparticles. Adv Drug Deliv Rev 64(2):129–137
107. Renero-Lecuna C, Iturrioz-Rodríguez N, González-Lavado E, Padín-González E, Navarro-Palomares E, Valdivia-Fernández L, García-Hevia L, Fanarraga ML, González-Legarreta L (2019) Effect of size, shape, and composition on the interaction of different nanomaterials with HeLa cells. J Nanomater
108. Chaicherd S, Killingsworth MC, Pissuwan D (2019) Toxicity of gold nanoparticles in a commercial dietary supplement drink on connective tissue fibroblast cells. SN Appl Sci 1:1–8
109. Jeevanandam J, Chan YS, Danquah MK (2016) Biosynthesis of metal and metal oxide nanoparticles. ChemBioEng Rev. 3:55–67
110. Naz S, Gul A, Zia M (2020) Toxicity of copper oxide nanoparticles: a review study. IET Nanobiotechnol 14:1–13
111. Roy S, Sadhukhan R, Ghosh U, Das TK (2015) Interaction studies between biosynthesized silver nanoparticle with calf thymus DNA and cytotoxicity of silver nanoparticles, Spectrochim. Acta Part A Mol Biomol Spectrosc 141:176–184
112. Johnston HJ, Hutchison G, Christensen FM, Peters S, Hankin S, Stone V (2010) A review of the in vivo and in vitro toxicity of silver and gold particulates: particle attributes and biological mechanisms responsible for the observed toxicity. Crit Rev Toxicol 40:328–346

113. Oberdörster G, Oberdörster E, Oberdörster J (2005) Nanotoxicology: an emerging discipline evolving from studies of ultrafine particles. Environ Health Perspect 113:823–839
114. Egbuna C, Parmar VK, Jeevanandam J, Ezzat SM, Patrick-Iwuanyanwu KC, Adetunji CO, Khan J, Onyeike EN, Uche CZ, Akram M, Ibrahim MS, El Mahdy NM, Awuchi CG, Saravanan K, Tijjani H, Odoh UE, Messaoudi M, Ifemeje JC, Olisah MC, Ezeofor NJ, Chikwendu CJ, Ibeabuchi CG (2021) Toxicity of nanoparticles in biomedical application: nanotoxicology. J Toxicol 2021:9954443. https://doi.org/10.1155/2021/9954443
115. Bacchetta R, Santo N, Valenti I, Maggioni D, Longhi M, Tremolada P (2018) Comparative toxicity of three differently shaped carbon nanomaterials on Daphnia magna: does a shape effect exist? Nanotoxicology 12:201–223
116. Abramenko NB, Demidova TB, Abkhalimov EV, Ershov BG, Krysanov EY, Kustov LM (2018) Ecotoxicity of different-shaped silver nanoparticles: case of zebrafish embryos. J Hazard Mater 347:89–94
117. Carnovale C, Bryant G, Shukla R, Bansal V (2019) Identifying trends in gold nanoparticle toxicity and uptake: size, shape, capping ligand, and biological corona. ACS Omega 4:242–256
118. Zhou H, Gong X, Lin H, Chen H, Huang D, Li D, Shan H, Gao J (2018) Gold nanoparticles impair autophagy flux through shape-dependent endocytosis and lysosomal dysfunction. J Mater Chem B 6:8127–8136
119. Steckiewicz KP, Barcinska E, Malankowska A, Zauszkiewicz–Pawlak A, Nowaczyk G, Zaleska-Medynska A, Inkielewicz-Stepniak I (2019) Impact of gold nanoparticles shape on their cytotoxicity against human osteoblast and osteosarcoma in in vitro model. Evaluation of the safety of use and anti-cancer potential. J Mater Sci Mater Med 30:1–15
120. Wang W-N, Tarafdar JC, Biswas P (2013) Nanoparticle synthesis and delivery by an aerosol route for watermelon plant foliar uptake. J Nanopart Res 15:1417. https://doi.org/10.1007/s11051-013-1417-8
121. Verma SK, Das AK, Patel MK, Shah A, Kumar V, Gantait S (2018) Engineered nanomaterials for plant growth and development: a perspective analysis. Sci Total Environ 630:1413–1435. https://doi.org/10.1016/j.scitotenv.2018.02.313
122. Oukarroum A, Bras S, Perreault F, Popovic R (2012) Inhibitory effects of silver nanoparticles in two green algae, Chlorella vulgaris and Dunaliella tertiolecta. Ecotoxicol Environ Saf 78:80–85. https://doi.org/10.1016/j.ecoenv.2011.11.012
123. Dash A, Singh AP, Chaudhary BR, Singh SK, Dash D (2012) Effect of silver nanoparticles on growth of eukaryotic green algae. Nano-Micro Lett 4:158–165. https://doi.org/10.1007/BF03353707
124. Barbafieri M, Giorgetti L (2016) Contaminant bioavailability in soil and phytotoxicity/genotoxicity tests in Vicia faba L.: a case study of boron contamination. Environ Sci Pollut Res 23:24327–24336. https://doi.org/10.1007/s11356-016-7653-6
125. Plechkova NV, Seddon KR (2008) Applications of ionic liquids in the chemical industry. Chem Soc Rev 37:123–150
126. Dupont J, Scholten JD (2010) On the structural and surface properties of transition-metal nanoparticles in ionic liquids. Chem Soc Rev 39:1780–1804
127. Ma Z, Yu J, Dai S (2010) Preparation of inorganic materials using ionic liquids. Adv Mater 22:261–285
128. Ryu HJ, Sanchez L, Keul HA, Raj A, Bockstaller MR (2008) Imidazolium-based ionic liquids as efficient shape-regulating solvents for the synthesis of gold nanorods. Angew Chemie Int Ed 47:7639–7643
129. Redel E, Thomann R, Janiak C (2008) First correlation of nanoparticle size-dependent formation with the ionic liquid anion molecular volume. Inorg Chem 47:14–16
130. Roy P, Lynch R, Schmuki P (2009) Electron beam induced in-vacuo Ag deposition on TiO_2 from ionic liquids. Electrochem Commun 11:1567–1570
131. Harada M, Kimura Y, Saijo K, Ogawa T, Isoda S (2009) Photochemical synthesis of silver particles in Tween 20/water/ionic liquid microemulsions. J Colloid Interface Sci 339:373–381
132. Dupont J, Suarez PAZ (2006) Physico-chemical processes in imidazolium ionic liquids. Phys Chem Chem Phys 8:2441–2452

133. Migowski P, Zanchet D, Machado G, Gelesky MA, Teixeira SR, Dupont J (2010) Nanostructures in ionic liquids: correlation of iridium nanoparticles' size and shape with imidazolium salts' structural organization and catalytic properties. Phys Chem Chem Phys 12:6826–6833
134. Gutel T, Santini CC, Philippot K, Padua A, Pelzer K, Chaudret B, Chauvin Y, Basset J-M (2009) Organized 3D-alkyl imidazolium ionic liquids could be used to control the size of in situ generated ruthenium nanoparticles? J Mater Chem 19:3624–3631
135. Smith DK, Miller NR, Korgel BA (2009) Iodide in CTAB prevents gold nanorod formation. Langmuir 25:9518–9524
136. Rayavarapu RG, Ungureanu C, Krystek P, Van Leeuwen TG, Manohar S (2010) Iodide impurities in hexadecyltrimethylammonium bromide (CTAB) products: lot−lot variations and influence on gold nanorod synthesis. Langmuir 26:5050–5055
137. Wang F, Tang R, Buhro WE (2008) The trouble with TOPO; identification of adventitious impurities beneficial to the growth of cadmium selenide quantum dots, rods, and wires. Nano Lett 8:3521–3524
138. Lazarus LL, Riche CT, Malmstadt N, Brutchey RL (2012) Effect of ionic liquid impurities on the synthesis of silver nanoparticles. Langmuir 28:15987–15993
139. de Oliveira PFM, Torresi RM, Emmerling F, Camargo PHC (2020) Challenges and opportunities in the bottom-up mechanochemical synthesis of noble metal nanoparticles. J Mater Chem A 8:16114–16141
140. Fan Z, Zhang H (2016) Crystal phase-controlled synthesis, properties and applications of noble metal nanomaterials. Chem Soc Rev 45:63–82
141. Zhou S, Zhao M, Yang T-H, Xia Y (2019) Decahedral nanocrystals of noble metals: synthesis, characterization, and applications. Mater Today 22:108–131. https://doi.org/10.1016/j.mattod.2018.04.003
142. Xia Y, Xiong Y, Lim B, Skrabalak SE (2009) Shape-controlled synthesis of metal nanocrystals: simple chemistry meets complex physics? Angew Chemie Int Ed 48:60–103. https://doi.org/10.1002/anie.200802248
143. Xia Y, Xia X, Peng H-C (2015) Shape-controlled synthesis of colloidal metal nanocrystals: thermodynamic versus kinetic products. J Am Chem Soc 137:7947–7966. https://doi.org/10.1021/jacs.5b04641
144. Scarabelli L, Sánchez-Iglesias A, Pérez-Juste J, Liz-Marzán LM (2015) A "tips and tricks" practical guide to the synthesis of gold nanorods. J Phys Chem Lett 6:4270–4279. https://doi.org/10.1021/acs.jpclett.5b02123
145. Fievet F, Lagier JP, Figlarz M (1989) Preparing monodisperse metal powders in micrometer and submicrometer sizes by the polyol process. MRS Bull 14:29–34. https://doi.org/10.1557/S0883769400060930
146. Fiévet F, Ammar-Merah S, Brayner R, Chau F, Giraud M, Mammeri F, Peron J, Piquemal J-Y, Sicard L, Viau G (2018) The polyol process: a unique method for easy access to metal nanoparticles with tailored sizes, shapes and compositions. Chem Soc Rev 47:5187–5233. https://doi.org/10.1039/C7CS00777A
147. Kimling J, Maier M, Okenve B, Kotaidis V, Ballot H, Plech A (2006) Turkevich method for gold nanoparticle synthesis revisited. J Phys Chem B 110:15700–15707. https://doi.org/10.1021/jp061667w
148. Wuithschick M, Birnbaum A, Witte S, Sztucki M, Vainio U, Pinna N, Rademann K, Emmerling F, Kraehnert R, Polte J (2015) Turkevich in new robes: key questions answered for the most common gold nanoparticle synthesis. ACS Nano 9:7052–7071. https://doi.org/10.1021/acsnano.5b01579
149. Turkevich J, Stevenson PC, Hillier J (1951) A study of the nucleation and growth processes in the synthesis of colloidal gold. Discuss Faraday Soc 11:55–75. https://doi.org/10.1039/DF9511100055
150. Frens G (1973) Controlled nucleation for the regulation of the particle size in monodisperse gold suspensions. Nat Phys Sci 241:20–22. https://doi.org/10.1038/physci241020a0

Recent Trends in Metallic Nanocomposites for Sensing and Electrochemical Devices

Beauty Pandey and Daya Shankar

Abstract Recent years have witnessed an exponential rise in the development of sensor devices with enhanced characteristics. The research arena of sensing has been devoted to increasing the sensitivity, stability, scalability, selectivity, responsiveness, reactivity, cost-effectiveness, and ease of operation. Nanocomposites, owing to their size and shape-dependent interatomic bond lengths, exceptional mechanical properties, high chemical reactivity, good optical and electronic features, specific biocompatibility, and other excellent physiochemical properties, have exhibited immense potential in the intensification of these properties in sensors. Electrochemical devices employed for sensing have emerged as one of the most significant types of sensors. Electrochemical sensors are characterized by their reactivity, which depends on electrochemical signals, and their functionality is determined by ion and electron transportation, which in turn is measured in the form of electrical responses. Metallic nanocomposites have triggered greater attention in electrochemical detection due to their ability to impart high strength, better stiffness, improved toughness, increased wear and corrosion resistance, high melting points, low density, increased catalytic activity, and other improved characteristics even at very low concentrations. The main thrust is to offer a comprehensive review highlighting the ongoing and upcoming research in the trajectory of metallic nanocomposite-based electrochemical sensors for different applications such as gas sensing, metal trace detection, biosensing, etc. The emphasis will be on the performance of metallic nanocomposites with different carbon-based nanomaterials such as graphene, carbon nanotubes (CNTs), etc. in electrochemical sensors. Carbon nanomaterial-based electrochemical sensors have an edge over other systems because they aid in better electron transfer, thereby promoting the adsorption of molecules. The key aspect here is to understand the mechanical attributes of nanocomposites; hence, the role of metallic nanocomposites-based flexible sensor actuators will also be concisely discussed. In this case, the emphasis would be on investigating the mechanical robustness and compliance of metallic nanocomposite-modified flexible sensors.

B. Pandey (✉) · D. Shankar
School of Technology, Woxsen University, Hyderabad 502345, India
e-mail: pandey.beauty@gmail.com; pandey.beauty@yahoo.com

S. K. Tiwari et al. (eds.), *Nanoparticles Reinforced Metal Nanocomposites*,
https://doi.org/10.1007/978-981-19-9729-7_8

Keywords Nanocomposites · Metals · Carbon · Graphene · Carbon-nanotube · Electrochemical · Sensor · Gas sensing · Heavy-metal · Biosensing

1 Introduction

The increasing infiltration of the global population within urban conurbations has made sensing an extraordinarily fascinating issue universally. Consequently, an ever-growing demand to resource, study, and utilize a vast array of materials for sensing has been witnessed both by the scientific and technological communities. It won't be wrong to state that in today's scenario, sensors represent a class of devices of extreme importance not only for research but also in terms of applications. The reason behind this is that, though the worldwide economic expansion and industrial developments essential to cater to the needs and the continued sustainability of our day-to-day living have been immensely appreciated, the by-products of human activities in the form of the releases from a multitude of chemical procedures, agricultural outputs, energy conversions, and various other applications have elevated pollution. Such anthropogenic processes have acted as major contributors to the contamination of the atmosphere, waterbodies, food products, and everything around us. The critical impact of modern living conditions on environmental safety is hazardous and elicits concern [1–5]. As a result of the indiscriminate release of a wide range of hazardous analytes and chemicals into the atmosphere from airborne, water, and solid waste as a result of escalating international development and manufacturing, various safety measures are being implemented globally. This has incited the development of new, highly efficient, sensitive methods and sensors. In simple terms, a sensor can be identified as a device that produces an alert in response to some stimulus. It is typically capable of detecting electronically variable quantities such as mass, electrical conductivity, or capacitance and converting the measurement into specific signals. Depending on the diversity, stability, selectivity, and accuracy of the information extracted, sensors empower us to not only sense the environment around us but also to utilize that information for varied purposes [6–8].

The widespread usage of sensors ranges from the detection of gases and chemicals to tracing heavy metals and biological species in the atmosphere and surroundings. The past few years have seen a surfeit of research in the field of sensing owing to its applicability in the analysis of environmental pollution, the monitoring of industrial emissions and process assessment, the investigation of medical conditions, the diagnosis of public security, the control of agriculture, and a variety of other industries. The major attributes of a good sensing system are its specificity, sensitivity, reliability, portability, possibility of miniaturization, real-time analysis, and user-friendly operation. The field of sensors has significantly evolved compared to the last decade, owing to an increasing demand for the development of screening methods that are simple, robust, fast, cost-effective, scalable, flexible, small, have quick responses, are capable of in situ analysis, show rapid recovery, and are field portable [1–5, 9–11]. Sensors denote a category of devices that have an active sensing element along with a signal transducer, which serve to transmit the signal from a specific material or from some kind of shift in a reaction without any amplification. These devices then generate the output signals in electrical, thermal, or optical form, which are translated into digital signals for further processing [12]. Over time, various sensing mechanisms and associated devices have been configured. Based upon sensing mechanisms, there are several types of sensors, such as chemical sensors, field-effect transistors (FET), biosensors, and micro-electromechanical systems (MEMS) [9, 10]. From the humongous range of chemical sensors that are available, electrochemical sensors are one of the simplest, widely explored and employed in practical applications.

Electrochemical sensors belong to a class of sensors where the electrode is the transducer element. As evident from their name, such sensors are electrochemical in nature and are generally based on redox reactions involving a particular chemical species in the electrolyte at the working electrode. Electrochemical sensors detect the target material quantitatively as an oxidation or reduction current, producing output as a varying electrical signal. Electrochemical sensors are among the most investigated and enjoy tremendous benefits over conventional analytics for their simple measurement procedure, adaptability, compactness, fast response, sufficient sensitivity, and selectivity. Though there are various other chemical sensor systems that utilize chemical reactions to convert the response from target species to measurable quantities with fairly high sensitivity, they are not equipped for in-place measurements since they are based on indirect detection of target analytes. On the contrary, an electrochemical sensor system can effortlessly track changes to facilitate real-time and in situ investigations of the materials present in the host without causing much damage to the host system. Additionally, they have innate accuracy and specificity, are flexible, and are easy to set up. Not to mention that they are portable and lightweight, making them simple to use and resulting in faster research data acquisition and arrangement [13–19]. Regardless, all of the desirable characteristics of electrochemical sensors are heavily dependent on the appropriate composition, design, and structure of the electrode, because it is the electrode that comes into direct contact with any chemical analyte. Hence, in order to establish the varsity and improve the performance of the electrochemical sensors, modifications to the electrode surface are a necessary obligation and are carried out consistently. To fulfil these criteria, researchers have been designing and developing sensitive materials with different compositions, structures, and morphologies for the working electrode in an electrochemical sensor.

Owing to this, the practice of applying nanomaterials and nanostructures as electrochemical sensing tools has been progressively growing [20–22]. Subsequently, being associated with nanotechnology, electrochemical sensors have displayed improved selection and volatility and are also becoming increasingly precise, specific, and highly sensitive. Nanoscale materials are materials or structures that lie on the nanometer scale in one, two, or three dimensions [23–25]. Their advent has led to the emergence of numerous innovative materials that have brought forward various technological, commercial, and scientific developments. Nanomaterials have obtained such great prominence on account of their exceptional size-dependent physicochemical properties such as large surface area, high chemical reactivity, unique melting points, thermal and electrical conductivities, improved properties of light absorption, scattering, and optical sensitivity, better wettability, and catalytic activity, making them exhibit remarkably excellent performance over the bulk [24, 26]. Figure 1 provides the classification of nanomaterials as per their dimensionality. A qualitative analysis of the characteristics of 0D, 1D, 2D, and 3D materials has been presented. In view of these extraordinary properties, nanomaterials have evolved as one of the most favourable candidates to meet the desired requirements for the construction of highly sensitive electrochemical sensors. With remarkable achievements in nanotechnology and nanoscience, metal nanoparticles have shown tremendous potential in the design of new and improved sensing devices, especially electrochemical sensors [27–31].

Depending upon the role and the system, nanoparticles of gold (Au), silver (Ag), platinum (Pt), palladium (Pd), copper (Cu), nickel (Ni), titanium (Ti), magnesium (Mg), and cobalt (Co) have been widely employed by the researchers while designing and fabricating electrochemical sensors. The metal nanoparticles exhibit diverse effects on the response of the electrochemical sensor besides improving their mechanical, thermal, and electrical properties. They aid in the immobilization of molecules, the improvement of electrochemical catalysis reactions, and the enhancement of electron transfer. Moreover, metal nanoparticles may also perform as analytical transducers as well as signal amplification elements [27, 32, 33]. Recent years have witnessed increasing interest in nanocomposites due to their improved properties as compared to single metal nanoparticles [34–36]. Generally, a nanocomposite material is defined as a mixture or matrix wherein different materials join to develop unique properties while also ensuring that at least one phase is in the range of 1–100 nm [37–39]. Consequently, as dimensions reach the nanometer scale, largely

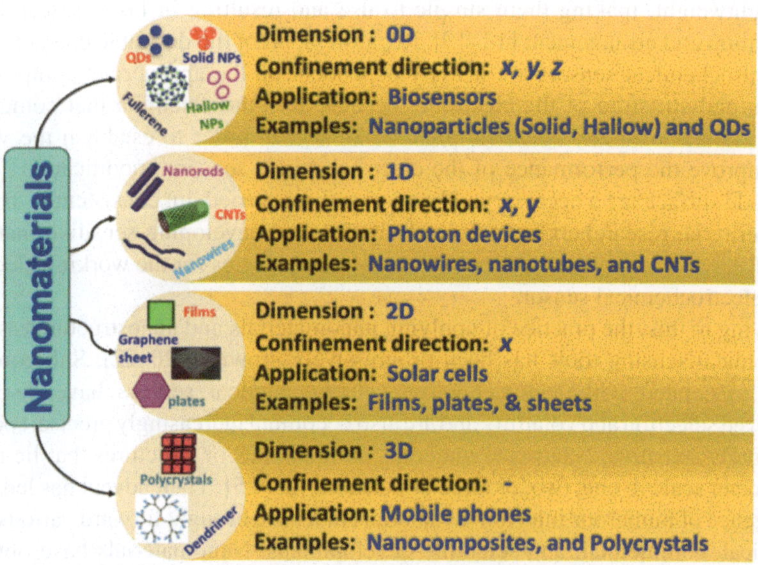

Fig. 1 Dimensional classification of nanomaterials [19]

improved interactions occur at the phase interfaces, which enhance materials' properties further. Thus, the metallic nanocomposites can have an amalgamation of strikingly different electrical, mechanical, optical, thermal, catalytic, and electrochemical properties from their components [38, 40–43]. Such novel nanocomposites with astonishing new and sometimes interesting properties that are created by combining different metal nanoparticles with different functional materials greatly enhance the range of sensing applications. Some of the interesting and excellent properties of nanocomposites are depicted in Fig. 2.

On account of its unique electrochemical properties, carbon has long been employed as an electrochemical sensing interface [44]. Following their discovery, carbon nanomaterials such as carbon quantum dots, carbon nanotubes (CNTs), graphene, and others became very popular sensing materials due to their extraordinary properties such as increased surface area, improved environmental stability, and exceptional mechanical, electrical, thermal, and chemical properties [10, 45–47]. They also have incredible conductivity, electrochemical properties, tunable surface functionality, and biocompatibility [48–54]. Also, these carbon nanomaterials offer superb adsorption, possess high electron transportation, enable fast recovery, have flexibility, and are very robust. Though these novel carbon nanomaterials appeared to be promising alternatives for achieving enhanced performance of electrochemical sensors, they also suffered from some serious disadvantages, such as low selectivity, poor repeatability, and non-homogeneity of the functional groups [55–57]. Thus,

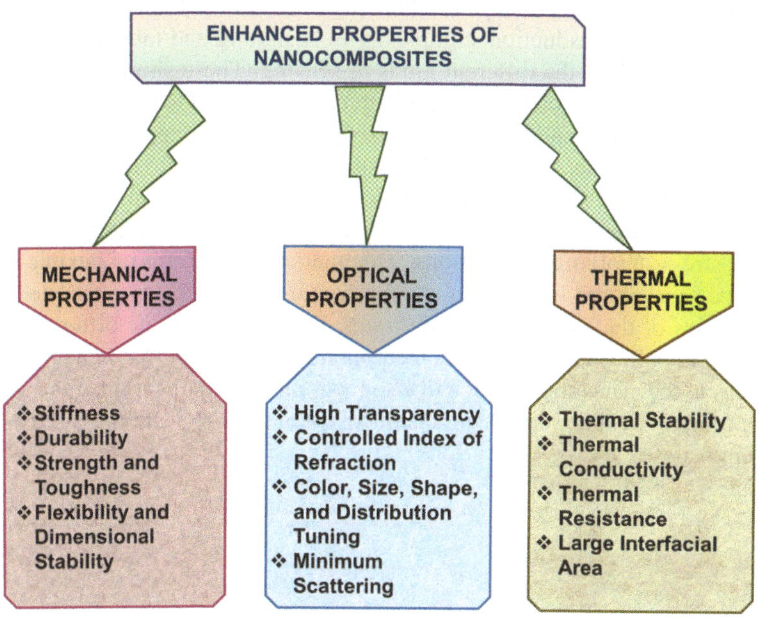

Fig. 2 Enhanced properties of nanocomposites. Adapted with Permission from [20] Copyright (2018) Elsevier

without forfeiting the advantages of carbon-based nanomaterials, their nanocomposites with metals have been extensively investigated as different sensing materials [10, 29, 45, 59]. These new classes of metallic nanocomposites with carbon-based nanomaterials have exhibited tremendous performances as electrochemical sensors because they were able to yield a synergistic effect among conductivity, compatibility, and catalytic activity of all the constituents, which not only accelerates the signal transduction but also amplifies the recognition of events that lead to highly sensitive sensing with quick outcomes and visual responses, suitable for being used in resource-controlled environments. To exploit this potential of carbon nanomaterials-based metallic nanocomposites as sensors, intensive research efforts have been made in recent years, resulting in a wide range of scientific breakthroughs in electrochemical sensing applications.

Notwithstanding anything, the intervention of sensors has largely revolutionized our world and helped in improving the day-to-day living standards of humans. As already discussed, the passage of time witnessed the rapid expansion of sensing technologies [57–61]. Sensors can now be broadly classified as non-flexible or rigid sensors and flexible sensors. Initially, non-flexible or rigid sensors were designed using conventional techniques. Although rigid sensors are widely popular owing to their ability to work in extreme environmental conditions, the ease with which they integrate into systems or modify them, their low power consumption, their small thermal constant, their low hysteresis, and their excellent signal-to-noise ratio, they have certain disadvantages that restrict their applications. Their limitations include their rigidity in capturing analytes efficiently, stiffness, intransigency, high cost, poor quality signal transduction, complexity in designing and fabricating, and their inability to perform all the different kinds of sensing. These shortcomings become even more prominent when the sensing system is employed to monitor any physiological parameter of an individual or is associated with any application that causes prominent stress on the sensor, thereby damaging the sensor. These discrepancies in non-flexible sensors led the scientists to look for an alternative where the sensor can be used dynamically to negate any personal inconvenience or any damage to the sensor in stress-involving applications. This marked the advent of flexible sensors that can be twisted and turned to a certain degree depending upon the mechanical characteristics of the electrodes or substrates. In this chapter, the different sensing activities of metallic nanocomposites incorporated with carbon-based nanomaterials will be discussed. The monograph will also focus on the modification of the mechanical properties of flexible sensors on account of the inclusion of carbon-based metallic nanocomposites.

2 Electrochemistry

The expansive interest in electrochemical sensors can be connected to their economical cost, relatively fast time-response, operational simplicity, and robustness of electrochemical measurements [61–63]. Electrochemical sensors are based on electrochemistry, which is an umbrella term to describe all typical dynamic processes occurring at the electrode–electrolyte interface. Basically, electrochemistry is a group of electroanalytical methods used to study the relationship between chemistry and electricity by monitoring different signal outputs in the form of electric current, charge, potential, phase, frequency, etc. derived from a chemical reaction [64]. Fundamentally, an electrochemical sensor is nothing but an anode and a cathode, immersed in an electrolytic solution to allow the transport of ions between both electrodes [16–19, 65–68]. Consecutively, the response from an electrochemical sensor is typically obtained from the electrical signal obtained in the presence of an analyte. The analyte of interest is either oxidized or reduced by the electrode. Basically, electrochemical sensors function by translating the information related to electrochemical reactions between an electrode and an analyte into a relevant qualitative or quantitative signal and can produce digital signals from these electronic outputs for further analysis. Figure 3 demonstrates the workings of a typical electrochemical sensor. An electrochemical sensor consists of a receptor that binds the sample, or analyte, and an electrode that converts the reaction into a measurable electrical output [67–73]. In principle, an electrochemical sensor can work with a minimum of two electrodes, which serve as the sensing electrode and counter electrode [16–19, 65–68, 73]. These two electrodes are connected by an electrically conducting medium in the form of an electrolyte and via an electric circuit externally. A general set-up of a two-electrode electrochemical cell is shown in Fig. 4. The electrodes are made with materials having catalytic properties such that chemical reactions take place and an effective three-phase boundary having an analyte, catalyst, and electrolyte is created. However, a two-electrode sensor system has many shortcomings.

An increase in the analyte concentrations leads to higher currents in the sensor, which further creates a voltage drop, changing the pre-set voltage of the sensor, leading to ambiguous measurements or, in some cases, stopping the chemical reaction occurring within the sensor and causing an error in the measurement. Henceforth, a third electrode that functions as a reference and is located away from the current flow is added to the sensor. Mainly, the reference electrode serves as a constant potential source, and with respect to its potential, the sensor voltage is measured continuously. An electrochemical sensor comprising a three-electrode configuration is illustrated in Fig. 5. This modification to the two-electrode sensor configuration with a third reference electrode improves measurement quality in terms of linearity and selectivity while also supporting a longer lifetime.

In addition to the reference electrode, there are two other conductive and chemically stable electrodes; one is a working electrode, and the other is a counter or auxiliary electrode. The working electrode is known as the sensing or redox electrode and acts as the transduction element in the reaction, whereas the counter electrode is used

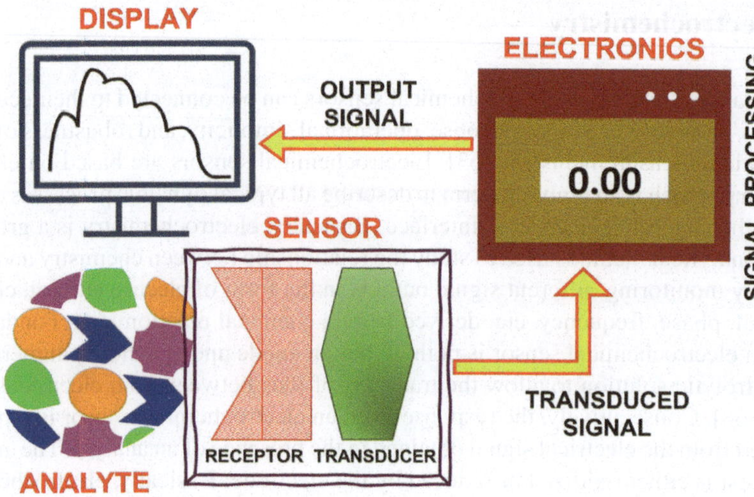

Fig. 3 Schematic representation of working of a typical electrochemical sensor. Adapted with Permission from [16] Copyright (2017) Elsevier

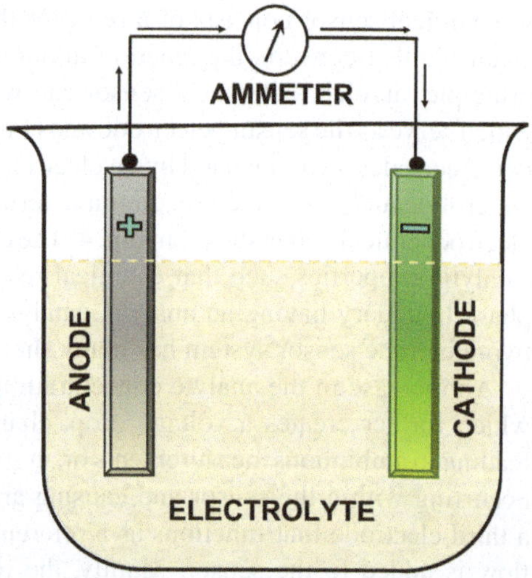

Fig. 4 Schematic representation of a two-electrode electrochemical cell. Adapted with Permission from [16] Copyright (2017) Elsevier

to establish a link with the electrolyte to maintain a current supply at the working electrode. Hence, all conventional electrochemical sensors are primarily designed as a three-electrode system. As the reactions are mainly sensed in the close vicinity of the electrode surface, the electrodes play a decisive role in the output performance of electrochemical sensors. Depending upon the chosen operation of a particular electrode, its material, modification of its surface, or dimensions greatly affect its detection quality.

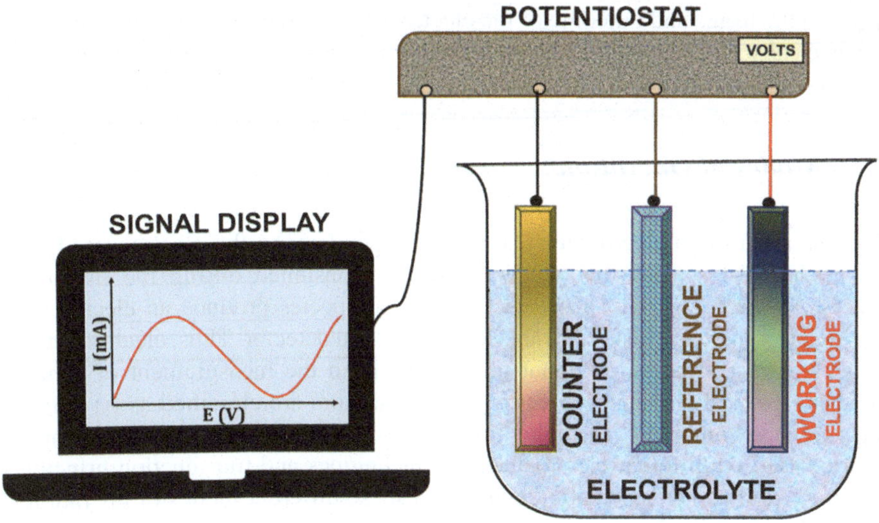

Fig. 5 Schematic representation of a three-electrode electrochemical cell. Adapted with Permission from [16] Copyright (2017) Elsevier

As previously mentioned, that in spite of the development in the electroanalytical techniques, there was a requirement for potential control. Fundamentally, as per the Nernst equation (Eq. 1) [16, 70], the potential of an electrochemical system is controlled by the activity of the analyte or its concentration.

$$E = E^0 + \frac{RT}{nF} \ln \frac{[RED]}{[OXI]} \quad (1)$$

where E = cell potential; E^0 = standard potential of a half-reaction; R = universal gas constant; T = temperature; n = number of electrons (eq. mol^{-1}) involved in the half-reaction; F = Faraday constant; [RED] = activity of the reduced species; and [OXI] = activity of the oxidized species. With the progress in the reaction, there is a decline in the analyte concentration that changes the cell potential. This kind of potential control also evades the complications associated with the application of overpotential, which is the sum of additional potential with that of the thermodynamic potential and is needed to overcome the influences of electric and kinetic resistance as well as activation energy [16, 70]. It's important to note that under the ideal conditions, reference electrodes must be based on a reversible reaction that satisfies the Nernst equation and should possibly have a low dependence on temperature [16, 65, 70]. Some of the common reference electrodes which meet these requirements are the normal (or standard) hydrogen electrode (NHE or SHE), saturated calomel electrode (SCE), and saturated silver/silver chloride electrode ($E_{Ag/AgCl}$). The metal electrodes (SCE and $E_{Ag/AgCl}$) are more frequently used and thus serve as an important factor in the usage and outcome of the electrochemical sensors, since they are more mechanically resilient and can be preserved.

2.1 Modes of Operation

The operation of an electrochemical sensor depends upon the electrons and other electrochemical species that are generated or consumed during the interaction process [19, 65, 71–74]. These electrochemical species produce an electrochemical signal, which is measured by an electrochemical detector. Thus, electrochemical sensors extract information about the sample from the measurement of some or another electrical parameter. Electrochemical sensors are classified as (i) amperometric, (ii) voltammetric, (iii) potentiometric, (iv) conductometric, and (v) impedimetric based on different electroanalytical techniques and transduction principles. Figure 6 exhibits the schematic diagrams of these different modes of operation of electrochemical sensors.

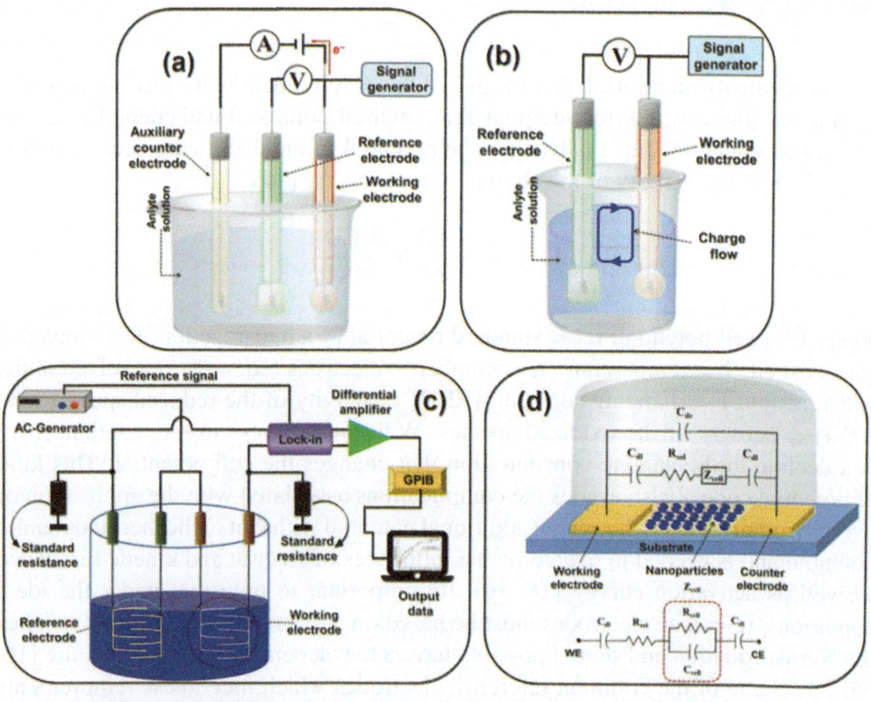

Fig. 6 Schematic representation of different modes of operations for electrochemical sensors; **a** amperometric/voltammetric, **b** potentiometric, **c** conductometric biosensors, and **d** equivalent circuit of the impedimetric sensor (Cdl = double-layer capacitance of the electrodes, Rsol = resistance of the solution, Cde = capacitance of the electrode, Zcell = impedance introduced by the bound nanoparticles, and Rcell and Ccell are the resistance and capacitance in parallel) [19]

(i) Amperometric Sensors: These sensors utilize an applied potential for causing oxidation or reduction of the analyte. The sensing is done by recording the change in current response with change in time and the concentration of the analyte [74, 75]. Basically, if during as electrochemical reaction a measurable current is to be detected between the two electrodes, amperometric technique is used. Their limit of detection can be as low as 10^{-5} M. They are one of the most popular electrochemical sensors owing to their sensitive detection and ease of use.

(ii) Voltammetric Sensors: These sensors operate by scanning a certain region of potential, and the current response is depicted in the form of a peak or a plateau [74, 76]. The resultant signal is proportional to the analyte concentration available in the system. The voltammetric techniques that are more generally used consist of linear sweep voltammetry (LSV), cyclic voltammetry (CV), differential pulse voltammetry (DPV), and square wave voltammetry (SWV) [74, 76, 77]. Although all these techniques work on the same principle, the potential region is scanned in differently. The detection limits respectively, for LSV, CV, DPV, and SWV are 10^{-5}, 10^{-5}, 10^{-7}, and 10^{-8} M [76]. Hence depending upon the scanning methods, DPV and SWV are the most sensitive voltammetry techniques [78].

(iii) Potentiometric Sensors: These sensors record a potential difference between the two electrodes in an electrochemical reaction. For this technique, as opposed to amperometry, the change in current is minimal while the change in potential is measurable [74, 79]. Since the potentiometric sensors are often constructed in such a manner as to produce signals with regard to changes in concentration of a particular ion, they are more commonly known as ion selective electrodes (ISEs) [79]. Apart from the working principle, potentiometric cells differ from amperometric or voltammetric cells on account of their set-ups. Potentiometric sensors use two reference electrodes to determine the potential change with changing target concentrations, as opposed to the one reference electrode employed in traditional electrochemical sensing techniques. These sensors enjoy widespread usage because of their small size, low limit of detection, rapid response, and ability to operate in conditions that are independent of sample volume.

(iv) Conductometric Sensors: These sensors function by detecting the conductance or resistance between the electrodes in an electrochemical cell. The conductance or resistance change in this technique is assessed from the response of interactions taking place in the analyte [74, 80]. They usually determine the changes in the ionic strength in the electrochemical cell arising due to catalytic modifications to estimate target molecules qualitatively and quantitatively [80, 81]. These sensors are preferred modes for conductivity detection as they can be employed for sensing electroactive as well as electroinactive species. This technique uses a conductivity measuring electrode that can either be in direct contact with the sample analyte or is covered with a thin insulating layer. This detection method can sense all the properties with respect to a particular class of compounds or even an individual analyte [82].

(v) Impedimetric Sensors: These sensors determine impedance (both resistance and reactance) between the electrodes in an electrochemical cell. They perform by monitoring the change in resistance and capacitance occurring at the interface of the working electrode with respect to the varying concentration of analyte [74]. The impedimetric technique is also called electrochemical impedance spectroscopy (EIS) technique. The impedimetric sensors carry out the measurements by applying an alternating current that affects the voltage, and the change in the resultant voltage signal is recorded as a function of frequency (1×10^{-5} to 0.7 V) [83]. Impedimetric sensors are stable and have good reproducibility in their responses. They possess the ability to respond to a wide linear range, have low power requirement, and are highly sensitive to slight changes in impedance [57, 74]. Additionally, few other types of electrochemical sensor but potentiometric, amperometric, and conductometric are the most widely used electrochemical detection techniques as the electrical signals recorded in these can be easily linked by Ohm's Law of electricity which states that "the product of current in amperes and the resistance in ohms equals the potential difference in volts" [71].

2.2 *Parameters Defining Performance of Nanomaterial-Based Sensors*

Ever since the incorporation of nanomaterials was reported to influence the performance of detection techniques, extended efforts have been undertaken to fabricate, design, and employ various nanomaterial-based sensors for different sensing applications [2, 84, 85]. As this review aims to discuss the impact of nanocomposites on the operation of sensors, an understanding of the important parameters that determine the working of the sensing devices is imperative before delving deep into the magnanimous expanse of sensors and actuators. An amalgamation of parameters that extended the improvement in sensor functionality has been widely investigated in relation to this cause. Some of these factors are shown in Fig. 7 and are discussed below.

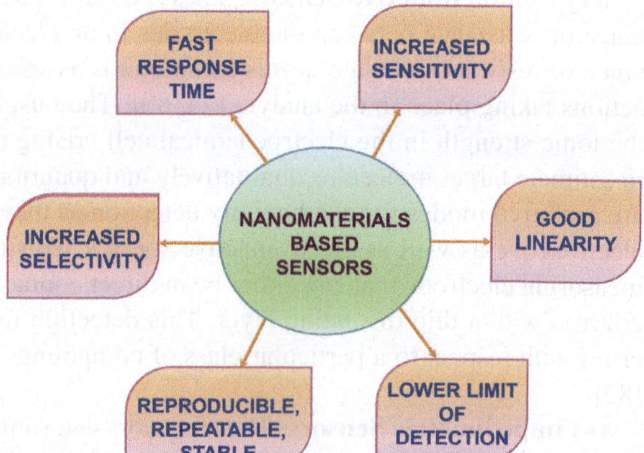

Fig. 7 Important performance parameters of nanomaterial-based sensors. Adapted from [2]

(a) Sensitivity

The sensitivity of a sensor is described as the ratio between minimum input of a physical parameter and a detectable output created due to that physical parameter [2]. In simple words, sensitivity (S) measures the output change (ΔY) with respect to the input change (ΔX). It has been noted that nanomaterial-based sensors have displayed an improved sensitivity that has been ascribed to the increased surface-to-volume ratio of the nanodimension [2, 86]. This improved sensitivity is accounted to an increase in conductivity which arises due to the controlled shapes and dimensions of the nanomaterials.

(b) Linearity

Linearity or non-linearity is a parameter in the operational window [2]. The linearity expression of the senor is defined by the extent of departure of the calculated curve of the sensor from its ideal one. Linearity or non-linearity is basically the deviation of the output curve of the sensor from a particular straight line. Nanorange sensors have demonstrated good linearity in their calibration curves owing to the reproducible and uniform nature of the adopted nanomaterials [2, 87].

(c) Limit of Detection [LoD]

The limit of detection (LoD) of the sensors is defined as the reliable detection of the lowest signal or the lowest concentrations to be determined from the signal by the sensor [2, 88]. It gives the smallest amount of the analyte present that can be distinguished confidently or statistically from zero and intertwines both the sensitivity and the resolution. Traditional sensors suffer greatly from poor linearity in calibration curves, which results in an overestimation or underestimation of analyte concentration [2, 84]. Owing to their reproducibility and their uniformity of shape and size, nanomaterial-based sensors have evolved as a feasible solution by enabling the generation of linear calibration curves that result in the enhanced sensitivity of the target species. Moreover, the miniaturization of materials provides an increased surface area in the nanorange, which lowers the limit of detection significantly [2, 85].

(d) Selectivity

The selectivity of a sensor refers to its capability to display the maximum response of a specific analyte while simultaneously discriminating the interference from other molecules. It is the ability of the sensor to generate the maximum output of the target species while distinguishing the response of other competing substances [2, 89]. Traditional sensors face a major drawback in terms of low selectivity. However, nanomaterial-based sensors with unique shapes have been fabricated that selectively target the analyte from a sea of competing substances present in the complex sample. Furthermore, due to the rise in the surface-to-volume ratio of the nanomaterials, the binding sites also amplify, resulting in the increased selectivity of the nanorange sensors [89, 90]. As a result, by fine-tuning the sensor's structure in the nanodimension, selectivity improves significantly.

(e) Response Time

The response time is typically the time required for a sensor to convert its output from a previous state to a final settled value, such that the final value should be within a tolerance band of the correct new value. The responsiveness of a sensor measures the time or frequency taken by an applied input parameter to produce an effective change [91]. The sensors do not change output immediately but take a certain period of time to indicate this effect with a change in an input parameter. This time is known as the response time and is basically the time that a sensor requires for its output to transform from its previous value to a final value under the action of an input change [2, 90]. Nanomaterial-based sensors have exhibited response times lower than those of traditional sensors [92]. This depicts a quicker response to the change in input for the nanomaterial-enabled sensors.

(f) Reproducibility, Repeatability, and Stability

Reproducibility and repeatability of sensors are parameters to measure precision on an intra- and inter-day basis, respectively. Reproducibility assesses whether an entire sensing experiment can be reproduced in its entirety, while repeatability refers to the variation in measurements from a single instrument or person under identical conditions [2, 93]. Together with reproducibility and repeatability, stability is also an important factor that determines the efficiency of a sensor. The sensor's stability is defined by its ability to withstand hostile environments and long operational hours [94]. It tells about the degree to which sensor characteristics remain constant over time. Nanomaterial-based sensors are found to hold their structure even after functioning repeatedly for a longer duration of time without significantly losing their precision or performance, in contrast to traditional sensors, whose performance often degrades if used for a long time in aggressive environments [95]. Hence, it has been widely acclaimed that sensors exhibit good reproducibility, repeatability, and stability in the nanorange [93–95]. Now that the major parameters determining the performance of the sensing mechanism have been identified and investigated, various rigid and flexible sensor actuators developed from metallic nanocomposites for different applications will be reviewed.

3 Review of Metal Nanocomposite-Based Electrochemical Sensors

3.1 Gold (Au) Nanocomposite-Based Electrochemical Sensors

Au Nanocomposite-Based Traditional Electrochemical Sensors: Gold nanoparticles (AuNPs), due to their exciting physiochemical properties have been momentously used in electrochemical sensors. The interesting properties of AuNPs include their high conductivity that allows an intensive electron exchange between the electrode and the analyte [96, 100], their small size, their stable and uniform surface that enables functionalization which further leads to improved selectivity and sensitivity of the electrochemical sensors [97]. Researchers have shown that combining AuNPs with rGO, nitrogen-doped graphene, and other conductive materials enhances their electrochemical detection capability [98, 99]. Ting et al. [100] demonstrated that cysteamine-capped AuNPs functionalized with graphene quantum dots (GQDs) exhibited LODs of 0.05 and 0.02 nM, and high sensitivities of 3.69 and 2.47 μA/nM for the electrochemical detection of Copper ($Cu^{2+)}$ and Mercury (Hg^{2+}) ions, respectively. They attributed the high activity of the electrochemical sensor to the synergetic effect between cysteamine-capped AuNPs and GQDs. Sakthivel et al. [101] fabricated an amperometric sensor using carbon black—polytyramine—AuNPs for the assessment of hazardous contaminant hydroquinone. Govindasamy et al. [102] assembled gold nanorods on multiwalled carbon nanotubes to determine the effect of dopamine neurotransmitter in biological and drug samples. They found that due to their speediness, selectivity, and sensitivity, this electrochemical sensor was well matched for the detection. Majhi et al. [103] prepared a p-type gas sensor from gold (Au) and nickel oxide (NiO) core–shell nanoparticles. They found that the response of p-type Au and NiO core–shell NPs enhanced as compared to pure NiO NPs and was 2.54 for 100 ppm of ethanol at working temperature of 200 °C. A sensing interface for finding serotonin using molecularly imprinted polymers (MIPs) incorporated with AuNPs anchored on reduced graphene oxide (rGO)/polyaniline (PANI) nanocomposites was created by Xue et al. [104]. This material was helpful in not only improving the selectivity and sensitivity for serotonin sensing, but also enhancing the conductivity of the membrane and was also able to prohibit interferences from endogenous complexes. Ma et al. [105] achieved sensitive detection of glucose by modifying glass carbon electrode with AuNPs—polypyrrole—two-dimensional MoS_2 nanosheets. Wang et al. [106] were able to develop a selective, sensitive, and exceptionally reusable sensor for electrochemical estimation of mercury ions from thymine-modified AuNPs/rGO nanocomposites. A versatile sensor designed by modifying of carbon paste electrodes with AuNPs was reported by Abdel-Raoof et al. [107] for voltammetric detection of brexpiprazole. Integrating AuNPs increased the sensitivity of the carbon paste electrodes towards brexpiprazole. The advancement in synthesis of non-enzymatic electrodes from high-performance Au-graphene nanocomposites for high reliability evaluation of glucose was comprehensively discussed by Shu et al. [108]. Jiao et al. [109] constructed high-sensitivity electrode from Au-reduced graphene oxide/poly (diallyldimethylammonium chloride) (PDDA) nanocomposites through a facile, cost-effective, one-pot technique. This sensor displayed a large linearity range for determination nitrite ions present in water supplies, meat, and dairy products. Meng et al. [110] investigated the ethanol gas sensing properties of gold/tin oxide/reduced graphene oxide ($Au/SnO_2/rGO$) nanocomposites. They exhibited that $Au/SnO_2/rGO$ nanocomposites sensors had low operating temperature as well as excellent reproducibility with linearity in wide range of 1 to 1000 ppm. Turcheniuk et al. [111] reviewed high sensing applications of graphene—graphene—Au NPs hybrid materials for DNA study, and also for analysing proteins, dopamine, and hydrogen peroxide (H_2O_2). They also reported fabrication of a non-enzymatic H_2O_2 electrodes for sensitively detecting H_2O_2 levels in human serum obtained from human cervical cancer cells. A highly sensitive nanotube hybrid developed from CNT/Au/SnO_2 nanotubes was applied by Du et al. [112] to detect carbon monoxide (CO) gas at ambient temperature. The present gas sensor indicated a high level of sensitivity of about 70 and with a recovery time of less than 20 s for 2500 ppm concentration of CO and showed superior response.

Au Nanocomposite-Based Flexible Electrochemical Sensors: Lee et al. [113] synthesized a flexible and transparent gas sensor from single-walled carbon nanotubes (SWCNT) decorated with AuNPs. They observed that this gas sensor functioned as a detector for determining even very low concentrations of ammonia gas (NH3) with a sensitivity of up to 255 ppb (parts per billion) at room temperature. This detection was done in terms of the deviation of electrical resistance of SWCNT films and is reported as one of the lowest values of concentration estimated for nanotube-derived sensors. Sershen et al. [114] investigated the working of a microfluidic device embedded with an optically controllable hydrogel valve prepared from AuNPs immobilized in thermosensitive polymers. These composite hydrogel actuators had a response time of less than 5 s. Shi et al. [115] constructed a photothermal hydrogel actuator that demonstrated a precise finger-like bending on light treatment. To obtain this one-by-one bending during illumination, the hydrogel actuator was prepared by combining the hydrogel with AuNPs and non-thermoresponsive polyacrylamide. They illustrated that the nanocomposite actuators produced by this method were flexible and achieved reversible bending as well as non-bending movements after being irradiated with light. Zhu et al. [116] designed a layered hydrogel nanocomposite consisting of AuNPs and AuNSs (Au nanosheets). They discovered that this hydrogel actuator underwent a spatially anisotropic deformation upon light irradiation. It was observed that the composite hydrogel continued to change its shape until it reached equilibrium after 15 min, and turning off the light restored it to its original state approximately after 5 s, indicating its high reversibility. Lim et al. [117] integrated gold–carbon nanotube–gold (Au/CNT/Au) sensors together with flexible thin-film circuits and implanted them on a soft elastomeric membrane to create a sodium detection system. The nanocomposite sensor had high stability and, even with continuous bending, exhibited mechanical robustness as well as reliability. This sensor-circuit integrated system demonstrates stable sodium measurements when mounted on the skin with minimized motion artefacts. The sensitivity of this fully flexible sodium sensor was found to be 55.5 0.3 mV/decade and showed less than 3% change when mounted on the skin. The present study revealed immense potential for the development of a wireless all-in-one wearable sensor system for monitoring health conditions. Figure 8 establishes the mechanical reliability and describes the performance characteristics of the Au/CNT/Au wearable sensor system.

A flexible non-enzymatic electrochemical device assembled from flexible AuNPs/PANI/carbon cloth (CC) for glucose sensing was reported by Xu et al. [118]. The electrochemical measurements revealed that that even after repetitive bending of the flexible electrode for 300 times, the response remain unchanged. The

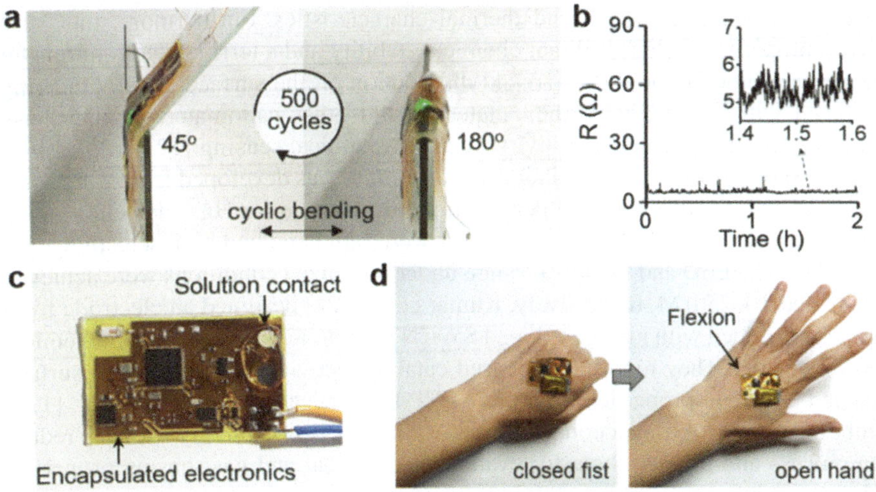

Fig. 8 Mechanical and performance reliability of the all-in-one flexible system. **a** Cyclic bending test to validate the mechanical reliability of the flexible system (500 cycles; the radius of curvature: 3 mm). **b** Change of electrical resistance of the system during the cyclic bending showing the negligible change. **c** Device that is encapsulated by an elastomeric membrane except the solution contact opening on the sensors. **d** Photos of the wireless thin-film electronics mounted on the hand with repetitive hand movements to make a closed fist and open hand. Reprinted with Permission from [117] Copyright (2021) Elsevier

experiments displayed broad linear range (10.26 μM to 10.0 mM), high sensitivity (150 μA cm^{-2} mM^{-1}), and a detection limit (3.08 μM), for the AuNPs/PANI/CC non-enzymatic glucose sensor. The accumulation of metal NPs and catalytic leaching during harsh processes hinders the catalytic and electrocatalytic operations of metal NPs. Consequently, high surface area catalysts that are stable and leaching free are not only extremely desirable but are also challenging. Patil et al. [119] investigated the sensing activities of AuNP-hosted mesoporous nitrogen-doped carbon matrix, fabricated from bovine serum albumin (BSA) through calcination (AuNPs@NBSAC). AuNPs@NBSAC-flexible sensor system indicated outstanding linear response towards dopamine throughout the concentration range from 1 to 50 μM with a detection limit of 0.05 μM. The nanohybrid sensor also demonstrated a brilliant response towards uric acid, for a wide detection range from 5 to 200 μM with its limit of detection being 0.1 μM.

3.2 Silver (Ag) Nanocomposite-Based Electrochemical Sensor

Ag Nanocomposite-Based Traditional Electrochemical Sensors: Silver nanoparticles (AgNPs) have emerged as one of the most essential nanoparticles due to distinctive electrical, optical, and thermal characteristics. Furthermore, they also possess properties such as brilliant chemical stability and a terrific electron transfer rate that are based upon their size and distribution on the surface [120]. On being combined with conductive carbon materials and other nanomaterials, they have demonstrated an enhanced sensitivity in electrochemical sensing [121, 122]. Eksin et al. [123] described an electrochemical sensor that was developed from folic acid-functionalized AgNP-integrated pencil graphite electrode for Hg^{2+} detection in tap water samples. These AgNP-based sensors were rapid and had high selectivity and sensitivity. The LoD and the linear range under optimized conditions were achieved as 8.43 and 10–25 μM, respectively. Kumar et al. [124] designed an electrode from AgNPs embedded with nitrogen-doped SWCNT for non-enzymatic electrochemical sensing of urea. They reported increased catalytic activity with significant surface area and electrical properties from the AgNP-based electrode. Mirzaei et al. [125] synthesized Ag/Fe_2O_3 nanocomposites have core–shell structure by two-step reduction process and utilized them for evaluating electrical and gas sensing properties of low concentrations of ethanol. It was found that unusual nanoarchitecture as well as the chemical and electronic sensitization effect caused by AgNPs in Ag/Fe_2O_3 sensors modulated the gas sensing properties of pristine Fe_2O_3 sensors. Zhang et al. [126] put forward a facile strategy for construction of an OTA aptasensor for electrochemical tracing of ochratoxin A(OTA) by depositing AgNPs on polydopamine nanospheres (PDANSs). The following AgNPs/PDANSs-based OTA aptasensor had a low detection limit of 0.57 pM and provided a wide linearity in the range of 1.0×10^{-6} to 1.0×10^{-2} μM. It also successfully detected OTA in red wine samples displaying its capability for electrochemical analysis of various toxic molecules with outstanding stability, specificity, and reproducibility. Yao et al. [127] created an electrochemical sensor from copper porphyrin metal organic frameworks (Cu-MOFs) embedded with multiwalled carbon nanotubes (MWCNT) and decorated with Au@Ag core–shell nanoparticles to determine the concentrations of acetaminophen and dopamine. Photochemical methods for synthesis of Ag nanoparticle anchored on porous graphitic C_3N_4 (p-g-C_3N_4) were reported by Jiang et al. [128]. These Ag/p-g-C_3N_4 nanocomposites indicated brilliant electrocatalytic reduction of H_2O_2. Additionally, the cyclic voltammetry and amperometry measurements of Ag/p-g-C_3N_4 nanocomposite electrochemical sensor yielded exceptional analytical response to H_2O_2, with rapid response, wide linear range, and low detection limit. Figure 9 describes cyclic voltammograms (CVs) of the bare GCE (glassy carbon electrode), p-g-C_3N_4/GCE (porous g-C_3N_4-modified GCE), a-Ag/p-g-C_3N_4/GCE (Ag-deposited p-g-C_3N_4 in the absence of N-GQDs), and Ag/p-g-C_3N_4/GCE (Ag-deposited p-g-C_3N_4 in the presence of N-GQDs) in N_2 saturated 0.2 M pH 7.4 PBS (phosphate-buffered saline) containing 5.0 mM H_2O_2. N-GQDs here stand for nitrogen-doped graphene quantum dots. The CVs reveal that the bare GCE and p-g-C_3N_4/GCE exhibit weak responses towards the reduction of H_2O_2, whereas the presence of Ag nanoparticles, in a-Ag/p-g-C_3N_4/GCE, results in a good H_2O_2 reduction response.

Nantaphol et al. [129] investigated the electrochemical performances of AgNP-modified glassy carbon. They determined enzymatic assay of cholesterol by carrying out cyclic voltammetry and chronoamperometry. The estimated results showed an

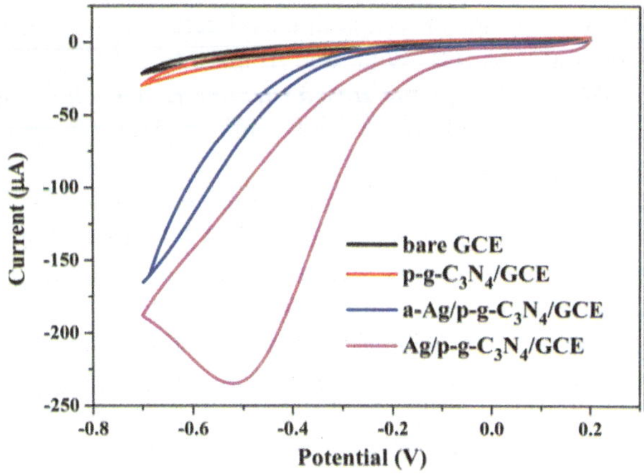

Fig. 9 Cyclic voltammograms (CVs) of bare GCE, p-g-C_3N_4/GCE, a-Ag/p-g-C_3N_4/GCE, and Ag/p-g-C_3N_4/GCE in N_2-saturated 0.2 M PBS at pH 7.4 in the presence of 5.0 mM H_2O_2, scan rate: 50 mV s^{-1}. Reprinted with Permission from [128] Copyright (2011) Royal Society of Chemistry

extensive linear range from 3.9 to 773.4 mg/dL with 0.99 mg/dL detection limit. The proposed sensor with high specificity indicated its potential application for other important biomarkers as well. Zahran et al. [130] prepared an AgNP-dependant sensor for electrochemical detection of atrazine (Atz), a toxic herbicide. The tracer was fabricated by immobilizing dissolved organic matter (DOM) with AgNPs which was then implanted on a glassy carbon electrode. The electrode traced Atz with a linear range of 10–140 μg/L and had acceptable recovery values. A highly sensitive electrochemical immunosensor based on AgNP-coated graphene (AgNPs-G) for the ultra-sensitive detection of avian influenza virus H7(AIV H7) was designed by Huang et al. [131]. This technique exhibited better signal amplification and showed a dynamic working range of 1.6 × 10(−3)~16 ng/mL, with a low limit of detection ~1.6 pg/mL. The novel immunosensor displayed high specificity and sensitivity towards AIV H7 and showed possibility for quick discovery of other pathogenic microorganisms. Ozcelikay et al. [132] described simple voltammetric sensor built on glassy carbon electrode from benzalkonium chloride, and AgNPs were the highly sensitive analysis of antiviral drug tenofovir. The nanocomposite sensor displayed a linearity range from 6.0×10^{-8} to 1.0×10^{-6} M, with a superb detection limit of 2.39×10^{-9} M. Traiwatcharanon et al. [133] produced an electrochemical sensor for estimating sodium ions (Na^+) by cyclic voltammetry. The electrode was developed by accumulating AgNPs with graphene oxide (GO) on a screen-printed silver electrode. The nanocomposite sensor was found to be highly sensitive with value of 0.269 mA/mM/cm^2, showed a low limit of detection of 9.344 mM, and exhibited linear relationship of 0–100 mM for food application. Moreover because of high selectivity, good stability, excellent reproducibility, they had real-time usage for Na^+ detection in fish sauce, seasoning powders, and other food samples. Imran et al. [134] explained the synthesis and working of sensitive electrochemical sensor for determining lipopolysaccharide (LPS) that provides a useful information regarding the indirect detection of various pathogens. The proposed sensor was based on AgNPs functionalized with positively charged chitosan-modified glassy carbon electrodes.

The nanocomposites accomplished detection over an extensive concentration range from 0.001 to 100 ng/mL and from 10 to 10^7 CFU/mL.

Ag Nanocomposite-Based Flexible Electrochemical Sensors: Amjadi et al. [135] reported the preparation and application of a highly flexible and sensitive strain sensor grown from silver nanowire (AgNW) embedded in a polydimethylsiloxane (PDMS) elastomer in one of the pioneering works on flexible and wearable electronic devices. The developed AgNW/PDMS elastomer strain sensor demonstrated strong piezoelectricity with gauge factors tunable in the ranges of 2–14, along with an increased stretchability of up to 70%. The extensive applicability of the proposed strain sensor was demonstrated by creating a smart glove integrated with five stretchable strain sensors that was utilized for the real-time detection of finger motions. The employability of the strain sensors in the smart glove was demonstrated by using the finger postures to control the finger motion of an avatar in the virtual environment. This AgNWs/PDMS nanocomposite strain sensor for detecting human motion is depicted in Fig. 10. As revealed by the figure, there is an exceptional agreement between the loading profile and the sensor response. Neella et al. [136] made a flexible temperature sensor based on Kapton sheets from RGO-Ag nanocomposite films. The RGO-Ag sensor's electrothermal performance in a hot–cold temperature set-up demonstrated outstanding temperature sensing characteristics, a quick response time of 470 ms, good stability, and repeatability.

Shengbo et al. [137] proposed a strain sensor constructed from AgNPs and AgNWs. This sensor demonstrated improved activity in terms of electronic skin and was found to possess high sensitivity and stretchability. The Ag NPs and NWs strain sensors exhibited strong piezoresistivity, a tunable gauge factor, and a linear increase in the strain rms of electronic skin and was found to possess high sensitivity and stretchability. The AgNPs and NWs strain sensors exhibited strong piezoresistivity, a tunable gauge factor, and a linear increase in the strain. The high gauge factor demonstrates the irreplaceable role of Ag NPs in the sensor. Furthermore, this strain sensor had an enhanced capacity to sense human movements like talking, finger bending, wrist raising, and walking. Ko et al. [138] reported the fabrication of a flexible, stretchable, and piezoresistive physical sensor from AgNPs/CNT nanocomposite films that can be used as a wearable medical device or electronic skin for continuous monitoring of human health. The AgNPs/CNT tracing material was highly sensitive, had a quick response, and revealed good electrical conductivity with remarkable sensing performances. They successfully operated at low voltages and displayed good mechanical stability under different conditions of pressure, loading, bending, and elongations. Zhang et al. [139] synthesized a flexible strain sensor based on carbon black and AgNPs embedded in a thermoplastic polyurethane matrix for human motion detection. The prepared wearable strain material had high stretchability and possessed brilliant static as well as dynamic stability. In contrast to a bare carbon black-based strain sensor, the carbon black/AgNPs nanocomposite strain sensor indicated an improvement of 18 times in sensitivity at 100% strain. Owing to its excellent strain-detecting response, this carbon black/AgNPs strain sensor can offer extensive possibilities in the sensing of finger bending, wrist rotation, elbow flexion, and various other human motions.

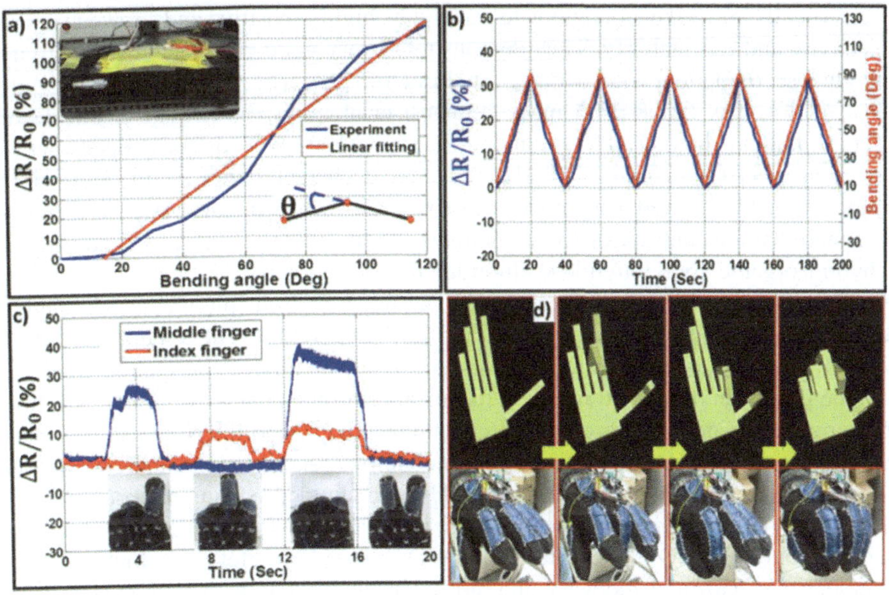

Fig. 10 Human motion detection by the sandwich-structured AgNWs_PDMS nanocomposite strain sensors: **a** response of the sandwich-structured strain sensor to the bending angles from 0° to 120° (inset: photograph of the artificial finger); **b** response of the strain sensor under repeated bending/relaxation cycles (10° to 90°): **c** motion detection of index and middle fingers; **d** control of avatar fingers in the virtual environment using wireless smart glove system. Reprinted with Permission from [135] Copyright (2014) American Chemical Society. Adapted from [137]

3.3 Platinum (Pt) Nanocomposite-Based Electrochemical Sensors

Pt Nanocomposite-Based Traditional Electrochemical Sensors: Niu et al. [140] presented an innovative biocompatible carrier for protein/enzyme synthesized from Platinum-3D graphene oxide aerogel (Pt-3DGA) nanocomposite. They aimed for improving electron transfer efficiency between redox protein/enzyme and electrode from this nanocomposite. Electrochemical measurements showed excellent electrocatalytic reduction with a broad range of detection and good recovery in real samples. Zhang et al. [141] constructed an aqueous ammonia sensor from PtNPs polypyrrole/nickel foam (Pt-PPy/Ni foam) for direct detection of ammonia in water. Electrochemical analysis revealed good electrocatalytic oxidation of ammonia and demonstrated a detection limit of 12.36 nM, with a sensitivity of 4.19 μA μm^{-1} and an extensive linear range from 0.5 to 400 μm. Cui et al. [142] prepared an electrode by implanting a novel ordered flower-like PtNPs on helical CNTs for electrochemical biosensing of bovine serum albumin. One-step synthesis of a non-enzymatic glucose sensing from an electrochemical sensor based on rGO supported Pt–nickel oxide nanoplate arrays (Pt–NiO/rGO) was reported by Wang et al. [143].

The vertically arrayed nanocomposite was stable, highly sensitive and showed superb selectivity and repeatability. They demonstrated very good electrocatalytic activity for glucose oxidation with a detection limit of 2.67 μM and indicated a wide range of linearity 0.008–14.5 mM. Leonardi et al. [144] fabricated electrode of Pt and titanium dioxide (TiO_2) nanoparticle-modified rGO for the amperometric detection of H_2O_2 in milk. The electrochemical characteristics of Pt–TiO_2–rGO nanocomposites showed very good sensitivity and broad linear response proving its suitability for analysis of toxic chemicals in milk. Ong et al. [145] developed a hybrid organic (polyvinylidene fluoride) PVDF–inorganic M–rGO–TiO_2 (M = Ag, Pt) nanocomposite for detecting H_2 gas released from photocatalytic degradation of volatile organic compounds, at room temperature. The synergistic effect of TiO_2 nanotubes, Ag/Pt nanoparticles and rGO, facilitates the charge transfer, and the presence of both promotes mechanical reinforcement as well as flexibility. The proposed nanocomposites have tremendous potential to introduce an economical, easy-to-maintain, and highly durable technology for environmental remediation and solving global energy crisis. Ma et al. [146] assembled sensing nanocomposites from PtNP-functionalized WO_3 (tungsten oxide) (Pt/WO_3) for tracing and measuring carbon monoxide (CO). The obtained Pt/WO_3 nanocomposites were highly sensitive, selective, had quick recovery, and showed brilliant catalytic response to CO even at low concentrations. Yang et al. [147] presented an electrochemical sensor based on PtNP-decorated rGO sheets (Pt–rGO) for analysing the electrocatalytic activity of nitrite ions. The cyclic voltammetry results indicated that the Pt–rGO nanocomposite demonstrated increased electron transfer rate and significant progress in the electrochemical activity for the nitrite oxidation. Amperometric studies revealed high sensitivity, good linearity, quick response, better stability, fine reproducibility, and a low detection limit in the performance of Pt–rGO nanocomposite towards nitrite sensing. Xu et al. [148] reported the electrochemical sensing of dopamine (DA) and uric acid (UA) in the presence of high concentration of ascorbic acid from Pt/rGO-modified GCE. The cyclic voltammetry and differential pulse voltammetry measurements exhibited good performance by the Pt/rGO-modified electrode towards detection of DA and UA individually as well as simultaneously in the presence of 1.0 mM ascorbic acid. Li et al. [149] built an electrochemical sensor on glassy carbon electrode (GCE) by electrodepositing PtNPs on mesoporous carbon (MC) surface and combining Nafion with Pt-MC composites. The electrochemical measurements of the obtained sensor displayed outstanding electrocatalytic oxidation of dopamine. The sensor was highly stable, reproducible, and selective as well as showed a sensitivity of 12.23 μA mM^{-1} with a limit of detection of 0.034 μM towards dopamine detection. Upan et al. [150] synthesized an electrochemical aptamer-based sensor from PtNP-functionalized carboxylated-graphene oxide (PtNPs/GO-COOH). (PtNPs/GO-COOH)-modified electrode was utilized for sensing alpha-fetoprotein (AFP). The fabricated aptasensor offered a linear range from 3.0 to 30 ng mL^{-1} with a detection limit of 1.22 ng mL^{-1}. Furthermore, it was highly selective, stable, and sensitive towards the sensing of AFP in human serum samples with good recovery rate. Peng et al. [151] constructed a simple and high-performance electrochemical sensor for determination of N-nitrosodiphenylamine (NDPhA), considered as a carcinogen that is used widely in daily human necessities as an additive. They fabricated poly(diallyldimethylammonium chloride)-stabilized graphene/PtNPs (PDDA–Gr/PtNPs) for NDPhA detection by wet-chemical approach. PDDA–Gr/PtNP-based sensing platform exhibited outstanding electrooxidation of NDPhA and demonstrated good linearity with a range of 1.0×10^{-7}–$5.0 \times 10 - 5$ mol L^{-1} and low detection with a limit of 3.3×10^{-8} mol L^{-1}. Mazzotta et al. [152] investigated electrochemical tracing of H_2O_2 by using glassy carbon electrode functionalized with tunable PtNPs. The amperometric measurements indicate effect of size on electrocatalytic properties of PtNPs demonstrating high performances by smaller NPs in the detection of H_2O_2 in the concentration range of 25–750 μM, with a limit of detection of 10 μM. The as-prepared electrodes reveal good sensing of H_2O_2 in tap water samples as displayed by their sensitivity, selectivity, repeatability, and stability analysis.

Pt Nanocomposite-Based Flexible Electrochemical Sensors: Tanner et al. [153] obtained a controllable strain gauge factor from a strain sensing device fabricated by depositing a two-dimensional layer of PtNPs between interdigitated Au electrodes on an oxidized silicon substrate. They found that the nanoparticle array demonstrated semiconducting characteristics when their density was less than a threshold value, beyond which they exhibited metallic behaviour. They reported that a PtNP-based device at room temperature revealed strain sensitivity that was more than an order of magnitude higher as compared to continuous metallic films. PtNP arrays have the potential to achieve highly sensitive strain gauges and can be easily integrated on silicon or flexible substrates. Tanner et al. [154] produced a strain sensor based on PtNPs with diameters of 4–5 nm synthesized by room temperature sputtering. The strain sensing devices made from these films have a high sensitivity that is controlled by the nanoparticle density. They also showed a good temperature sensitivity that can also be controlled by the nanoparticle density, providing an additional degree of design flexibility. Skotadis et al. [155] reported on the electrochemical tracing of ethanol using a sensor made of PtNPs and poly(2-hydroxyethyl methacrylate) (PHEMA) deposited on flexible polyimide substrates. On being exposed to the vapours of ethanol or humidity, the resistance or capacitance of the sensor changes, indicating their applicability in the detection of such gases. Nguyen et al. [156] presented the fabrication of amperometric biosensor electrodes synthesized from PtNPs-MWCNTs and a conductive polymer deposited on a flexible substrate for the in vivo determination of glutamate. The proposed sensor was designed to monitor the extracellular dynamics of glutamate and other potential biomarkers that may arise in the event of a spinal cord injury. The results obtained revealed that the implantable biosensors are highly sensitive, selective, flexible, stable, and can be easily inserted into the spinal cord for in vivo evaluation. Liu et al. [157] investigated a flexible temperature sensor developed from platinum and indium oxide (In_2O_3). The Pt/In_2O_3 temperature sensors implanted on flexible polyimide substrate exhibited a maximum voltage of 34.33 mV as output with an ultra-high sensitivity of 204.35 μV/°C, for the low-temperature recordings. The sensor was found to have good repeatability, stability, and the temperature drift, allowing for its probable usage in the areas of industrial robot and biomedicine.

3.4 Palladium (Pd) Nanocomposite-Based Electrochemical Sensors

Pd Nanocomposite-Based Traditional Electrochemical Sensors: Johnson et al. [158] built hydrogen sensor based on palladium (Pd)-functionalized multilayer graphene nanoribbon networks. The sensors reflect a good sensitivity to hydrogen even at low concentration levels with a quick response rate and recovery time. Li et al. [159] proposed a hydrogen sensor prepared by depositing PdNP-elaborated PMMA (polymethyl methacrylate) organic sol on silica microfiber. Owing to the synergistic effect of silica microfiber and amorphous structure of PMMA film, the PdNPs effectively absorb H_2 molecules causing a shift in resonance wavelength and a change in refractive index. The electrochemical sensing measurements of these sensors demonstrate an average sensitivity of 5.58 nm/%. Lu et al. [160] developed an amperometric sensor for bromate detection. For sensor fabrication, glassy carbon electrode was modified by incorporating PdNPs in graphene oxide nanosheets. Gao et al. [161] presented gas sensor for H_2 detection by decorating silicon (Si) nanomesh structure PdNPs. The PdNP-modified H_2 sensor reveals excellent sensitivity for H_2 gas as compared to an Si thin-film sensor. Furthermore, the sensing device displays good stability, quick response, and improved selectivity towards H_2 gas determination. He et al. [162] achieved H_2O_2 sensing by preparing PdNPs assembled on multiporphyrin array ultra-thin films. These films demonstrated a shift of 6 nm in their absorption band on being exposed to an aqueous solution of H_2O_2. They were also found to act as light-harvesting units by generation of photocurrent. Hu et al. [163] constructed a sensing device for sensing chromium [Cr(VI)] in water. The sensor was prepared by depositing AuPdNP-functionalized electrochemically reduced graphene oxide (AuPdNPs/ERGO) nanocomposite on glass carbon electrode. The fabricated AuPdNPs/ERGO nanocomposite-modified electrode demonstrates good detection of 0.013 μM in two concentration ranges from 0.05 to 5 μM and 5 to 1000 μM. Furthermore, the proposed sensor is highly selective, sensitive, and stable and displays a high rate of recovery towards Cr(VI) detection. Phan et al. [164] presented the synthesis and working of Pd-graphene nanocomposite-based H_2 sensors. These sensors were resistivity type and were utilized to evaluate different concentrations of H_2 and interfering gases at various temperatures. Pd-graphene sensors displayed good sensitivity, selectivity, and linear response and detected H_2 with very low limit of 0.2 ppm in concentrations ranging from 1 to 1000 ppm. The H_2 sensor exhibited robust mechanical nature, structural stability, long-term durability, and high resistance to oxidation.

Pd Nanocomposite-Based Flexible Electrochemical Sensors: Sun et al. [165] fabricated high-performance H_2 sensors with outstanding mechanical flexibility from PdNPs embellished CNTs implanted on thin plastic substrates. These sensors showed sensitivity of ~75% for 0.05% hydrogen and with a quick time of response of ~3 s. The H_2 sensor exhibits unchanged performance even after being bent to a radius of 7.5 mm and after being operated for 2000 cycles of bending/relaxing. Chung et al. [166] prepared flexible sensors for hydrogen gas determination by PdNP-decorated graphene. The Pd-graphene sensor provides 33% response on being exposed to 1000 ppm H_2 and can detect low concentrations H_2 at room temperature. The flexibility of the sensor is excellent and can withstand bending up to a radius of 3 mm making it suitable for a wide-ranging application that involves mechanical durability and high gas response. Electrochemical bending tests as illustrated in Fig. 11 were carried out to investigate the flexible sensing application of Pd-graphene H_2 sensor. Zhang et al. [167] fabricated pH sensor from a composite PdNPs/nitrogen-doped carbon (NC) attached to the polyethylene terephthalate film. This proposed device from PdNPs/NC functions as strip-based solid-state economic pH sensor and exhibits flexibility, transparency, disposability, and user-friendliness. The prepared pH sensor demonstrated high selectivity and specificity and possessed a sensitivity of 55 mV/pH unit with a relative standard deviation of 0.79%. The sensor showed great potential for its usage in investigation of biological samples, such as saliva and gastric juices. Su et al. [168] reported the construction of a novel and flexible H_2 gas sensor that was achieved by assembling Pd-based complex on multiwalled CNTs (MWCNT-Pd) which was further erected on a polyester (PET) substrate. The MWCNT-Pd-based H_2 gas sensor displayed gas sensing response that was greater than the rigid sensors at room temperature. Nuthalapati et al. [169] developed a flexible sensor based on a nanocomposite of rGO and Pd (rGO-Pd) that combined with a polydimethylsiloxane (PDMS) substrate. The rGO-Pd nanocomposite device with high sensitivity and quick response was tested as a strain sensor as well as a temperature sensor. The strain sensor demonstrated high durability even beyond 1000 cycles, quick response of ~39 ms, and a gauge factor of ~22 to ~198 in the strain range of 0.05–0.625%. The temperature sensing activity by the rGO-Pd nanocomposite was characterized with excellent sensitivity and linearity for both negative and positive temperature coefficients.

3.5 Other Metal Nanocomposite-Based Electrochemical Sensors

Rikhari et al. [170] presented the corrosion resistance and biocompatibility studies of titanium (Ti) coated with polypyrrole/chitosan (PPy/CHI) composite. Ti-PPy/CHI composite displayed improved microhardness and adhesion strength. The corrosion protection ability as measured by dynamic electrochemical impedance spectroscopy demonstrated enhanced impedance PPy/CHI-coated Ti. Keerthi et al. [171] developed titanium nanoparticles (TiNPs) embedded in functionalized multiwalled carbon nanotube-based disposable electrochemical sensor for detecting ractopamine (RAC). The proposed sensor exhibited excellent electrochemical sensing ability with an extensive range of linear response (0.01–185 μM) and possessed ultra-low limit of detection (0.0038 μM). Additionally, this electrochemical sensing device showed satisfactory stability, reproducibility, repeatability, and demonstrated effective practicability with substantial recovery rate. You et al. [172] fabricated a detection electrode

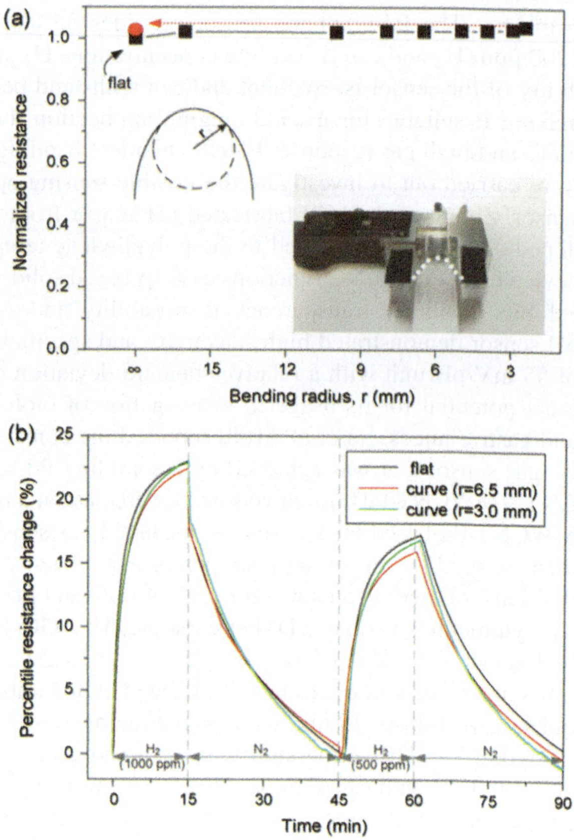

Fig. 11 Bending test for flexible sensing application. **a** The normalized resistance of the sensor with respect to bending radius. The red plot presents the normalized resistance when the sensor returned to the release state. **b** The sensing responses exposed to 1000 and 500 ppm hydrogen with respect to the bending radius of the sensor. Reprinted with Permission from [166] Copyright (2022) Elsevier

from nickel nanoparticles (NiNPs) dispersed in disordered graphite-like carbon (Ni-NDC) for using it in high-performance liquid chromatography (HPLC). The modified electrode possessed superb electrocatalytic ability, good stability to the electrooxidation of sugars and accomplished advanced detection limits for sugars. The electrode on being employed in HPLC was able to provide a good separation of four sugars (glucose, fructose, sucrose, and lactose) at low constant detection potential with a linearity of over 3 orders of magnitude. Comparative HPLC tests using Ni-NDC film electrode and a Ni-bulk electrode for detection of glucose, fructose, sucrose, and lactose are given in Fig. 12. Jia et al. [173] constructed a sensing device based on NiNP-implanted nanoporous carbon nanorods (Ni/NCNs) for glucose detection. The as-prepared Ni/NCN glucose sensor exhibited outstanding electrocatalytic behaviour with ultra-low detection limit, a broad linear detection, quick response time, and excellent stability and was utilized to examine glucose concentrations in human blood samples.

Wang et al. [174] built an electrochemical sensor from a nanocomposite comprising of cylinder-like NiNPs/nitrogen-doped carbon nanosheet/chitosan

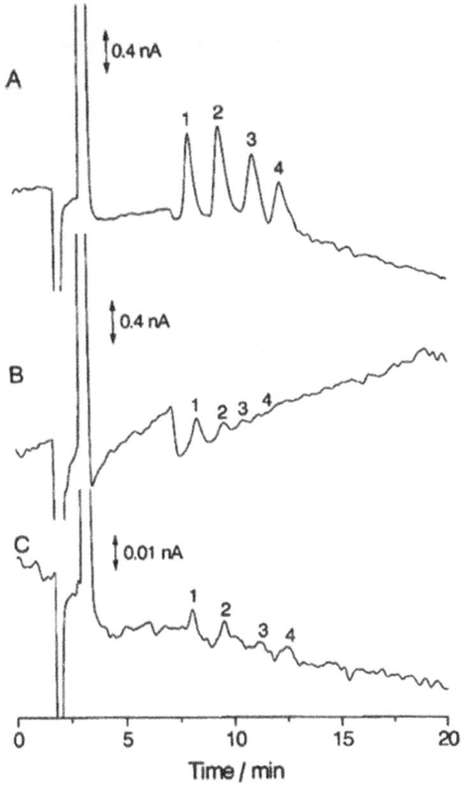

Fig. 12 Liquid chromatogram separation of mixture of glucose (1), fructose (2), sucrose (3), and lactose (4) at 0.8% Ni-NDC (A, C) and Ni-bulk (B) electrodes carried out with 0.03 M NaOH as mobile phase. Reprinted with Permission from [172] Copyright (2003) American Chemical Society

nanocomposite (NiNP/NCN/CS). NiNP/NCN/CS nanocomposite modified the glassy carbon electrode and was used for finding bisphenol A (BPA). The BPA sensor was stable and produced two linear differential pulse responses in the concentration ranges of 0.1–2.5 mM and 2.5–15.0 mM with a detection limit of 45 nM. Cui et al. [175] synthesized NiNP-designed graphene–diamond hybrid electrodes for electrochemical glucose detection. The NiNP-modified electrodes show good selectivity in glucose sensing with a linear range of 2 μM–1 mM and a detection limit of 2 μM. Riaz et al. [176] investigated the application a highly active electrocatalyst made from a nanocomposite of cobalt (CoNPs) and carbon for efficient H_2O_2 sensing on different sensing platforms. The CoNPs/carbon-modified glassy carbon electrode demonstrated an extremely high sensitivity of 300 $\mu A/cm^2$ mM with a detection limit of 10 μM. Due to its satisfactory performance in multiple sensing interfaces, this CoNP-based nanocomposite electrocatalyst can lead to numerous portable applications for practical sensing of H_2O_2. Hammud et al. [177] fabricated sensors for detecting fructose based on nanocomposites of hierarchical porous carbon incorporated with CoNPs. The cyclic voltammetry measurements using CoNP-modified carbon electrode possessed a limit of detection of 0.5 mM, whereas impedimetric measurements displayed broad dynamic range of 8.0–53.0 mM with a sensitivity of 24.87 Ω mM^{-1} for fructose sensing. Švorc et al. [178] applied an electroanalytical approach for the estimation of ibuprofen in pharmaceuticals and human urine samples by using bare boron-doped diamond electrode as sensor. The elaborated electrochemical protocol provided low detection limits of 4.1×10^{-7} and 9.3×10^{-7} mol L^{-1} in with satisfactory intra-day repeatability. The modified electrodes possessed reasonable selectivity and was also used to analyse the effect of possible endogenous compounds such as ascorbic acid, dopamine, caffeine, uric acid, and glucose present in the samples. In another work, Švorc et al. [179] investigated an electrochemical sensor to be utilized in practical applications of food quality control. They prepared the sensor from a miniaturized boron-doped diamond electrode for reliable determination of theobromine in chocolates. The electrochemical voltammetric measurements demonstrated low limits of detection of 0.42 and 0.51 μM with sufficient intra-day repeatability.

4 Conclusion

This chapter discusses reports on electrochemical sensors modified with different metallic nanocomposites to obtain an understanding of the recent trends in the arena of sensing devices. The efforts made to improve electrochemical sensing by incorporating metallic nanocomposites with various carbon-based nanomaterials were reviewed. The focus has been on metallic nanocomposites-based rigid and traditional electrochemical sensing systems as well as on innovative flexible sensors and actuators. These electrochemical sensors developed from metallic nanocomposites have widespread applications ranging from gas sensing, metal detection, biosensing, healthcare, wearable human–computer interface devices, food control, and environmental monitoring. The nanocomposite-based detection system has the capability to strengthen essential requirements that reinforce and advance the transition from traditional sensors towards fully integrated electrochemical sensing. Depending upon our literature analysis, it would not be wrong to say that the field of new electrochemical tracing techniques built upon metallic nanocomposites is continuously growing. It is expected that the rapidly evolving area of flexible and stretchable electrochemical sensing systems will become highly imperative in the future with their significant usage in different utilities related to human–machine interfaces in different segments of our day-to-day lives. Nonetheless, additional efforts are still needed to be undertaken to increase the functionalities and modalities while simultaneously addressing the existing challenges to achieve further developments in the realm of electrochemical sensing. The past few decades have witnessed rapid progress and increased applications of electrochemical sensing technology owing to their cost-effectiveness, quick response, simplicity, and mechanical strength. We have summarized the electrochemistry and the different electroanalytical techniques on which the workings of electrochemical sensors are based. Electrochemistry is a broad term that encompasses all of the typical dynamic operations that occur at the electrode–electrolyte interface. Electroanalytical techniques are employed to investigate outputs resulting from variations in electric current, charge, potential, phase, frequency, etc. due to chemical reactions occurring at the electrode–electrolyte interface. An electrochemical sensor comprises electrodes immersed in the analyte solution and operates by monitoring the electrical responses obtained at the electrodes. As the electrodes play a major role in recording the signals, their composition, morphology, surface functionalization, and mode of operation significantly influence the sensing potential of the electrochemical sensors.

Based upon the different electrical parameters recorded during the electroanalytical measurements, electrochemical sensors can be broadly categorized as amperometric sensors, potentiometric sensors, voltammetric sensors, conductometric sensors, and impedimetric sensors. The performance of sensors is investigated in terms of their sensitivity, limit of detection, repeatability, reproducibility, stability, selectivity, response time, etc. The impact of metallic nanocomposites on the different sensor parameters has been reviewed to demonstrate their effectiveness for different applications. Carbon nanomaterials like graphene, CNTs, and their derivatives, on account of their low density, easy functionalization, higher surface area, abundant adsorption sites, high charge mobility, electrical tunability, etc., have become one of the preferred choices for sensing applications, as reflected by the research trends. To further extend their applicability and to improve their response time, selectivity, and sensitivity, composites of carbon nanomaterials with metal nanoparticles and nanostructures are widely being used all over the world. The present work has been focussed on evaluating the electrochemical sensing capabilities of nanocomposites of metals with carbon nanomaterials, in both rigid and flexible configurations. Also, considerations about the sensors being mechanically compliant along with technological efficiency have been of paramount importance for this chapter. The presence of strong adhesion and reversibility is one of the most vital characteristics that needs to be targeted for flexible and stretchable sensors to ensure their high performance.

Our literature study has been aimed at providing a comprehensive review of the enhanced detection characteristics of nanocomposites of metals such as Au, Ag, Pt, Pd, Ti, Co, etc. with carbon nanomaterials. AuNP-based sensors have usage in the sensing of toxic gases and biological compounds, monitoring the quality of food and dairy products, glucose, and copper ion detection, etc. Flexible sensor actuators fabricated from AuNPs have displaced enhanced reversibility and excellent bending responses to light irradiation. On the other hand, Ag nanocomposites are used for the detection of mercury ions, hazardous chemicals, pathogens, cholesterol, etc. AgNP-modified flexible strain sensors have demonstrated extremely high stretchability and strong piezoresistivity. They have shown a remarkable ability to detect any kind of human movement and have been successfully employed in wearable electronics. Likewise, the effectiveness of nanocomposites PtNPs and PdNPs as electrochemical sensors was investigated. Pt nanocomposite sensors have been reported to exhibit increased sensitivity towards CO sensing, protein assays, and the tracing of toxic chemicals. Pt-functionalized flexible sensors have better strain as well as temperature sensitivity and can serve as biomarkers. Moreover, the Pd nanocomposites show good responses to H2 and chromium. Flexible PD sensors are applied for the investigation of biological samples, such as saliva and gastric juices.

In summary, metal nanocomposite-based electrochemical sensors have distinctive merits and have great potential for an extensive range of various applications, making them immensely popular. However, various aspects of the diverse areas of their usage should be brought together to facilitate the generation of new ideas and the eradication of prevailing limitations. Owing to technical revolutions and material evolutions, new avenues for exciting applications are continuously emerging in this field. Consequently, tremendous prospects for the further development of metallic nanocomposites-based electrochemical sensors exist. The constant progress in this fast-paced arena is expected to yield numerous opportunities as well as provide exciting breakthroughs.

Acknowledgements We are very grateful to Dr. Santosh Kumar Tiwary, Scientist (NAWA) of Warsaw University, Poland, for his feedback and suggestions that have proved immensely beneficial towards the improvement of this book chapter.

References

1. Said RAM, Hasan MA, Abdelzaher AM, Abdel-R, Ahmed M (2020) J. Electrochem. Soc.167:037549.
2. Munonde TS, Nomngongo PN (2021) Sensors 21:131
3. Fazio E, Spadaro S, Corsaro C, Neri G, Leonardi SG, Neri F, Lavanya N, Sekar C, Donato N, Neri G (2021) Sensors 21:2494
4. Sihlahla M, Mouri H, Nomngongo PN (2019) J Afr Earth Sci 160:103635
5. Embaby A, Redwan M (2019) Egypt. Environ Monit Assess 191:686
6. Cheng ZX, Ren XH, Xu JQ, Pan QY (2011) J Nanomater 2011:654715
7. Ayesh Ahmad I (2016) J Nanomater 2016:2359019
8. Ahammad AJS, Lee J-J, Rahman MA (2009) Sensors 9(4):2289–2319
9. Zhang T, Mubeen S, Myung NV, Deshusses MA (2008) Nanotechnology 19:332001
10. Dariyal P, Sharma S, Chauhan GS, Singh BP, Dhakate SR (2021) Nanoscale Adv 3:6514–6544
11. Reddy KK, Bandal H, Satyanarayana M, Goud KY, Gobi KV, Jayaramudu T, Amalraj J, Kim H (2020) Adv Sci 7:1902980
12. Ohashi T, Dai L (2006) In: Dai L (ed) Carbon nanotechnology. Elsevier, pp 525–575
13. Li H, Zhang J, Li G, Tan F, Liu R, Li R, Zhang T, Jin H, Li Q (2014) Carbon 66:369–376
14. Sacco L, Forel S, Florea I, Cojocaru C-S (2020) Carbon 157:631–639
15. Hou Z, Wu J, Zhou W, Wei X, Xu D, Zhang Y, Cai B (2007) IEEE Trans Electron Devices 54:1545–1548
16. Simões FR, Xavier MG (2017) In: Da Róz AL, Ferreira M, Leite F deLima, Oliveira ON (eds) Micro and nano technologies, nanoscience and its applications. William Andrew Publishing, pp 155–178
17. Miyake M (2003) In: Yasuda E, Inagaki M, Kaneko K, Endo M, Oya A, Tanabe Y (eds) Carbon alloys. pp 435–445
18. Choudhury S (2020) In: Prasad MNV, Grobelak A(eds) Waterborne pathogens. Butterworth-Heinemann, pp 219–235
19. Naresh V, Lee N (2021) Sensors 21:1109
20. Kumar S, Sarita NM, Dilbaghi N, Tankeshwar K, Kim K-H (2018) Prog Polym Sci 80:1–38
21. Jadon N, Jain R, Sharma S, Singh K (2016) Talanta 161:894–916
22. Zhu C, Yang G, Li H, Du D, Lin Y (2015) Anal Chem 87:230–249
23. Cheng X (2014) In: Feldman M (ed) Nanolithography. Woodhead Publishing

24. Kreyling WG, Behnke MS, Chaudhry Q (2010) Nano Today 5(3):165–168
25. Bellah MM, Christensen SM, Iqbal SM (2012) J Nanomater 2012:486301
26. Nasrollahzadeh M, Issaabadi Z, Sajjadi M, Sajadi SM, Atarod M (2019) Interface. Sci Technol 28:29–80
27. Luo X, Morrin A, Killard AJ, Smyth MR (2006) Electroanalysis 18(4):319–326
28. Maduraiveeran G, Jin W (2017) Trends Environ. Anal Chem 13:10–23
29. Malaki M, Xu W, Kasar AK, Menezes PL, Dieringa H, Varma RS, Gupta M (2019) Metals 9:330
30. Huang X, Qi X, Boey F, Zhang H (2012) Chem Soc Rev 41:666–686
31. Riedel R (2012) Chem Soc Rev 41:5029–5031
32. Pallares M, Thanh NTK, Su X (2019) Nanoscale 11(46):22152–22171
33. Azamia M, Arkanb E (2015) J Rep Pharm Sci 4(1):111–124
34. Saleh TA, Shetti NP, Shanbhag MM, Reddy KR, Aminabhavi TM (2020) Mater Sci Energy Technol 3:515–525
35. Camargo PHC, Satyanarayana KG, Wypych F (2009) Mater Res 12(10):1–39
36. Lateef A, Nazir R (2017) In: Sia PD (ed) Science and applications of tailored nanostructures. One Central Press, pp 239–255
37. Asmatulu R, Khan WS, Reddy RJ, Ceylan M (2015) Polym Compos 36(9):1565–1573
38. Twardowski TE (2007) Introduction to nanocomposite materials: properties, processing. Destech Publications, Characterization
39. Di Ventra M, Evoy S, Heflin JR (2004) Introduction to nanoscale science and technology. Springer
40. Pina S, Oliveira JM, Reis RL (2015) Adv Mater 27(7):1143–1169
41. Rafiee MA, Rafiee J, Wang Z, Song H, Yu Z-Z, Koratkar N (2009) ACS Nano 3(12):3884–3890
42. Mariano M, El Kissi N, Dufresne A (2014) J Polym Sci B Polym Phys. 52(12):791–806
43. Hu H, Onyebueke L, Abatan A (2010) J Miner Mater Charact Eng 9(04):275
44. Power AC, Gorey B, Chandra S, Chapman J (2018) Nanotechnol Rev 7(1):19–41
45. Yang C, Denno ME, Pyakurel P, Venton BJ (2015) Anal Chim Acta 887:17–37
46. Yang N, Swain GM, Jiang X (2016) Electroanalysis 28:27–34
47. Aykaç A, Gergeroglu H, Beşli B, Akkaş EÖ, Yavaş A, Güler S, Güneş F, Erol M (2021) Nanoscale Res Lett 16:65
48. Tagmatarchis N (2012) Advances in carbon nanomaterials: science and applications. CRC, Boca Raton
49. Navarro-Pardo F, Martínez-Hernández AL, Velasco-Santos C (2015). In: Mohanty S, Nayak SK, Kaith BS, Kalia S (eds) Polymer nanocomposites based on inorganic and organic nanomaterials. Wiley, Oxford, pp 347–399
50. Sattler KD (2016) Carbon nanomaterials sourcebook: graphene, fullerenes, nanotubes, and nanodiamonds. CRC Press
51. Lee SW, Lee W, Hong Y, Lee G, Yoon DS (2018) Sens Actuators B Chem 255:1788–1804
52. Xu K, Fu C, Gao Z, Wei F, Ying Y, Xu C, Fu G (2018) Instrum Sci Technol 46:115–145
53. Wang Y, Yeow JTW (2009) J Sens 2009:493904
54. Toda K, Furue R, Hayami S (2015) Anal Chim Acta 878:43–53
55. Han T, Nag A, Mukhopadhyay SC, Xu Y (2019) Sens Actuators A Phys 291:107–143
56. Demon SZN, Kamisan AI, Abdullah N, Noor SAM, Khim OK, Kasim NAM, Yahya MZA, Manaf NAA, Azmi AFM, Halim NA (2020) Sens Mater 32:759–777
57. He S, Yuan Y, Nag A, Feng S, Afsarimanesh N, Han T, Mukhopadhyay SC, Organ DR (2020) Int J Environ Res Public Health 17(14):5220
58. Nag A, Mukhopadhyay SC, Kosel J (2017) IEEE Sens J 17:3949–3960
59. Han ST, Peng H, Sun Q, Venkatesh S, Chung KS, Lau SC, Zhou Y, Roy VAL (2017) Adv Mater 29:1700375
60. Algamili AS, Khir MHM, Dennis JO, Ahmed AY, Alabsi SS, Hashwan SSB, Junaid MMA (2021) Nanoscale Res Lett 16:16
61. Ferrari AG-M, Carrington P, Rowley-Neale SJ, Banks CE (2020) Sci Water Res Technol 6:2676–2690

62. Bard A, Faulkner L (2021) Electrochemical methods: fundamentals and applications. Wiley
63. Compton RG, Banks CE (2010) Understanding voltammetry. Imperial College Press, London
64. Wang J (2006) Analytical electrochemistry. Wiley
65. Grieshaber D, MacKenzie R, Vörös J, Reimhult E (2008) Sensors 8:1400–1458
66. Manjavacas G, Nieto B (2016) In: Ball M, Basile A, Veziroglu TN (eds) Compendium of hydrogen energy, vol 4. Woodhead Publishing Series in Energy, pp 215–234
67. Chaubey A, Malhotra BD (2002) Biosens Bioelectron 17(6–7):441–456
68. Harper A, Anderson MR (2010) Sensors 10:8248–8274
69. Ben AM, Korpan Y, Gonchar M, El'skaya A, Maaref MA, Jaffrezic-Renault N, Martelet C (2006) Biosens Bioelectron 22(5):575–581
70. Holler FJ, Skoog DA, Crouch SR (2007) Principles of instrumental analysis (6th edn). Thomson Brooks, Cole
71. Janata J (2010) Electrochemical Sensors, Electrochemistry Encyclopedia
72. Stradiotto NR, Yamanaka H, Zanoni MVB (2003) J Braz Chem Soc 14(2):159–173
73. Electrochemical sensors (www.intlsensor.com) pp 27–35
74. Islam T, Hasan MM, Awal A, Nurunnabi M, Ahammad AJS (2020) Molecules 25(24):5787
75. Tang Z, Ma Z (2017) Biosens Bioelectron 98:100–112
76. Meirinho SG, Dias LG, Peres AM, Rodrigues LR (2016) Biotechnol Adv 34:941–953
77. Joshi A, Kim K (2020) Biosens Bioelectron 153:112046
78. Scholz F (2015) Voltammetric techniques of analysis: the essentials. ChemTexts 1:17
79. Ding J, Qin W (2020) Anal Chem 124:115803
80. Muhammad-Tahir Z, Alocilja EC (2003) Biosens Bioelectron 18:813–819
81. Kamel S, Khattab T (2020) Biosensors 10:67
82. Mukherji S, Mondal D (2017) In: Narayan RJ (eds) Medical biosensors for point of care (POC) applications. Woodhead Publishing, pp 99–131
83. Bahadir EB, Sezgintürk MK (2016) Biotechnol 44:248–262
84. Liu J, Pan L, Shang C, Lu B, Wu R, Feng Y, Chen W, Zhang R, Bu J, Xiong Z, Bu W, Du J, Shi J (2020) Sci Adv 6(16):eaax9757
85. Shoaie N, Daneshpour M, Azimzadeh M, Mahshid S, Khoshfetrat SM, Jahanpeyma F, Gholaminejad A, Omidfar K, Foruzandeh M (2019) Microchim Acta 186(7):465
86. Leal-Junior AG, Frizera A, Pontes MJ (2018) Opt Laser Technol 100:272–281
87. Tiwari JN, Vij V, Kemp KC, Kim KS (2016) ACS Nano 10:46–80
88. Bernal E, Guo X (2014) Adv Gas Chromatogr 3:57–63
89. Schroeder V, Savagatrup S, He M, Lin S, Swager TM (2018) Chem Rev 119:599–663
90. O'Riordan A, Barry S (2016) Rep Electrochem 6:1
91. De A, Chen S, Carlen ET (2014) In: Coffer JL (ed) Semiconducting silicon nanowires for biomedical. Woodhead Publishing, pp 229–265
92. Rowland CE, Brown CW III, Delehanty JB, Medintz IL (2016) Mater Today 19:464–477
93. Mnyipika SH, Nomngongo PN (2017) Int J Electrochem Sci 12:4811–4827
94. Ghosh A, Zhang C, Shi SQ, Zhang H (2019) Clean-Soil Air Water 47:1800491
95. Arduini F, Cinti S, Scognamiglio V, Moscone D (2020). Nanomaterial-based sensors. In: Handbook of nanomaterials in analytical chemistry. Elsevier, pp 329–359
96. Dutta S, Strack G, Kurup P (2019) Sens Actuators B Chem 281:383–391
97. Zhou M, Han L, Deng D, Zhang Z, He H, Zhang L, Luo L (2019) Sens Actuators B Chem 291:164–169
98. Zhu Y, Pan D, Hu X, Han H, Lin M, Wang C (2017) Sens Actuators B Chem 243:1–7
99. Cheng Y, Fa H, Yin W, Hou C, Huo D, Liu F, Zhang Y, Chen C (2016) J Solid State Electrochem 20:327–335
100. Ting SL, Ee SJ, Ananthanarayanan A, Leong KC, Chen P (2015) Electrochim Acta 172:7–11
101. Sakthivel R, Annalakshmi M, Chen SM, Kubendhiran S (2019) J Electrochem Soc 166(2):B680–B689
102. Govindasamy M, Manavalan S, Chen SM, Rajaji U, Chen TW, Al-Hemaid FM, Ali MA, Elshikh MS (2018) J Electrochem Soc 165:B370–B377

103. Majhi SM, Naik GK, Lee HJ, Song HG, Lee CR, Lee IH, Yu YT (2018) Sens Actuators B Chem 268:223–231
104. Xue C, Wang X, Zhu W, Han Q, Zhu C, Hong J, Zhou X, Jiang H (2014) Sens Actuators B Chem 196:57–63
105. Ma K, Sinha A, Dang X, Zhao H (2019) J Electrochem Soc 166(2):B147–B154
106. Wang N, Lin M, Dai H, Ma H (2016) Biosens Bioelectron 79:320–326
107. Abdel-Raoof AM, El-Shal MA, Said RA, Abostate MH, Morshedy S, Emara MS (2019) J Electrochem Soc 166(12):B948–B945
108. Shu H, Chang G, Su J, Cao L, Huang Q, Zhang Y, Xia T, He Y (2015) Sens Actuators B Chem 220:331–339
109. Jiao S, Jin J, Wang L (2015) Sens Actuators B Chem 208:36–42
110. Meng F, Zheng H, Chang Y, Zhao Y, Li M, Wang C, Sun Y, Liu J (2018) IEEE Trans Nanotechnol 17(2):212–219
111. Turcheniuk K, Boukherroub R, Szunerits S (2015) J Mater Chem B 3:4301–4324
112. Du N, Zhang H, Ma X, Yang D (2008) Chem Commun 46:6182–6184
113. Lee K, Scardaci V, Kim H-Y, Hallam T, Nolan H, Bolf BE, Maltbie GS, Abbott JE, Duesberg GS (2013) Sens Actuators B Chem 188:571–575
114. Sershen SR, Mensing GA, Ng M, Halas NJ, Beebe DJ, West JL (2005) Adv Mater 17(11):1366–1368
115. Shi Q, Xia H, Li P, Wang Y-S, Wang L, Li S-X, Wang G, Lv C, Niu L-G, Sun H-B (2017) Adv Opt Mater 5(22):1700442
116. Zhu Z, Senses E, Akcora P, Sukhishvili SA (2012) ACS Nano 6(4):3152–3162
117. Lim H-R, Lee Y, Jones KA, Kwon Y-T, Kwon S, Mahmood M, Lee SM, Yeo W-H (2021) Sens Actuators B Chem 331:129416
118. Xu M, Song Y, Ye Y, Gong C, Shen Y, Wang L, Wang L (2017) Sens Actuators B Chem 252:1187–1193
119. Patil AB, Zheng C, Ma L, Wu R, Mengane SK, Zhang Y, Liu X, Meng Z, Zhang W, Xu Z, Chen C, Huang J, Liu XY (2021) Nanotechnology 32(6):065502
120. Zhang X-F, Liu Z-G, Shen W, Gurunathan S (2016) Int J Mol Sci 17(9):1534
121. Renedo OD, Martínez MJA (2007) Electrochem Commun 9:820–826
122. Han T, Jin J, Wang C, Sun Y, Zhang Y, Liu Y (2017) Nanomaterials (Basel) 7(2):40
123. Eksin E, Erdem A, Fafal T, Kıvçak B (2019) Electroanalysis 31(6):1075–1082
124. Kumar TV, Sundramoorthy K (2018) J Electrochem Soc 165(8):B3006–B3016
125. Mirzaei A, Janghorban K, Hashemi B, Bonavita A, Bonyani M, Leonardi S, Neri G (2015) Nanomaterials 5(2):737–749
126. Zhang J, Yang K, Chen L (2019) J Electrochem Soc 166(6):H182–H186
127. Yao W, Guo H, Liu H, Li Q, Xue R, Wu N, Li L, Wang M, Yang W (2019) J Electrochem Soc 166(14):B1258–B1267
128. Jiang D, Zhang Y, Chu H, Liu J, Wan J, Chen M (2014) RSC Adv 4:16163–16171
129. Nantaphol S, Chailapakul O, Siangproh W (2015) Sens Actuators B Chem 207(Part A):193–198
130. Zahran M, Khalifa Z, Zahran MAH, Azzem MA (2020) ACS Appl Nano Mater 3:3868–3875
131. Huang J, Xie Z, Xie Z, Luo S, Xie L, Huang L, Fan Q, Zhang Y, Wang S, Zeng T (2016) Anal Chim Acta 913:121–127
132. Ozcelikay G, Dogan-Topal B, Ozkan SA (2018) Electroanalysis 30:943–954
133. Traiwatcharanon P, Siriwatcharapiboon W, Wongchoosuk C (2020) Chemosensors 8(3):58
134. Imran M, Ehrhardt CJ, Bertino MF, Shah MR, Yadavalli VK (2020) Micromachines 11:413
135. Amjadi M, Pichitpajongkit A, Lee S, Ryu S, Park I (2014) ACS Nano 8(5):5154–5163
136. Neella N, Gaddam V, Nayak MM, Dinesh NS, Rajannaa K (2017) Sens Actuators A Phys 268:173–182
137. Shengbo S, Lihua L, Aoqun J, Qianqian D, Jianlong J, Qiang Z, Wendong Z (2018) Nanotechnology 29(25):255202
138. Ko W-Y, Huang L-T, Lin K-J (2021) Sens Actuators A Phys 317:112437
139. Zhang W, Liu Q, Chen P (2018) Materials 11(10):1836

140. Niu X, Xie H, Luo G, Men Y, Zhang W, Sun W (2018) J Electrochem Soc 165:B713–B719
141. Zhang L, Wan J, Li J, Cui Q, He D, Zhao C, Suo H (2020) J Electrochem Soc 167(2):027537
142. Cui M, Zhang Q, Fu M, Fan X, Lu H, Wang H, Zhang Y, Wang H (2019) J Electrochem Soc 166(2):B117–B124
143. Wang L, Lu X, Wen C, Xie Y, Miao L, Chen S, Li H, Li P, Song Y (2015) J Mater Chem A 3:608–616
144. Leonardi SG, Aloisio D, Donato N, Russo PA, Ferro MC, Pinna N, Neri G (2014) Chem Electro Chem 1:617–624
145. Ong WL, Gao M, Ho GW (2013) Nanoscale 5:11283–11290
146. Ma J, Ren Y, Zhou X, Liu L, Zhu Y, Cheng X, Xu P, Li X, Deng Y, Zhao D (2018) Pt Adv Funct Mater 28:1705268
147. Yang B, Bin D, Wang H, Zhu M, Yang P, Du Y (2015) Colloids Surf A Physicochem Eng Asp 481:43–50
148. Xu TQ, Zhang QL, Zheng JN, Lv ZY, Wei J, Wang AJ, Feng JJ (2014) Electrochim Acta 115:109–115
149. Li X, Tian A, Wang Q, Huang D, Fan S, Wu H, Zhang H (2019) Int J Electrochem Sci 14:1082–1091
150. Upan J, Youngvises N, Tuantranont A, Karuwan C, Banet P, Aubert P-H, Jakmunee J (2021) Sci Rep 11:13969
151. Peng X, Zou J, Liua Z, Guo Y (2019) New J Chem 43:820–826
152. Mazzotta E, Di Giulio T, Mastronardi V, Pompa PP, Moglianetti M, Malitesta C (2021) ACS Appl Nano Mater 4:7650–7662
153. Tanner JL, Mousadakos D, Broutas P, Chatzandroulis S, Raptis YS, Tsoukalas D (2011) Nanoparticle strain sensor. Procedia Eng 25:635–638
154. Tanner JL, Mousadakos D, Giannakopoulos K, Skotadis E, Tsoukalas D (2012) Nanotechnology 23:285501
155. Skotadis E, Mousadakos D, Katsabrokou K, Stathopoulos S, Tsoukalas D (2013) Sens Actuators B Chem 189:106–112
156. Nguyen TNH, Nolan JK, Park H, Lam S, Fattah M, Page JC, Joe HE, Jun MBG, Lee H, Kim SJ, Shi R, Lee H (2019) Biosens Bioelectron 131:257–266
157. Liu Z, Tian B, Fan X, Liu J, Zhang Z, Luo Y, Zhao L, Lin Q, Han F, Jiang Z (2020) Sens Actuators A Phys 315:112341
158. Johnson JL, Behnam A, Pearton SJ, Ural A (2010) Adv Mater 22:4877–4880
159. Li J, Fan R, Hu H, Yao C (2018) Mater Lett 212:211–213
160. Lu L, Hu X, Zhu Z, Li D, Tian S, Chen Z (2020) J Electrochem Soc 167(3):037512
161. Gao M, Cho M, Han H-J, Jung YS, Park I (2018) Small 14:1703691
162. He WL, Fang F, Ma DM, Chen M, Qian DJ, Liu MH (2018) Appl Surf Sci 427:1003–1010
163. Hu J, Liu Y, Gao G, Zou X (2018) J Electrochem Soc 165:B893–B899
164. Phan DT, Chung GS (2014) Int J Hydrog Energy 39:620–629
165. Sun Y, Hau Wang H (2007) Appl Phys Lett 90:213107
166. Chung MG, Kim D-H, Seo DK, Kim T, Im HU, Lee HM, Yoo J-B, Hong S-H, Kang TJ, Kim YH (2012) Sens Actuators B Chem 169:387–392
167. Zhang W, Liu X, Lin Y, Ma L, Kong L, Min G, Wu R, Mengane SK, Yang L, Patil AB, Liu XY (2022) Chin Phys B 31:028201
168. Su P-G, Chuang Y-S (2010) Sens Actuators B Chem 145(1):521–526
169. Nuthalapati S, Shirhatti V, Kedambaimoole V, Pandi NV, Takao H, Nayak MM, Rajanna K (2022) Sens Actuators A Phys 334:113314
170. Rikhari B, Pugal Mani S, Rajendran N (2018) Carbohydr Polym 189:126–137
171. Keerthi M, Panda AK, Wang Y-H, Liu X, He J-H, Chung R-J (2022) Food Chem 378:132083
172. You T, Niwa O, Chen Z, Hayashi K, Tomita M, Hirono S (2003) Anal Chem 75:5191–5196
173. Jia H, Shang N, Feng Y, Ye H, Zhao J, Wang H, Wang C, Zhang Y (2021) J Colloid Interface Sci 583:310–320
174. Wang Y, Yin C, Zhuang Q (2020) J Alloys Compd 827:154335

175. Cui N, Guo P, Yuan Q, Ye C, Yang M, Yang M, Chee KWA, Wang F, Fu L, Wei Q, Lin CT, Gao J (2019) Sensors (Basel) 19(13):2979
176. Riaz MA, Yuan Z, Mahmood A, Liu F, Sui X, Chen J, Huang Q, Liao X, Wei L, Chen Y (2020) Sens Actuators B Chem 319:128243
177. Hammud HH, Alotaibi N, Otaibi NA, Aljaafari A, Ahmed F, Azam A, Prakasam T (2021) Chemosensors 9(1):6
178. Švorc Ľ, Strežová I, Kianičková K, Stanković DM, Otřísal P, Samphao A (2018) J Electroanal Chem 822:144–152
179. Švorc Ľ, Haššo M, Sarakhman O, Kianičková K, Stanković DM, Otřísal P (2018) Microchem J 142:297–304

Carbon–Metal Hybrid Nanomaterials for High Technologies

Priyambada Mallick, Ankita Subhrasmita Gadtya, Debajani Tripathy, Santosh Ku. Satpathy, and Srikanta Moharana

Abstract Carbon–metal-based hybrid materials have recently received encouraging results from scientists, researchers, and industrialists. Hybrid materials are distinct as their attributes are not only the sum of their individual characteristics but rather their synergy. These advanced hybrid nanomaterials provide an extra degree of freedom (DOF) that might result in the emergence of advanced or enhanced mechanical, electrical, magnetic, catalytic, and optical properties when building novel materials. Carbon nanotube and carbon nanofiber hybrids, in addition to metal, metal oxide, or inorganic nanoparticles, can have the ability to overcome energy storage device challenges. As a result, this research area has tremendous potential for the evolution and growth of high-efficiency materials. In the meantime, further research into the similarity and affinity among carbon-based nanomaterials and metals is required to enhance the usage of these hybrids in electrical, magnetic, optical, mechanical, environmental, and biological applications. Furthermore, a deeper interpretation of the characteristics of producing carbon–metal-based hybrids will allow for the progress of novel procedures that will produce innovative purposes for more cost-effective and dependable approaches for the manufacturing of advanced carbon–metal-based hybrid materials.

Keywords Nanomaterial · Carbon nanotubes · Graphene · Composites · Supercapacitor · Energy storage

1 Introduction

Efforts to investigate nanostructured materials and their derivatives are critical for developing innovative materials with exceptional qualities. It has been proven that by employing various manufacturing procedures, it is possible to generate new sophisticated nanomaterials with amazing properties for a diverse set of applications [1]. Nanomaterials have extraordinary physiochemical performances, including melting

P. Mallick · A. S. Gadtya · D. Tripathy · S. Ku. Satpathy · S. Moharana (✉)
School of Applied Sciences, Centurion University of Technology and Management, Bhubaneswar, Odisha, India
e-mail: srikantanit@gmail.com; srikanta.moharana@cutm.ac.in

S. K. Tiwari et al. (eds.), *Nanoparticles Reinforced Metal Nanocomposites*,
https://doi.org/10.1007/978-981-19-9729-7_9

point, thermoelectric activities, photo-absorbing properties, reactivity, scattering, and optical activities, with enhanced catalytic properties in comparison with their polycrystalline equivalents. These structural, electrical, and magnetic properties of the material are greatly affected by the content of nanomaterials, the process of fabrication, grain size, and grain boundary structures. Researchers and scientists have recently investigated the thermomechanical nature of nanomaterials based on these characteristics. In the era of nanostructured materials, carbon-based hybrid materials have several diverse performances because of the availability of various allotropies such as diamond, graphene, carbon nanotubes (CNTs), fullerenes, carbon nanofibers (CNFs), and their derivatives such as graphene oxide, graphene quantum dots, carbon-based quantum dots, and nanodiamond ranging from 1 to 3D structures [2]. Single carbon and its multiple derivatives are very adaptable substances with a wide range of distinct qualities that have been employed in technology and human life for ages. Each allotropy of carbon has been used as an energy material due to its tremendous properties, including structural diversity, surface functionalization, mechanical, thermal, electrical, biological, optical, etc. These hybrid materials can also be used in several sectors, including metallurgy, energy, industry, medicine [3], environmental protection, and technological applications [4]. However, the rapid progress of industries requires finer materials with novel mechanical, electrical, electrochemical, thermal, catalytic, optical, and biological properties for better use in the future. The solution was discovered in the development of carbon–metal hybrid nanomaterials, which involve the incorporation of two or more carbon allotropes with a possible combination of appropriate metals or metal oxides into advanced hybrid materials that not only exhibit emergent features but also synergistic effects. All the superior characteristics of carbon-based nanomaterials open up advanced avenues for the creation of innovative, effective, and multifunctional nanostructured composites by a possible combination of metals, including Si, Au, Ni, Cu, Pt, and others [1].

Furthermore, the design and synthesis procedures of carbon-based hybrid materials (e.g., carbon nanotubes, carbon nanofibers, graphene, graphene oxide, and fullerenes) have opened advanced avenues for the study and application of these materials in the fields of electronic and chemical sensors, supercapacitors, batteries, and other energy storage devices [5]. For this, the synthesis and design approach must be manageable and quotable, enabling technique scale and device downsizing [2]. These hybrid nanomaterials have the potential to provide a viable interface between the worlds of living and non-living [6–8].

With the improvement of the global economy, one of the greatest challenges in the world is to meet energy requirements, which are exponentially rising and are projected to triple by 2050. To solve the problems, the energy supply must be increased. According to researchers and scientists, carbon–metal hybrid nanomaterials are hopeful and suitable (when employed as energy materials) for mitigating the threat. Carbon–metal hybrid nanomaterials have recently been used in a wide range of applications in many diverse areas, including photovoltaic technology, field emission transistors, electronics [1], sensors, scanning probe microscopy, fuel cells, supercapacitors, batteries, and other energy storage, etc. As a result, the incredible characteristics of these hybrid materials, such as their various preparation techniques, industrial-scale production, and greener and more environmentally friendly nature, are unquestionably essential. Thus, the amazing future of these smart carbon–metal hybrid nanomaterials can be considered the central point of research for many scientists and technologists in the twenty-first century [8]. This chapter aims to provide a summary of new improvements in carbon-based nanomaterials, their synthesis techniques, characteristics, and applications.

2 Overview of Carbon-Based Hybrid Nanomaterials

Carbon-based nanomaterials span the entire dimensional spectrum, including zero-dimensional (0D) fullerenes, one-dimensional (1D) carbon nanotubes (CNTs) and carbon nanofibers (CNFs), two-dimensional (2D) graphene and its derivatives, three-dimensional (3D) nanodiamonds (NDs), and fullerenes. CNTs, CNFs, graphene, fullerene, graphene quantum dots (GQTs), and nanodiamonds [9] have piqued the scientific community's interest as promising building blocks for creating a wide range of hybrid nanostructures with applications in electronics, optics, supercapacitors [10], energy storage applications, and biomedicine [11]. True hybrids and pseudo-hybrids are the two types of carbon-based hybrid materials [6]. A structure in which carbon nanofibers are formed on top of a nickel-deposited diamond thin film as a functional substrate is an example of a real carbon-based hybrid nanomaterial. In this case, the diamond-like carbon film transforms the CNF structure from tubular to platelet by functioning as extra carbon during the early phases of growth and providing good electrical and mechanical connections. Carbon nanotubes placed on the surface of an aqueous solution, on the other hand, are not actual hybrids but rather pseudo-hybrids. Electrochemical energy storage devices (EESD) and renewable energy sources (RES) are necessary for meeting the energy demands of the future and mitigating the associated risks. The EES devices aid in the provision of consistent power distribution. Furthermore, there is an increasing demand for portable energy devices, which can be met by producing hybrid energy storage devices (ESDs) [12]. The solar cells and hybrid supercapacitors, along with Li-ion batteries, which are part of hybrid renewable energy producing devices (HREPDs), can enhance the performance of the hybrid ESDs (as shown in Fig. 1). To protect the life cycle and capacitance of the ESDs, core nanostructure electrodes containing carbon-based materials like graphene, graphene oxide, carbon nanotubes are important for them. The passage of electrons and charges is relatively easy in these carbon structures, which makes them more useful.

Researchers have become more interested in 2D materials such as graphene due to their excellent properties such as maximum surface-to-bulk ratio, high density of surface-active sites, high thermal durability, and thermal conductivity. Apart from these properties, a few more electrical properties also help in enhancing their performance. Batteries and supercapacitors (SC) are most useful for energy storage purposes [12–14]. The SC has advantageous properties like greater power density and a higher charge–discharge rate. Li-ion batteries are one of them, having a more

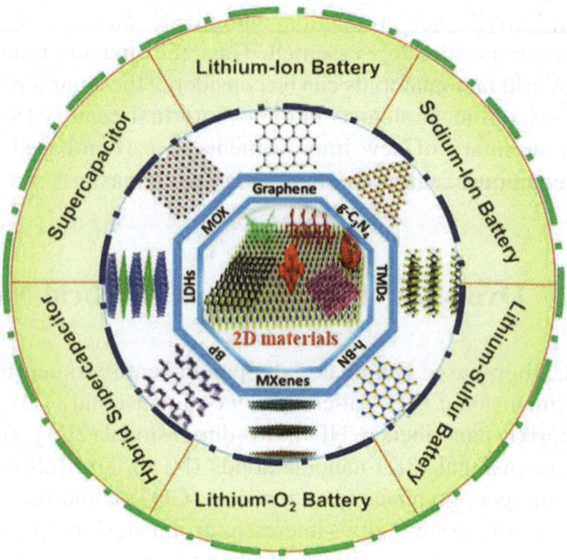

Fig. 1 2D nanomaterial architectures showing nano-sheet network structures incorporating carbon-based nanomaterials [12]

specific capacitance. Li-ion batteries are mostly used as portable energy storage devices, like in the case of smartphones, which require more energy density because of their large display [15]. These carbon-based hybrid nanomaterials are the most appropriate materials for hybridization with an external suitable metal, metal oxide, or other ceramic for improved device performance and multi-functionality in the fields of electronics, optics, photovoltaic cells, thermal devices, and drug delivery because of their well-established tremendous characteristics such as high electric and thermal conductivity, superior optical properties, and strong mechanical properties. Some of the carbon-based nanoparticles that were attracted are described in detail below.

2.1 Carbon Nanotubes (CNTs)

Carbon nanotubes (CNTs) have piqued the interest of researchers as a new generation of nanoprobes because of their tremendous structural, mechanical, electrocatalytic, electrical, and optical capabilities. Because of their high aspect ratio, unique surface roughness, high electron mobility, high conductivity, chemical stability, and sensitivity, they are ideal for energy storage applications. CNTs have a reputation for being promising materials for enhancing electron transmission, making them ideal for merging electrochemical and electrical sensors [4]. Following Iijima's discovery of CNTs in 1991 [16, 17], interest in carbon-related nanomaterials increased. They are composed of a one-atom-thick layer of hexagonally bonded carbon atoms rolled up as cylinders. CNTs are an excellent starting point for the development of donor–acceptor (D-A) hybrid structures including photo- or electro-active organic components because of their exceptional electrical conductivity and chemical durability [18]. CNTs, in particular, were discovered to considerably improve heat transmission in polymer composites because of their structure, high aspect ratio, and high electrical and thermal conductivity [19]. Furthermore, most research found that incorporating CNTs with metals or metal oxide materials increased the selectivity and sensitivity of the hybrid materials [20]. Carbon nanotubes are classified into three types: (i) single-walled carbon nanotubes (SWCNT), which are composed of rolling up a single graphene sheet having a high aspect ratio; (ii) multi-walled carbon nanotube (MWCNT), which has a few graphene sheet layers in the rolling pattern and has nearly 3.4 Å interlayer spacing due to van der Waals forces; and (iii) double-walled carbon nanotube (DWCNT), which has properties of both SWCNTs and MWCNTs [18]. These nanotubes can be classified into three types based on their electrical conductivity: armchair, zigzag, and chiral. Armchair carbon nanotubes are conductors, but zigzag and chiral carbon nanotubes can be either conductors or semiconductors. Another essential attribute is the insolubility of carbon nanotubes in most liquids, including water, polymer resins, and practically all solvents. To facilitate and standardize nanotube dispersion in liquids, functional groups or polar compounds can be inserted onto the walls without materially changing their characteristics [21].

2.2 *Graphene*

Graphene is a two-dimensional (2D) material composed of a sp2-hybridized carbon network with a carbon–carbon covalent bond with a distance of approximately 1.42 and an interlayer spacing of 3.4. It possesses several exceptional properties, including high electronic conductivity, good thermal stability, high mechanical strength, high elasticity, a primarily large surface area [21], and the absence of mechanical impurities [22], all of which contribute to its potential suitability for energy storage applications. Besides these properties, graphene possesses semiconductor behavior with a zero band gap, which favors battery, PV cell, and other energy storage applications. This single-atom-thick honeycomb-patterned sheet of carbon is the strongest, finest, and stiffest material [23], as well as an efficient heat and electrical conductor [11]. Wallace analyzed the electrical structure of graphene in 1947 [24], and McClure determined the associated wave equation in 1956 [25]. Mouras and colleagues coined the term "graphene" in 1987 as "graphitic intercalation compounds (GIC)" [26]. Graphene research has exploded in the past two decades, with scientists discovering a slew of novel features. Its exceptionally huge surface area, chemical purity, hydrophilic characteristics, and free p electrons make it an excellent option for energy storage applications as well as for the adsorption or detection of heavy metal ions [20].

2.3 Graphene Quantum Dots (GQDs)

GQDs, a zero-dimensional (0D) graphene sheet with less than 100 nm of lateral dimension in one or few layers, are another recently created and advanced smart material from the carbon family. Due to quantum confinement, these materials have exceptional photoluminescence behavior. GQD has a better chance in biomedical applications than graphene or graphene oxide due to its small size. However, before designing GQDs for practical applications, biocompatibility and toxicity remain major concerns. The size of a GQD has a significant impact on its toxicity, surface functionalization, and ability to cross biological barriers. More systematic studies involving the size of GQDs are still required in the future. Unlike graphene, GQDs show a nonzero band gap, which extends their application beyond energy storage. Surprisingly, GQDs have greater biocompatibility and photobleaching resistance. Furthermore, GQDs have graphene-like properties such as strong photoluminescence, a large surface area, available p electrons, low toxicity, cost-effectiveness, excellent solubility, and environmental friendliness, making them one of the best smart nanomaterials for a wide range of applications such as lithium-ion batteries, PV cells, sensors, and energy storage applications [27], imaging, cancer therapy-targeted drug delivery, sensors, photothermal therapy, and environmental friendliness. GQDs can be prepared by using several techniques, but hydrothermal and microwave techniques are very popular because they can enhance physical, chemical, optical, electrical, and other properties for use in high-tech applications. However, the increased availability of high-quality GQDs to the scientific community will encourage more in-depth studies of their unique properties as well as the development of new applications.

2.4 Nano-diamonds

Nanodiamonds have grown in importance in science and technology because of their superior properties, like extreme hardness, excellent mechanical properties, high surface areas, chemical inertness, excellent photo stability, high thermal conductivities, good biocompatibility, wide optical transparency, and tunable surface structures. Besides these properties, nanodiamonds show excellent fluorescent, biocompatibility, and surface modification properties. Furthermore, when stimulated by a laser, defect centers inside the ND release photons capable of penetrating tissue, making them ideal for biological imaging and therapeutic applications [11]. As these materials exhibit excellent mechanical and electrical/thermal conductivity, they can be applied in the field of energy storage devices. As nanodiamonds are non-toxic in nature, they can be widely used in mass spectrometry, biosensors, magnetic resonance imaging, and energy storage devices such as EDLCs (electrochemical double-layer capacitors), PV cells, and batteries. Although several reviews have been published on the synthesis and applications of nanodiamonds, research on the high-tech aspects of NDs is still in its infancy, and some challenges must be overcome before practical applications can be implemented.

2.5 *Carbon Nanofibers*

Carbon nanofibers (CNFs) are defined as non-continuous one-dimensional (1D) materials of cylindrical or conical shape made up of stacked and curved graphene sheets organized in a variety of ways [28]. They are commonly sp2-based linear filaments with diameters within the 50–200 nm range and aspect ratios greater than 100 [29]. CNFs offer unique surface shapes, stable structure characteristics, and surface attributes that can be altered chemically to achieve a specific aim, and they are widely available [17]. In comparison with CNTs, CNFs are novel carbon-based nanomaterials. CNFs exhibit low toxicity when compared to other traditional adsorbents, such as CNTs [30]. Furthermore, carbon nanofibers have several advantages over CNTs, including more active functional sites, a relatively large surface area, ease of functionalization, and tightly bound metal ions. As a result, they are widely utilized in a variety of energy storage applications such as fuel cells, sensors, batteries, and field effect transistors, among others. Zhang created bimetallic (Pt-Au)-based CNFs for heavy metal ion detection utilizing an electrospinning approach in 2017 [31]. The high conductivity of CNFs resulted in the great sensitivity of the Pt-Cu-CNF-based sensor [32]. The selectivity and sensitivity of carbon–metal-based hybrid materials can be increased because of the remarkable conductivity and sorption ability of nanoparticles such as sulfur, nitrogen, platinum, and copper [20].

2.6 *Fullerenes*

A fullerene can show a variety of shapes, including hollow spheres, ellipsoids, tubes, or a variety of other configurations. Fullerenes have a different structure than CNTs, CNFs, and graphene. Fullerenes are closed, hollow cages made of five or six sp^2-hybridized carbon rings, with each atom connected to three other carbon atoms in the x–y plane and by a weakly delocalized electron cloud along the z-axis, and arranged into pentagons and hexagons based on the total number of carbon atoms [33]. Because of the presence of both pentagonal and hexagonal carbon, fullerenes have distinct electrical and morphological properties. The existence of pentagons is required because they introduce curvature and so allow the cage to be closed [17]. Although the presence of carbon nanoallotrope fullerenes (C60, C70, and C84) with hybridization between sp^2 and sp^3 was predicted in 1970, C60 was only discovered in 1985. Most fullerenes (e.g., C60) have spherical shapes; however, oblong shapes like a rugby ball are also possible (e.g., C70) [3]. C60 is the most prevalent and has received the most attention to date. In 2018, Ciotta and his colleagues created UF-FQDs (unfolded fullerene quantum dots) by oxidation techniques and discovered their great sensitivity to heavy metal ions, including Cu, As, Pb, and Cd ions [34]. Due to their outstanding electrical and structural features, these investigations revealed that fullerenes are efficiently exploited for the detection of heavy metal ions. However, a lot of study and research is required to understand the properties of fullerenes, which can benefit numerous applications, such as the detection of heavy metal ions, energy storage, and biological sensors [20].

3 Synthesis of Carbon-Based Hybrid Nanomaterials

The multiplicity of synthesis pathways and the relative simplicity of hybrid materials are key advantages. Their surface structures are exceedingly complex and are determined by the raw materials as well as the manufacturing method and pre-treatment process [21]. Ex situ and in situ technologies can be used to create carbon-based hybrid nanomaterials. The ex situ approach entails preparing the inorganic materials separately in the appropriate size and morphology (often spherical nanoparticles), then attaching this component to the carbon surface via covalent, non-covalent, or electrostatic interactions. On the other hand, the in situ technique entails the production of an inorganic component in the presence of initial or functionalized CNTs and CNFs, on which the component develops as particles, nanowires, or thin films. Carbon–metal-based hybrid nanomaterials can also be manufactured in two ways: bottom-up and top-down. These nanomaterials can be synthesized at low temperatures via sol–gel and hydrothermal reactions and in a variety of morphologies, including three-dimensional structures, thin films, and nanoparticles. The kind and purity of carbon-based nanomaterials, as well as their surface functionalization, influence the synthesis process of carbon-based hybrids and the degree of their synergistic impact [2]. Covalent and non-covalent procedures as well as in situ and ex situ synthesis techniques such as chemical vapor deposition (CVD), the hydrothermal method, reduction, electrostatic-force guided assembly, and Hummer's techniques are commonly used to combine carbon-based nanomaterials with appropriate metals, metal oxides, or other ceramics. The primary benefits of covalent and non-covalent techniques are their ability to customize the shape, size, structure, and size of suitable metals. Before attaching to the surface of carbon-based nanomaterials, the desired hybrids can be synthesized using developed or well-defined procedures that are free of the effect of carbon materials. Meanwhile, utilizing the covalent approach, external nanocrystals can be precisely positioned on a specific region of carbon materials. In situ synthesis is thought to be important for creating hybrid nanomaterials with innovative structural designs, optimal compatibility, and numerous capabilities. Chemical vapor deposition (CVD) development of graphene shells on the surface of Fe, Cu, and Au nanoparticles is one example [35].

3.1 Sol–Gel Method

Inorganic and organic/inorganic hybrid material creation and processing are both generically referred to as "sol-gel" in the literature. Sol-gel processing generally involves the creation of colloidal suspensions (referred to as "sol"), which are then transformed into gels and finally 10 into solid material. The formation of a sol, a colloidal suspension, is the first step in the sol-gel process. Colloid precursors feature a metal atom surrounded by ligands, which can be either inorganic anions or organic alkoxides. In this context, "sol" refers to dispersed solid colloidal particles in a solution solvent, whereas "gel" refers to a 3D porous, interconnected network structure in the liquid phase. In this process, sol is transformed into a gel by following several steps. Dispersed solid nanoparticles (sols with a diameter of 1-100 nm) are mixed in a homogeneous liquid medium and agglomerated to form a continuous three-dimensional network (gel) with pore diameters in the sub-micrometer domain in the liquid phase. In sol-gel synthesis, monomeric alkoxides are typically utilized [36, 37]. Materials that are a composite of carbon and metal have remarkable mechanical, electrical, thermal, and optical properties. In addition, composites made from a combination of carbon nanotubes, nanofibers, and inorganic nanoparticles are receiving increasing interest. The synergistic impacts of design and morphology-driven charge and energy transport processes make it feasible to include less material while improving output. This is made possible by the fact that these processes are driven by morphology. As a result of their fundamental nanostructure, high surface region, substantial strength, and adsorption limit, carbon-metal nanomaterials are relevant to a wide range of fields [38, 39]. The sol-gel method is widely used, so it is reasonable to assume that it represents some novel type of combination. With this combined approach, inorganic particles are uniformly dispersed on the surface of carbon nanomaterials. This cycle is controlled by dispersion, and when the pH changes, the precursors start to bond together and shape the inorganic particles. Ahmed et al. [40] have reported the use of the sol-gel technique to load silver-doped titanium dioxide (TiO_2/Ag) with single-wall nanotubes (SWNTs) and multi-wall nanotubes (MWNTs). By decorating SWNTs and MWNTs with Ag-TiO_2 nanoparticles, it was hoped that a novel hybrid would be produced, one that combines the antibacterial capabilities and cytotoxicity of two functional resources to achieve a more potent inhibitory impact on bacteria and cancer cells. Further, neither the cytotoxicity of CNTs against uterine cancer (SiHa) cells nor that against normal human (WRL68) cells have been reported in the literature. Ag/TiO_2-CNT conjugates' cytotoxicity and biocompatibility are discussed. Similarly, through the sol-gel method, Mohammad et al. [41] have created agarose-doped TiO_2 nanoparticles covered in carbon nanotubes. A nanocomposite with strong antibacterial activity against E. coli and S. aureus has been examined for its various morphological features. Yang et al. [42] used the sol-gel process to create a carbonaceous material, ZnS nanoparticles, by combining nitrogen and sulfur co-doped with carbon nanosheets (ZnS@NSC). Carbon nanosheets are employed for 11 sodium ion storage due to their most active locations, and they are also used to improve electrical conductivity. The composite

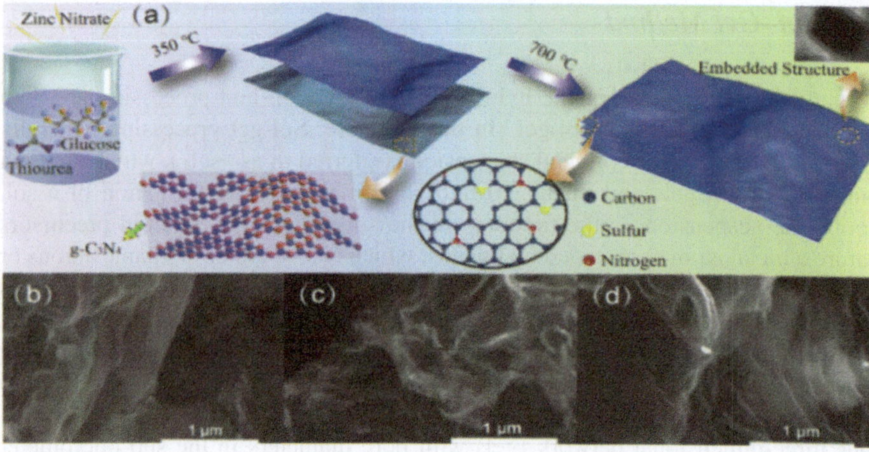

Fig. 2 **a** Schematic diagram for the synthesis of ZnS@NSC-x-700 and **b–d** SEM micrographs of ZnS@NSC-x-700 [42]

CeSeZn bonds provide good structural stability for ZnS particles to pin on the 3D carbon framework. The bridging effect gives electrons a way to move from the conductive framework to the active materials. It can also make the material more conductive by allowing free electrons to move directly between C, S, and Zn atoms (as shown in Fig. 2).

Based on the TG results, Liu et al. [43] made $Ti_2Nb_{10}O_{29}$/C nanoparticles with 13% carbon. This was done with an in situ sol–gel strategy (as shown in Fig. 3). Small grain size and carbon variation can influence the pseudo-capacitive impact of $Ti_2Nb_{10}O_{29}$/C nanoparticles, boosting their exceptional rate capability, particularly at high current rates. $Ti_2Nb_{10}O_{29}$/C nanoparticles, in contrast to $Ti_2Nb_{10}O_{29}$ nanoparticles, are blended using a simple and effective process in a sol–gel scheme due to the carbon layer. The carbon layer not only works on the pseudo-capacitive effect and electric conductivity, but also acts as a flexible obstruction to support the pressure of the volume change caused by the release and charge processes. This results in spectacular rate limitation and massive cycle implementation.

3.2 *Hydrothermal Treatment*

Hydrothermal synthesis (also known as solvothermal synthesis) is a single-step process to prepare the ultrafine nanomaterials in a hydrothermal environment (i.e., neither the solution is non-aqueous nor aqueous) at a low temperature in the range of 100–374 °C without the use of a calcination process. The basic principle of the technique is a reaction in an aqueous solution or suspension of the precursors at high temperature and pressure. Under fixed pressure and temperature, the hydrothermal technique is carried out in specific equipment known as an autoclave. The autoclave

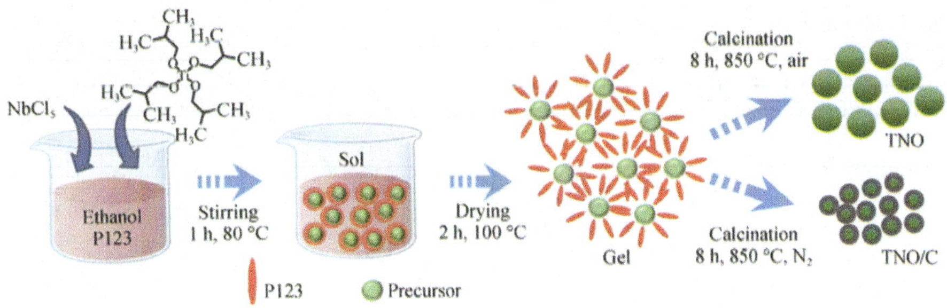

Fig. 3 $Ti_2Nb_{10}O_{29}$/C nanoparticles (TNO/C) and $Ti_2Nb_{10}O_{29}$ nanoparticles' growth mechanisms (TNO) [43]

is loaded with reagents and then placed in the oven for some time, allowing the reaction to take place without direct supervision. This procedure is performed at high temperatures. This method allows for hydrolysis. Navrotskaya and his colleagues recently used this technique to create core–shell-structured carbon nanofiber-Titanate nanotubes (CNF-TiNT) and observed an increase in surface area, which is responsible for CNF-TiNTs' effective photocatalytic activity [2]. The addition of elemental oxide to the surfaces of CNTs or CNFs alters the properties of hybrid materials. Using a simple acid-based hydrothermal technique, Li and his co-workers [44] have created a composite that includes a heteroatom-doped nanostructured hollow graphitic carbonaceous (Fig. 4). This approach is used to break down the remaining indistinct carbon and functionalize the resulting graphitic hybrid carbon. According to their findings, N-doped hollow graphitic carbon hybrid material shows tremendous promise as being superior to carbon-based anodes for lithium-ion batteries (LIB).

He et al. [45] have described in their study the one-step hydrothermal synthesis of a hybrid material made of MoO_3 nanobelts and oxide multi-walled carbon nanotubes (MoO_3-CNTs). To plan uniform MoO_3-CNT films, an altered vacuum

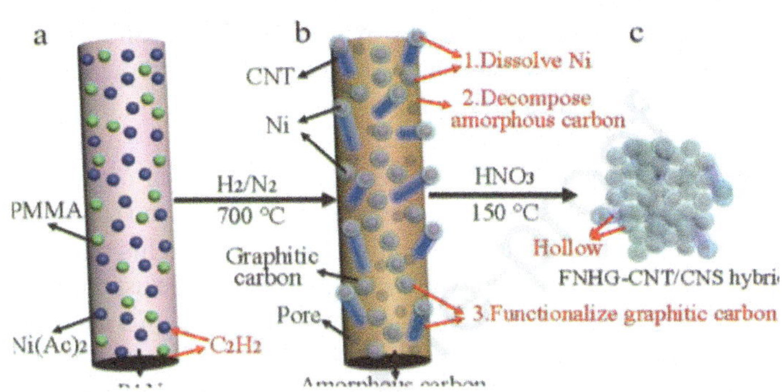

Fig. 4 Schematic diagram of the formation of the functionalized N-doped hollow graphitic FNHG-CNT/CNS hybrids [44]

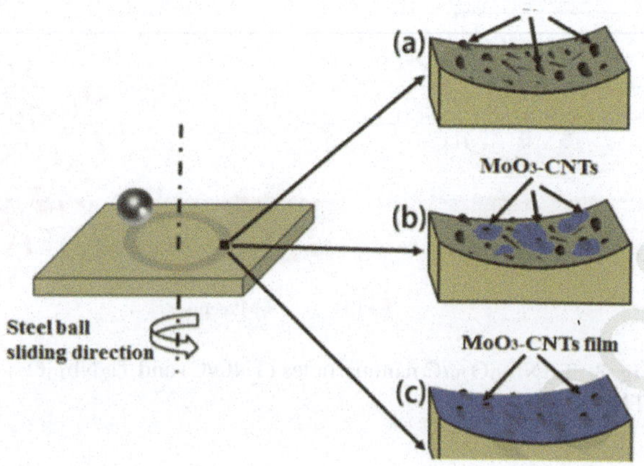

Fig. 5 Graphical representation of friction and wear process of **a** GF/EP, **b** MoO_3-CNTs/GF/EP, **c** f-MoO_3-CNTs/GF/EP composites [45]

filtration approach was used (f-MoO_3-CNTs) (as shown in Fig. 5). These MoO_3-CNTs and f-MoO_3-CNTs were used to improve the tribological characteristics of a glass-textured epoxy (GF/EP) composite constructed using the vacuum-assisted resin transfer molding (VARTM) technique. They discuss how distinct organic and inorganic nanofiller structures with carbon-based fillers and metallic oxide-based carbon nanotubes (CNTs) are regularly employed to increase matrix performance and generate high electrical conductivity in the study. In comparison to the ideal GF/EP composite, the MoO_3-CNTs- and MoO_3-CNTs film (f-MoO_3-CNTs) modified GF/EP composites show exceptional tribological performance, notably the f-MoO_3-CNTs/GF/EP composite.

Bakhtiarzadeh and his co-workers [46] have developed a novel process for constructing a composite containing core–shell carbon coated with Fe_3O_4 nanoparticles (Fig. 6). By using hydrothermal treatment, these nanoparticles are combined with MnO_2 nanosheets to form polymorphous and crystalline-type structures of MnO_2 nanosheets. In their study, they discuss the use of a newly created Fe_3O_4@Cg/MnO_2 nanostructure catalyst for the selective oxidation of BzOH (phenol) to BzCHO (benzaldehyde) using oxygen as a green oxidant treatment, and these nanoparticles are combined with MnO_2 nanosheets to form polymorphous and crystalline-type structures of MnO_2 nanosheets. In their study, they discuss the use of a newly created Fe_3O_4@Cg/MnO_2 nanostructure catalyst for the selective oxidation of BzOH (phenol) to BzCHO (benzaldehyde) using oxygen as a green oxidant. The newly created nanostructured catalyst could also be recycled and utilized for six runs without losing substantial catalytic activity or stability. The main benefits of this method are that it is easy to use, quick to set up, very selective, and can be used more than once.

Sabeeh et al. [47] have created a carbonaceous material with a nanostructure that has many unique qualities, such as strong conductivity, a greater surface area, good chemical stability, and a distinctive structure. As a result, these materials have a high likelihood of improving the electrochemical presentation of transition metal-based pseudo-capacitive resources. They discuss the investigation of the hydrothermal process, which is utilized to organize copper sulfide nanochips. By putting together,

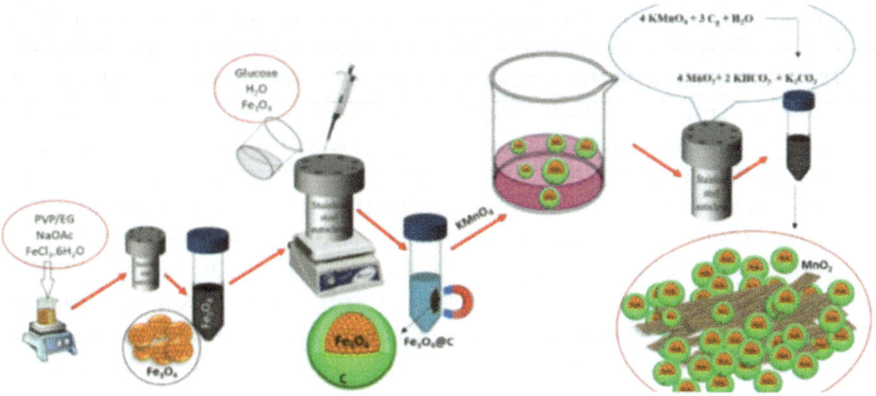

Fig. 6 Schematic diagram of MnO_2 on CHO substrate through hydrothermal technique [46]

a nano-hybrid with both conductive and capacitive CNTs, the sample was made to have better electrical conductivity and capacitance.

3.3 Chemical Vapor Deposition (CVD)

In the semiconductor sector, this process is commonly employed to generate high-clearance solid materials or thin films. In this technique, the substrate (i.e., catalyst) is typically immersed in the vapor precursor, after which the reaction creates the required chemical. This method is utilized to produce clean carbon nanomaterials by first producing CNTs and CNFs. Catalyst particles serve as both a substrate beneath the developing carbon nanomaterials and an arming dopant in this case. Catalysts must be non-toxic or reduce the toxicity of carbon-based nanoparticles before they may be used in ecology or medicine. The position and growth of carbon nanomaterials on the plane can be regulated using this technology. This procedure is straightforward, low cost, and precise. The concept of hybridizing CNTs with Al_2O_3 is based on CNT agglomeration prevention owing to Van der Waals interaction. Epoxide compounds, including CNTs-Al_2O_3, exhibit compressive strength and Young's modulus amplification greater than 100%. The insertion of nanocatalysts using this method is intended to improve the thermal characteristics of CNMs [2]. Chemical vapor deposition (CVD) is an important technology for creating a thin, high-purity coating. At lower temperatures, different types of catalysts are used for the deposition of metallic coatings. Rashid et al. [48] discuss in their review various acceptable transition metals such as palladium, platinum, gold, and nickel, which have been proven to be excellent catalysts for metal deposition on polymers. Hoyos-Palacio et al. [49] used chemical vapor deposition to create multi-walled carbon nanotubes (MWCNTs) by combining carbon nanotubes coated with silver nanoparticles. SEM, TEM, Raman, XRD, and XPS were used to characterize the MWCNTs-Ag. The results showed that the silver nanoparticles were evenly distributed across the MWCNTs' outer surface. Finally, the produced CNT-Ag may find usage in improved nanocomposites, antimicrobial products, and sensors. The investigation of multi-walled carbon nanotubes (CNTs) formation utilizing two different chemical vapor deposition (CVD) techniques—low pressure plasma-boosted and atmospheric pressure thermal CVD—is discussed by Moshkalyov and his co-workers [50]. Here, the utilization of long, straight carbon nanotubes is coupled with thermal CVD (chemical vapor deposition), employing methane (CH4)-based gas mixtures and thinner Ni films, as a catalyst, which have been created to produce nickel nanoparticles that perform better. Bachmatiuk and his co-workers [51] investigate the graphitic nanostructure, which is made up of carbon nanotubes and graphene. In the study, SiO_2 is used as a catalyst to create the SiC compound. SiO_2 nanoparticles are employed in the chemical vapor deposition (CVD) of graphitic carbon nanostructures. This work enables a better understanding of the growth mechanisms at work in the production of carbon nanotubes and carbon nanofibers utilizing SiO_2 catalysts. Ni et al. [52] have created pony-size Cu nanoparticle-based catalysts utilizing a chemical vapor deposition (CVD) technique with floating Cu atoms and mesoporous C-N support. By modifying the electronic structure of Cu, mesoporous C-N with a large surface area is used to stabilize scattered Cu nanoparticles and boost the electrocatalytic activity of Cu. Furthermore, their paper mentions the significant catalytic stability and remarkable methanol-tolerant characteristics that encourage the use of Cu-based oxygen reduction reaction (ORR) electrocatalysts in fuel cells. Similarly, Yadav et al. [53] investigated the structure of carbon nanotubes cotton (CNT-c) using two distinct sulfur precursors via floating catalyst chemical vapor deposition (FC-CVD) (Fig. 7). It has been discovered that single-walled carbon nanotubes (SWCNTs) and multi-walled carbon nanotubes (MWCNTs) can be formed using carbon disulfide and thiophene, respectively.

3.4 Hummer's Method

Recently, the Hummer's method has attracted much more attention from scientists, researchers, and industrialists for the preparation of carbon–metal-based hybrid nanomaterials. This technique possesses high efficiency and reaction safety. This synthesis takes place basically in two steps: (i) Toxic gases are first removed through an oxidation process, and then residual materials are purified. This technique increases the reaction yield and reduces the toxicity of the carbon–metal-based hybrid nanomaterials. The enhanced Hummer's process is less hazardous and has various advantages in terms of the synthesized product. In 2013, Chen and his groups successfully prepared graphene oxide by using the improved Hummer's method to decrease the

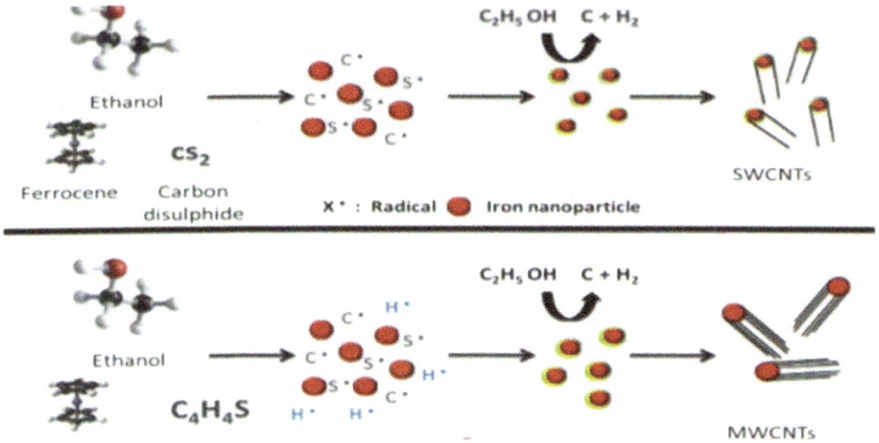

Fig. 7 Diagram illustrating the growth mechanisms of SWCNTs and MWCNTs using thiophene and carbon disulfide as sources of sulfur, respectively [53]

cost of graphene oxide [48]. Later in 2017, Zaaba and his co-workers also successfully synthesized graphene oxide with low cost and low toxicity using this technique [49].

4 Polymer-Supported Carbon–Metal Hybrid Nanocomposites for Energy Storage Applications

Current research and development have primarily concentrated on energy storage devices [50–52], including supercapacitors, electronics, fuel cells, sensors [19], lithium-ion batteries, etc. Supercapacitors have a high power density, Columbia efficiency, long cyclic stability, and a shorter interval for full charge–discharge cycles than batteries. Carbon-based nanomaterials, including CNTs, graphene, and mesoporous carbon electrodes, gained much more interest in the field of energy storage devices due to their large specific surface area and excellent electrical conductivity. Carbon-based nanomaterials, combined with other suitable metals, are excellent candidates for batteries, supercapacitors, and other applications [53]. Carbon nanomaterials, such as 1D carbon nanotubes, 2D graphene, 3D mesoporous carbon, and their metal oxide composites, have been greatly used as electrodes in supercapacitors, batteries, and PV cells to enhance capacitance, electrochemical performance, energy, and power density. The specific capacitance of carbon–metal hybrid nanomaterials can be raised by an order of magnitude: typically, 100–1000 F/g. The energy density of carbon nanomaterials can be greatly increased by combining them with other metals. Although combining graphene and carbon nanotubes (CNTs) with appropriate metals is a viable strategy for supercapacitors (SCs) with high flexibility and strain resistance [54], these characteristics make the battery, SCs, and other energy storage devices appropriate for diverse applications, including hybrid electric cars, renewable energy storage devices, and portable electronics. There are, however, still some issues that must be solved to enhance the performance, characteristics, and applications of carbon–metal-based nanomaterials [53].

Carbon nanoparticles have been shown to successfully inhibit tin expansion, resulting in good cycling capabilities. Carbon-tin-based or other carbon–metal-based nanomaterials ranging from zero to multi-dimensional have been employed as anode materials for Li-ion batteries. The role and contribution of various carbon components (e.g., graphene, carbon nanotubes, etc.), as well as their interaction with other metals (e.g., Sn, Si, Ni, Au, etc.), were introduced [35]. Because of the superior characteristics of silver nanoparticles, the performance of CNT-Si hybrid nanoparticles for energy storage applications can be improved [54]. It can also boost the antibacterial activity of textile fibers. CNTs coated with highly conductive and catalytic metals like Pt, Pd, and Ru have the potential to develop the mechanical, electrical, thermal, and catalytic properties of electrode materials for use in the fuel cell, Li-ion batteries, and supercapacitors [11]. These successful applications demonstrated carbon–metal hybrid nanostructures' significant potential as innovative materials. Certain significant problems limit the performance of carbon–metal-based hybrid nanomaterials, such as uneven architectures, agglomerations, limited lifetimes, and low dependability. Importantly, the low reliability and limited lifetime are mostly driven by the weak bonding strength between carbon and metals. To solve these issues and increase the performance of carbon–metal-based hybrid nanomaterials, it is critical to boost bonding strength [55].

The primary goal of developing energy storage supercapacitors is to increase energy density without sacrificing high power density. These two measures assess supercapacitor performance. The quality of electrode materials can be improved by rigorous development at a higher level. Because of their superior electrical and mechanical properties, various carbon-related materials, such as three-dimensional porous carbons, two-dimensional graphene, and its derivatives, and others, have been used in supercapacitors. Based on the storage mechanisms, these materials have fast discharge and charge as well as high cycle capacities [56].

Zhang et al. have worked on carbon-based nanocomposite interface engineering for enhanced electrochemical energy storage. They found that when nanocomposites interact, there are synergistic effects that make the electrochemical performance better. By employing carbon nanocomposites and combining them with other materials and chemicals, such as metal oxides, conducting polymers, and various metals, they created a variety of composites with improved super-capacitive properties. For instance, they created electrodes with high reversible capacity, excellent rate capability, and increased cycle stability using core–shell structured graphene that was combined with Fe_2O_3 nanocomposite [57]. Electrical double-layer capacitors (EDLCs) and pseudo-capacitors, which are both supercapacitors, are mostly employed for this energy storage purpose. By building up charge at the double-layered interface between the electrode and electrolyte, electrochemical energy is stored in EDLCs. Therefore, the capacitive performance is investigated by computing

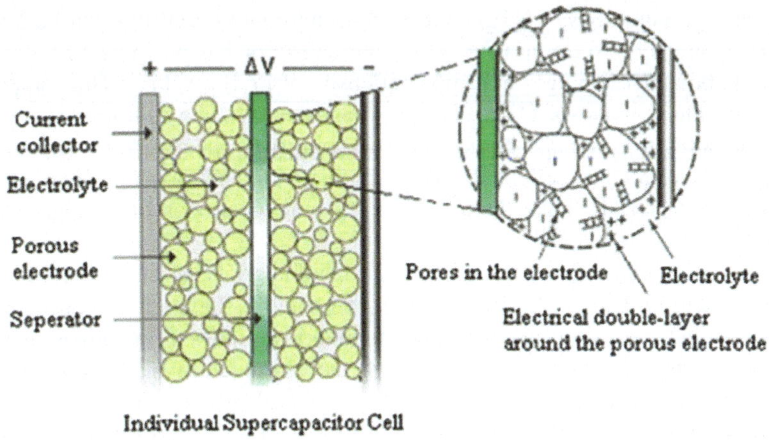

Fig. 8 A Supercapacitor presentation [60]

the electrode's surface area. However, in pseudo-capacitors, the Faraday reaction occurs at the surface of electrically active species to store electrochemical energy [58, 59]. Carbon nanotube (CNT)-related supercapacitor performance (Fig. 8) is related to several physical factors in EDLCs, like surface area, pore size, conductivity, and so on. Its performance is determined by the CNTs' mix, pre-treatment, and post-treatment. Different techniques, such as functionalization and doping, can improve supercapacitor performance, resulting in increased diffusivity, conductivity, and capacitance values. The introduction of functional groups can also increase stability [60].

To demonstrate the relationship between void and lithium-ion for storage performance, Cui and co-workers constructed a pomegranate-like silicon-carbon nanocomposite with varying void space. For the larger energy storage goal to be maintained, there must be a clearly defined void space between the silicon and the carbon composite [61]. Heydari et al. [62] have synthesized a nanocomposite of $CoFe_2O_4$-carbon microspheres for electrochemical energy storage applications by the precipitation method. In this work, they have found out that the carbon/cobalt ferrite electrode has shown high specific conductance along with great cycling stability and high-rate capability, which shows the great capability of the electrode to become a supercapacitor. The cycling life test shows 81% retention, which is very good for storage purposes. They have also concluded that nanocomposite shows good energy and power density, which are helpful in the energy storage system. To store energy efficiently, Siwal and his groups have described a carbon-based polymer nanocomposite [63, 64]. Supercapacitors have made considerable use of carbon nanocomposites as the active electrode due to their higher surface area and improved electrical and mechanical qualities. These supercapacitors have electrodes created through stacking, a 3D pillared graphene-CNT system, and edge functionalities. For instance, they have taken polypyrrole (PY) and enhanced its electrical performance by doping it with carbon nanotubes (CNTs). Extreme strength and durability can be attained from the manufactured composite. The supercapacitor has an increased electrical density of about 101.0 Wkg^{-1} and a power density of 17,186 Wkg^{-1}. The compounds based on graphene make excellent energy storage devices. The addition of polymeric nanocomposites to graphene-based materials improves their specific capacitance. The large surface area, excellent electrical conductivity, and robust mechanical qualities of these materials make them ideal for use as electrodes in a supercapacitor [65].

The development and design of complex nanomaterials for energy storage have been suggested by Liu et al. [66]. Lithium-ion batteries have great significance in electrochemical energy storage. The Li-ion batteries require a fundamental shift in electrodes to aid lithium integration, where the expandability is accompanied by remarkable bond cleavage, low ionic conductivity, high volumetric variation, and so on. Carbon-based composites are being used in Li-ion batteries due to their better electrical properties and good coulomb force proficiency. Because of their chemical resistivity, electrical properties, and lightweight, carbon-based polymeric composites such as CNTs and graphene increase the rate of Li-ion batteries as well as the integration of different polymers such as polypyrrole. Philip et al. [67] doped polyaniline, a conducting polymer, with CNT to examine the electronic interaction and improve the chemical and electrical properties. The interaction of CNT and polyaniline not only widened the applicability but also improved the nanocomposite's homogeneity. They observed that the nanocomposite had higher conductivity than the polymer due to greater uniformity and charge delocalization power. This contributes to effective delocalization on the polymer rather than the main chain. They used a phenyl-amino group for functionalization to distribute CNTs uniformly, and this dispersion resulted in a uniformly formed polyaniline shell. The produced nanocomposite has been covalently functionalized with the polymer polyaniline. Naoi and his co-workers created a new generation nano-hybrid supercapacitor using an in situ material processing technique known as "Ultra Centrifugation (UC) treatment" to generate an ultrafast electrode material known as LTO ($Li_4T_5O_{12}$). They found that the UC treatment boosts cohesion power, resulting in the created supercapacitor and battery not only satisfying the power needs but going beyond the capacity of storing current energy. They discovered that the nano-hybrid supercapacitor has a much higher power density (approximately 16 kWL^{-1}), and it can store three to five times as much energy. The composite showed 107 mAh g^{-1} at 3000 C and 78 mAh g^{-1} at 120 C, proving that the nano-hybrid composite has a high retention capacity rather than a rapid charging rate [68]. Frackowiak and his team have developed a nanocomposite material by combining the polymer poly (3,4-ethylene dioxythiophene) known as PEDOT with CNT by electrochemical polymerization, which has resulted in the highest capacitance for the supercapacitor. The manufactured nanocomposite material exhibits excellent cycle performance and a high rate of stability. The volumetric energy is greatly increased because of the high density of PEDOT. Since CNTs have an open mesoporous network, the interface between the electrode and the composite material is easily accessible for charge propagation. While the charging and discharging cycles proceeded, the supercapacitor stored a large quantity of energy in the PEDOT.

By boosting the voltage of the supercapacitor using activated carbon as the negative electrode, they also improved its power and energy density. In this scenario, the researchers have seen a voltage of 1.8 V and a rotational frequency of almost 10,000 cycles when the measurements were made in an aqueous medium [69–74].

Asen et al. integrated graphene oxide (GO), polypyrrole (PPy), and V_2O_5 using the electrochemical deposition process to create a ternary phase nanocomposite [75]. The specific capacitances of the PPy, V2O5, and GO drop as the scan rate increases since the relationship between the two is inverse. But because of the low scan rate, the ions and electrode interact more effectively, which contributes to increased capacitive behavior. An excellent specific capacitance of 750 F/g at 5 A/g, or around 83%, and 3000 charge–discharge cycles were displayed by the produced nanocomposite. The greatest power density and current density of the nanocomposite device were 27.6 Wh/kg and 13,680 W/kg, respectively. All of these results demonstrated the supercapacitor's excellent performance rate. Rao et al. [76] created the nanocomposite material CNT/metal sulfides as an electrode for energy storage and conversion applications. The CNT network provides an electron channel for the metal sulfide. The nanocomposites produced have a power conversion efficiency of 6.41%. The supercapacitor demonstrated an astounding specific capacitance value of 398.16 F/g and an energy density of 35.39 Wh/Kg, as well as an improved cycling stability of 98% capacity retention after 1000 cycles with better flexibility [77, 78]. Heli and Yadegari used an electrochemical synthesis approach to create an effective supercapacitor electrode using graphene and the conducting polymer poly (ortho-aminophenol). The charge of the Pt/GNS/POAP electrode is substantially higher than the sum of the charges of the Pt/POAP and Pt/GNS electrodes for all potential sweep rates [as shown in Fig. 9a]. [79]. The changes in specific capacity by cycle number are depicted in Fig. 9b. The polymer amino phenol was chosen because it has two oxidizable functional groups, –NH2 and –OH. The nanocomposite developed exhibits exceptional super-capacitive behavior, with a specific capacitance value of 281.1 F/g. This specific capacitance value is three times greater than graphene's typical specific capacitance value. This increased value is owing to the EDLC's synergistic impact. Because of hydrogen bonding and stacking, the nanocomposite has a higher specific capacitance value, as well as rate capability and cycling stability. After 1200 cycles, the cycle's retention capacity was found to be greater than 99% [80].

5 Conclusions

The formation of unique multifunctional hybrid nanomaterials having advanced mechanical, chemical, electrical, optical, and thermal behavior can be achieved by combining carbon-based nanomaterials, including CNFs, CNTs, fullerenes, and graphene and their derivatives, with metal, metal oxide, or other ceramics. This chapter summarized several methodologies for the preparation of carbon–metal-based hybrid nanomaterials with varied structures ranging from zero to high dimensions and their practical application in supercapacitors, Li-ion batteries, PV cells,

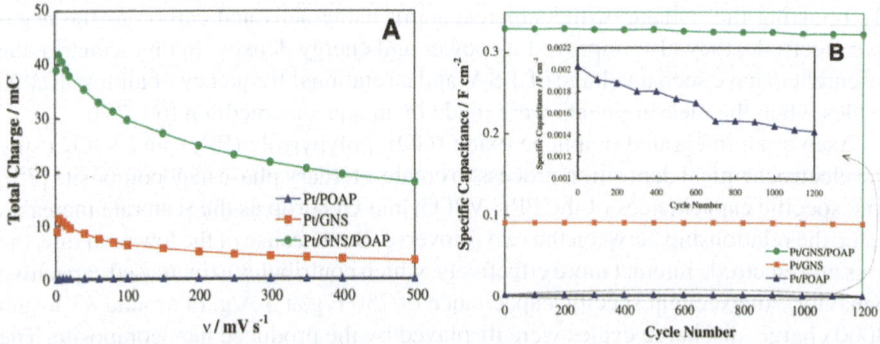

Fig. 9 **a** Variation of total charge value and **b** specific capacitance of the electrode as a function of cycle number [79]

and other energy storage devices. The primary goals and benefits of hybridizing carbon with metal hybrid nanomaterials are: (1) to obtain hybrid nanomaterials with complementary and combined features or performance; (2) to improve mechanical, chemical, electrical, and thermal stability for the application in energy storage applications; and (3) to produce vast quantities of composite nanomaterials at a minimal cost (it is still a major challenge at present). However, the majority of the hybrid nanomaterials from 0 to 3D were made using in situ approaches due to their simplicity, low production cost, and feasibility for large-scale production. Furthermore, in situ synthesis usually results in direct interaction between carbon-based nanomaterials and suitable metals, whereas non-covalent and covalent techniques can result in ligands or residues at the surface of carbon-based nanomaterials and metals or metal oxides.

As a result, the potential improvement in favorable features of these carbon-based hybrids could have a wide-ranging impact, including on solar energy, batteries, light-driven catalysis, supercapacitors, other energy storage devices, electrical/biosensors, drug administration, and other biocompatible devices. Even though considerable progress in structural design and production for carbon–metal hybrid nanomaterials has been made in recent years, the advancement has not yet reached a point where these nanocomposites may be used in practical applications. As a result, additional effort and research are required in the future. First, perfect control of the size, shape, and location of external nanocrystals on the surface of carbon materials remains a challenge. Although covalent and in situ techniques can partially accomplish this, researchers prefer to develop simple methods to achieve these objectives. Second, the underlying mechanism of hybrid nanomaterials is still poorly understood. Third, because nanoscale heterostructure research has typically been undertaken in scientific labs, large-scale repeatability and productivity are major barriers. Fourth, the present research is mostly based on the investigation of individual hybrid nanomaterials. However, to be useful, these hybrid nanomaterials must be packed or integrated into a bulk material.

Acknowledgements The authors thankfully acknowledge the support provided by Centurion University of Technology and Management, Odisha, India, for carrying out the current research.

Conflicts of Interest The authors declared no conflict of interest.

References

1. Degirmenci U, Kirca M (2018) Comput Mater Sci 154:122–131
2. Navrotskaya AG, Aleksandrova DD, Krivoshapkina EF, Sillanpää M, Krivoshapkin PV (2020) Front Chem 8:546
3. Patel KD, Singh RK, Kim HW (2019) Mater Horiz 6:434–469
4. Maiti D, Tong X, Mou X, Yang K (2019) Front Pharmacol 9:1401
5. Eder D (2010) Chem Rev 110:1348–1385
6. Laurila T, Sainio S, Caro MA (2017) Prog Mater Sci 88:499–594
7. Peltola E (2019) Front Mater Sci 6:202
8. Nasir S, Hussein MZ, Zainal Z, Yusof NA (2018) Materials 11:295
9. Jawaid M, Ahmad A, Ismail N, Rafatullah M (2021) Environmental remediation through carbon based nano composites. Springer
10. Celasco E, Chaika AN, Stauber T, Zhang M, Ozkan CS, Ozkan US, Harun SW (2019) Handbook of graphene. Wiley
11. Li H, Song SI, Song GY, Kim I (2014) J Nanosci Nanotechnol 14:1425–1440
12. Khan K, Tareen AK, Aslam M, Mahmood A, Zhang Y, Ouyang Z, Zhang H (2020) Prog Solid State Ch 58:100254
13. Preethika M, Shetty BH, Govindasamy M, Sundramoorthy AK (2021) Nanoscale Adv
14. Kumbhakar P, Gowda CC, Mahapatra PL, Mukherjee M, Malviya KD, Chaker M, Tiwary CS (2021) Mater Today 45:142–168
15. Zhong C, Deng Y, Hu W, Qiao J, Zhang L, Zhang J (2015) Chem Soc Rev 44:7484–7539
16. O'regan B, Grätzel M (1991) Nature 353:737–740
17. Tulaphol S, Bunsan S, Kanchanatip E, Miao H Y, Grisdanurak N, Den W (2016) Int J Environ Sci Technol 13:1465–1474
18. Barrejón M, Arellano LM, D'Souza F, Langa F (2019) Nanoscale 11:14978–14992
19. Ayranci R, Başkaya G, Güzel M, Bozkurt S, Şen F, Ak M (2017) ChemistrySelect 2:1548–1555
20. Afreen S, Talreja N, Chauhan D, Ashfaq M (2020) Multifunctional hybrid nanomaterials for sustainable agri-food and ecosystems. Elsevier 335–353
21. Cardoso CE, Almeida JC, Lopes CB, Trindade T, Vale C, Pereira E (2019) Nanomaterials 9:814
22. Scida K, Stege PW, Haby G, Messina GA, García CD (2011) Anal Chim Acta 691:6–17
23. Khan A, Jawaid M, Neppolian B, Asiri AM (2019) Graphene functionalization strategies. Springer Singapore
24. Morichi S, Komoda Y, Inagaki H, Itami H, Itami H (2008) Doboku Gakkai Ronbunshuu A 64:452–457
25. Sharma MP, Johnson LG, McClure JW (1973) Phys Lett A 44:445–446
26. Hérold A (1987) ChemPhys of Intercalation NATO ASI Series, B Phys 17
27. Lin Z, Yang Y, Zhang A (2017) Polymer-engineered nanostructures for advanced energy applications. Springer, pp 173–304
28. Pérez-Mayoral E, Calvino-Casilda V, Soriano E (2016) Catal Sci Technol 6:1265–1291
29. Georgakilas V, Perman JA, Tucek J, Zboril R (2015) Chem Rev 115:4744–4822
30. Singh S, Ashfaq M, Singh RK, Joshi HC, Srivastava A, Sharma A, Verma N (2013) New Biotechnol 30:656–665
31. Zhang B, Zhu H, Zou M, Liu X, Yang H, Zhang M, Du M (2017) J Mater Sci 52:8207–8218

32. Garcia-Cardona J, Sirés I, Alcaide F, Brillas E, Centellas F, Cabot PL (2020) Int J Hydrog Energy 45:20582–20593
33. Martínez LMT, Kharissova OV, Kharisov BI (2019) Handbook of eco-materials. Springer International Publishing, pp 1–22
34. Ciotta E, Prosposito P, Tagliatesta P, Lorecchio C, Stella L, Kaciulis S, Pizzoferrato R (2018) Sensors 18:1496
35. Li Y, Wu J, Chopra N (2015) J Mater Sci 50:7843–7865
36. TC Moreira F, P Moreira-Tavares A, Sales GF (2015) Curr Top Med Chem 15:245–255
37. Dey T, Naughton D (2016) J Solgel Sci Technol 77(1):1–27
38. Hossain MS, Alam MB, Shahjahan M, Begum MHA, Hossain MM, Islam S, Al-Mamun M (2018) J Adv Dielectr 8:1850030
39. Shen H, Yin Y, Tian K, Baskaran K, Duan L, Zhao X, Tiwari A (2018) J Alloys Compd 766:601–608
40. Ahmed DS, Mohammed MK, Mohammad MR (2020) Chem Papers 74:197–208
41. Mohammad MR, Ahmed DS, Mohammed MK (2019) J Solgel Sci Technol 90:498–509
42. Yang K, Guo Q, Li H, Hao X, Ma Y, Yang M, Xia H (2018) J Power Sources 402:340–344
43. Liu GY, Zhao YY, Tang YF, Liu XD, Liu M, Wu PJ (2020) Rare Met 39:1063–1071
44. Li X, Wang M, Li R, Li C, He J, Qiu M, Chen Y (2021) Compos Commun 23:100578
45. He Y, Wu D, Zhou M, Liu H, Zhang L, Chen Q, Guo Z (2020) Appl Surf Sci 506:144946
46. Bakhtiarzadeh Z, Rouhani S, Karimi Z, Rostamnia S, Msagati TA, Kim D, Shokouhimehr M (2021) Mol Catal 509:111603
47. Sabeeh H, Aadil M, Zulfiqar S, Rasheed A, Al-Khalli NF, Agboola PO, Shakir I (2021) Ceram Int 47:13613–13621
48. Chen J, Yao B, Li C, Shi G (2013) Carbon 64:225–229
49. Zaaba NI, Foo KL, Hashim U, Tan SJ, Liu WW, Voon CH (2017) Procedia Eng 184:469–477
50. Mallick P, Satpathy SK, Behera B (2022) Bull Mater Sci 45:198
51. Satpathy SK, Sen S, Behera B (2017) J Mater Sci Mater Electron 28:9102
52. Mallick P, Patra R, Mohanty D, Satpathy SK (2022) Sadhana 47:134
53. Chen X, Paul R, Dai L (2017) Natl Sci Rev 4:453–489
54. Iro ZS, Subramani C, Dash SS (2016) Int J Electrochem Sci 11:10628–10643
55. Wang N, Pandit S, Ye L, Edwards M, Mokkapati VRSS, Murugesan M, Liu J (2017) Carbon 111:402–410
56. Yadav MD, Dasgupta K (2020) Chem Phys Lett 748:137391
57. Liang Y, Zhang W, Wu D, Ni QQ, Zhang MQ (2018) Adv Mater Inter 5:1800430
58. Soneda Y (2013) Handbook Adv Ceram 211–222
59. Fang B, Binder L (2006) J Phys Chem B 110:7877–7882
60. Pan H, Li J, Feng Y (2010) Nanoscale Res Lett 5:654–668
61. Liu N, Lu Z, Zhao J, McDowell MT, Lee HW, Zhao W, Cui Y (2014) Nature Nanotechn 9:187–192
62. Heydari N, Kheirmand M, Heli H (2019) Int J Green Energy 16:476–482
63. Siwal SS, Zhang Q, Devi N, Thakur VK (2020) Polymers 12:505
64. Patra R, Mohanty D, Nayak PK (2022) Polym Sci Ser B 7
65. Dang ZM, Yuan JK, Yao SH, Liao RJ (2013) Adv Mate 25:6334–6365
66. Liu Y, Zhou G, Liu K, Cui Y (2017) Acc Chem Res 50:2895–2905
67. Philip B, Xie J, Abraham JK, Varadan VK (2005) Polym Bull 53:127–138
68. Naoi K, Naoi W, Aoyagi S, Miyamoto JI, Kamino T (2013) AccChem Res 46:1075–1083
69. Frackowiak E, Khomenko V, Jurewicz K, Lota K, Béguin F (2006) J Power Sources 153:413–418
70. Ryu KS, Kim KM, Park NG, Park YJ, Chang SH (2002) J Power Sources 103(2):305–309
71. Frackowiak E, Beguin F (2002) Carbon 40:1775–1787
72. Patra S, Munichandraiah N (2007) J Appl Polym Sci 106:1160–1171
73. Marchioni F, Yang J, Walker W, Wudl F (2006) J Phys Chem B 110:22202–22206
74. Omar FS, Numan A, Duraisamy N, Ramly MM, Ramesh K, Ramesh S (2017) Electrochim Acta 227:41–48

75. Asen P, Shahrokhian S (2017) Int J Hydrog Energy 42:21073–21085
76. MuraleeGopi CV, Ravi S, Rao SS, Eswar Reddy A, Kim HJ (2017) Sci Rep 7:1–12
77. Yang Z, Chen CY, Liu CW, Chang HT (2010) Chem Comm 46:5485–5487
78. Joshi P, Zhang L, Chen Q, Galipeau D, Fong H, Qiao Q (2010) ACS Appl Mater Inter 2:3572–3577
79. Heli H, Yadegari H (2014) J Electroanal Chem 713:103–111
80. Gonçalves D, Faria R C, Yonashiro M, Bulhoes LOS (2000) J Electroanal Chem 487:90–99

Nanomaterials for Fabrication of Thermomechanical Robust Composite

Priyambada Mallick, Santosh Ku. Satpathy, and Srikanta Moharana

Abstract Nanomaterials have received a lot of interest as an emerging material because of their small size, surface effect, and tunneling effect, along with their potential utilization in traditional materials, electronic devices, energy storage devices, and other industries. Nanoparticles are nanomaterials with dimensions ranging from 1 to 100 nm. Nanomaterials with remarkable structural, mechanical, catalytic, optical, electrical, and magnetic characteristics that differ significantly from the bulk materials can be created. They can be categorized differently based on their qualities, forms, and sizes. There are different nanomaterials, including metals and ceramics are reinforced in the polymeric matrix to obtain composites with improved physical and chemical characteristics. A lot of research has been extensively reported on the impact of introducing nanomaterials into the polymeric matrix. The effect of nanomaterial selection, synthesis technique, grain size, and boundary structures on the mechanical characteristics of nanomaterials is presented in this chapter. Hybrid polymeric composites have undergone significant development and utilization for energy applications in recent times. However, future applications in the fields of engineering, industry, and medicine can be made possible by progressing further research on the molding technique of nanomaterials. Therefore, scientists and researchers put efforts into investigating the molding and fabrication technique as well as strengthening the process of an advanced, robust nanocomposite to fulfill the essential needs in future application. This chapter provides an overview of nanostructured materials along with composite preparation processes and discusses the effect that these approaches have on the thermomechanical performance of nanomaterial-based robust composites. However, future applications in the fields of engineering, industry, and medicine can be made possible by progressing further research on the molding technique of nanomaterials. As a result, scientists and researchers are investigating the molding and fabrication technique, as well as strengthening the process of an advanced robust nanocomposite to meet the critical needs in a future application.

Keywords Nanomaterial · Polymers · Mechanical properties · Composites

---

P. Mallick · S. Ku. Satpathy · S. Moharana (✉)
School of Applied Sciences, Centurion University of Technology and Management, Paralakhemundi, Odisha, India
e-mail: srikantanit@gmail.com; srikanta.moharana@cutm.ac.in

S. K. Tiwari et al. (eds.), *Nanoparticles Reinforced Metal Nanocomposites*,
https://doi.org/10.1007/978-981-19-9729-7_10

1 Introduction

Nanotechnology has piqued the interest of many over the last two decades due to its potential for the development of novel nanomaterials and device structures with exceptional physical and chemical characteristics. Recently, nanomaterials having a size of one dimension with a scale of 1–100 nm in 3D space are gaining much more attention as emerging materials. Due to its small size, surface morphology, different phases, and quantum tunneling effect, it has a lot of remarkable potential applications in the fields of electronics, automotive, electrochemical, biomedical, photochemical, electrical, medical, and industrial. Nanomaterials have extraordinary physiochemical performances, including melting point, thermoelectric activities, photo-absorbing properties, reactivity, scattering, and optical activities, with enhanced catalytic properties in comparison with their polycrystalline equivalents. Efforts to investigate nanostructured materials and their derivatives are critical for developing innovative materials with exceptional qualities. It has been proven that by employing various manufacturing procedures, it is possible to generate new sophisticated nanomaterials with amazing properties for a diverse set of applications. These structural, electrical, and magnetic properties of the material are greatly affected by the content of nanomaterials, the process of fabrication, grain size, and grain boundary structures. A lot of research has been extensively reported on the impact of introducing nanomaterials into the polymeric matrix. The effect of nanomaterial selection, synthesis technique, grain size, and boundary structures on the mechanical characteristics of nanomaterials is presented in this chapter. Researchers and scientists have recently investigated the thermomechanical nature of nanomaterials based on these characteristics [1]. But the problem and issues related to the improvement of the thermomechanical properties of advanced robust nanocomposites have not been completely solved till now [2]. The most important thermomechanical parameters in nanocomposites include stiffness, glass transition, storage and loss modulus, coefficient of damping (tan), distortion heat, temperature, and coefficient of thermal expansion, among others [3]. To improve their function, the thermomechanical and mechanical properties must be thoroughly investigated. Herein, we have only focused on the thermomechanical properties of a robust composite. Also introduced is the current progress in research and application range of nanomaterials, which is remarkable for the development of thermomechanically robust composites [4]. To improve mechanical and thermophysical properties, nanoparticles are dispersed in a matrix material such as metals, ceramics, or polymers. Polymer nanocomposite materials have found use in critical domains such as the automobile and aerospace industries. A material's mechanical properties define how it behaves under a variety of circumstances and stresses. Mechanical characteristics such as brittleness, strength, plasticity, toughness, hardness, ductility, yield stress, rigidity, and elasticity are the ten standard components of conventional materials' mechanical properties in metals. Most inorganic and non-metallic materials are brittle and lack desirable mechanical qualities, including plasticity, toughness, elasticity, ductility, etc. In addition, unlike inorganic materials, organic ones can be pliable without exhibiting traits like brittleness or rigidity. Because of the size, shape, and quantum nature of nanoparticles, nanomaterials exhibit remarkable mechanical capabilities. When nanoparticles are incorporated into a polymeric material, the grain boundary is strengthened, and the material's mechanical characteristics are enhanced due to the formation of an intra- and intergranular structure [4–7]. For example, adding 3 wt/% nano-SiO_2 to concrete can improve its compressive strength, bending strength, and splitting tensile strength [8].

It is feasible, for example, to increase the compressive strength, bending strength, and splitting tensile strength of concrete by adding 3 wt/% nano-SiO_2 [9]. The tensile strength, elongation at break, and impact strength of kenaf epoxy composites are significantly enhanced by the addition of 3% nano-sized oil palm empty fruit string filler. Because of their outstanding mechanical properties and unique traits not seen in macroscopic materials, nanomaterials have a wide range of potential uses. To identify possible technical applications and industrial productions, we must first determine the mechanical properties of various nanomaterials and composite materials. In this book chapter, an overview of nanomaterials, synthesis methodologies, and the mechanical properties of nanomaterials and their composites is presented.

2 Overview of Nanostructured Materials

In recent years, researchers and industrialists have shown a lot of interest in the preparation and development of advanced nanostructured materials [10] with excellent performance. The high surface areas and enlarged chemical reactivity, combined with the improved mechanical strength of the nanostructured materials, have gained them worldwide attention. In general, nanomaterials are categorized as natural (obtained from nature and obey all the laws of nature), incidental (by-products obtained from industry like coal dust), and engineering nanomaterials (obtained from advanced synthesis techniques with complex shapes) [11]. In general, nanostructured materials are materials that look mostly like crystallites and have at least one nanoscale dimension in terms of grain size and thickness (layer) that is less than 100 nm. Based on the dimensions of the features, these nanomaterials can be categorized into four types: [12]:

(i) Zero-dimensional (0D) nanomaterials having nanoscale dimensions in all directions, e.g., nanospheres, nanoparticles, quantum dots, etc.
(ii) One-dimensional (1D) nanomaterials have comparatively large-scale dimensions in one direction than others, e.g., nanorods, nanotubes, nanowires, nanobelts, nanoribbons, nanostars, etc.
(iii) Two-dimensional (2D) nanomaterials having comparatively large-scale dimensions in any two directions than others, e.g., graphene nanosheets, nanoplates, nanodisks, etc. [13]

(iv) Three-dimensional (3D) nanomaterials have comparatively large-scale dimensions in three directions than others, e.g., nanotetrapods, nanoflowers, nanocombs, etc.

Nanostructured materials have become a fascinating domain in the diverse fields of biotechnology, bioengineering, the medical field, condensed matter physics, material chemistry, and ionics engineering, as well as the academic, industrial, and commercial sectors [14]. Nanostructured (NS) materials usually exhibit various excellent characteristics such as high strength and hardness, increased diffusivity, and useful sintering properties [15, 16]. On the basis of composition, nanomaterials may be classified into four different types [11].

(i) Carbon-based nanomaterials: Carbon nanotubes, graphene, porous carbon, etc., with high conductivity and stability play a crucial role in the practical applicability and advancement of multifunctional interdisciplinary areas.
(ii) Metal-based nanomaterials: These materials are made up of various metals such as gold (Au), silver (Ag), platinum (Pt), and copper (Cu) and have excellent physical, chemical, and catalytic properties. They can be applied for sensors, paints, cosmetics (sunscreen), and dental care. [17] Also, metal oxide nanomaterials such as zinc oxide (ZnO), tin oxide (SnO_2), and copper oxide (CuO) have a wide range of applications in the fields of electronics and photonics, optoelectronics, solar cells, energy, etc.
(iii) Polymer-based nanomaterials: These are composed of a matrix and a filler, such as polyaniline, polypyrrole, poly(dopamine), and others, and have a high sensitivity for direct application in sensors and bioengineering [18].
(iv) Composite nanomaterials: These are mixtures of simple nanoparticles or compounds such as nanoclays and nanoflowers. There are several nanostructured materials found in the form of composites [19–21], capsules, porous materials [22], and fibers on a nanometric scale.

The other types of nanostructured materials (ceramics, optical materials, polymers, and metal nanocoats) and various nanodevices related to sensing switches, etc., have been reported [23]. Various applications of nanostructured materials include a wide range in pharmaceutical, industrial chemistry, electronics, space applications, energy storage applications [24–26], materials and metallurgy, biological fields, and the food and medical industries [27].

Recently, nanostructured materials used in the field of nanomedicine may lead to an important contribution to the fields of drug delivery, nanomedicine, and medical imaging [28]. Siwick and his team [29] have developed a nanostructured material with a special emphasis on the structure of a photonic crystal for application in 3D optical data storage with high density. Vaqueiro et al. [30] have reported on thermoelectric materials for the better utility of earlier resources of energy like building power manufacturing systems, which extract valuable electrical power from the wasted heat. It has paved the way for improvements in thermoelectric performance achieved through nanostructured materials. Ozturk and his group have synthesized a carbonaceous material [sandwiched graphene-fullerene composites (SGFC)], which is formed via a covalent bond (junction) among layers of graphene and non-homogeneously distributed fullerene. The outcome of this study opened a new path for potential candidates for application in hydrogen storage applications (ultra-lightweight) due to their superior surface ratio and adaptable porous microstructure [31]. It has been reported that nanostructured materials have a great influence on tissue engineering, with distinctive performances demanding significant applicability in rigid and flexible tissue engineering [32]. The polymeric nanofibers from the domain of organic nanomaterials might be utilized as an aid in the cultivation of cells [33] and also have more opportunities for various inorganic nanomaterials [34–36]. Recently, polymeric nanostructured materials (PNMs) have played a considerable role in the diagnosis and treatment of diseases. These nanostructured materials have a broad range of applications in the food sector, such as nanofood, probiotics, nanocoating in edible form, and modern packaging [34]. Kumar and his colleagues discovered that silver epoxy nanocomposite [37] had the highest thermal conductivity, Young's modulus, and tensile strength values when compared to sole epoxy nanocomposite.

3 Nanomaterial Fabrication Techniques

The enhancement of thermomechanical properties such as thermal energy transport, melting point, thermoelectric activities, photo-absorbing property, reactivity, scattering, and optical activities in nanomaterials does not depend only on the nature of individual components but also on the various fabrication techniques, the morphology, and the nature of the interface. One of the major critical issues with nanomaterials is the loss of their original properties during fabrication, which limits their applications. Various fabrication techniques, including physical and chemical methods, use surfactants to prevent the agglomeration [38] of nanomaterials, which is crucial for developing the mechanical and thermomechanical properties. A lot of nanomaterials have been prepared over the last century by the old technique, which comprises grinding and mixing ingredient powders and then calcining them in a furnace at a high temperature. But chemical methods are more considerable because of the strong covalent bonds under different conditions [39]. Some of the most suitable chemical techniques for the fabrication of nanomaterials are chemical vapor deposition (CVD), molecular beam epitaxy (MBE), the sol–gel technique, hydrothermal synthesis, molecular self-assembly, lithography, etc., which are discussed in detail below.

(i) **Chemical Vapor Deposition (CVD)**

Chemical vapor deposition (CVD) is a very promising technique that is mostly used in the semiconductor industry for the synthesis of nanometric layers of inorganic materials on the surface of 3D substrates and for depositing thin films of various materials [38]. Chemical vapor condensation or chemical vapor synthesis occurs when solid films are deposited on surfaces. There are four successive steps involved in this process: (a) introduction of the volatile or vaporized precursors (which may be solid, liquid, or gas under ambient conditions) by carrier gas to the reactor chamber; (b) adsorption of the substrate to one or more volatiles at a high temperature, which favors homogeneous nucleation to form by-products; (c) breakdown of these products on a heated substrate, followed by heterogeneous nucleation and solid layer or grain development; and (d) the creation and extraction of volatile products under certain conditions (temperature, pressure, substrate, etc.) from the chamber by the carrier gas [39].

The quality of the by-products produced through CVD techniques is influenced by various factors, such as type and quality of precursors, desired volatility, thermal ability, temperature and pressure in the chamber, chemical properties of the substrate and gas carrier, and time and rate of deposition. Based on the type of chamber and precursors, the CVD techniques are categorized into various types, which are given in Fig. 1.

One of the most widely used precursors for the fabrication of nanoparticles in CVD techniques is silver nitrate ($AgNO_3$). In 2018, Piotr Piszczek synthesized silver nanoparticles using CVD techniques, which reduce the cytotoxicity of silver-based nanoparticles [40]. Wang and his colleagues recently used this technique to create pure, structurally uniform, single-crystalline semiconducting oxides free of defects and dislocations, such as ZnO, In_2O_3, Ga_2O_3, CdO, PbO_2, and SnO_2. Zhao discovered in 2019 that the CVD technique is the most advantageous synthesis technique for surface modification of nanomaterials [38]. The CVD process enables control over the structure, shape, and development of the robust nanocomposite that is formed. However, the CVD method has some disadvantages, such as long reaction times (1 min to hours) and low-temperature processes (in the range of 700–1473 K) [41], which can be overcome by using modern technology.

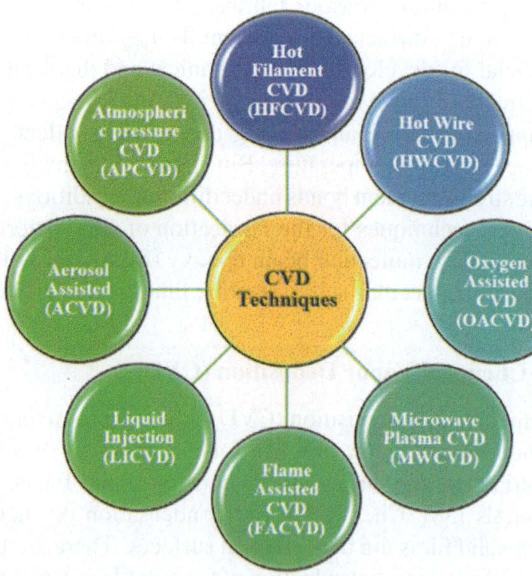

Fig. 1 Different types of chemical vapor deposition techniques

(ii) **Molecular Beam Epitaxy**

One of the most time-consuming, technically demanding, and challenging physical evaporation experimental techniques is molecular beam epitaxy (MBE) [42], which allows for the layer-by-layer expansion of thin films of various novel nanomaterials without the innovation of any chemical reactions. It can also be used to deposit a wide range of materials, including metals, semiconductors, magnetic materials, oxides, organic molecules, chalcogenide layers, etc. For the development of chemical sensors, the nanomaterials should be synthesized using a novel technique that can control the structure, composition, and morphology of the surface. Molecular beam epitaxy (MBE) is the most favorable process for the preparation of nanomaterials.

This method works on the principle of vacuum evaporation, where thermal molecular and atomic beams directly impinge on a heated substrate under ultra-high vacuum (UHV) conditions [43]. The UHV condition helps with minimal impurities and surface modification to extract high-purity nanomaterials. MBE, as a low-temperature process, reduces autodoping, allows for precise control of the doping process, and keeps the growth rate between 0.01 and 0.3 m per minute. A typical MBE experimental setup consists of two or more Knudsen effusion cells (K-Cells) containing pure solid elements like selenium, silicon, bismuth, gelinium, etc., a UHV chamber (where the growth of materials takes place), and a sample holder with a substrate. The first step in the MBE growth process is to heat the K-cells to the appropriate temperatures until the elements in each cell sublimate. The shutters are then opened, allowing physical vapor from each K-cell to permeate into the chamber until it reaches the substrate, where it is deposited, and the thin film gets formed. The final composition and stoichiometry of the film will be determined by the temperature and surface atomic structure of the substrate, as well as the flux ratios of individual components reaching the substrate. The substrate can be continually turned at modest rotation speeds (1–2 revolutions per minute) using a stepper motor coupled to the magnetic manipulator for more uniform development. In 2013, Lorenzo Morresi concluded that the evaporation and growth of materials can be influenced by temperature controllers, shutters, beam flux monitors, mass analyzers, and reflection high-energy electron diffraction (RHEED) systems [43]. Ishikawa reported that the MBE technique is a novel process for the preparation of robust composites [44]. Asghar and his team successfully prepared zinc oxide (ZnO) nanoparticles by using this technique [45]. Growth of porous nanostructures in GaN with low dimension can be obtained by synthesizing the nanomaterials using molecular beam epitaxy (MBE) techniques [46]. Later, E. Fadaly and his team synthesized InAs nanowires on silicon (Si) by using this technique and observed catalyst-free growth of the nanowires [47].

(iii) **Sol–gel Synthesis**

Although the sol–gel process is one of the outdated chemical synthesis techniques that was developed in 1940 [47], the importance of this process in material fabrication has been growing rapidly. Sol–gel processes have a remarkable advantage for preparing superfine nanopowders of metal oxides (MO) as well as non-oxide materials. The mechanical and chemical stability of the materials can be improved by fabricating with this technique, which is very useful for the development of the sensor. Inorganic and organic/inorganic hybrid material creation and processing are both generically referred to as "sol–gel" in the literature. Sol–gel processing generally involves the creation of colloidal suspensions (referred to as "sol"), which are then transformed into gels and finally into solid material. The formation of a sol, a colloidal suspension, is the first step in the sol–gel process. Colloid precursors feature a metal atom surrounded by ligands, which can be either inorganic anions or organic alkoxides. Sol denotes dispersed solid colloidal particles in a solution solvent, whereas gel denotes a 3D porous, interconnected network structure in the liquid phase [48]. In this process, sol is transformed into a gel by following several steps. Dispersed solid nanoparticles (sols with a diameter of 1–100 nm) are mixed in a homogeneous liquid medium and agglomerated to form a continuous three-dimensional network (gel) with pore diameters in the sub-micrometer domain in the liquid phase (Fig. 2).

The properties of sol-gels depend on important parameters such as pH, type of solvent, temperature, time, catalysts, and agitation mechanisms [38, 41]. Rahman used the sol–gel technique to create silica nanoparticles with improved mechanical, thermal, physical, and chemical properties [49]. In a recent study [29], a sol–gel processing technology was developed to generate a wide range of ceramic materials, including Al2O3, Fe2O3, SiO2, TiO2, and others. In 2021, Kumar and his team prepared TiO2 nanoparticles using this technique and observed excellent interfacial bonds of the nanoparticles with epoxy materials and glass fibers [50].

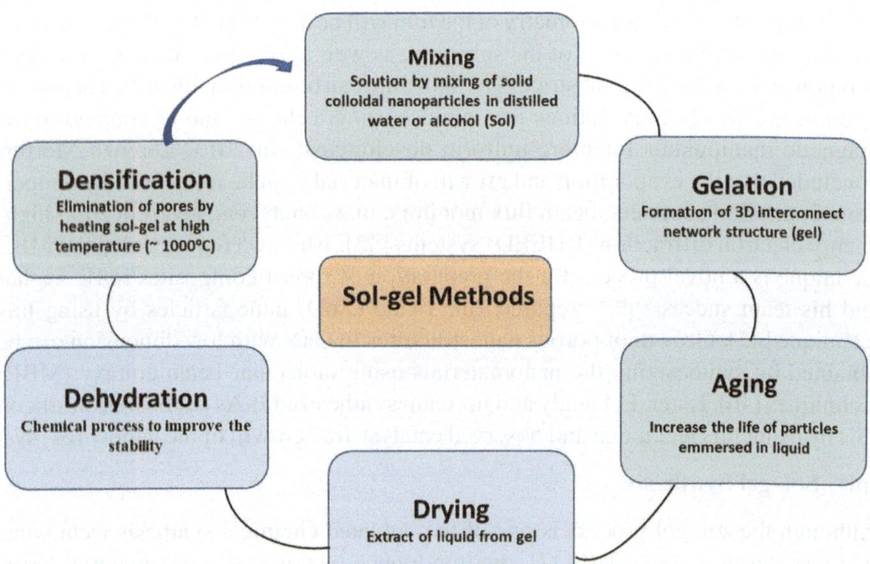

Fig. 2 Flow diagram for the different steps of the sol–gel technique

(iv) **Hydrothermal Synthesis**

Hydrothermal synthesis (also known as solvothermal synthesis) is a single-step process to prepare the ultrafine nanomaterials in a hydrothermal environment (i.e., neither the solution is non-aqueous nor aqueous) at a low temperature in the range of 100–3740 °C without the use of a calcination process. The basic principle of the technique is a reaction in an aqueous solution or suspension of the precursors at high temperature and pressure [51, 52]. Recently, the hydrothermal process has gained much more attention for the synthesis of nanomaterials as grain size, shape, crystalline state, and surface morphology of materials can be changed by controlling various factors such as the property of the solvent, concentration of reactants, temperature, aging, pH time, and additives in this technique [36]. Under fixed pressure and temperature, the hydrothermal technique is carried out in specific equipment known as an autoclave. The autoclave is loaded with reagents and then placed in the oven for some time, allowing the reaction to take place without direct supervision. This procedure is performed at high temperatures. This process produces chemically synthesized nanostructured powders with excellent consistency and particle uniformity and an appropriate particle size that has higher mechanical or electrical properties. The nanomaterials that are synthesized by using this process have worldwide applications in the fields of solar cells, batteries, MLCCs, etc. [18]. Benega and his colleagues reported in 2021 that carbon-based nanocomposites prepared by hydrothermal synthesis had superior physiochemical properties than others [53]. Kigozi observed that the metal oxide-based nanocomposite that is prepared using this technique can be used in energy storage applications with enhanced stability [54].

(v) **Molecular self-assembly**

One of the most successful chemical synthesis routes for designing complex nanostructures in the range of 1–100 nm is molecular self-assembly (MSA), in which atoms or molecules assemble in equilibrium conditions to form a stable and well-defined nanophase via non-covalent bonds [55]. It is the most attractive bottom-up process because it is technologically feasible and cost-effective and provides well-defined and functional geometries with structural freedom under specific, controllable thermodynamic conditions. All natural organic and inorganic nanomaterials that are prepared using this technique are thermodynamically stable, relatively defect-free, and self-healing [56]. But the important challenge of the MSA process is the lack of knowledge on the development of molecular shape, the nature of non-covalent forces, the interplay between enthalpy and entropy, etc. Besides these, the most promising avenues for self-assembly are presently based on organic compounds, and organic compounds, as a group (although with exceptions), are electrical insulators. As a result, the methods used in the self-assembled system must be redesigned or developed to produce advanced types of organic molecules with appropriate properties for information processing and electrical or mechanical transduction. Ozin reported that the molecular self-assembly process is scientifically popular for the fabrication of nanocomposite by providing enhanced properties such as electrical, magnetic, and optical, which can develop the future of nanotechnologies as well as the daily lives of human beings and our surrounding environment [56]. Recently, Arul and his team prepared different types of nanostructures successfully to provide structural freedom for application in intracellular drug delivery [57].

(vi) **Lithography**

Lithography is the process where a substrate (like glass, silicon, gallium arsenide, etc.) coated with a photosensitive or radiation-sensitive polymer called resist is illuminated. When this resist is radiated, its physical properties change to form structures [58]. The size of formed structures is affected by the choice and thickness of the resist. Lithography is used for prototyping in electronics, microfluidics, optics, lab-on-a-chip, etc. The performance of a lithography can be determined from three parameters: (i) resolution (the minimum feature dimension that can be transferred with high fidelity to a resist film on a semiconductor wafer), (ii) registration (a measure of how accurately patterns on successive masks can be aligned or overlaid with respect to previously defined patterns on the same wafer), and (iii) throughput (the number of wafers that can be exposed per hour for a given mask level and is thus a measure of the efficiency of the lithographic process) [53]. We can produce nano and microstructures with different resolutions by using different lithography techniques, which are given below [59] (Fig. 3).

4 Robust Composite Fabrication Techniques

Robust composites can provide a variety of benefits, making them desirable in many high-performance applications. As a result, composites are increasingly being used

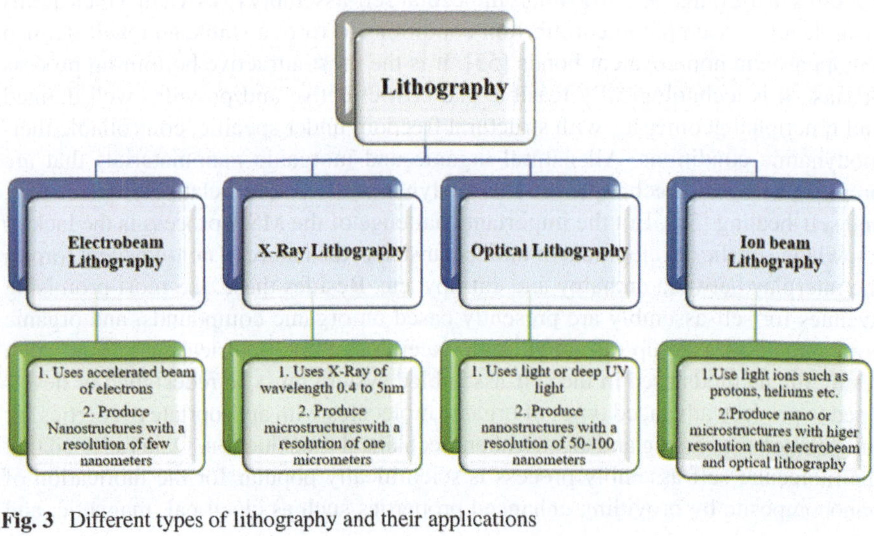

Fig. 3 Different types of lithography and their applications

not only in high-performance applications like racing cars, airplane components, and sporting goods, but also in lower-cost, higher-volume industries like automotive. With increased use comes an increased demand on manufacturing processes to maintain high quality while combining higher volumes and lower costs. Robust composite materials have many desirable properties that make them suitable for various applications [60]. To fabricate the robust composite, the researchers should handle the material in a safe and healthy environment by using protective gear and following policies. There are various fabrication techniques to shape resins and reinforcements, which are classified into two categories: (i) open molding and (ii) closed molding techniques (Fig. 4).

(i) **Open Molding technique**

Contact molding, or "open molding," is a low-cost process used for the production of large, robust composites where the raw materials are exposed to air before the fabrication. First, a one-sided mold is applied with a release agent and gel coat. Then, the molding materials or reinforcements are placed on top of the mold using either the spray-up process or the hand layup process. Additional layers of laminate are added to build thickness and strength as desired. The air is then rolled out of the laminate by hand, and the part is left to cure. In addition to reinforcements, low-density core materials such as balsa wood, foam, or honeycomb can be added to stiffen the laminate without adding significant weight. Open molding utilizes different processes where hand layup, spray-up, and filament winding techniques are mostly used, which are discussed below in detail.

(a) **Hand layup**

Hand layup is the most basic, simplest, and oldest technique of robust composite fabrication. The process involves laying the molding materials or reinforcements on top of the mold by hand and then acquiring the desired thickness of the laminated materials by stacking them up layer by layer. If the resin is applied to the layer of laminated material by using a brush or roller, then the process is known as wet layup, which is a different version of hand layup. A resin having high molecular weight and

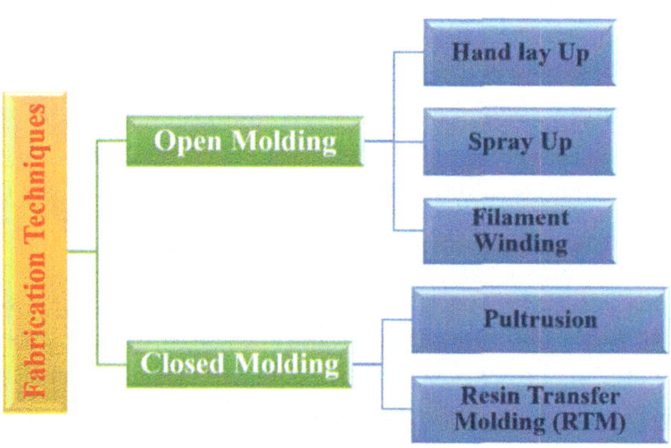

Fig. 4 Various fabrication techniques for robust composite

low viscosity can be used in this process to enhance its thermal and mechanical properties. Also, these properties are greatly influenced by factors such as resin mixing and the quality and content of the laminate resin. The hand layup fabrication process is mainly used in marine and aerospace structures. Although this method is successful for fabricating the robust composite, it has various disadvantages. It requires a large amount of time and a lot of skilled employees. Recently, Abas et al. have demonstrated the effect of silicon carbide (SiC) concentration in polyester nanocomposite on the thermomechanical properties like hardness, distortion of bending, thermal conductivity, coefficient of thermal expansion, etc., that are prepared by using the hand layup technique [61].

(b) **Spray Layup**

Spray-up, also known as "chopping," is another open-mold automated technique that is faster to produce chopped laminate using multiple molds. This process depends more on the operator than the hand layup process to control the thickness and consistency of the process. Although the production volume per mold is low, it is feasible to produce substantial production quantities using multiple molds. This process uses simple, low-cost tooling, and simple processing. Portable equipment permits on-site fabrication with virtually no part-size limitations. The process involves three important stages such as (i) application of gel coat to the double-sided mold; (ii) deposition of resin on the mold through a chopper gun; (iii) addition of chopped laminate layers to acquire the desired thickness. Roll stock reinforcements, such as woven roving or knitted fabrics, can be used in conjunction with chopped laminates. The same core materials and molds as in hand layup techniques are used in this process.

(c) **Filament winding**

Another open molding technique is filament winding, which is a continuous, low-cost, highly automated, and computer-controlled fabrication method. This technique is controlled by a computer, which reduces the number of employees. Filament winding uses a rotating mandrel (made up of steel or aluminum) as the mold, which helps in the production of a laminate surface with a high strength-to-weight ratio on the outside of the product. The uniformity of by-products and fiber orientation is highly controlled by this technique. The process of filament winding follows three steps to produce highly engineered structures with maximum tolerances.—(i) provide strand roving in a resin bath and keep it on a rotating mandrel; (ii) add filament on the mandrel to achieve the desired strength; and (iii) cure the laminate on the mandrel after the addition of appropriate layers. If the chopping or spray-up process is involved in filament winding, then that process is called the "hoop chop process."

(ii) **Closed Molding Technique**

According to Composites World, closed molding is also the low-cost process used for faster and more consistent production of large, robust composites where the raw materials are exposed to air before the fabrication. But a two-sided mold is applied with a release agent and gel coat. This technique has various advantages, such as

(i) less waste production, (ii) better surface cosmetics, and (iii) suppressing post-work. Besides these, it meets the needs of local and state manufacturers by emitting less radiation. These processes are automated, reducing the number of dependent workers. This technique utilizes different processes such as vacuum bag molding, vacuum infusion process, and compression molding, where resin transfer molding (RTM) and pultrusion techniques are mostly used, which are discussed below in detail.

(a) **Resin Transfer Molding (RTM)**

One of the widely used intermediate-volume closed molding processes is resin transfer molding (RTM) for the fabrication of nanocomposite materials. This process is significantly more reliable because it provides a faster rate of production, high consistency even at room temperature, and is also cheaper than the open molding processes. The steps involved in these techniques are (i) injection of resin inside a mold that is coated with a gel under low pressure, (ii) laying up the raw materials and orientation (like 3D reinforcements) inside the mold cavity, and (iii) control of thickness and temperature by different tooling. Vacuum assist can be used to increase the flow of resin inside the mold cavity. RTM can utilize either hard (such as aluminum, electroformed nickel shells, or machined steel molds) or soft tooling (such as polyester or epoxy molds) depending upon the expected duration of the run. Tooling can range from very low-cost to high-cost, life-long molds.

(b) **Pultrusion**

The pultrusion method is a smooth, continuous, cheap, and automated technique without any post-processing methods for the fabrication of nanocomposites with both simple and complex shapes having high structural, mechanical, and thermal properties. The portmanteau term "pultrusion" is the combination of "pull" and "extrusion," which means pulling of the material. A die pulls inserted fiber reinforcement in a hot bath of resin with specific shapes in this process. Then, that die (which may be hot oil or electric) is continuously heated to cure the resin and control the ratio of resin. The resin is inserted directly inside the die in the most recent technology, eliminating the need for an external resin bath.

5 Different Factors Influencing Mechanical Performance of Nanomaterials

The mechanical, thermal, as well as structural properties of nanomaterials are greatly affected by different factors such as their size, the quality of their raw materials, different fabrication techniques, and temperature. Nanomaterials exhibit various mechanical properties, including brittleness, strength, plasticity, hardness, toughness, fatigue strength, elasticity, ductility, rigidity, yield stress, etc.

(i) **Selection of Nanoparticle**

Different nanoparticles exhibit excellent mechanical properties. Thus, for the enhancement of the mechanical properties of nanomaterials, the selection of nanoparticles plays a vital role. Xu investigated whether the mechanical properties of robust composites improved first and then decreased as the amount of titanium diboride increased. Further, he increases the mechanical properties by adding aluminum trioxide (Al_2O_3) [62]. Thus, the addition of a large number of nanoparticles reduces the mechanical nature of nanomaterials. So, for the modification of nanomaterials with improved mechanical properties, we should focus on the amount, ratio, nanoparticle size, etc.

(ii) **Production process**

Besides the selection of nanoparticles, the production process also influences the mechanical nature of nanomaterials. Processing parameters such as sintering and calcination temperatures, synthesis techniques, processing time, choice, and concentration of nanoparticles can control the mechanical properties like hardness, rigidity, elasticity, ductility, etc., of nanomaterials. Karimzadehet al. reported in 2014 that the value of Young's modulus and hardness initially increases with increasing sintering temperature and then decreases after 1200 °C [63]. Different sintering temperatures have different influences on the mechanical properties of materials.

(iii) **Grain size**

Because nanomaterials are made up of grains and grain boundaries, we should concentrate on the various shapes and sizes of grain sizes in order to improve the mechanical properties of nanomaterials. Recently, a large number of scientists and researchers [64] investigated the effect of grain size on the mechanical properties of nanomaterials and concluded that grain size enhances the mechanical strength [65] and toughness [66] and also influences the fracture resistance [67] of nanomaterials.

(iv) **Grain boundary structure**

In robust composites, besides grain size, the structure of the grain boundary is a vital factor for the enhancement of mechanical properties. However, the density, chemical bond, and structure of the grain boundary influence the mechanical properties of nanomaterials. It has been discovered that the parameters influencing the structure of grain boundaries can have an effect on the mechanical properties of robust composites [68].

6 Effect of Thermomechanical Performance Nanomaterial-Based Robust Composites

The robust composite is a combination of matrix and nanomaterials with an improved thermomechanical nature. A large number of articles about the thermomechanical properties of nanocomposites containing nanoplatelets, nanospheres, and nanocylinders are available in the literature [69]. Surface modification of nanoparticles and adjusting the properties of the interfacial polymer layer by using proper chemistry and physics are crucial issues to enhance the thermomechanical properties of nanocomposites [70]. Recently, scientists and researchers have mainly focused on the thermomechanical factors such as coefficient of damping, modulus loss, storage, and coefficient of thermal expansion to enhance these properties by modifying the surface morphology of nanoparticles [71]. There has been a lot of research on 3D hybrid nanostructures with the goal of incorporating their exceptional qualities into polymeric matrices to create high-performance nanocomposites. It has been reported that the material's mechanical characteristics, including hardness, tensile, and flexural strength, have improved [72–75]. Usually, thermomechanical properties can be enhanced if the nanoclay is intercalated or exfoliated. In 2016, Yasmin et al. observed that the intercalation of nanoclay in epoxy composites enhanced the elastic and storage modulus and also reduced the coefficient of thermal expansion [76]. Rafieian et al. observed that the addition of cellulose nanofibrils (CNF) in the robust composite can greatly affect the mechanical properties rather than the thermal properties [77]. The addition of a carbon tube in reinforced nanocomposite can influence and obtain the values of thermal conductivity and other thermomechanical property parameters as needed [78]. Similarly, Araby et al. [79] have demonstrated the construction of conductive three-dimensional networks. They suggest that 1D nanostructures act as nanowires that transmit electrons and stress to 2D nanostructures (graphene nanoplatelets, GNPT). Therefore, 1D–2D interconnected nanostructures function as conductive channels. In addition, the hybrid reinforcement dispersion in the elastomeric matrix has also improved.

7 Mechanical Properties of Nanomaterial-Based Robust Composites

Different nanomaterials exhibit excellent mechanical properties such as elasticity, plasticity, tensile strength, stress, strain, Young's modulus, rigidity, hardness, and toughness under different external forces and environments due to the volume, surface, and quantum effects of nanoparticles [80]. Hence, by the addition of nanomaterials such as SiO2 and nanooil, the grain boundary as well as the mechanical properties of the robust composite can be improved [81]. Ajeesh et al. investigated in 2016 that the content of carbon nanofiber helps to increase the storage modulus as well as thermal conductivity in polyetherketone [82], which can be applied for high-temperature applications. Zhang et al. have demonstrated that the value of thermal conductivities increases with the addition of silica to epoxy nanocomposite, but the value of shear and elastic modulus reduces, which helps in the improvement of the thermomechanical nature of nanocomposite [83]. Due to the unique mechanical properties of nanomaterials, they will have a wide range of applications in future. With progress in further research on nanomaterials, future applications in the fields of engineering, industry, and medicine can be possible. Methods for modifying the surface of graphene using both covalent and non-covalent functionalization were reported by Kulkarni and his co-workers [84]. They explored how the electrical, mechanical, and thermal properties of graphene-epoxy composites change depending on manufacturing techniques, filler dispersion, and filler surface modification. In their analysis of graphene, carbon nanotubes (CNTs), and hybrid graphene-CNT-reinforced epoxy composites, Singh et al. [85] have fabricated epoxy-based composites with mechanical, thermal, electrical, and flame-retardant properties that were examined along with their sensitivity to the filler type and functioning. These composites are used in the application of airplane bodies, electromagnetic shielding, corrosion-resistant coatings, etc. Szeluga et al. [86] investigated the influence of graphene fillers on the mechanical, thermal, electrical, and flame-retardant properties of epoxy composites. Epoxy composite qualities were summarized in relation to filler size, exfoliation level, and functioning. Epoxy composites enhanced with graphene were examined by Atif et al. [87]. The mechanical, thermal, and electrical properties of the resulting epoxy composites were linked to the filler size, morphology, and level of functionalization. In order to enhance the damping qualities, Jin and his co-workers [84] have fabricated an IPN based on epoxy resin and polyurethane pre-polymers. However, this material has a lower tensile modulus and strength than pure epoxy. Epoxy matrix toughening is also demonstrated. By providing a wide surface area, nanoparticles facilitate the establishment of primary valence bonds at the filler/resin contact and efficient stress transfer across the interface [84].

8 Conclusions

Nanostructured materials are widely investigated for their utmost potential as the material of electrodes, especially owing to their thermomechanical performance in the domain of energy storage applications in recent times. In this chapter, an overview of nanostructured materials with their various fabrication techniques and thermomechanical properties is discussed in detail for a better understanding and future perspective. Also, this chapter focuses on important factors such as the selection of nanomaterials, processing parameters, grain size, and grain boundary structure that influence the thermomechanical behavior of nanomaterials. A robust composite having advanced thermal and mechanical properties can be obtained by using advanced fabrication techniques such as CVD, molecular beam epitaxy (MBE), the sol–gel technique, hydrothermal synthesis, molecular self-assembly, and lithography. The various open and closed molding processes are also described minutely for understanding the structural stability of robust composites for intended future applications. It has been proven that by employing various manufacturing procedures, it is possible to generate new sophisticated nanomaterials with amazing properties for a diverse set of applications. These structural, electrical, and magnetic properties of the material are greatly affected by the content of nanomaterials, the process of fabrication, grain size, and grain boundary structures. The thermomechanical properties of nanomaterials endow them with broad application prospects and huge potential value in future. But the problem and issues related to the improvement of the thermomechanical properties of advanced robust nanocomposites have not yet been completely solved. The most important thermomechanical parameters in nanocomposites include stiffness, glass transition, storage, and loss modulus, coefficient of damping (tan), distortion heat, temperature, and coefficient of thermal expansion, among others. A lot of research has been extensively reported on the impact of introducing nanomaterials into the polymeric matrix. The effect of nanomaterial selection, synthesis technique, grain size, and boundary structures on the mechanical characteristics of nanomaterials is presented in this chapter. This chapter provides an overview of nanostructured materials along with composite preparation processes and discusses the effect that these approaches have on the thermomechanical performance of nanomaterial-based robust composites. Hybrid polymeric composites have undergone significant development and utilization for energy applications in recent times. However, future applications in the fields of engineering, industry, and medicine can be made possible by progressing further research on the molding technique of nanomaterials. As a result, scientists and researchers are investigating the molding and fabrication technique, as well as strengthening the process of an advanced robust nanocomposite to meet the critical needs in a future application.

Acknowledgements The authors thankfully acknowledge the support provided by Centurion University of Technology and Management, Odisha, India for carrying out the current research.

Conflicts of Interest The authors declared that no conflict of interest.

References

1. Nikzad M, Masood SH, Sbarski I (2011) Mater Des 32:3448
2. Wang F, Zhou S, Yang M, Chen Z, Ran S (2018) Polymers 10:401
3. Huang X, Zhi C (2016) Springer 10:978
4. Wu Q, Miao WS, Gao HJ, Hui D (2020) Nanotechnol Rev 9:259–273
5. Zou B, Huang CZ, Wang J, Liu BQ (2006) Trans Tech Publications Ltd 315:154
6. Wang XH, Xu CH, Yi MD, Zhang HF (2011) Adv Mater Res 154:1319
7. Wang JL, Meng LJ (2014) Appl Mech Mater 535:785
8. Ghabban A, Zubaidi A B, Jafar M, Fakhri Z (2018) IOP ConfSer: Mater Sci Eng 454:012016
9. Saba N, Paridah MT, Abdan K, Ibrahim NA (2016) Constr Build Mater 123:15
10. Bavykin DV, Friedrich JM, Walsh FC (2006) Adv Mater 18:2807
11. Abdel Karim R, Reda Y, Abdel-Fattah A (2020) J Electrochem Soc 167:037554
12. Maglia F, Tredici IG, Anselmi-Tamburini U (2013) J Eur Ceram Soc 33:1045

13. Bag A, Lee NE (2019) J Mater Chem C 7:13367
14. Kanatzidis MG, Poeppelmeier KR, Bobev S, Guloy AM, Hwu SJ, Lachgar A, Seshadri R (2008) Prog Solid State Ch 36:133
15. Pal K, Asthana N, Aljabali AA, Bhardwaj SK, Kralj S, Penkova A, Gomes de Souza F (2021) Crit Rev Solid State Mater Sci 17
16. Bhoi NK, Singh H, Pratap S (2020) J Compos Mater 54:813
17. Iqbal M, Usanase G, Oulmi K, Aberkane F, Bendaikha T, Fessi H, Elaissari A (2016) Mater Res Bul 79:97
18. Sahay R, Reddy VJ, Ramakrishna S (2014) Int J Mech Eng 9:13
19. Patra R, Mohanty D, Nayak PK (2022) Polym Sci Ser B 7
20. Sahu AK, Satpathy SK, Rout SK (2020) Trans Electr Electron Mater 21:217
21. Mallick P, Satpathy SK, Behera B (2022) Braz J Phys 52:187
22. Wei Q, Fu Y, Zhang G, Yang D, Meng G, Sun S (2019) Nano Energy 55:234–259
23. Jeevanandam J, Barhoum A, Chan YS, Dufresne A, Danquah MK (2018) Beilstein J Nanotechnol 9:1050
24. Mallick P, Satpathy SK, Behera B (2022) Bull Mater Sci 45:198
25. Satpathy SK, Sen S, Behera B (2017) J Mater Sci Mater Electron 28:9102
26. Mallick P, Patra R, Mohanty D, Satpathy SK (2022) Sadhana 47:134
27. Nasiri N, Clarke C (2019) Biosensors 9:43
28. Mudedla SK, Singam EA, Vijay Sundar J, Pedersen MN, Murugan NA, Kongsted J, Subramanian V (2015) J PhysChem C 119:653
29. Siwick BJ, Kalinina O, Kumacheva E, Miller RD, Noolandi J (2001) J Appl Phys 90:5328
30. Vaqueiro P, Powell AV (2010) J Mater Chem 20:9577
31. Ozturk Z, Baykasoglu C, Kirca M (2016) Int J Hydrog Energy 41:6403
32. Pryjmakova J, Kaimlova M, Hubacek T, Svorcik V, Siegel J (2020) Int J MolSci 21:2521
33. Jean-Gilles R, Soscia D, Sequeira S, Melfi M, Gadre A, Castracane J, Larse M (2010) J Nanotechnol Eng Med 1
34. Sharma S, Jaiswal S, Duffy B, Jaiswal AK (2019) Bioeng 6:26
35. Mohanty NK, Pradhan RN, Satpathy SK (2014) J Mater Sci Mater Electron 25:117
36. Mohanty D, Satpathy SK, Behera B, Mohapatra RK (2020) Mater Today Proc 33:5226
37. Siva Kumar M, Kumar S, Gouda K, Bhowmik S (2021) Polym Polym Compos 29:1551
38. Ganachari SV, Banapurmath NR, Salimath B, Yaradoddi JS, Shettar AS, Hunashyal AM, Hiremath GB (2017) Handbook of ecomaterials, p 83
39. Zhao X, Wei C, Gai Z, Yu S, Ren X (2020) Chem Pap 74:767
40. Piszczek P, Radtke A (2018) Noble Precious Metals 187
41. Manawi YM, Samara A, Al-Ansari T, Atieh MA (2018) Mater 11:822
42. Henini M (Ed.) (2012) Elsevier, Newnes
43. Morresi L (2013) Bussum: Bentham Science Publishers, p 81
44. Ishikawa Y, Shibata N, Fukatsu S (1996) Appl Phys Lett 68:2249
45. Asghar M, Mahmood K, Raja MY, Hasan MA (2013) Adv Mater Res 622:919
46. Soopy AKK, Li Z, Tang T, Sun J, Xu B, Zhao C, Najar A (2021) Nanomater 11:126
47. Fadaly E(2015) Master's thesis, Molecular beam epitaxy of catalyst-free InAs nanowires on Si
48. Rao BG, Mukherjee D, Reddy BM (2017) Nanostructures for novel therapy. Elsevier
49. Sumida K, Liang K, Reboul J, Ibarra IA, Furukawa S, Falcaro P (2017) Chem Mater 29:2626
50. Ab Rahman Ismail, Padavettan Vejayakumaran (2012) J Nanomater 2012
51. Tan WK, Muto H, Kawamura G, Lockman Z, Matsuda A (2021) Nanomater 11:181
52. Holliday S, Stanishevsky A (2004) Surf Coat Technol 188:741
53. Özbay N, Şahin RY (2017) AIP Conf Proc 1809:020040
54. Benega MAG, Silva WM, Schnitzler MC, Andrade RJE, Ribeiro H (2021) Polym Test 98:107180
55. Kigozi M, Ezealigo BN, Onwualu AP, Dzade NY (2021) Chemically Deposited nanocrystalline metal oxide thin films. Springer, Cham
56. Ozin GA, Hou K, Lotsch BV, Cademartiri L, Puzzo DP, Scotognella F, Ghadimi A, Thomson J (2009) Mater Today 12:12

57. Arul A, Rana P, Das K, Pan I, Mandal D, Stewart A, Maity B, Ghosh S, Das P (2021) Nanoscale Adv 3:6176
58. Lombardo D, Calandra P, Pasqua L, Magazù S (2020) Mater 13:1048
59. Kajbafvala A, Bahmanpour H, Maneshian MH, Li M (2013)J Nanomater
60. Venugopal G, Kim SJ (2013)Advances in micro/nano electromechanical systems and fabrication technologies. IntechOpen
61. De Teresa JM (2020) Nanofabrication: nanolithography techniques and their applications. IOP Publishing
62. Li J, Zhang C, Liang R, Wang B (2008) Int J Prod Res 46:2087
63. Abas FO, Abass RU (2018) MATEC Web of Conf 225:01021
64. Xu C, Yi M, Zhang J, Fang B, Wei G (2012) Ceramic materials-progress in modern ceramics, IntechOpen
65. Karimzadeh A, Ayatollahi MR, Bushroa AR, Herliansyah MK (2014) Ceram Int 40:9159
66. Acchar W, Cairo CAA, Chiberio P (2019) Compos Struct 225:111109
67. Trombini V, Tonello KPS, Santos T, Bressiani JC, Bressiani AHDA (2012) Mater Sci Forum 727:597
68. Wang D, Zhao J, Zhou Y, Chen X, Li A, Gong Z (2013) Comput Mater Sci 77:236
69. Zhou T, Huang C (2015) Comput Mater Sci 104:177
70. Khorshidi B, Biswas I, Ghosh T, Thundat T, Sadrzadeh M (2018) Sci Rep 8:10
71. Zeng H, Wu J, Pei H, Zhang Y, Ye Y, Liao Y, Xie X (2021) Chem Eng J 405:126865
72. Subhani T, Latif M, Ahmad I, Rakha SA, Ali N, Khurram AA (2015) Mater Des 87:436
73. Chang HP, Liu HC, Tan CS (2015) Polymer 75:125
74. Li W, Dichiara A, Bai J (2013) Compos Sci Technol 74:221
75. Zhang S, Yin S, Rong C, Huo P, Jiang Z, Wang G (2013) Eur Polym J 49:3125
76. Yasmin A, Luo JJ, Abot JL, Daniel IM (2006) Compos Sci Technol 66:2415
77. Rafieian F, Shahedi M, Keramat J, Simonsen J (2014) J Food Sci 79:N100
78. Dittrich B, Wartig KA, Mülhaupt R, Schartel B (2014) Polymers 6:2875
79. Araby S, Saber N, Ma X, Kawashima N, Kang H, Shen H, Ma J (2015) Mater Des 65:690
80. Güler Ö, Bağcı N (2020) J Mater Res Technol 9:6808
81. Ajeesh G, Bhowmik S, Sivakumar V, Varshney L, Kumar V, Abraham M (2017) J Compos Mater 51:1057–1072
82. Zhang X, Wen H, Wu Y (2017) Polymers 9:430
83. Kulkarni HB, Tambe P, M Joshi G (2018) Compos Interfaces 25:381
84. Singh NP, Gupta VK, Singh AP (2019) Polymer 180:121724
85. Szeluga U, Pusz S, Kumanek B, Olszowska K, Kobyliukh A, Trzebicka B (2021) Crit Rev Solid State Mater Sci 46:152
86. Atif R, Shyha I, Inam F (2016) Polymers 8:281
87. Jin H, Zhang Y, Wang C, Sun Y, Yuan Z, Pan Y, Cheng R (2014) J Therm Anal Calorim 117:773

Progress in Metal Nanoparticles-Based Elastic Materials

Rakesh Shrestha, Sagar Ban, Gaurav Khatiwada, Saroj Raj Kafle, Santosh K. Tiwari, and Rajendra Joshi

Abstract Metal nanoparticles incorporation in several materials have greatly enhanced their properties and also resulted in some unique features that can be very beneficial for the future. These nanoparticles show potential for devices such as polymers, biomedical devices, hydrogels, glass composites, printable electronics, and superhydrophobic materials. The materials, thus formed, have favorable physical as well as chemical properties and have shown enhancements in modulus of elasticity, impact strength, hardness, optical properties, thermomechanical properties, low dielectric loss, self-healing properties, biocompatibility, excellent durability, and more. The advancements achieved in several elastic-material applications over the last decade, with the incorporation of several metal nanoparticles, have been discussed in detail in this chapter.

Keywords Metal nanoparticles · Nanocomposites · Elastic modulus · Hydrogels · Superhydrophobic materials · Printable elastic electrodes

1 Introduction

Nanoparticles and the scope surrounding them are extremely broad, with numerous research goals and applications. Nanoparticles, of metal in particular, have found applications in the development of elastic materials that are applicable

R. Shrestha · S. Ban · G. Khatiwada · R. Joshi (✉)
Department of Chemical Science and Engineering, Kathmandu University, PO Box 6250, Dhulikhel, Kavre, Nepal
e-mail: rajendra.joshi@ku.edu.np

S. R. Kafle
Department of Chemical Engineering, Chungbuk National University, Chungbuk 28644, Korea

S. K. Tiwari
Department of Chemistry, University of Warsaw, Warsaw, Poland

R. Joshi
Department of Pharmacy, School of Science, Kathmandu University, PO Box 6250, Dhulikhel, Kavre, Nepal

S. K. Tiwari et al. (eds.), *Nanoparticles Reinforced Metal Nanocomposites*,
https://doi.org/10.1007/978-981-19-9729-7_11

in devices such as polymers, biomedical materials, hydrogels, glass, printed boards, and electronics [1]. At present, the preparation of materials with enhanced functionality is highly desired; for instance, the nanoparticles adhering to the surface of the materials, the deformation behavior, and enhanced elasticity without the loss of strength [2]. The significant difference that exists between the biological tissues and the elastic modulus of the metals has resulted in the resorption phenomena occurring in the carbon nanostructured implants (CNI) [3]. With favorable physical as well as chemical properties, metal nanoparticles are gaining widespread popularity [4]. The most significant feature of the metallic nanoparticles is their ratio of surface area to volume, which allows their interaction with other particles. The change in the shape, size, and composition of the metal nanoparticles, when incorporated in the material to be prepared, can change their properties significantly to meet the required properties in applications like biomedical, coating, storage materials, polymer, superhydrophobic materials, and much more. Moreover, the metal nanoparticles improve the modulus of elasticity, impact strength, hardness, and optical properties of the host materials.

Different noble metal nanoparticles have been discovered and practiced in industrial processes, like hydrogels for catalysis, and non-industrial processes, like bio-applications. Besides the improvements in the properties, the nanoparticles also aid in the preservation and increasing the life cycle of the devices [5]. The metallic nanoparticles are also utilized in the preparation of products with antimicrobial activity. This implies that the shelf life of the materials is improved and which will attract the attention of researchers to combine the biopolymer with the metal nanoparticles and prepare a hybrid system comprising a multi-functional property. These recent advancements in the application of the elastic materials incorporated by the metal nanoparticles are discussed in the following sections.

2 Stretchable Conductive Composites and Dielectrics

The need for health care sensors, flexible and electronic displays, and electronic skins is increasing, resulting in increased research interest in stretchable materials. With low-loss and a high dielectric constant, stretchable conductive composites have also been used in solar cells, epidermal electronics [6], health monitoring, robotics, and prosthetics [7]. Several recent advances on stretchable electronic systems that show desirable properties such as high conductance and its retention even with mechanical deformation, and a strong adhesive tendency to the composite matrix, without compensation for any of the properties, have been reported. The dielectric permittivity of polymeric materials can be significantly enhanced by incorporating conductive fillers such as metal nanoparticles [8]. With 20% silver nanoparticles (AgNPs) added to the polydimethylsiloxane (PDMS) matrix, the permittivity was 5.9 with a dielectric strength of 13.4 $V \cdot m^{-1}$. Likewise, with a strain at a break of 800% and, at 100% strain, a low elastic modulus of 350 kPa was also observed. Furthermore, the silver flakes showed independence from the polymer matrix and provided flexibility in matrix selection [9]. Likewise, the dielectric permittivity further increased to 21 at 31% vol. filler, and these materials with their increase in permittivity, and large strain, led to their application in energy harvesting devices that operate with a low electric field [10]. The permittivity can be further increased with the addition of AgNPs and nickel nanoparticles (NiNPs) to the PDMS matrix, and when compared to the PDMS alone, there was a low dielectric loss of 0.009 while the dielectric constant significantly increased by 1146% [11]. They also stated that the low elastic modulus of 3.57 kPa combined with a strain at break of 139.68% made it suitable for use in wearable electronics.

Carbon black (CB) particle incorporation in the PDMS matrix led to the condensation of Si–OH groups with the remaining Si–OH group from the elastomer surface, resulting in a large deformation of 200% [12]. Likewise, there was no loss in the conductivity even after 10,000 cycles, which resulted in stretchable electrodes that are highly conductive and suitable for dielectric elastomers. The resistivity of conventional stretchable electrically conductive composite materials increased with continuous stretching [13]. The silver-dendrites incorporation at high-volume ratio in the composite showed a 78% decrement in resistivity from 5×10^{-4} Ω·cm at 100% elongation owing to the lower percolation threshold of the silver dendrites as compared to conventional electrically conductive composites [14]. Materials with functional resistive sensing properties could be created by coating flexible capacitive sensors with conductive nanoparticles and polymers. Figure 1 shows graphene oxide (GO)-coated polyester fibers on a polyurethane fiber at 0% strain. Under 50% strain, the fiber angle and distance increased. This sensor has the potential to be attached to a human body and measure physiological signals [15].

The liquid metal nanoparticles possess properties like low viscosity and vapor pressure, self-repairing ability, and large surface tension [16]. At 500% strain, the printable and stretchable electrodes, which are highly conductive with a conductivity of 2.4×10^4 S·cm^{-1}, were formed from the polymers incorporated with liquid metal nanoparticles such as eutectic gallium indium (EGaIn) particles. Liquid metal nanoparticles that are gallium-based possess several abilities, such as the ability to

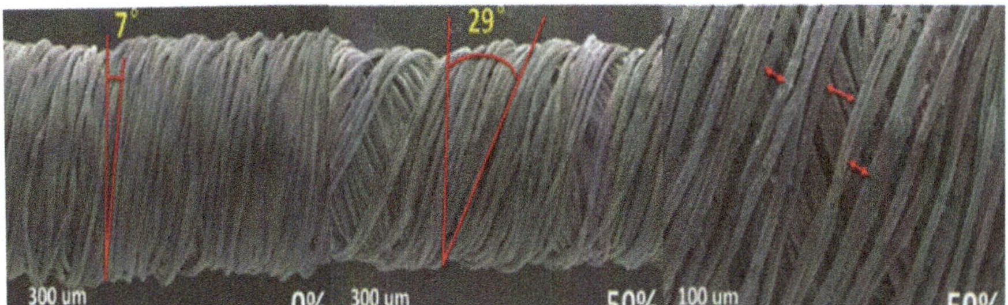

Fig. 1 Fiber for flexible and wearable respiratory sensors [15]

self-heal, change shape, and form a liquid–solid core shell structure [17]. The permittivity can also be controlled by altering the surface hardness of elastomers and the diameters of liquid metal nanoparticles. These stretchable materials are extensively attractive for application in deformable electronics like bio-integrated devices and robotic sensory skins [16]. The enhancement in thermal transport phenomena was observed when surface-modified EGaIn nanoparticles were incorporated in a PDMS matrix. Even with the composite containing EGaIn particles having a high volume ratio, it possessed a 6.91 kPa modulus of elasticity, which is very low, and under thermal and mechanical stress, it also showed electrical insulating properties. The resulting thermoelectric performance of the composite has a promising application in thermo-sensing devices [18].

The study on nanocomposites and amorphous metals is gaining growing research interest, with the conductors showing improvements in strength as well as strain. With a conductivity of 1.25×10^4 S·cm^{-1}, the highly flexible conductors derived from armored nanofibers and gold nanoparticles (AuNPs) showed toughness as well as Young's modulus of 1.3 MJ·m^{-3} and 5.29 GPa, respectively. The conductor showed properties comparable to amorphous metals, which also exceeded most of the composite materials [19]. The AuNPs incorporated PDMS to form a nanocomposite termed a piezoelectric nanogenerator (PNG), which showed 1.8% efficiency in energy conversion. The PNG delivered 8.34 mW·m^{-2} peak power density [20]. This indicated the potential for application in energy conversion. The application of dielectric polymer nanocomposites is rapidly emerging in the fields of solar cells, electronic transistors, light emitting diode applications, and energy storage devices [21]. A flexible polytetrafluoroethylene (PTFE) and polyethylene terephthalate (PET) composite with grown thin films of copper using an ion beam sputtering technique was observed [22], with reported electrical conductivity increments of 9.65×10^{-7} S·cm^{-1} for Cu/PET and 9.24×10^{-7} S·cm^{-1} for Cu/PTFE composites. It was also reported that the dielectric constant also increased, to 19 for Cu/PET and to 2.25 for Cu/PTFE composite. This property of the flexible nanocomposites could find application in modern electronic devices that could resist harsh environments and have widespread lifetimes.

3 Shape Memory Application

Alloys for shape memory applications (SMA) should possess unique thermomechanical characteristics. The shape memory effect (SME) usually occurs when a material is subjected to a substantial amount of plastic strain and which recovers its previous shape by applying heat. Heat applied internally or externally raises the temperature over the phase transition point, causing SMA to revert to its original structure [23]. The superelasticity of SMA is another important property owing to which, when a load or stress is removed, the SMA recovers its original shape from nonlinear strain quickly [24, 25]. Some rapid progress has been observed in material science and nanotechnology, where nanoparticles have been largely incorporated. Temperature sensitivity has also been discovered in polymer nanocomposites with shape memory properties. Generally, there is a decrease in the elastic modulus of an alloy with a rise in temperature, while when SiO_2 nanoparticles were added, the modulus of elasticity of the nanocomposite showed a nonlinear rise and a decrease in Poisson's ratio [26, 27].

Over the recent years, NiTi alloys have been the most commonly used SMAs in various engineering and medical applications [28, 29]. The binary Ni–Ti shape memory alloy possesses unique mechanical–functional properties that include superelasticity and a shape memory effect [30]. The shape memory alloy based on NiTi, which was formed via laser powder bed fusion (LPBF) technology, has coarse columnar B2 grains that possess 70 GPa elastic modulus with higher stiffness, and fine equiaxed B19' grains that show 30 GPa elastic modulus, showing that the alloy exhibits good mechanical deformation behavior [31]. Superelasticity degradation has also been observed in nano-scaled NiTi alloys with some amorphous regions. The main reason behind the observed degradation is the plastic deformation accumulation and the martensite that originate from the crystalline and shape memory alloy regions, which could be resolved by heating the alloy at a higher temperature and at a lower maximum stress as it avoids the phase transformation [32]. There was a significant increment in the strain recovery when LaB_6 was increased to 1.5 wt%. However, strain related to the shape memory effect of 2.04% and 4.14% elastic recovery added up to 6.18% of the maximum recoverable strain in the composites of NiTi when the LaB_6 amount was adjusted to 0.97 wt% [33].

Antimicrobial properties of biomedical implants have been significantly preferred to reduce problems associated with infection during implantation [34, 35]. Silver ions are typically released within the toxic limit in Ag-containing alloys [36]. As a result of silver's antiseptic properties against infection after surgery and wound therapy, treated Ti-16Nb-Ag alloy with different martensite variants demonstrated a good combination for biomedical applications. With the increase in Ag content, the alloy showed a high strength of 840 MPa as well as a low modulus of elasticity, as shown in Fig. 2 [37]. However, brittleness presented an issue in the metallic alloys composed of Ag. Embrittlement was observed during a tensile test with cyclic loading and unloading with the alloy with the most phases, which eventually failed in the early elastic region, whereas strain deformation performance was 29% when the alloy contained a significant amount of the one phase. This biocompatible alloy has, thus, attracted increased interest due to the requirement for biomedical shape memory alloys [38].

4 Polymer Composite

Metal nanoparticles have found increased application in technologies for vibration reduction with the need for damping materials that have high performance [39]. In a study by Jiang et al., metal nanoparticles were deposited over the carbon nanomaterials' surfaces, and it was found that the modulus of storage of CuAlMn/polymer

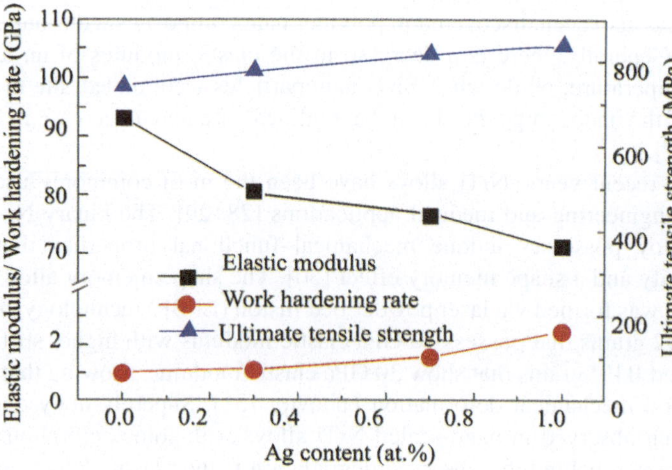

Fig. 2 Modulus of elasticity, work hardening rate, and ultimate tensile strength dependence on Ag content [37]

composite was above 3500 MPa, whereas for CuAlMn/polymer/(CNT + CNF + GNS) composite, it was about 1000 MPa [40]. Similarly, Yang et al. 2019 deposited CNT (1.5%) and Gr (1%) over the bitumen binder and found an enhancement of elastic recovery properties [41]. The mixture of CNT, alumina, and glass fibers were deposited over the polymer-based composite along with consideration for the crosslinking of epoxy molecules, and it was reported that the addition of fibers to pure polymers increased the elasticity by 89% [42].

The elastic modulus of the polymer composite is also affected by inclination. Joshi et al. discovered that the effective modulus of elasticity in CNT-based nanocomposite increased with lateral and axial inclination. A multiwall carbon nanotube with three-dimensional nitrogen doping (N-MWCNT) was synthesized using chemical vapor deposition (CVD) [43]. Experimentally, the modulus of elasticity increased with an increase in the diameter of N-MWCNT [44].

CuO nanoparticles were incorporated as nanofillers in a polyvinyl alcohol using the film-casting method, and surface morphology and elastic modulus were studied. The effect on mechanical properties of different loadings of MWCNTs in polyvinyl alcohol-based hydrogels has been shown in Fig. 3 [45]. Furthermore, a vacuum-assisted resin transfer molding process was used to create polyester composites based on woven carbon fiber and a combination of copper and graphene oxide. The composite showed 61.2% enhancement in strength and 57.5% in moduli [46].

The ultrasonication dispersion technique has also been used to prepare the reinforced epoxy nanocomposite by using TiO_2 spherical nanoparticles, and it was found that a maximum enhancement of 27% in modulus of elasticity was observed [47]. Lakshmikandhan et al. developed a polybenzoxazine nanocomposite by varying the weight percentages of TiO_2 and Al_2O_3 and found a higher value for its elastic properties [48].

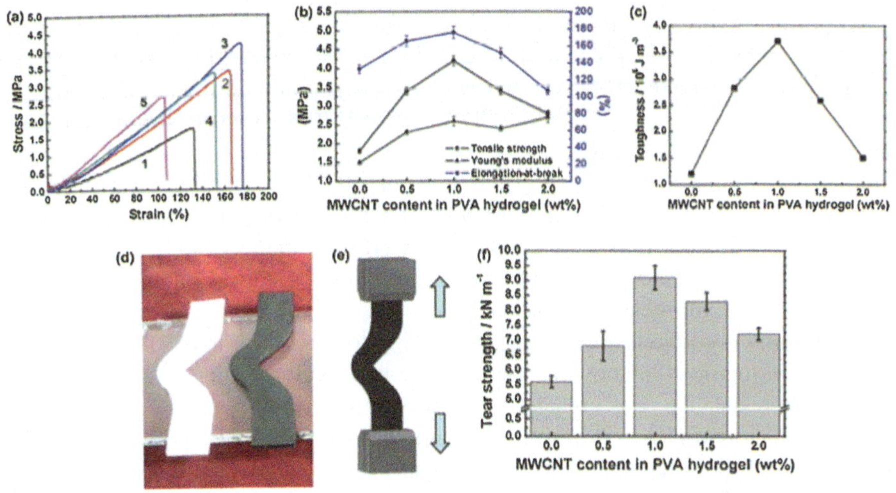

Fig. 3 Mechanical properties of hydrogels made of MWCNTs—PVP/PVA. **a** Graphs showing tensile stress and tensile strain for neat 0.5%, 1%, 1.5%, and 2% by weight loading. **b**, **c**, and **f** Comparison of different loading on mechanical properties. **d** and **e** Actual and schematic representation of measuring tear strength [45]

5 Flexible Hydrogels

Hydrogel products possess the capability to store water in their three-dimensional networks, allowing them to be applied in industries, medicine, and environmental-related applications [49]. Owing to their flexible nature, hydrogels are mostly used as human motion sensors, for tissue engineering in the medical field, in robotics engineering, and in electric circuits [50, 51]. Rafieian et al. experimented with different nanoparticles to form a hydrogel network and prepared different nanocomposite hydrogels to improve the drawbacks of hydrogels and increase their flexibility and self-healing capability [52]. Some of the advancements related to the metal nanocomposite hydrogels are presented.

Reduced graphene oxide coated with dopamine (rGO@PDA) was added to prepare hydrogels that are highly flexible and also possess self-healing properties, and it was found that the interaction of boron–oxygen and metal ligands formed reversible dynamic networks [53]. This significantly improved the mechanical strength of the biomass as well as its self-healing efficiency. rGO@PDA showed enhanced conductivity and flexible properties, and these self-healing hydrogels showed a large strain of 728%. Polydopamine (PDA), graphene oxide (GO), and polyvinyl alcohol (PVA)-based nanocomposite hydrogels showed enhanced mechanical and electrical properties with 146.5 kPa tensile strength and 2580% fracture strain [54]. This hydrogel could continuously monitor human body motion as well as parameters related to physiology with the aid of its self-adhesive, conductive, and self-healing properties.

A porous hydrogel prepared from GO-stabilized pickering aqueous foam templates consisted of PVA and a copolymer of acrylamide (AM) and *N*-(3-sulfopropyl)-*N*-methacroyloxyethyl-*N*,*N*-dimethylammonium betaine (SBMA) [55]. This porous composite reportedly exhibited 400% tensile strain and 80% composite strain. This strain-sensitive composite hydrogel could be used to detect various human motions as well as to develop flexible wearable strain-sensing materials. Strain sensors that are flexible and also have self-repairing abilities were developed using poly (AA-co-SMA)/c-CNF/Fe^{3+} nanocomposite hydrogels. With hydrophobic association and ionic interaction in supramolecular cross-linked nanocomposite hydrogels, there is the possibility for self-repair, if damaged, and nanocomposite hydrogels also possess the ability to recover sensing [56].

A highly ductile conductive nanocomposite hydrogel, with increased toughness and fatigue resistance, was created by immersing gelatin and silver-coated copper (Ag@Cu) nanoparticles in Na_2SO_4 solution in a single step [57]. With the uniform distribution of these nanoparticles in hydrogel over a wide range of strains, it was found that the electrical conductivity was excellent, and it also showed high tensile strength sensitivity which was also reported to be stable. Due to its sensitive and stable nature, it can successfully be used for fabricating electronic skin for bionic robots. Silver flakes with sizes suspended in a polyacrylamide–alginate hydrogel matrix have electrical conductivity and mechanical deformation resistance despite being partially dehydrated. The electrical conductivity was greater than 300 $S{\cdot}cm^{-1}$ and was also capable of maintaining soft compliance and deformability while delivering a direct current [58]. The SEM images in cross-sectional view are shown in Fig. 4, for different volumes of *N*,*N*′-methylenebisacrylamide (MBA) on a nanocomposite of poly (acrylamide) and poly (ethylene glycol) and silver (PAM–PEG–Ag) [59].

A one-step in-situ polymerization technique was used to fabricate a self-healing, highly stretchable, and strain-sensitive hydrogel from acrylic acid, GO, ammonium persulfate, and iron ions without the use of a chemical crosslinker. The hydrogel showed improved stretchability of 1185.53% break elongation due to the dual crosslinking effect of polyacrylic acid–GO, which also showed self-healing efficiency of 88.64% [60]. This strain-conductive and strain-sensitive hydrogel showed potential to be used for monitoring the human body as a flexible sensor. An experiment was conducted using deformable liquid metal droplets in combination with sonication to initiate and gel acrylic acid, this LM-induced hydrogel provided stretchability, softness, and notch-insensitive toughness while also offering modality and exceptional self-healing properties, showing a new hydrogel application. The hydrogel shows potential for wearable devices and in energy storage devices as a solid electrolyte, as well as in tissue engineering [61].

6 Polymeric Membrane

The nanoparticle Al_2O_3 addition on membranes has enhanced membrane elasticity, but excessive amounts of this particle on membranes cause elasticity to decrease with

Fig. 4 **a–d** SEM images of nanocomposite of PAM–PEG hydrogel and silver with different volumes, i.e., 0.2, 0.4, 0.8, and 1.6 ml of MBA

a decrease in membrane elongation [62]. Zhang et al. added carbon nanotubes (CNT) to Al_2O_3, and large elastic deformation, that is tunable, was observed [63]. However, when Al_2O_3 nanoparticles were combined with polyurethane (PU) additives and added to the polycarbonate (PC) membrane, the elasticity improved [64]. Chemical modification of TiO_2 was carried out by Bet-Moushoul et al. in order to produce a hollow fiber PES membrane in which good thermal resistance was seen, but elasticity was low [65]. A phase inversion technique was used to prepare a new nanowire ultrafiltration membrane by dispersing TiO_2, and thermal gravitational and FTIR analysis were performed; a slight increase in the elasticity of polymeric chains was reported [66]. Lin et al. also added Ag to the TiO_2 membrane, which resulted in a greater improvement in elastic properties [67].

A wet chemical method was carried out in a polymer membrane by the addition of ZrO_2 nanomaterials with a structure-controlling agent, poly (ethylene oxide), poly (phenylene oxide), and poly (ethylene oxide) (PEO-PPO-PEO, P123). The method showed that the interparticle elasticity could be enhanced by P123 [67]. Furthermore, a recast and impregnation method was carried out in which Nafion®/Zr-0 was doped, and it was found that the elastic modulus of the formed membrane was higher [68]. Sigwadi et al. reported enhanced elasticity with a 10 to 50 mm/min strain rate with a 10 mm/min increment after adding ZrO_2-CNT nanofiller within the Nafion® membrane [69]. Electrospinning techniques have been carried out for the fabrication of novel polyvinylidene fluoride (PVDF)—(0.5–2%) Ag and PVDF—(0.5–2%) Ag–1% graphene oxide (GO). 0.5–2% AgNPs were added to PVDF, which markedly improved the elastic modulus [70]. Furthermore, Benavente et al. incorporated AgNPs in the cellulosic membrane, which further improved the mechanical strength [71].

7 Glass Composites

Platinum ions were implanted over the silica glass matrix, and nanoindentation tests were carried out by Torres-Torres et al., and it was found to increase the elasticity [72]. Gutiérrez-Menchaca et al. distributed platinum nanoparticles in the SG substrate with ion implantation and its subsequent annealing, and the effective elastic modulus was reportedly increased [73]. The distribution of Ag particles over SiO_2 glass particles showed that even a 1.4 vol% addition of silver particles changed the elasticity [74]. Similarly, sol–gel techniques were carried out for the production of agglomeration-free silica nanoparticles in SiO_2, and with 15% silica nanoparticle loading, a greater elastic modulus was obtained than the neat resin [75]. Single fiber fragmentation tests were carried out on polypropylene glass, and the addition of various types and amounts of nanoparticles led to an incredible improvement in the elastic modulus in that matrix [76]. Conventional melt quenching techniques were carried out to make the ternary phosphate glasses, and pulse-echo techniques measured the elastic properties. When PbO was added, it was discovered that the elastic properties improved [77]. Furthermore, a series of binary $(PbO)_x(P_2O_5)_{1-x}$ lead phosphate glasses were created using the same technique that resulted in the increase in Young's modulus of elasticity [78]. Sayyed et al. added both lead (Pb) and bismuth (Bi) to the six different glasses, and an ultrasonic wave showed enhancement in elastic properties [79].

Spraying and electrophoretic deposition were carried out to make the antibacterial system, and silica gentamicin nanoparticles were implanted in glass, and it was found to increase the elastic modulus of the elastic system [80]. The mechanical and thermal properties of glass fiber-reinforced polymer (GFRP) embedded with Al_2O_3 nanoparticles have been improved for use as composites in marine environments. When 0.1 wt% Al_2O_3 nanoparticles were included in the GFRP composites, there was a 12% boost in flexural strength, an 11% increment in the interlaminar shear strength in dry conditions, and the saltwater diffusion coefficient was lowered by 17%, which creates an opportunity for this nanocomposite to be used in marine environments [81]. Also, nano-TiO_2 particles of various concentrations have been combined to increase the mechanical capabilities of the epoxy polymer matrix. Even when nano-TiO_2 particles were added at 0.1 wt% to the matrix, a 15% increment in seawater diffusivity was observed. The nanocomposites that were seawater-aged showed 15% and 23% increments in flexural and interlaminar shear strengths, respectively [82]. It can be concluded that the mechanical properties of glass fiber-reinforced polymer (GFRP) composites could be improved and used in marine environments by embedding nanofillers in the epoxy polymer matrix.

The low-cost sodium silicate was used in a low-ambient-pressure drying process to create the water–glass-based silica aerogel, which was then used as an epoxy system filler. There was 80% gain in flexural modulus when compared to pure epoxy, and a strength gain of 40% was also reportedly attained due to the fact that silica aerogel based on water–glass held great application as the low-cost filler in polymer composites [83]. The addition of silica nanoparticles to polymers reinforced with kenaf fiber (KFRP) and woven glass (GKFRP) showed that tensile as well as flexible characteristics improved greatly even for the 25 wt% nano-silica loading. When epoxy polymer was mixed with nano-silica, a stiffer matrix was formed, which improved load transfer capability to the fibers as well as mechanical properties [84].

Yttria-stabilized tetragonal zirconia polycrystals (Y-TZP) are commonly used in fixed partial prostheses and dental applications, such as crowns. However, because of faults in its microstructure generated during the fabrication, Y-TZP have restricted applicability. A powder metallurgy method was used to fabricate SiO_2 nanocomposites to reduce the limitations of Y-TZP, and it was found that enhanced thermal properties along with micro-shear bonding and strength of bending were observed for the sample with 15% silica, indicating that it could be used in dental applications [85].

Despite aluminum alloys showing potential to be used as ideal lightweight engineering materials having low density, high specific strength, and strong corrosion resistance, their low strength and stiffness, when compared to Ti alloys and steel, still limit their use. A combination of powder metallurgy and hot extrusion was used to create a novel aluminum-based composite that is reinforced by metallic glass nanoparticles that are Ti-based for increased strength. Because the metallic glass nanoparticles were dispersed densely and uniformly in the Al matrix, the strength was greatly increased without compromising the plasticity, resulting in a super-high specific yield strength [86].

Metallic glass films containing nanostructured $Zr_{50}Cu_{50}$ with exceptionally tunable mechanical characteristics were deposited with pulsed laser deposition. In these films, a hardness of 10 and a 140 GPa modulus of elasticity were observed along with exceptional total elongation to failure greater than 9%. These characteristics aided in obtaining a balance between ductility and strength that may also be modified by manipulating film design. Thus, they show potential to be used in microelectronics as well as in formulating coatings [87].

A high-strength metallic glass based on CoFe was used for making strong, and ductile in tension, Co-Fe-Ta-B-O oxide glass at room temperature. The dual-phase structure within supra-nanometer sizes was present in the developed metallic-glass-reinforced oxide glass matrix nanocomposite. The nanocomposite showed 29% greater tensile strength than glasses made of single-phase oxide due to dispersion strengthening effects. Furthermore, enhancement of tensile plasticity to 2.7% was observed with a mixture of dual-phases having continuous glass/glass interfaces in the formed nanocomposite [88].

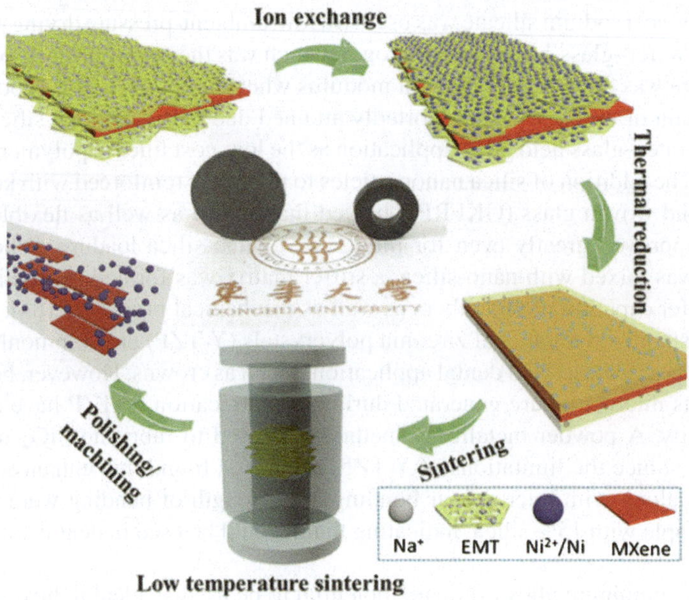

Fig. 5 Fabrication scheme of $Ti_3C_2T_x$/Ni/AS composite [89]

Titanium carbide MXene, which show both elasticity and electromagnetic attenuation propeties, could be used to make microwave-absorbing composites. The mesoporous $Ti_3C_2T_x$/Ni/AS glass composite was created by sintering a hybrid Ni ion-exchanged EMT zeolite/MXene composite at ultra-low temperatures, as shown in Fig. 5. When MXene was added to the composite, the hardness was improved by 40% to 700 MPa for AS glass, which also corresponded to the high modulus of elasticity of $Ti_3C_2T_x$ [89].

8 Superhydrophobic Materials

At present, electrodes that can withstand large mechanical deformations with retained activity and which show electrical stability have been highly desirable. For this, largely scalable metallic nanowires which are coated on a stretchable fabric showing wide range of compatibility have been developed. AgNPs which are percolated inside the electro-spun poly (styrene-block-butadiene-block-styrene) (SBS) rubber fibers have contributed to the high conductivity of the mat. When the ε was at 0.2 for a single fiber, the conductivity showed large reduction. Likewise, the conductivity of 2.2×10^3 S·cm^{-1} was attained, which was also reported to be proportional to the fiber mat thickness for a 150 μm thick mat at 100% strain [90].

Wearable electronic sensors with real-time human activities detection, which have shown enhanced flexibility for use as an apparatus for power storage and are corrosion-resistant for operating under humid conditions, have also shown significant progresses. A facile-spraying of polystyrene-b-poly(ethylene-co-butylene)-b-polystyrene (SEBS) and 1-octadecanethiol modified silver nanoparticles (M-AgNPs) on a natural rubber, which is pre-stretched, have shown superhydrophobicity with sensitive response to bending as well as stretching [91]. Under the application of mechanical forces, strong acid/alkali and heat, the coatings showed exceptional durability. The coating provided more than 160° of contact angle and a 10-Ω resistance as well as high conductivity. However, at strain of 200%, 100%, and 0%, the resistance was shown as 10, 58.5, and 165 Ω with gradual increments for the coating. The sputter-deposited NiNPs on a fluorine-containing coating onto cellulose filter papers coated with silver nanowires (AgNWs) for developing electromagnetic interference (EMI) shielding papers were reported [92]. These shielding papers showed excellent electrical conductivity, EMI shielding performance, and magnetic properties. At 0.013 mg·cm^{-2} loading of 0.109 vol% AgNWs, EMI shielding effectiveness of 88.4 dB was shown by the resultant papers. For $CS_{30}N_O$ papers, the AgNWs' volume content was 0.38% while the corresponding wt% was 8.72 wt%. The electrical conductivity of CS_xN_O papers increased by more than 200%, to 6331.4 S·m^{-1} when the cycles of dip-coating was increased by three times. Similarly, with increased conductivity, the shielding effectiveness also increased to 69.6 dB for $CS_{10}N_O$ paper and 76.1 dB for $CS_{20}N_O$ paper.

In another study by Xu et al., CuO spheres were in situ incorporated on polyimide (PI), following hydrothermal method for developing an acid/alkali corrosion resistant surface, which could withstand high as well as low temperatures. The composite showed high Young's modulus and good flexibility. The superhydrophobicity was retained, even after bending, exhibiting a water contact angle of 153° and a rolling angle of 6.5° [93]. The PDMS surface was placed on the top of the silicon stamp to obtain the adhesion as well as the conformal contact under the application of high vacuum of 9.5×10^{-5} Pa. The NPs were moved to the PDMS surface after the composite was baked at 60 °C for 6 h. Figure 6 shows the AFM images of the PDMS surface with and without nanoparticles. The droplet motion was observed after examining the contact angle hysteresis by the deposition of droplet over a tilted surface. The superhydrophobic surface with nanoparticles presented an extraordinary water repellency when the droplet was subjected at a low roll-off angle of 3° as shown in Fig. 7 [94].

9 Printable Elastic Electrodes and Sensors

Sensors and electrodes that have high conductivity, and are intrinsically stretchable, show large potential in applications related to stretchable transistors, actuators, and LED arrays, as well as other devices used for energy harvesting [95]. Apart from this, these electrodes and sensors are lightweight, and they provide comfort, curvilinear-fit flexibility for body shapes of different types, and breathability as breakthrough

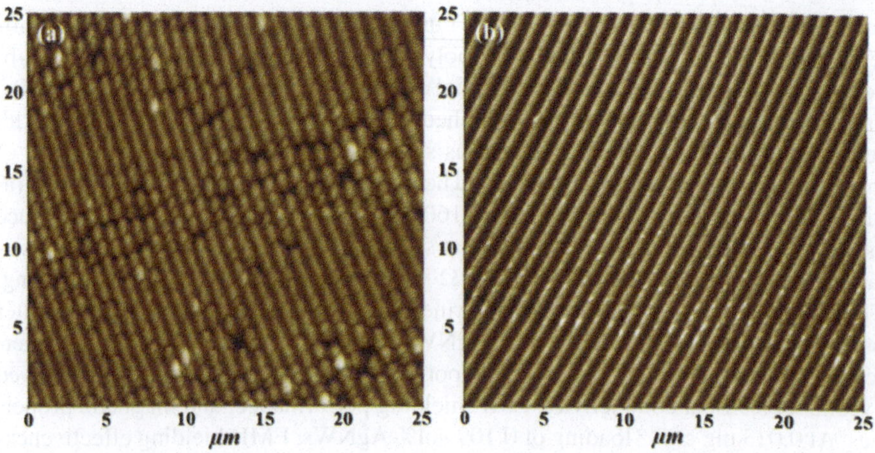

Fig. 6 AFM showing PDMS surface **a** without nanoparticles and **b** with nanoparticles [94]

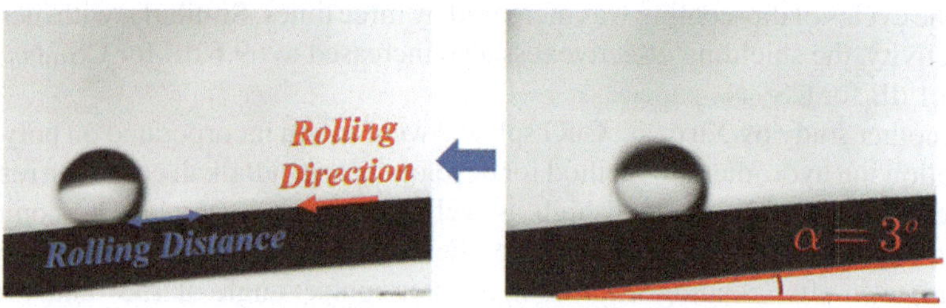

Fig. 7 Optical images showing mobility of water droplet at a low roll off angle of 3° on the superhydrophobic surface [94]

developments in wearable health monitoring devices [96]. In order to develop high-performing conductors as well as electronic devices that show stretchable properties, it is vital to understand the features of these nanocomposites and determine the feasibility of their manufacture. Some recent advances in the manufacture of printable elastic electrodes and sensors, as well as their applications, are discussed.

Single-walled carbon nanotubes (SWCNTs) containing AgNPs were produced using $NaBH_4$ as a reductant. With these stretchy electrodes, a conductivity of 4.9 $\times 10^3$ S·cm^{-1} was obtained, which also exhibited stability over 10,000 cycles at a strain of 20%, when exposed to high irradiation energy. The sensors and electrodes formed with silver nanoparticles embedded in CNTs could find application in soft as well as wearable electronics [97].

Polystyrene spheres coated with silver (PS@Ag) were mixed with liquid polydimethylsiloxane (PDMS) and printed on a screen with geometry of desirable dimensions to fabricate a high-performing stretchable and printable conductive elastic composite based on a strain sensor. This strain sensor shows the possibility for versatile application with robust mechanical properties in flexible and wearable electronics due to its potential to be fabricated at a large-area scale [98]. Low-cost and flexible silver NPs electrodes on polyurethane were created using a simple screen-printing technique for use in wearable and stretchable electronics. At 3-mm linewidth, the resistance of the AgNP-electrodes, which are screen-printed, remained constant until strains of 20% (wavy pattern) and 15% (plain pattern) were applied (horseshoe pattern). Similarly, with less than 1 mm inner as well as outer bending radii, the electrodes showed high flexibility in applications. Furthermore, they could interconnect LEDs and develop strain sensors [99].

A sandwich structured hybrid of single-walled carbon nanotubes (SWCNTs) and silver nanowires (AgNWs) was used to fabricate electrode films that showed high stretchability, flexibility, and transparency. These electrodes, which were produced by embedding AgNWs–SWCNTs, between layers of PU films and PDMS, showed 20% stretchability and also showed environmental durability when exposed to 85 °C and a humidity of 85%, with no appreciable performance loss [100]. Strain sensors with good stretchability were created by printing multi-walled carbon nanotubes (MWCNTs) and AgNPs conductive patterns with microscale pores onto PDMS. This highly sensitive printed sensor showed a maximum strain limit of 74%, which could be applicable in soft robotics, wearable electronics, and also human–machine interfaces [101].

Despite their increased conductivity and stretchability, metal–elastomer nanocomposites have limited practical applications due to their low cyclic stability and long-term durability. These shortcomings were improved by a composite consisting of an elastomeric binder matrix containing metal fillers and nanofibers of polyvinylidene fluoride that are electrospun. The resulting electrode showed outstanding cyclic endurance and was also highly conductive with a stretchability property of nearly 800% [102].

A nanocomposite with the ability to detect as well as differentiate temperature-induced strain could be developed with titanium carbide (MXene), AgNW–PEDOT: PSS tellurium nanowire. The sensitivity as well as stretch properties of the sensing devices were considerably improved by the synergistic impact between MXene nanosheets and PEDOT: PSS. This stretchable physical sensor could detect stimuli such as temperature variation and mechanical deformation and could play a critical role in skin-mountable electronics [103].

Due to the limited area electrochemical performance of stretchable micro-supercapacitors (MSCs), their application has been largely limited. With the application of 3D printing and pseudoplastic nanocomposite gels that are unidirectionally frozen, an intrinsically stretchable MSCs can be fabricated, where the gel is composed of $Ti_3C_2T_x$ MXene nanosheets, MnO_2 nanowire, AgNWs, and fullerene [104]. At 10 $mV{\cdot}s^{-1}$ scan rate, this MSC which is 3D-printed, achieved an unparalleled 216.2 $mF{\cdot}cm^{-2}$ areal capacitance when impregnated with a polymer gel electrolyte. Further, the electrolyte showed stability after 1000 stretch/release cycles and when stretched to 50%. This stretchy MSC also had a high-rate capability and a high areal energy capacity and outperformed all other reported stretchable MSCs [105].

10 Conclusion

The incorporation of several metal nanoparticles—AgNPs, NiNPs, EGaIn particles, liquid metal nanoparticles, AuNPs, and CuNPs—has found application in stretchable conductive composites and dielectrics as health care sensors, flexible electronic displays, electronic skins, solar cells, epidermal electronics, and more. These stretchable materials are extensively attractive for application in deformable electronics with properties such as a high dielectric constant, low dielectric loss, high conductivity, the capability to maintain good conductivity, and strong adhesion to the composite matrix. Furthermore, significant improvements in dielectric permittivity, lower dielectric loss, and low elastic modulus at high strain have been achieved for use in energy harvesting. Likewise, several coatings of nanomaterials have transformed flexible capacitive sensors into functional resistive sensing materials, and liquid metal nanoparticles have enhanced abilities to self-heal, change shape, and form a liquid–solid core shell structure with good energy conversion efficiency.

SMAs incorporated with nano-scaled Ni–Ti alloys, Ti-16Nb-Ag alloys, and AgNPs with enhanced thermomechanical characteristics, superelasticity, and good mechanical deformation have also been reported for their application in the field of material science and their biocompatibility for medical applications. However, problems related to brittleness still limit their widespread applications.

Likewise, polymer composites of Cu, Al, Mn, CuO nanofillers, spherical TiO_2 nanoparticles, and Al_2O_3 have been used in vibration reduction technologies with their enhanced storage modulus, increased elasticity, and composite strengths. On the other hand, iron nanoparticles, silver-coated copper nanoparticles, and liquid metals containing hydrogels have shown enhanced flexibility and self-healing capability for applications in robotic engineering, electrical circuits, monitoring human body motion, and much more. Superhydrophobic materials containing AgNPs, NiNPs, AgNWs, and CuO nanoparticles have been developed that could withstand large mechanical deformations, provide high conductivity, show excellent durability, and exhibit outstanding magnetic properties with the ability to withstand high as well as low temperatures. They have found applications in electrodes with retained activity and electrical stability and that are corrosion resistant under humid conditions for wearable electronic sensors that are sensitive to bending as well as stretching. The presented progress of metal nanoparticles for several elastic material applications makes research in the related field very appealing to future researchers.

References

1. Hartland GV (2004) Measurements of the material properties of metal nanoparticles by time-resolved spectroscopy. Phys Chem Chem Phys 6(23):5263–5274. https://doi.org/10.1039/B413368D
2. Mahross HZ, Baroudi K (2015) Effect of silver nanoparticles incorporation on viscoelastic properties of acrylic resin denture base material. Eur J Dent 09(02):207–212. https://doi.org/10.4103/1305-7456.156821

3. Basova TV et al (2021) The use of noble metal coatings and nanoparticles for the modification of medical implant materials. Mater Des 204:109672. https://doi.org/10.1016/j.matdes.2021.109672
4. Kumar HK et al (2018) Metallic nanoparticle: a review. Biomed J Sci Tech Res **4**(2):3765–3775. https://doi.org/10.26717/BJSTR.2018.04.00
5. dos Santos CA, Ingle AP, Rai M (2020) The emerging role of metallic nanoparticles in food. Appl Microbiol Biotechnol 104(6):2373–2383. https://doi.org/10.1007/s00253-020-10372-x
6. Vural M et al (2015) Sprayable elastic conductors based on block copolymer silver nanoparticle composites. ACS Nano 9(1):336–344. https://doi.org/10.1021/nn505306h
7. You I, Kong M, Jeong U (2019) Block copolymer elastomers for stretchable electronics. Acc Chem Res 52(1):63–72. https://doi.org/10.1021/acs.accounts.8b00488
8. Sengwa RJ, Dhatarwal P (2021) Polymer nanocomposites comprising PMMA matrix and ZnO, SnO_2, and TiO_2 nanofillers: a comparative study of structural, optical, and dielectric properties for multifunctional technological applications. Opt Mater 113:110837. https://doi.org/10.1016/j.optmat.2021.110837
9. Guo W et al (2019) Matrix-independent highly conductive composites for electrodes and interconnects in stretchable electronics. ACS Appl Mater Interfaces 11(8):8567–8575. https://doi.org/10.1021/acsami.8b21836
10. Quinsaat JEQ et al (2015) Highly stretchable dielectric elastomer composites containing high volume fractions of silver nanoparticles. J Mater Chem A 3(28):14675–14685. https://doi.org/10.1039/C5TA03122B
11. Feng P, Zhong M, Zhao W (2019) Stretchable multifunctional dielectric nanocomposites based on polydimethylsiloxane mixed with metal nanoparticles. Mater Res Express 7(1): 015007. https://doi.org/10.1088/2053-1591/ab5b4b
12. Bele A et al (2018) Conductive stretchable composites properly engineered to develop highly compliant electrodes for dielectric elastomer actuators. Smart Mater Struct 27(10):105005. https://doi.org/10.1088/1361-665x/aad977
13. Yoon IS et al (2020) Ag flake/silicone rubber composite with high stability and stretching speed insensitive resistance via conductive bridge formation. Sci Rep 10(1):5036. https://doi.org/10.1038/s41598-020-61752-2
14. Tang M et al (2020) Silver dendrites based electrically conductive composites, towards the application of stretchable conductors. Compos Commun 19:121–126. https://doi.org/10.1016/j.coco.2020.03.010
15. Dinh T et al (2020) Stretchable respiration sensors: advanced designs and multifunctional platforms for wearable physiological monitoring. Biosens Bioelectron 166:112460. https://doi.org/10.1016/j.bios.2020.112460
16. Liu Y, Ji X, Liang J (2021) Rupture stress of liquid metal nanoparticles and their applications in stretchable conductors and dielectrics. NPJ Flex Electron 5(1): 11. https://doi.org/10.1038/s41528-021-00108-w
17. Lin Y, Genzer J, Dickey MD (2020) Attributes, fabrication, and applications of gallium-based liquid metal particles. Adv Sci 7(12):2000192. https://doi.org/10.1002/advs.202000192
18. Bark H et al (2021) Deformable high loading liquid metal nanoparticles composites for thermal energy management. Adv Energy Mater 11(35):2101387. https://doi.org/10.1002/aenm.202101387
19. Lyu J et al (2016) High strength conductive composites with plasmonic nanoparticles aligned on aramid nanofibers. Adv Func Mater 26(46):8435–8445. https://doi.org/10.1002/adfm.201603230
20. Pusty M, Shirage PM (2020) Gold nanoparticle–cellulose/PDMS nanocomposite: a flexible dielectric material for harvesting mechanical energy. RSC Adv 10(17):10097–10112. https://doi.org/10.1039/C9RA10811D
21. Kausar A (2019) Polymeric nanocomposites reinforced with nanowires: opening doors to future applications. J Plastic Film Sheeting 35(1):65–98. https://doi.org/10.1177/8756087918794009

22. Atta A (2020) Enhanced dielectric properties of flexible Cu/polymer nanocomposite films. Surf Innovations 9(1):17–24. https://doi.org/10.1680/jsuin.20.00020
23. Antonucci V, Martone A (2015) Chapter 2—phenomenology of shape memory alloys. In: Lecce L, Concilio A (eds) Shape memory alloy engineering. Butterworth-Heinemann, Boston, pp 33–56. https://doi.org/10.1016/B978-0-08-099920-3.00002-4
24. Zareie S et al (2020) Recent advances in the applications of shape memory alloys in civil infrastructures: a review. Structures 27:1535–1550. https://doi.org/10.1016/j.istruc.2020.05.058
25. Van Humbeeck J (2001) Shape memory alloys: a material and a technology. Adv Eng Mater 3(11):837–850. https://doi.org/10.1002/1527-2648(200111)3:11%3c837::AID-ADEM837%3e3.0.CO;2-0
26. Hassanzadeh-Aghdam MK, Mahmoodi MJ (2018) Micromechanics-based characterization of elastic properties of shape memory polymer nanocomposites containing SiO_2 nanoparticles. J Intell Mater Syst Struct 29(11):2392–2405. https://doi.org/10.1177/1045389X18770862
27. Ma A et al (2018) Surface-initiated metal-free photoinduced ATRP of 4-vinylpyridine from SiO_2 via visible light photocatalysis for self-healing hydrogels. Ind Eng Chem Res 57(51):17417–17429. https://doi.org/10.1021/acs.iecr.8b05020
28. Van Humbeeck J (1999) Non-medical applications of shape memory alloys. Mater Sci Eng, A 273–275:134–148. https://doi.org/10.1016/S0921-5093(99)00293-2
29. Karaca HE et al (2014) NiTiHf-based shape memory alloys. Mater Sci Technol 30(13):1530–1544. https://doi.org/10.1179/1743284714Y.0000000598
30. Kang G, Song D (2015) Review on structural fatigue of NiTi shape memory alloys: pure mechanical and thermo-mechanical ones. Theor Appl Mech Lett 5(6):245–254. https://doi.org/10.1016/j.taml.2015.11.004
31. Gu D et al (2021) Additively manufacturing-enabled hierarchical NiTi-based shape memory alloys with high strength and toughness. Virtual Phys Prototyping 16(sup1):S19–S38. https://doi.org/10.1080/17452759.2021.1892389
32. Ko W-S et al (2021) Dissecting functional degradation in NiTi shape memory alloys containing amorphous regions via atomistic simulations. Acta Mater 202:331–349. https://doi.org/10.1016/j.actamat.2020.10.070
33. Yi X et al (2021) The higher compressive strength (TiB+La_2O_3)/Ti–Ni shape memory alloy composite with the larger recoverable strain. Compos Commun 23:100583. https://doi.org/10.1016/j.coco.2020.100583
34. Melaiye A, Youngs WJ (2005) Silver and its application as an antimicrobial agent. Expert Opin Ther Pat 15(2):125–130. https://doi.org/10.1517/13543776.15.2.125
35. Zhang E et al (2021) Antibacterial metals and alloys for potential biomedical implants. Bioactive Mater 6(8):2569–2612. https://doi.org/10.1016/j.bioactmat.2021.01.030
36. Jiao J et al (2021) Recent advances in research on antibacterial metals and alloys as implant materials. Front Cell Infect Microbiol 11. https://doi.org/10.3389/fcimb.2021.693939
37. Sun B et al (2018) Martensite structure and mechanical property of Ti-Nb-Ag shape memory alloys for biomedical applications. Vacuum 156:181–186. https://doi.org/10.1016/j.vacuum.2018.07.029
38. Toriyabe A et al (2021) Mechanical property enhancement of the Ag–tailored Au–Cu–Al shape memory alloy via the ductile phase toughening. Intermetallics 139:107349. https://doi.org/10.1016/j.intermet.2021.107349
39. Shafique M, Luo X (2019) Nanotechnology in transportation vehicles: an overview of its applications, environmental, health and safety concerns. Materials (Basel, Switzerland) 12(15):2493. https://doi.org/10.3390/ma12152493
40. Jiang Z et al (2020) Vibration damping mechanism of CuAlMn/polymer/carbon nanomaterials multi-scale composites. Compos B Eng 199:108266. https://doi.org/10.1016/j.compositesb.2020.108266
41. Yang Q et al (2019) Rheological and micro-structural characterization of bitumen modified with carbon nanomaterials. Constr Build Mater 201:580–589. https://doi.org/10.1016/j.conbuildmat.2018.12.173

42. Roustazadeh D, Aghadavoudi F, Khandan A (2020) A synergic effect of CNT/Al_2O_3 reinforcements on multiscale epoxy-based glass fiber composite: fabrication and molecular dynamics modeling. Mol Simul 46(16):1308–1319. https://doi.org/10.1080/08927022.2020.1815729
43. Joshi UA, Sharma SC, Harsha SP (2012) Effect of carbon nanotube orientation on the mechanical properties of nanocomposites. Compos B Eng 43(4):2063–2071. https://doi.org/10.1016/j.compositesb.2012.01.063
44. Shan C et al (2013) Three-Dimensional Nitrogen-Doped Multiwall Carbon Nanotube Sponges with Tunable Properties. Nano Lett 13(11):5514–5520. https://doi.org/10.1021/nl403109g
45. Alam A et al (2018) Polymer composite hydrogels containing carbon nanomaterials—morphology and mechanical and functional performance. Prog Polym Sci 77:1–18. https://doi.org/10.1016/j.progpolymsci.2017.09.001
46. Deka BK et al (2016) Interfacial resistive heating and mechanical properties of graphene oxide assisted CuO nanoparticles in woven carbon fiber/polyester composite. Compos A Appl Sci Manuf 80:159–170. https://doi.org/10.1016/j.compositesa.2015.10.023
47. Balguri PK, Samuel DGH, Thumu U (2021) A review on mechanical properties of epoxy nanocomposites. Mater Today: Proc 44:346–355. https://doi.org/10.1016/j.matpr.2020.09.742
48. Lakshmikandhan T et al (2016) Development and characterization of functionalized Al_2O_3 and TiO_2-reinforced polybenzoxazine nanocomposites. Des Monomers Polym 19(1):67–76. https://doi.org/10.1080/15685551.2015.1092014
49. Ahmed EM (2015) Hydrogel: preparation, characterization, and applications: a review. J Adv Res 6(2):105–121. https://doi.org/10.1016/j.jare.2013.07.006
50. Mantha S et al (2019) Smart hydrogels in tissue engineering and regenerative medicine. Materials (Basel, Switzerland) 12(20):3323. https://doi.org/10.3390/ma12203323
51. Peng Q et al (2020) Recent advances in designing conductive hydrogels for flexible electronics. InfoMat 2(5):843–865. https://doi.org/10.1002/inf2.12113
52. Rafieian S et al (2019) A review on nanocomposite hydrogels and their biomedical applications. Sci Eng Compos Mater 26(1):154–174. https://doi.org/10.1515/secm-2017-0161
53. Sun Z et al (2020) Self-healing, sensitive and antifreezing biomass nanocomposite hydrogels based on hydroxypropyl guar gum and application in flexible sensors. Int J Biol Macromol 155:1569–1577. https://doi.org/10.1016/j.ijbiomac.2019.11.134
54. Zhang Y et al (2020) Flexible and wearable sensor based on graphene nanocomposite hydrogels. Smart Mater Struct 29(7):075027. https://doi.org/10.1088/1361-665X/ab89ff
55. Zhang Z et al (2021) Highly stretchable porous composite hydrogels with stable conductivity for strain sensing. Compos Sci Technol 213:108968. https://doi.org/10.1016/j.compscitech.2021.108968
56. Zhou H et al (2020) Self-repairing flexible strain sensors based on nanocomposite hydrogels for whole-body monitoring. Colloids Surf, A 592:124587. https://doi.org/10.1016/j.colsurfa.2020.124587
57. Chen K et al (2021) Highly stretchable, tough, and conductive Ag@Cu nanocomposite hydrogels for flexible wearable sensors and bionic electronic skins. Macromol Mater Eng 306(10):2100341. https://doi.org/10.1002/mame.202100341
58. Ohm Y et al (2021) Publisher correction: an electrically conductive silver–polyacrylamide–alginate hydrogel composite for soft electronics. Nat Electron 4(4):313–313. https://doi.org/10.1038/s41928-021-00571-3
59. Murali Mohan Y et al (2010) Controlling of silver nanoparticles structure by hydrogel networks. J Colloid Interface Sci 342(1):73–82. https://doi.org/10.1016/j.jcis.2009.10.008
60. Mao J et al (2020) Highly stretchable, self-healing, and strain-sensitive based on double-crosslinked nanocomposite hydrogel. Compos Commun 17:22–27. https://doi.org/10.1016/j.coco.2019.10.007
61. Xu J et al (2020) Polymerization of moldable self-healing hydrogel with liquid metal nanodroplets for flexible strain-sensing devices. Chem Eng J 392:123788. https://doi.org/10.1016/j.cej.2019.123788

62. Ng LY et al (2013) Polymeric membranes incorporated with metal/metal oxide nanoparticles: a comprehensive review. Desalination 308:15–33. https://doi.org/10.1016/j.desal.2010.11.033
63. Zhang J et al (2020) Improving actuation strain and breakdown strength of dielectric elastomers using core-shell structured CNT-Al_2O_3. Compos Sci Technol 200:108393. https://doi.org/10.1016/j.compscitech.2020.108393
64. Etemadi H et al (2021) Effect of alumina nanoparticles on the antifouling properties of polycarbonate-polyurethane blend ultrafiltration membrane for water treatment. Polym Eng Sci 61(9):2364–2375. https://doi.org/10.1002/pen.25764
65. Bet-moushoul E et al (2016) TiO_2 nanocomposite based polymeric membranes: a review on performance improvement for various applications in chemical engineering processes. Chem Eng J 283:29–46. https://doi.org/10.1016/j.cej.2015.06.124
66. Wei Y et al (2011) Effect of TiO_2 nanowire addition on PVDF ultrafiltration membrane performance. Desalination 272(1):90–97. https://doi.org/10.1016/j.desal.2011.01.013
67. Lin Y et al (2015) Enhancing mechanical and photocatalytic performances on TiO_2/Ti composite ultrafiltration membranes via Ag doping method. Sep Purif Technol 145:29–38. https://doi.org/10.1016/j.seppur.2015.02.024
68. Sigwadi R et al (2018) Mechanical strength of Nafion®/ZrO_2 nano-composite membrane. Int J Manuf Mater Mech Eng (IJMMME) 8(1):54–65. https://doi.org/10.4018/IJMMME.2018010104
69. Sigwadi R et al (2019) Enhancing the mechanical properties of zirconia/Nafion® nanocomposite membrane through carbon nanotubes for fuel cell application. Heliyon 5(7):e02112. https://doi.org/10.1016/j.heliyon.2019.e02112
70. Liu C et al (2018) Novel electrospun polyvinylidene fluoride-graphene oxide-silver nanocomposite membranes with protein and bacterial antifouling characteristics. Express Polym Lett 12(4)
71. Benavente J et al (2017) Inclusion of silver nanoparticles for improving regenerated cellulose membrane performance and reduction of biofouling. Int J Biol Macromol 103:758–763. https://doi.org/10.1016/j.ijbiomac.2017.05.133
72. Torres-Torres D et al (2020) Magnetic force microscopy study of multiscale ion-implanted platinum in silica glass, recorded by an ultrafast two-wave mixing configuration. Microsc Microanal 26(1):53–62. https://doi.org/10.1017/S1431927619015204
73. Gutiérrez-Menchaca J et al (2021) Enhanced fracture toughness of silica glass by ion-implanted platinum nanoparticles. Fatigue Fract Eng Mater Struct 44(6):1423–1438. https://doi.org/10.1111/ffe.13437
74. Liu L, Shinozaki K (2021) Toughening silica glass by imparting ductility using a small amount of silver nanoparticles. Mater Sci Eng, A 817:141372. https://doi.org/10.1016/j.msea.2021.141372
75. Uddin MF, Sun CT (2008) Strength of unidirectional glass/epoxy composite with silica nanoparticle-enhanced matrix. Compos Sci Technol 68(7):1637–1643. https://doi.org/10.1016/j.compscitech.2008.02.026
76. Pedrazzoli D, Pegoretti A (2013) Silica nanoparticles as coupling agents for polypropylene/glass composites. Compos Sci Technol 76:77–83. https://doi.org/10.1016/j.compscitech.2012.12.016
77. Matori KA et al (2017) Comprehensive study on physical, elastic and shielding properties of lead zinc phosphate glasses. J Non-Cryst Solids 457:97–103. https://doi.org/10.1016/j.jnoncrysol.2016.11.029
78. Matori KA et al (2013) Study of the elastic properties of (PbO)x(P2O5)1–x lead phosphate glass using an ultrasonic technique. J Non-Cryst Solids 361:78–81. https://doi.org/10.1016/j.jnoncrysol.2012.10.022
79. Sayyed MI et al (2020) The influence of PbO and Bi_2O_3 on the radiation shielding and elastic features for different glasses. J Market Res 9(4):8429–8438. https://doi.org/10.1016/j.jmrt.2020.05.113
80. Aydemir T et al (2021) Morphological and mechanical characterization of chitosan/gelatin/silica-gentamicin/bioactive glass coatings on orthopaedic metallic implant materials. Thin Solid Films 732:138780. https://doi.org/10.1016/j.tsf.2021.138780

81. Nayak RK (2019) Influence of seawater aging on mechanical properties of nano-Al_2O_3 embedded glass fiber reinforced polymer nanocomposites. Constr Build Mater 221:12–19. https://doi.org/10.1016/j.conbuildmat.2019.06.043
82. Nayak RK, Ray BC (2018) Influence of seawater absorption on retention of mechanical properties of nano-TiO_2 embedded glass fiber reinforced epoxy polymer matrix composites. Arch Civil Mech Eng 18(4):1597–1607. https://doi.org/10.1016/j.acme.2018.07.002
83. Salimian S, Zadhoush A (2019) Water-glass based silica aerogel: unique nanostructured filler for epoxy nanocomposites. J Porous Mater 26(6):1755–1765. https://doi.org/10.1007/s10934-019-00757-3
84. Sapiai N et al (2020) Tensile and flexural properties of silica nanoparticles modified unidirectional kenaf and hybrid glass/kenaf epoxy composites. Polymers 12(11). https://doi.org/10.3390/polym12112733
85. Rahimi S et al (2020) Effect of SiO_2 content on Y-TZP/Al_2O_3 ceramic-nanocomposite properties as potential dental applications. Ceram Int 46(8, Part A):10910–10916. https://doi.org/10.1016/j.ceramint.2020.01.105
86. Zhang WW et al (2018) A novel high-strength Al-based nanocomposite reinforced with Ti-based metallic glass nanoparticles produced by powder metallurgy. Mater Sci Eng, A 734:34–41. https://doi.org/10.1016/j.msea.2018.07.082
87. Ghidelli M et al (2021) Novel class of nanostructured metallic glass films with superior and tunable mechanical properties. Acta Mater 213:116955. https://doi.org/10.1016/j.actamat.2021.116955
88. Huang L et al (2020) A high-strength Co–Fe–Ta–B metallic-glass phase enabled tensile plasticity in Co–Fe–Ta–B–O oxide glass matrix nanocomposites. Appl Phys Lett 116(8):081903. https://doi.org/10.1063/1.5143598
89. Luo W et al (2022) A robust hierarchical MXene/Ni/aluminosilicate glass composite for high-performance microwave absorption. Adv Sci 9(4):2104163. https://doi.org/10.1002/advs.202104163
90. Park M et al (2012) Highly stretchable electric circuits from a composite material of silver nanoparticles and elastomeric fibres. Nat Nanotechnol 7(12):803–809. https://doi.org/10.1038/nnano.2012.206
91. Su X et al (2018) Highly stretchable and conductive superhydrophobic coating for flexible electronics. ACS Appl Mater Interfaces 10(12):10587–10597. https://doi.org/10.1021/acsami.8b01382
92. Zhan Y et al (2021) Superhydrophobic and flexible silver nanowire-coated cellulose filter papers with sputter-deposited nickel nanoparticles for ultrahigh electromagnetic interference shielding. ACS Appl Mater Interfaces 13(12):14623–14633. https://doi.org/10.1021/acsami.1c03692
93. Xu C-L et al (2017) Surface modification with hierarchical CuO arrays toward a flexible, durable superhydrophobic and self-cleaning material. Chem Eng J 313:1328–1334. https://doi.org/10.1016/j.cej.2016.11.024
94. Zhai S, Zhao H (2019) Silica-coated metallic nanoparticle-based hierarchical superhydrophobic surfaces fabricated by spin-coating and inverse nanotransfer printing. Appl Phys Lett 114(23):233702. https://doi.org/10.1063/1.5098780
95. Choi S et al (2019) High-performance stretchable conductive nanocomposites: materials, processes, and device applications. Chem Soc Rev 48(6):1566–1595. https://doi.org/10.1039/C8CS00706C
96. Hasan MM, Hossain MM (2021) Nanomaterials-patterned flexible electrodes for wearable health monitoring: a review. J Mater Sci 56(27):14900–14942. https://doi.org/10.1007/s10853-021-06248-8
97. Lee J-W et al (2021) Synthesis of silver nanoparticles embedded with single-walled carbon nanotubes for printable elastic electrodes and sensors with high stability. Sci Rep 11(1):5140. https://doi.org/10.1038/s41598-021-84386-4
98. Hu Y et al (2018) A low-cost, printable, and stretchable strain sensor based on highly conductive elastic composites with tunable sensitivity for human motion monitoring. Nano Res 11(4):1938–1955. https://doi.org/10.1007/s12274-017-1811-0

99. Yoon S, Kim H-K (2020) Cost-effective stretchable Ag nanoparticles electrodes fabrication by screen printing for wearable strain sensors. Surf Coat Technol 384:125308. https://doi.org/10.1016/j.surfcoat.2019.125308
100. Hwang B-Y et al (2018) Highly stretchable and transparent electrode film based on SWCNT/Silver nanowire hybrid nanocomposite. Compos B Eng 151:1–7. https://doi.org/10.1016/j.compositesb.2018.06.004
101. Min S-H, Lee G-Y, Ahn S-H (2019) Direct printing of highly sensitive, stretchable, and durable strain sensor based on silver nanoparticles/multi-walled carbon nanotubes composites. Compos B Eng 161:395–401. https://doi.org/10.1016/j.compositesb.2018.12.107
102. Jin H et al (2019) Highly durable nanofiber-reinforced elastic conductors for skin-tight electronic textiles. ACS Nano 13(7):7905–7912. https://doi.org/10.1021/acsnano.9b02297
103. Gruber B et al (2020) Mechanism of low temperature deformation in aluminium alloys. Mater Sci Eng A 795:139935. https://doi.org/10.1016/j.msea.2020.139935
104. Li X et al (2020) 3D-printed stretchable micro-supercapacitor with remarkable areal performance. Adv Energy Mater 10:1903794. https://doi.org/10.1002/aenm.201903794
105. Duan M, Luo L, Liu Y (2020) Microstructural evolution of AZ31 Mg alloy with surface mechanical attrition treatment: grain and texture gradient. J Alloy Compd 823:153691. https://doi.org/10.1016/j.jallcom.2020.153691

Application of Nanomaterials to Enhance Mechanical Properties of Metallic Alloys: Status and Prospects

Sagar Ban, Rakesh Shrestha, Gaurav Khatiwada, Saroj Raj Kafle, Santosh K. Tiwari, and Rajendra Joshi

Abstract During the last three decades, several groundbreaking research projects linked to materials research have been carried out, especially in the field of the fabrication of new materials with excellent thermo-mechanical properties. Innovative lightweight metallic alloys with high strength and toughness have been in high demand for use in aerospace, automotive parts, structures, and the 3C industries (computer, communication, consumer electronics). While previous methods for enhancing metallic mechanical properties included alloying with other metals and related materials, and developing new heat treatment and molding processes, the property enhancements were exclusive. Recent research trends have favored optimum nanoparticles incorporation in the metal matrix for yielding high strength while retaining ductility, increased hardness as well as toughness, and improved fatigue and creep. Furthermore, several studies on the mechanism of mechanical property enhancement have paved the way for a more comprehensive understanding and more detailed future research. This chapter provides a brief overview of the recent advances in improving the thermo-mechanical properties of nanomaterials incorporated into metal alloys for a variety of applications.

Keywords Ceramic nanoparticles · Thermo-mechanical properties · Metal matrix · Grain refinement · Failure · Deformation · Metallic alloys

---

S. Ban · R. Shrestha · G. Khatiwada · R. Joshi (✉)
Department of Chemical Science and Engineering, Kathmandu University, PO Box 6250, Dhulikhel, Kavre, Nepal
e-mail: rajendra.joshi@ku.edu.np

S. R. Kafle
Department of Chemical Engineering, Chungbuk National University, Chungbuk 28644, Korea

S. K. Tiwari
Department of Chemistry, University of Warsaw, Warsaw, Poland

Present Address:
R. Joshi
Department of Pharmacy, School of Science, Kathmandu University, PO Box 6250, Dhulikhel, Kavre, Nepal

S. K. Tiwari et al. (eds.), *Nanoparticles Reinforced Metal Nanocomposites*,
https://doi.org/10.1007/978-981-19-9729-7_12

1 Introduction

Alloys of metals such as Al and Mg serve as lightweight components showing good isotropic mechanical properties, great castability, excellent corrosion resistance, high-strength and for low-cost applications in automotive, aerospace applications, electronics, and 3C (computer, communication, and consumer electronics) industries [1–4]. Steels also have obvious advantages of rich resource, providing stable performance at low cost and have found wide applications. However, the uses of these alloys have been limited by the inadequate mechanical properties at elevated temperature [5]. In addition, problems related to low modulus of elasticities, ductility, poor creep, abrasion resistance, and high corrosion rate reduce their applicability since these are the core properties of the metallic alloys as shown in Fig. 1. Even when hardened, there have been problems associated with corrosion, cracking, and loss of mechanical credibility with time [6]. Therefore, there has been a need for developing lightweight composites to counter these inadequacies.

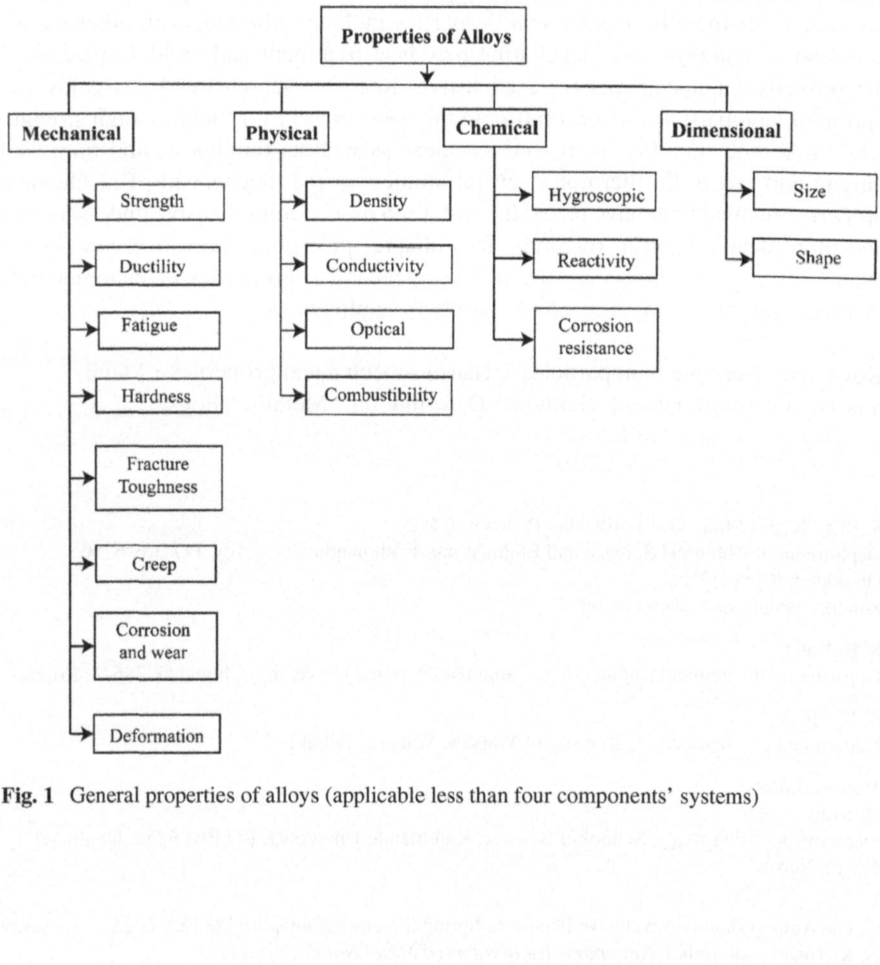

Fig. 1 General properties of alloys (applicable less than four components' systems)

Previous efforts toward enhancing the mechanical properties of the metals included adding metal elements to metal matrices for alloying as well as optimizing composition, and developing noble heat treatment processes along with newer processes for manufacture and molding [7]. In spite of these developments, simultaneous improvement of strength and ductility was not attained. A new approach to enhancing mechanical properties of nanomaterials has surfaced where grain refinement and increased toughness have been observed when nanoparticles served as a site for heterogeneous nucleation during solidification for metal grains. The grain coarsening was inhibited by these particles that occurs during the heat treatment owing to pinning effect of the grain boundary as a result of these nanosized particles. In other words, enhancement in mechanical strength of metallic matrices owing to incorporation of the small amount of nanomaterials attributed to nanosizing effect [8].

Research in the area has suggested incorporating a small trace of ceramic particles in the matrices for effective improvement of the mechanical properties at a low cost and for high performance, without the loss in weight so as to create metal nanocomposites [9, 10]. Uniform distributions of some trace nanoparticles such as oxides (TiO_2, Al_2O_3, Y_2O_3, ZrO_2), borides (TiB_2, ZrB_2, SiB_6), carbides (SiC, TiC, B_4C, VC), nitrides (AlN, TiN, BN), and carbonaceous materials have provided more heterogeneous nucleation sites, and these trace nanoparticles restrain dendritic growth by allowing other nanoparticles to adsorb onto the solid–liquid interface (a tree-like structure morphology of crystals growing during the metal solidification) and thus significantly refining the grain sizes of the primary γ- and δ-Fe phases [11]. The addition of particulate reinforcements has yielded high strength, high modulus, and wear properties, maintained toughness, and also exhibited high strengthening effect. Even 3% addition of these reinforcements significantly show superior mechanical properties while striking a balance between strength and plasticity [1]. The reinforcements usually contribute to less than 3 vol%, for avoiding agglomeration which can often have negative effect on strength and ductility. The general principle is to avoid surpassing certain optimum loading in the matrix, which would otherwise have a deleterious effect on the mechanical properties. The following sections give a comprehensive view toward incorporation of nanomaterials in the metal matrix and their effect on the mechanical properties.

2 Strength and Ductility

The strength increment of the alloys with ceramic nanoparticles is attributed to strengthening effect of the grain refinement, Orowan strengthening (strength owing to the dislocations passing resistance provided by the hard particles that are closely spaced) effect, and thermal expansion coefficient strengthening. The nanocomposite yield strength increment has been explained with the classical Hall–Petch relationship relating the reduction in the grain size by the relation (1) where k represents the material constant and $d_{nanocomposte}$ and d_{matrix} are the average sizes of the nanocomposite grains extruded and homogenized alloy matrix, respectively [12]. Figure 3 illustrates the mechanism for grain refinement by TiC nanoparticles addition in as-cast steel [7]. The TiC nanoparticles incorporation to as-cast steel led to the grain refinement, thus affecting both mechanical properties and microstructure of the steel. While process of nucleation was improved by some trace TiC nanoparticles and also refining the primary austenite grains, others adsorbed onto interface separating solid–liquid and restricted dendritic growth [7]. Ceramic nanoparticles addition to metal composites also stabilizes against grain growth, and the strengthening is then attributed to Orowan mechanism [13, 14], where the presence of second-phase particles impedes the dislocation movement resulting in the increment of shear stress that are resolved for the basal slip. Thus, increased yield strength is as stated in the Orowan relation (2) where G_m is the matrix shear modules, b represents Burger vector, d_p is the second-phase particle's average size, and V_p is its volume fraction. The enhancement of the yield strength also occurs due to differences in thermal expansion coefficient (CTE) as given by the relation (3) where β is a constant, $\Delta\alpha$ is the thermal expansion coefficient difference of the matrix from the nanoparticle phase, and the difference in the temperature of the hot deformation and tensile test is given as ΔT. Finally, the presence of the hard ceramic nanoparticles and matrix also shows the load transferring effect and improves the yield strength as given by the relation (4), where σ_m represents yield strength of the alloy and reinforcement's aspect ratio is represented by s [15].

$$\Delta\sigma_{\text{Hall - Petch}} = k \cdot \left(d_{\text{nanocomposite}}^{-1/2} - d_{\text{matrix}}^{-1/2}\right) \tag{1}$$

$$\Delta\sigma_{\text{Orowan}} = \frac{0.13 \cdot G_m \cdot b}{d_p \cdot \left[\left(\frac{1}{2V_p}\right)^{\frac{1}{3}} - 1\right]} \ln \frac{d_p}{2b} \tag{2}$$

$$\Delta\sigma_{\text{CTE}} = \beta \cdot G_m \cdot b\sqrt{\frac{12 \cdot \Delta\alpha \cdot \Delta T \cdot V_p}{b \cdot d_p \cdot (1 - V_p)}} \tag{3}$$

$$\Delta\sigma_{\text{load}} = \frac{1}{2}\sigma_m \cdot V_p \cdot s \tag{4}$$

The combined strengthening effects of the above-mentioned relations have been cited. 2 wt% addition of TiN in Al7075 alloys increased the ultimate compressive strength (UCS) by 71% and strain by 38% [16]. Similarly, reinforcing aluminum metal matrix composite (AA6061) with 0.2 wt% inorganic nanotubes (INT) and fullerene-like (IF) nanoparticles of WS_2 [17] improved tensile yielding strength (TYS), ultimate tensile strength (UTS) as well as ductility by up to 68%. Meanwhile, deleterious effect on the tensile strength was observed when WS_2 nanotubes were increased by 0.5 wt% resulting in INT agglomeration. The effect is attributed, through metallography microstructural analysis, to grain size reduction by 48.4% and increasing wt% of WS_2 INT or IF nanoparticles in the Al metal matrix composite as compared to neat AA6061 alloy ingot.

The grain refinement can also be illustrated as shown in Fig. 2, where the fine particles of second phase act as the crystalline primary phase nucleation sites in the molten steel which increases the rate of nucleation. On the other hand, those nanoparticles that do not get involved in the nucleation will get trapped in the solid–liquid interfacial dendrites (Fig. 3). The dendritic growth will be further hindered and result in the refinement of grains of the solidification microstructure [11].

Al_2O_3 also reduces the grain sizes, hinders the dislocation motion [19] and has the potential for inducing greater hardness than enhancing tensile properties only. Adding Al_2O_3 and other oxides such as TiO_2 and ZrO_2, as reinforcements on A356 Al alloy, via casting route [20] showed maximum UTS and ductility at 2 wt% Al_2O_3 with 3% TiO_2 addition. Similarly, Al_2O_3 enhanced the strength and ductility (tensile as well

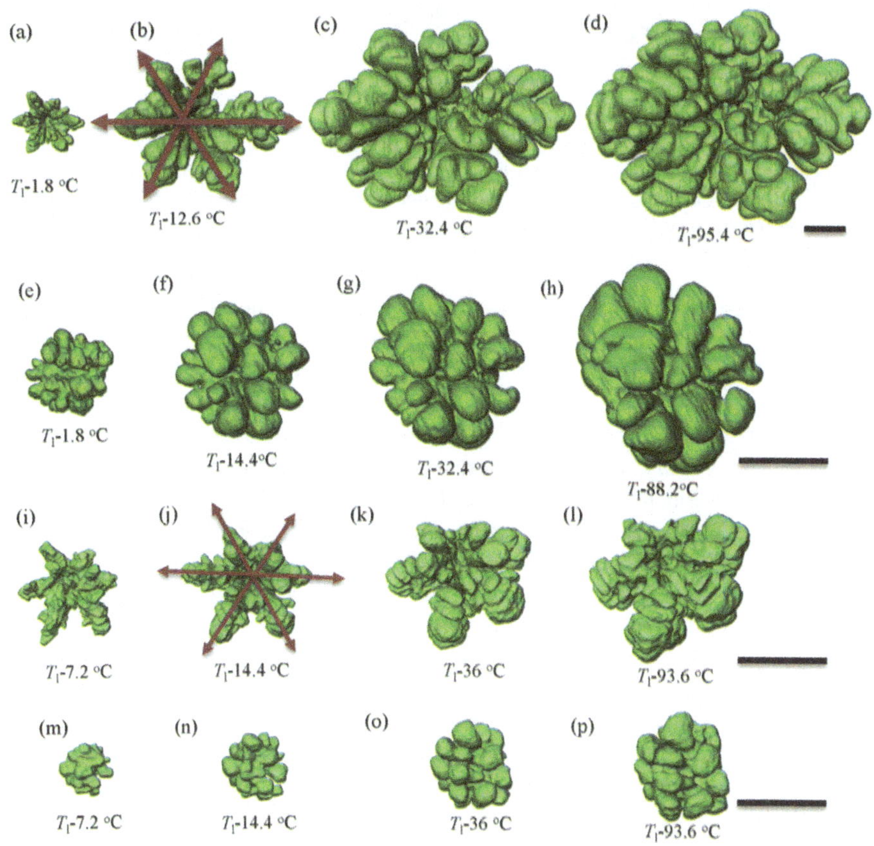

Fig. 2 Dendritic growth effect of nanoparticles of SiC NPs in Mg alloys. **a–d**, **i–l** represents evolution of dendrites in Mg alloys without nanoparticles and **e–h**, **m–p** evolution of dendrites containing nanoparticles [18]

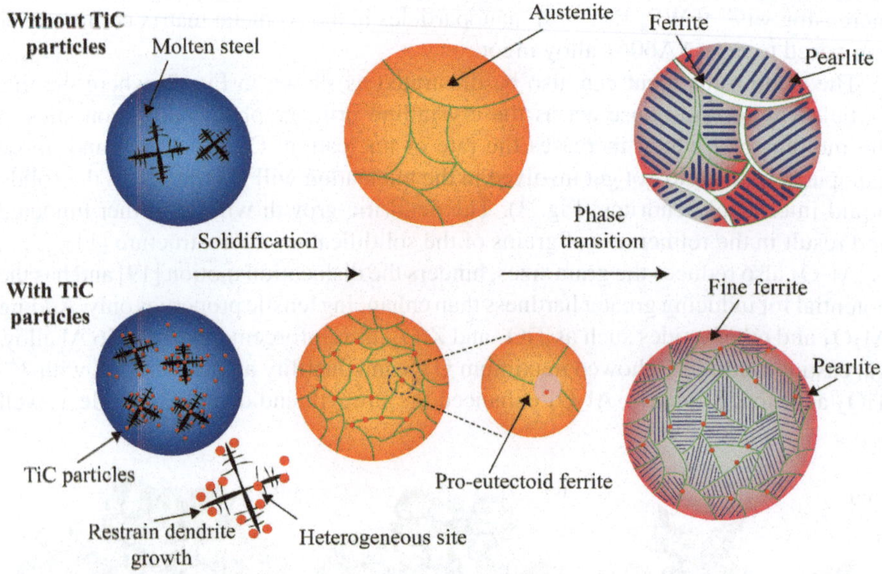

Fig. 3 Grain refining mechanism of the as-cast steel by the TiC nanoparticles addition [7]. While the process of nucleation was enhanced by some trace TiC nanoparticles alongside the primary austenite grains refinement, others restricted the dendritic growth by getting adsorbed onto the solid–liquid interface

as compression levels) of cast magnesium alloys (AZ31 and ZK60A). For tension level, 19% higher TYS, +21% UTS, +113% failure strain, and +162% work of failure (WOF) for Al_2O_3 containing AZ31 as compared to monolithic AZ31 and for compression level, Al_2O_3 containing AZ31 nanocomposite exhibited +5, +5, −4, and +11% change in the compressive yield strength (CYS), UCS, lower failure strain, and WOF, respectively, have been reported [2]. Similarly, as compared to monolithic ZK60A, for tension level, −4% lower TYS, +13% UTS, +170% failure strain, and +200% WOF for Al_2O_3 containing ZK60A and for compression level, Al_2O_3 containing ZK60A nanocomposite exhibited −10, +7, +15, and +26% change in the CYS, UCS, lower failure strain, and WOF, were reported, respectively.

Al_2O_3/7075 alloy fabricated using the solid–liquid mixed casting with loading increased to 1.5 wt% increased the tensile properties but, these properties, began to decline beyond 1.5 wt%. The increase in the UTS was only 29% while the hardness increased by 67% as compared to as-cast 7075 Al alloy [19]. Likewise, the UTS and yield strength were enhanced by 18.1% and 21.68%, respectively, when Al_2O_3 at 0.4 wt% was reinforced in 2024 Al alloy by stir casting technique [21].

Similarly, increased hardness, chemical stability, and melting point have been provided by other ceramic nanoparticles such as TiC while increased strength, Young's modulus, and resistance to thermal shock have been provided by SiC nanoparticles. TiN, an inorganic nitride ceramic material, possess good microhardness, superior wear and corrosion performances, high melting point, and good thermal as well as chemical stability. TiC nanoparticles addition in trace amount to 40Cr steel showed increased yield strength and enhanced tensile strength than unreinforced 40Cr steel by 19.7 and 30.4%, respectively [7]. The enhanced properties were attributed to strengthening due to fine grain, second phase, and thermal mismatch as previously stated. Likely, SiC nanoparticles (<500 nm) addition to 6061 Al alloy had significant improvement in Young's modulus and strength while having remarkable ductility [22]. The particle volume fractions increment also increased the yield stress, ultimate tensile strength, and Young's modulus. Likewise, its powder metallurgical addition to Mg has also shown largest flow stress and lowest creep rates [23]. TiN has promoted the solution of the alloying elements, retained Al–Zn–Mg–Cu alloy fine grains during the process of solid solution, and showed enhancement in the Vickers hardness of the alloy that is solution treated [24]. The TiN nanoparticles have enhanced the yield stress, ultimate tensile strength, and elongation better than with only Ti or without refiners [24]. The TiN nanoparticles incorporation to Ti–7Al–1Mo ternary alloy, at 3 wt%, increased CYS by 323 MPa [25]. For ZrO_2 addition, the maximum tensile strength and ductility were attained with 3% and 5% loading, respectively [20].

TiB_2 addition to Mg matrix composites showed reinforcement with alternating fiber-like nanoparticle-rich (NPR) and nanoparticle-free (NPF) zones and consequently showed bimodal properties of enhanced hardness and modulus of elasticity, in fiber-like NPR zones, and high plasticity, in NPF zones as can be seen in Fig. 4. It showed combination of high tensile strength of 388 MPa and ductility of 10.1% [12]. Even when 1.5% TiB_2 addition to A356 Al matrix increased the tensile properties and elongation, Akbari et al. found that an increase in the porosity content tangibly decreased the tensile properties with increasing casting temperatures [26]. Meanwhile, Al–Mn–Mg 3004 alloy at elevated temperature of 300 °C with 3 wt% TiB_2 has shown 13% improved yield strength [27].

Similarly, B_4C possesses properties of low thermal expansivity coefficient, high hardness, wear resistance, hardness, and good chemical stability at low density. Furthermore, a very good bonding characteristic was observed with aluminum alloys [28]. AA7075 alloy with 6 wt% B_4C nanoparticles has shown highest tensile and compressive strength of 240 and 329 MPa which declined on further increment of the loading [29]. Higher tensile yield along with good ductility was observed when TiO(C) nanoparticles were non-uniformly distributed on CoCrFeNiMn high entropy alloy [30]. The ultra-fine-grained region (UFG) and coarse-grained regions showed enhancements with structural heterogeneity to multiple levels. Young's modulus differences and strain gradient provided by the nanoparticles enhanced the back stress. Furthermore, non-uniformly distributed nanoparticles also showed resistance to forward stress as illustrated in Fig. 4.

However, there have been reported problems of performance deterioration of steel associated with ex-situ addition of ceramic nanoparticles with poor wettability between the ceramic reinforcement and matrix phase under harsh conditions. Furthermore, during casting, reduced strengthening effect may also be observed of the molten steel with the ceramic nanoparticles floating and agglomerating on the surface. Indeed, wettability may be improved by coating the nanoparticles with

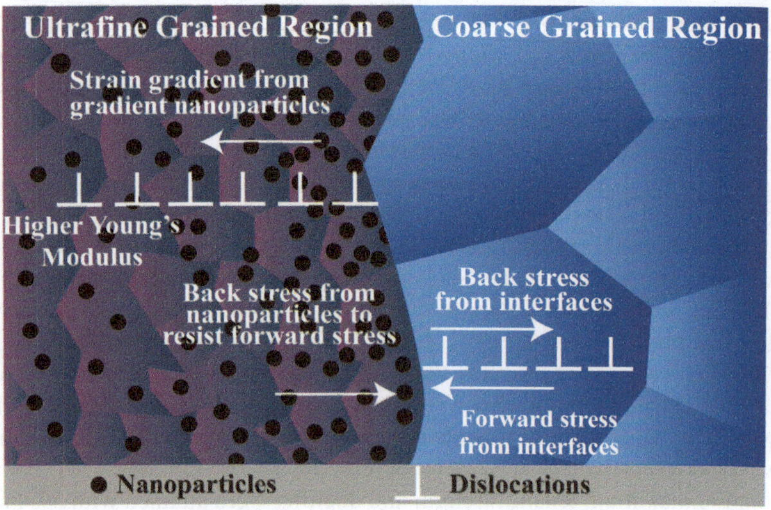

Fig. 4 Mechanism for enhancement in back stress due to the presence of strain gradient, higher Young's modulus, and non-uniformly distributed nanoparticles' resistance to forward stress

surface tension lowering agents. Besides, in-situ addition has been favored with the method providing cleaner interface, better thermal stability, fine sizes, and more uniform particle distribution at the interfaces of filler and matrices.

3 Fatigue

Most structural materials are a subject to repetitive amount of stresses that may produce permanent damage to the materials, starting from the surface and then propagating to the interior of the material, known as fatigue failure. Fatigue can be encountered in a large number of service-providing components such as aircraft, automobiles, turbines, pumps, compressors, and others [31]. Material selection, thus, plays an indispensable role in the component application. Besides the effects of mechanical loading of homogeneous materials, increased attention should be paid to microstructure alterations due to brazing and welding, coating techniques, or any kind of degradation that is environmentally assisted such as cracking due to stress-corrosion and corrosion-creep interactions [32]. Figure 5 gives a simple step-wise overview of designing a material that begins with a material selection and finally provides a safe service life. The challenge lies on future research, in the field of material fatigue, to include fatigue-life prediction methods.

At present, it has been estimated that 80–90% of the mechanical failures in materials are attributed to fatigue [33]. Nevertheless, fatigue resistance can be significantly enhanced with gradient nanostructures in the material surface for both the low- and high-cycle fatigue regimes [34]. In both low-cycle and high-cycle fatigue regimes,

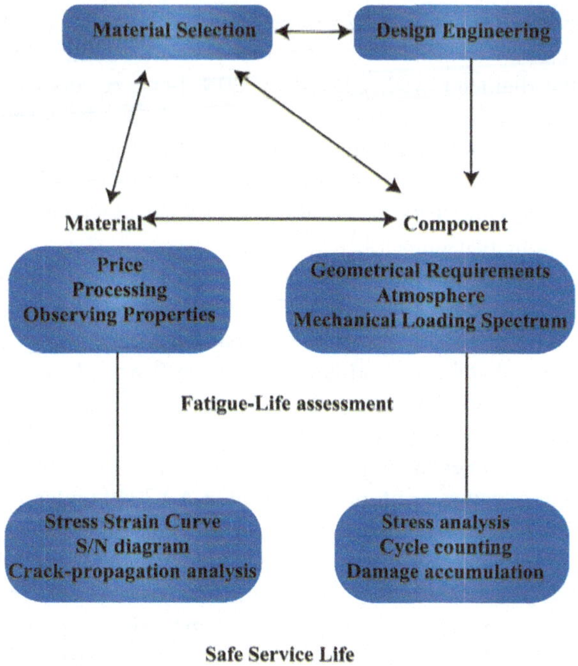

Fig. 5 Fatigue-life prediction in the engineering design process

the nano-grained sample's fatigue life is significantly longer as compared to conventional grain samples under the same stress and strain amplitude application. The limit for fatigue endurance was observed to be more than 10^7 cycles with maximum stress of 100 MPa for nano-grained Cu [35]. A similar phenomenon was observed, when surface mechanical rolling treated (SMRT) 316l austenitic stainless steel with controlled temperature was prepared. Here, at 175 °C, a surface layer of gradient nanostructure was formed with a controllable martensite fraction. The fatigue life was increased by three times as compared to conventional grain samples [36]. The grain size reduction of the steel, after heat treatment, can also decrease the fatigue-crack growth rate [37]. Similarly, when ZnO and Al_2O_3 nanoparticle were doped in AISI-1020 (low carbon steel), the fatigue strength improved by 14.1% for Al_2O_3 and 12.2% for ZnO. This showed that Al_2O_3 coated samples exhibited better performance than ZnO coated samples or uncoated samples [38]. In a low-cycle fatigue regime, equal channel angular pressing (ECAP) prepared ultra-fine-grained (UFG) alloys exhibited fatigue lives that were shorter. However, if ECAP followed a suitable heat treatment method, the fatigue life could be further enhanced. At high stress level, a strong reduced cycling softening with higher fatigue-life enhancement was observed when AA6061 aluminum alloy which was ultra-fine-grained was treated at 165 °C following ECAP [39].

Nanostructured Ti and NiTi alloys, prepared by the means of severe plastic deformation technique (SPD), are excellent choices for the engineering of implants applicable in medical devices that include orthopedics and dentistry [40]. The applied loading scheme, 67.75 N force in the nano-Ti, allowed diameter reduction of the implant by 10%; however, the device's fatigue strength was unacceptable when the diameter was reduced by 20% [41]. A super elastic NiTi nanocomposite, in a crystalline and amorphous form, can also be formed via SPD followed by low-temperature annealing. The produced-nanostructured alloy possesses the endurance of 10^8 reversible-phase-transition cycles under the application of 1.8 GPa stress. The synergistic properties of the two phases showed an effective way for improving the fatigue resistance [42].

Aluminum-based alloys are primarily used in structural applications such as aircraft and automobiles. Therefore, fatigue and tensile properties are some of the major concerns. As TiO_2 nanoparticles have properties like good wear resistance, thermal stability, and hardness, when 9 wt% TiO_2 was added in Al 7075 alloy via stir casting technique, a uniform distribution of TiO_2 in the matrix with an improvement in fatigue properties were observed. The endurance of the alloy reached 10^8 cycles under the application of 81.7 MPa stress [43]. Similarly, when 4 wt% ZrO_2 was added as reinforced material in AA7049 under a constant loading of 10^8 cycles, the fatigue-life factor increased by 66% [4]. When 9 wt% Al_2O_3 was added and 90 MPa stress was applied, the fatigue life increased by 17.4% than when the stirring temperature was 800 °C as compared to those with 0, 5 or 7 wt% Al_2O_3 [44]. Upon comparison, the AA6061/Al_2O_3 nanocomposite, prepared at a stirring temperature of 850 °C, has provided the best fatigue properties.

Enhanced physicochemical and mechanical properties such as high strength, abnormal thermal, and superior catalytic properties were exhibited by nanocrystalline alloys. The Joule heating technique, when applied for the formation of nanocrystalline NiCoP alloy under the application of 200 °C, provided the fatigue endurance limit in the range of 750–825 MPa. The material exhibited a high yield and tensile strength that did not degrade significantly [45]. Similarly, the nanocrystalline Cu and Cu–Al alloys, processed by high-pressure torsion, possessed a significantly improved tensile strength and reduced strain localization within stress bands. However, the fatigue limits did not show an appreciable improvement with their monotonic strength owing to a decrease in stacking fault energy [46].

4 Hardness

Hardness is the resistance of a metallic alloy to permanent indentation under static or dynamic load. If the alloy is uniform in structure and composition, hardness measured is the hardness of the bulk of the alloy on the surface layer. The hardness measurement can lead to determining the tensile or compressive flow curve upto certain plasticity. In the years between 1900 and 1939, for metallic systems, there were some widely used hardness tests such as Rockwell, Shore, Vickers, Knoop, and Brinell. As the Brinell hardness (HB) test is accurate and applicable for most metals, it has been the most preferable one. Although hardness measurement cannot substitute the measurement of test associated with different mechanical properties, it may give a rough idea about other mechanical properties too [47]. The determination of hardness (H) obtained from hardness data can be calculated by a relation (5) introduced by Tabor, where the yield stress is represented by σ and an influence of the sharp indenter geometry is included with a constant factor, C. Under Vickers indentation test, a value near 3 is obtained for ductile alloys [48, 49].

$$H = C \cdot \sigma \tag{5}$$

The hardness value, thus obtained, is significant for analyzing the resistance against wear mechanisms such as corrosion, scratching, and mechanism [50]. Similarly, strength of alloys can be analyzed alongside fatigue-life prediction and can, also, give the idea of the phase transformation.

The demand for high-strength alloys has been increasing in various engineering sectors. For this, a modification in the physical properties of an alloy is required. During cooling after heat treatment, high dislocations present around nanoparticles have also partially increased the hardness of the alloys [51]. These are generated due to differences in the thermal expansion coefficient. The stresses, generated due to thermal expansion coefficient mismatch, are relaxed which leads to punching out of the dislocations at the reinforcement–matrix interface. Thus, the composites may show peak hardness than the unreinforced alloy.

A gradient in hardness from the edge to the center of the surface was observed when alloying cobalt or cerium in aluminum by laser gazing technique [52]. The hardness of the metal can be enhanced by reinforcing different amounts of Al_2O_3 in the metal matrices. The HB hardness was improved by 15.78% when 0.4 wt% Al_2O_3 was reinforced in 2024 Al alloy by stir casting technique. The casting temperature obtained was 850 °C [21]. Similarly, increasing the amount of TiO_2 in the metal matrices of AA6061 and AA6082 showed significant improvement in the hardness of the composites, where a similar casting technique was employed. The hardness tended to be improved owing to the increased cohesion between the alloy components and TiO_2 and a high dislocation density after heat treatment. When 1.5 wt% TiO_2 was used, the HB hardness increased by 32% for AA6061 and 12.1% for AA6082 [53]. Al_2O_3 addition on A356 alloy showed increased hardness with increased loading, while the value reached a maximum and declined in the case of TiO_2 and ZrO_2 [20]. Similarly, TiN addition on Ni/W alloys has had significant effects on microhardness, wear resistance, and friction coefficient. Fine metal grain strengthening effect at 8 g/L TiN loading in Ni/W-TiN composite showed hardness of 897.6 HV. Similarly, the composite produced at 8 g/L possibly decreased the friction coefficient resulting in fine surface morphology [54]. Similarly, the hardness of the composite formed with adding B_4C nanoparticles on AA7075 alloys, as shown in the SEM image of B_4C in Fig. 6, improved with increasing the wt% of the B_4C loading, and reached 89 HV at 7 wt% loading [29]. The increase in the hardness has been highly attributed to Orowan strengthening mechanism.

As Tungsten Carbide (WC) nanoparticles possess advanced properties like high strength, melting point, and excellent wear and corrosion resistance, the addition of WC to the metal matrices can enhance hardness [55]. In a study by Borodianskiy

Fig. 6 SEM image of B_4C powders [29]

and Zinigrad, the hardness was increased by 7.3% when WC nanoparticles were incorporated into Al A356 alloy followed by gas dynamic treatment (GDT). The elongation of the metal was also enhanced by 60%, while the tensile and yield strength remained unchanged [56]. The performance of an aluminum alloy can be improved when a protective layer with excellent adhesive properties is deposited on the surface of alloy, providing resistance to corrosion and also high strength. A similar phenomenon was observed when Zr was coated on the aluminum alloy surface by high current pulsed electron beam (HCPEB) equipment. The surface hardness and corrosion resistance were significantly improved and were highest after 30 pulse irradiations [57]. Likewise, the hardness can also be improved by altering the pulse radiation from HCPEB. The increment was 14% when Cr was alloyed with a Ni substrate forming a nanocrystalline Cr–Ni alloy, and the irradiation was 20 pulses. The wear properties were also enhanced as the HCPEB pulse number increased [58].

The properties such as yield strength, Young's modulus, and ultimate tensile strength could be enhanced by nanostructured carbonaceous materials such as nanodiamond, fullerene, carbon black, and graphene reinforced in the Al matrices. The hardness was also significantly increased. The hardness was 210% more for nanodiamond-aluminum nanocomposite than pure aluminum powder. Similarly, the value increased to 79 HV from 59 HV for fullerene-aluminum nanocomposite compared to pure aluminum powder. This is due to the ease of homogeneous dispersion of these reinforcements in the metal matrices [59]. Also, carbon nanotubes filled aluminum prepared using mechanical ball milling and hot-pressing techniques exhibited five times higher hardness in comparison with base aluminum. Graphene incorporation via chemical vapor deposition showed a 39% improvement in the hardness of the composite. The HV was also 23.4% higher for covetic material in the presence of carbon nanomaterials [60].

5 Fracture and Toughness

Alloys are subjected to several stresses during their service which may be tensile, compressive, shear, or torsional. These stresses may result in an undesirable separation of the alloys to two or more pieces known as fracture. Temperature, during application of stress, is relatively low, below the melting temperature of the metallic alloy. The mechanical performance of materials can be also improved by fracture mechanics application which relies on the physics of stress and strain and also helps to predict the macroscopic mechanical failure through microscopic examination [61]. During fracture, the failure occurs by the initiation of crack and its extension followed by either opening, sliding, or tearing mode as shown in Fig. 7.

Brittle fracture is unstable and can be catastrophic, where the crack propagation direction is perpendicular to the applied tensile stress direction with no plastic deformation taking place. Meanwhile, extensive plastic deformation takes high energy absorption before fracture, during ductile fracture as shown in Fig. 8. So far, surface treatment and heat treatment have been the best tools for the protection against fracture [61]. Fracture of the metallic alloys can be improved by incorporating nanomaterials into the metal matrix as these nanomaterials hinder the expansion dislocation. Similarly, at elevated sintering temperature, an optimum nanomaterial loading within the matrix has also shown highest fracture toughness. Nano-Al_2O_3 incorporation has improved fracture along with the ductility of the alloys.

Hydrothermal and powder metallurgy processes were carried out by Cui et al., where fine ZrO_2 was distributed inside Mo grain, while slightly larger particles were distributed on the grain boundary. As a result, ZrO_2 hindered the expansion dislocation and improved Mo fracture [62]. Likewise, nano-ZrO_2 particulates of 3, 6, 9, 12, and 15 wt% were dispersed in the aluminum alloy by melt deposition method, and NMMC containing 12 wt% ZrO_2 showed the highest fracture toughness [63]. By mechanical milling method, Al_2O_3–ZrO_2–Ni nanopowder was prepared in which Ni particles were uniformly dispersed in the Al_2O_3–ZrO_2 matrix and sintering temperature was maintained at 1550 °C. Increasing the Ni content increased the fracture

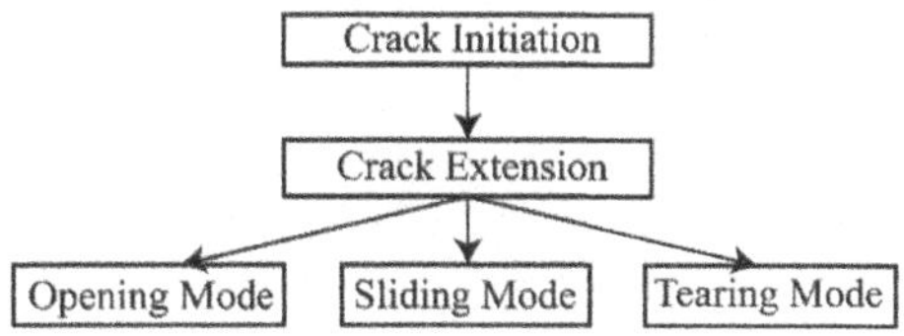

Fig. 7 Various modes of fracture

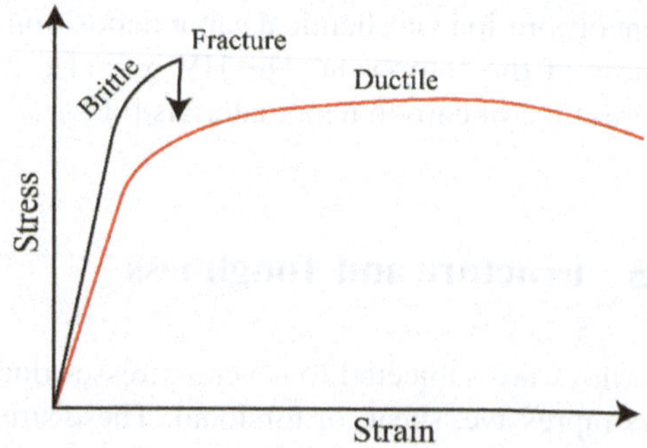

Fig. 8 Stress and strain relation for brittle and ductile materials

toughness of the alloy [64]. A conventional stir casting method was applied to fabricate AA7475 matrix composite containing nano-Al_2O_3, and the surface morphology as seen in Fig. 9 was observed. It was found that in AA7475 alloy, fracture mode is ductile with dimples shaped structure where fracture was improved [65]. Furthermore, mechanical stirring method was carried out with micro and nano-TiB_2 powders with average sizes of 20 and 5 nm into molten A356 aluminum matrix at various casting temperatures of 750, 850, and 900 °C. It was found that toughness increased with an increase in the reinforcement in comparison with non-reinforced alloy [26].

Heat treatment of Al–10Si–Mg alloy was carried out, and supersaturated silicon solid was dissolved into the Al matrix forming Mg_2Si nano-precipitate. It was found that heat treatment enhanced the ductility and formed Mg_2Si nano-precipitates which improved the wear resistance. The maximum and minimum erosion rates occurred at 30° and 90°, which decreased the impact toughness [66]. Upon the solution heat treatment of Al–10Si–Mg, Si gets precipitated from the supersaturated Al matrix. Sizes of precipitated Si increased with temperature, and thus, variation of Si sizes had great impact on mechanical properties of the Al–10Si–Mg alloy. After treatment, the fracture strain increased from 5 to 24% [67]. Heat treatment is one of the

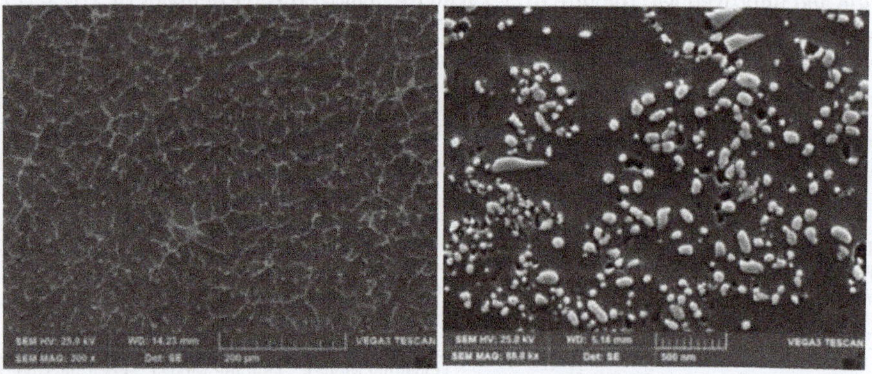

Fig. 9 SEM images of as-cast AA7475 alloy on the left and with 5% Al_2O_3 loading on AA7475 alloy on the right [65]

best methods for the reduction of matrix crack and shear bands of the nanocomposites [68]. But for copper, model based on segregating dopants over grain boundary was best for the improvement of fracture toughness [69]. Delamination toughening alongside transformation-induced plasticity has improved fracture resistance and has also provided higher yield strength of nearly 2 GPas [70]. It was found that controlled cracking at Mn-enriched prior-austenite grain boundaries normal to the primary fracture surface dramatically improved the overall fracture resistance.

6 Corrosion and Wear

There are several strategies for the prevention of corrosion in the case of metallic systems. One of them, to prevent metal alloys from wear and corrosion, is by adding nanofillers [71]. Many nanoparticles also simultaneously give self-healing properties to the alloys applicable for places, where regular maintenance is difficult, and metals are prone to wear and corrosion. From the recent developments, these kinds of alloy could find application from the depth of the ocean to the satellites. The AA 2024 alloy which has high durability and easy processing can be used on a wide range of applications such as for architecture and aviation but is subjected to a range of different environmental factors that can cause corrosion. To overcome this issue, polydimethylsiloxane (PDMS) matrix, fabricated with TiO_2 nanoparticles, formed a nanocomposite coating on AA 2024 via the spin coating method and improved the anticorrosion ability of metal. The electrochemical test showed the composite coating with 8 wt% TiO_2 nanofillers had excellent anticorrosion properties [62]. The graphite nanoparticles addition in electrolyte used for producing plasma electrolytic oxidation (PEO) coatings for AZ91 and AZ80 magnesium alloys showed improvement in corrosion and wear resistance [72]. The graphite nanoparticles provided increased thickness, producing dense coating which sealed the pores on the surface. Also, when silicon nitrite (Si_3N_4) nanoparticles were added into the reaction electrolyte of PEO coatings on AZ31 Mg alloy, a Mg_2SiO_4 phase was observed. A 2 g/L Si_3N_4 nanoparticles addition in electrolyte for PEO coating had the best resistance to corrosion, with increased hardness, good adhesive property, and low friction coefficient [73].

With good resistance to corrosion, strength, good magnetic properties, and electrical conductivity, CuFe alloys have received much attention from researchers. The properties of a mechanically milled mixture of Cu-25wt% Fe powder and Al_2O_3 at 5 vol% were studied, and it was found that the presence of Al_2O_3 caused the wear mechanism to change to abrasive from adhesive, due to which a good wear resistance in the nanocomposite was obtained. Also, a 75% reduction in corrosion rate was observed with the addition of Al_2O_3 nanoparticles [74]. Carbon steel has found application in petrochemical, transportation, pipelines, marine, etc., with very good mechanical property and at a low price. But these types of steel show low corrosion resistance in outdoor environments especially under the harsh conditions (high temperature–pressure). A low carbon steel was surface modified by Nd-PAA-BI (neodymium-polyacrylic acid-benzimidazole), uniformly precipitated Nd oxide/hydroxide, and

Nd complexes film over the steel surface, and it resulted in increased roughness and resistance to corrosion and at highest contact angle; the SEM image is shown in Fig. 10 [75].

Lignin nanoparticles (LNP) which are fabricated from micro-lignin, obtained from industrial and agricultural wastes, show anticorrosive behavior. The lignin nanoparticles with the size ranging 15–20 nm can be used in carbon steel as anticorrosive nanofillers for protection against stringent corrosive conditions. The studies from physio-mechanical and electrochemical impedance spectrometry suggest that epoxy matrix which is dispersed with LNP, to form epoxy-nanocomposite coatings, shows better protection against corrosion in base materials [76].

When Elaeis guineensis (a palm oil leaf) doped with silver nanoparticles (AgNPs) acted as a novel green corrosion inhibitor (EG/AgNPs) in steel-reinforced concretes. It was found that the steel-reinforced concrete with 5% EG/AgNPs incorporation formed a protective thin layer of extra calcium silicate hydrate (C–S–H) gel over the steel-reinforced surface providing enhanced corrosion resistance. The maximum inhibition efficiency was 95% and could be used as an optimum corrosion inhibitor to achieve durable concrete structures [77]. The organically modified silicate (ORMOSIL) coating well dispersed with reduced graphite oxide/silver nanoparticles (AgNPs@rGO) nanocomposite can be used as excellent corrosion and fouling inhibitor for copper metals. With strong binding strength between AgNPs and thiol functionality of the sol precursor, AgNPs/rGO nanocomposites were well-dispersed into the ORMOSIL sol. This would fill the original defect of the coating and effectively improve the corrosion resistance. Apart from this, the antifouling test showed these nanocomposites filled ORMOSIL coatings had excellent antibacterial and antialgae properties [78].

With the help of covalent grafting of graphene oxide (GO) with N,N-dimethylethanolamine and isophorone diisocyanate followed by a non-covalent grafting of GO with dodecylbenzenesulfonate, an amphiphilic graphene derivative was prepared [79]. This modified graphene oxide (IP-GO) coating showed excellent performance against corrosion in metal surfaces, particularly used in oil pipelines, coastal transmission towers, ships, and other metal equipment. The study suggested that the superior mechanical properties were observed with the composite material of waterborne polyurethane (WPU) coating with 0.3 wt% IP-GO nanosheets along with the lowest water absorption and a significant increase in anticorrosion performance. The IP-GO nanosheet can form a perfect network structure to block the intrusion of corrosive media in the polymer.

7 Creep

At elevated temperatures under stresses much lower than the yield stress, a form of time-dependent plastic deformation, creep, is observed in metallic alloys which limits their widespread applicability. Nanoparticles, when incorporated on the metal matrix, hinder the dislocation movement by pinning the grain boundary sliding which

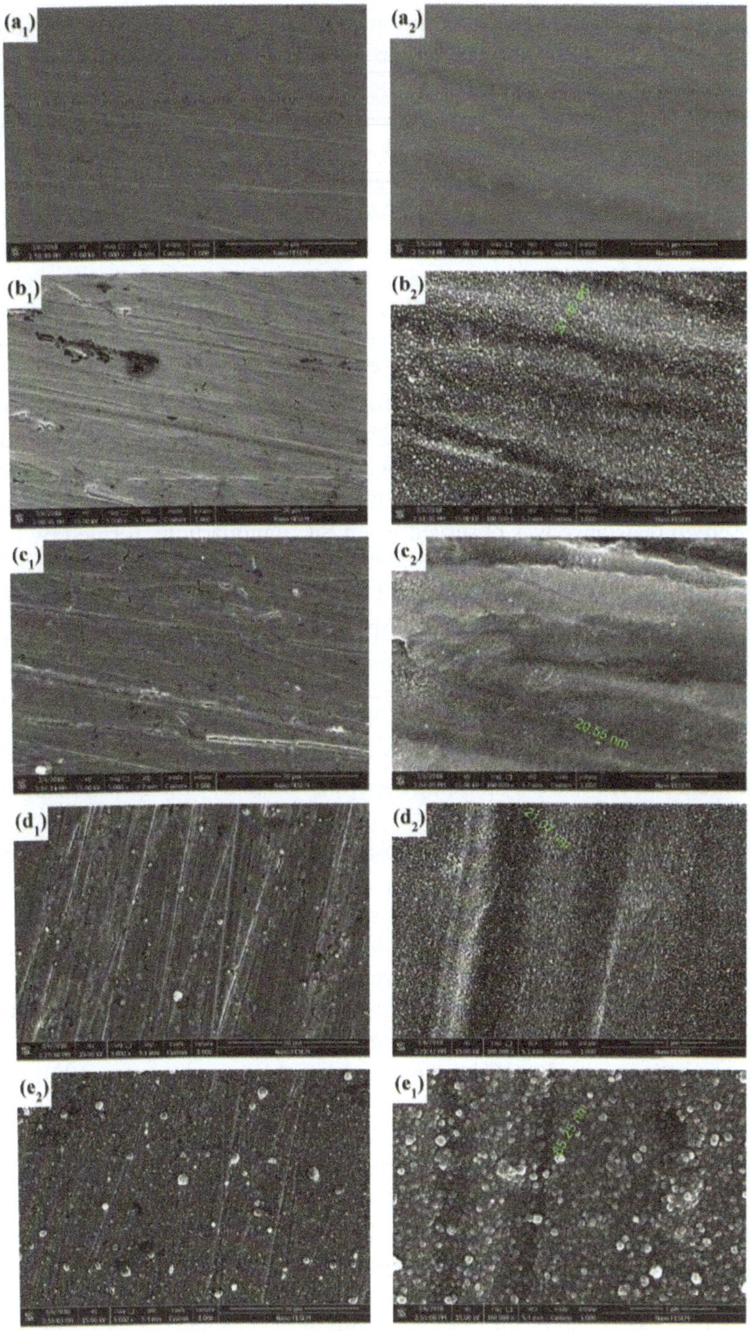

Fig. 10 FESEM micrographs showing $\mathbf{a}_1$–$\mathbf{a}_2$ untreated samples and $\mathbf{b}_1$–$\mathbf{b}_2$ Nd, $\mathbf{c}_1$–$\mathbf{c}_2$ Nd-PAA, $\mathbf{d}_1$–$\mathbf{d}_2$ Nd-BI and $\mathbf{e}_1$–$\mathbf{e}_2$ Nd-PAA-BI treated samples [75]

enhances the creep resistance. The addition of ZnO and Ag nanoparticles in the metal matrix has led to an increase in the creep lifetime. Similarly, SiC, AlN/Al, Zr, Sc, and TiB_2 nanoparticles incorporation has led to improved creep resistance at elevated temperature.

With the absence of lead, a potential substitute for Pb–Sn solder alloy is Sn–Ag–Cu solders. A strong adsorption effect was observed with the incorporation of ZnO nanoparticles (ZnO NPs) into Sn–3.0Ag–0.5Cu (SAC305) at 900 °C, when ZnO powders were melted and mixed mechanically for 2 h into SAC305. Furthermore, the modification with ZnO NPs, hindering dislocation slipping, is following a standard dispersion strengthening mechanism. The creep test result after the addition of 0.7 wt% ZnO NPs showed a decrease in the steady-state creep rate, while there was an increase in the creep lifetime of the plain SAC305 solder [80].

An Mg–Al-based alloy, AZ91, has found application in the interior parts of vehicles with properties of high mechanical strength along with retained ductility, providing corrosion resistance, and also with die-castability features. But, due to its reduced creep resistance above 130 °C, its application is restricted. Nevertheless, superior creep resistance was observed when SiC nanoparticle, formed by utilizing squeeze-casting method, was added in AZ91–2.0Ca–0.3Sb alloy [81].

AlN/Al nanoparticles were fabricated in Mg–2.85Nd–0.92Gd–0.41Zr–0.29Zn (Elektron21) alloy by a high shearing dispersion technique (HSDT) to increase its creep resistance in elevated temperatures. A reported improvement of one order magnitude in creep resistance of Elektron21 alloy was observed when 0.5% AlN/Al NPs were reinforced within the alloy at 240 °C and with an applied stress of 70–140 MPa as compared to unreinforced alloy. The creep resistance was improved due to the influence of AlN NPs by pinning the grain boundary sliding which hinders the dislocation movement [82].

Age hardenable aluminum alloys have different applications in modern engineering, but it is necessary to carefully perform the heat treatment at a controlled manner in order obtain an increase in their strength. However, heat treatment changes limit the tunability gradually. Recently, it has also been found that nanostructural Sc addition to Al alloys can provide resistance to both coarsening and high-temperature creep. The dislocation motion was likely resisted by nanometer-sized coherent Al_3Sc (L12 structure) dispersoids in the alloys. Further, these dispersoids provided stability against coarsening up to 300 °C by pinning down the grain boundary [83].

An alloy of the Al–Si–Mg, A356, even though having great casting and fluidity, low thermal expansion coefficient, and good resistance to corrosion resistance, has degraded mechanical properties at high temperatures, due to microstructural instabilities. Under the applied stress of 480–680 MPa, the impression creep behavior of the cast A356 alloy with the addition of 0.25 and 0.5 wt% Zr at the temperature range of 220–280 °C was studied. It was found that the Zr refined the α-Al grains, and also it was found that the eutectic structure was also modified, providing higher creep resistance [84]. Also, the result from the addition of 0.25 and 0.5 wt% rare earth Er showed similar creep resisting behavior in A356 alloy [85]. It was finally found that creep mechanism that was dominant in all the alloys was the lattice self-diffusion climb controlled creep.

Sn–9Zn is a cost-effective material with good mechanical properties considered as one of the best choices for application in green electronics replacing the traditional Sn–37Pb alloy. Although the use was limited by its poor resistance to oxidation and formation of micro-void, it can be fixed by the influence of Ag NPs. There was significant alteration of morphology of the Zn-rich phase by Ag NPs addition and increased microhardness, elastic properties that are temperature dependent, damping capacity, and finally creep resistance [86].

With the dispersoids formation that is thermally stable at 250–350 °C, Al–Mn–Mg 3xxx alloys are among the most promising candidates in light alloys for elevated temperature application. But during solidification, the Mn gets depleted and resulting in dispersoid free zones (DFZ) and thus decreasing the properties of the alloy. There has been reported reduction in volume fraction of the dispersoid free zones when TiB_2 NPs are introduced in Al–Mn–Mg 3004 alloy with the nanoparticles being distributed in the interdendritic grain boundaries. The creep rate then reduced by inhibiting the grain boundaries motion and dislocation motion at elevated temperatures [27].

8 Deformation

The main reason for the defect on the grain boundaries is due to irregular arrangement of atoms on the grain boundaries, which directly leads to high energy storage in the materials. As a result, the boundaries become the source of defects for the alloy. The presence of these defects leads to the deformation of the alloy. A study on deformation mechanism directly relates to the better understanding of mechanical properties. In recent years, nanomaterials incorporation into the metal matrix has highly resolved these problems [87]. Magnitude of the imposed stress and its effect on strain of the alloy, as given by the Hooke's relation (6), has great impact on the alloys' deformation.

Mathematically,

$$\sigma = \mathrm{E} \cdot \varepsilon \tag{6}$$

The proportionality constant E is called Young's modulus of elasticity, which is different for different metal alloy. The slope of curve of interatomic force separation at equilibrium spacing was found proportional to this modulus as given by relation (7).

$$\mathrm{E} \propto \left(\frac{\mathrm{dF}}{\mathrm{dr}}\right)_{\mathrm{at\ r = r0}} \tag{7}$$

Tungsten has the highest value of Young's modulus of elasticity as compared to other metals, namely magnesium, aluminum, brass, copper, or steel. In the case of gray cast iron, concrete, and many polymers, the stress–strain curve is not linear, and hence, tangent and secant modulus are generally used. However, for elastic deformation, stress is proportional to the strain. It is also called non-permanent types of deformation and time dependent. Elastic deformation remains only to strain of about 0.005. Beyond this point, stress is not proportional to strain and plastic deformation occurs [61].

Magnesium alloy is considered a light engineering material with broad application prospectives. However, the strength of its alloy is very low and also difficult to improve. The grain boundary strengthening based deformation process and the deformation mechanism transition are among the best techniques for preparation of high-strength Mg alloys [88]. Surface mechanical attrition (SMAT) method has produced A231 Mg alloy with nanocrystalline surface layer and gradient nanostructure wherein the grain size in the surface is refined to nanoscale. After SMAT process, SEM photograph has shown better plasticity of the matrix [89]. Various tensile stages of one-sided gradient structured AZ31 Mg alloy including initial stage, plastic deformation, elastic, and plastic deformation were observed. It was reported that internal granular cracks initiated with grain deformation and crack between the grains. Further, these were followed by continuous formation of parallel slip lines, initiation, and growth of trans-granular cracks. It was also found that fine-grained microstructure was alleviated of AZ31 Mg alloy along with effective obscuration of localized deformation and strain concentration [90].

Al alloy, as a lightweight material, has found application as a structural component in automobile and aerospace. Various microstructure modification processes and plastic deformation based manufacturing routes are performed to develop UFG materials Al matrix nanocomposites [91]. The Al alloying effect on the hot deformation behavior of bearing steel was studied by developing $GCr_{15}SiMo$ and $GCr_{15}SiMoAl$ steels, where the Al alloying increased the activation energy of deformation. After the hot deformation process, blemishes were observed for most of the $GCr_{15}SiMoAl$ specimens, indicating reduction in the high-temperature plasticity of the steel by Al alloy. Furthermore, the precipitated AIN particles, even at small amount, increased the strength and decreased the high-temperature plasticity [92]. In alloys EN AW 5182, EN AW 1085, and EN AW 6016, various mechanisms for deformation between room temperature and cryogenic temperature of −196 °C are already present, and when compared, an overall increase in UTS, elongation, and yield strength was observed [93].

Various electrodeless deposition, spark plasma sintering, and hot rolling process have strengthened carbon nanotubes with copper. Enhancement of properties such as high strength, high plasticity, and good electrical conductance was witnessed [94]. When cold drawn at room temperature, Ni_2Si nanofiber had abnormal plastic deformation in Cu matrix, which ultimately increased the strength and conductivity of the Cu alloy [95]. Rotary electroplating has been used to obtain nano-twin copper and Cu ion concentration effect on microstructure, and mechanical properties of nano-twin cu foil showed increased elongation by 22%, whereas UTS and yield tensile strength decreased by 8.3% and 3.9% [96].

9 Conclusion

Mechanical properties of alloys have been improved by incorporation of ceramic nanoparticles, which have led to grain refinement with grain size reduction and hindered motion of dislocations and grain growth stabilization. Likewise, the yield strength has also been improved with the presence of hard ceramic nanoparticles along with increased hardness, chemical stability, and melting point. Similarly, gradient nanostructures have enhanced fatigue resistance for both the low- and high-cycle regimes. There has also been enhancement in physicochemical and mechanical properties such as thermal and superior catalytic properties with nanocrystalline alloys. Some nanocrystalline alloys also showed reduced strain localization within stress bands. The hardness of the resultant alloys consisting nanoparticles has improved owing to the increased cohesion. Similarly, nanomaterials hindered the expansion dislocation and thus improved the fracture as well as creep of the metallic alloys. Nanoparticles have given self-healing properties which have resulted in increased resistance of metallic alloys to wear and corrosion. The presence of nano-reinforcements has resulted in excellent mechanical properties combinations in the alloyed metals which are far superior to their homogeneous equivalent. Inhomogeneous plastic deformation mechanisms with gradients in strain, new dislocations, and grain coarsening, which are mechanically driven, have been attributed for the improved mechanical properties. The high-strength and crack propagation resistance make these nano-reinforced metal alloys suitable for application where critical safety is required such as automobile, aerospace, microelectronics, and more. However, there are still studies required for exploring physical and chemical properties of these nano-reinforced alloys to realize their safe and robust application.

References

1. Nie K et al (2021) Magnesium matrix composite reinforced by nanoparticles—a review. J Magnes Alloys 9(1):57–77. https://doi.org/10.1016/j.jma.2020.08.018
2. Paramsothy M et al (2012) Al2O3 nanoparticle addition to commercial magnesium alloys: multiple beneficial effects. Nanomaterials 2(2):147–162. https://doi.org/10.3390/nano2020147
3. Jayasathyakawin S et al (2020) Mechanical properties and applications of magnesium alloy—review. Mater Today Proc 27:909–913. https://doi.org/10.1016/j.matpr.2020.01.255
4. Alalkawi H, Khenyab AYK, Ali AH (2019) Improvement of mechanical and fatigue properties for aluminum alloy 7049 by using nano composites technique. Al-Khwarizmi Eng J 15(1):1–9. https://doi.org/10.22153/kej.2019.08.001
5. Summers PT et al (2015) Overview of aluminum alloy mechanical properties during and after fires. Fire Sci Rev 4(1):3. https://doi.org/10.1186/s40038-015-0007-5
6. Li S et al (2020) A review on thermal conductivity of magnesium and its alloys. J Magnes Alloys 8(1):78–90. https://doi.org/10.1016/j.jma.2019.08.002
7. Zhang H et al (2021) Microstructure manipulation and strengthening mechanisms of 40Cr steel via trace TiC nanoparticles. Mater Sci Eng A 822:141693. https://doi.org/10.1016/j.msea.2021.141693

8. Yang H et al (2004) Tailoring, structure, and activity of carbon-supported nanosized Pt−Cr alloy electrocatalysts for oxygen reduction in pure and methanol-containing electrolytes. J Phys Chem B 108(6):1938–1947. https://doi.org/10.1021/jp030948q
9. Tjong SC (2007) Novel nanoparticle-reinforced metal matrix composites with enhanced mechanical properties. Adv Eng Mater 9(8):639–652. https://doi.org/10.1002/adem.200700106
10. Tjong SC (2014) Processing and deformation characteristics of metals reinforced with ceramic nanoparticles. In: Tjong S-C (ed) Nanocrystalline materials, 2nd edn. Elsevier, Oxford, pp 269–304. https://doi.org/10.1016/B978-0-12-407796-6.00008-7
11. Qiu F et al (2020) Application of nanoparticles in cast steel: an overview. China Foundry 17(2):111–126. https://doi.org/10.1007/s41230-020-0037-z
12. Xiao P et al (2020) Strengthening and toughening mechanisms of Mg matrix composites reinforced with specific spatial arrangement of in-situ TiB2 nanoparticles. Compos B Eng 198:108174. https://doi.org/10.1016/j.compositesb.2020.108174
13. Kandemir S (2017) Effects of TiB2 nanoparticle content on the microstructure and mechanical properties of aluminum matrix nanocomposites. Mater Test 59(10):844–852. https://doi.org/10.3139/120.111079
14. Viswanathan V et al (2006) Challenges and advances in nanocomposite processing techniques. Mater Sci Eng R Rep 54(5):121–285. https://doi.org/10.1016/j.mser.2006.11.002
15. Zhang Z et al (2011) Mechanical behavior of ultrafine-grained Al composites reinforced with B4C nanoparticles. Scripta Mater 65(8):652–655. https://doi.org/10.1016/j.scriptamat.2011.06.037
16. Wu W et al (2021) Laser powder bed fusion of crack-free TiN/Al7075 composites with enhanced mechanical properties. Mater Lett 282:128625. https://doi.org/10.1016/j.matlet.2020.128625
17. Huang S-J et al (2018) Al alloy metal matrix composites reinforced by WS2 inorganic nanomaterials. Mater Sci Eng A 709:290–300. https://doi.org/10.1016/j.msea.2017.10.041
18. Guo E et al (2018) The influence of nanoparticles on dendritic grain growth in Mg alloys. Acta Mater 152:127–137. https://doi.org/10.1016/j.actamat.2018.04.023
19. Ruirui W, Zheng Y, Qiushu L (2017) Microstructure and mechanical properties of 7075 Al alloy based composites with Al2O3 nanoparticles. Int J Cast Metals Res 30(6):337–340. https://doi.org/10.1080/13640461.2017.1316954
20. El-Mahallawi IS, Shash AY, Amer AE (2015) Nanoreinforced cast Al-Si alloys with Al2O3, TiO2 and ZrO2 nanoparticles. Metals 5:802–821. https://doi.org/10.3390/met5020802
21. Mohammed A-AHJ, Abd Al-Rasiaq A, Al-Jaafari MA (2018) Studying the effect of different wt% AL2O3 nanoparticles of 2024Al alloy/AL2O3 composites on mechanical properties. Al-Khwarizmi Eng J 14(2):147–153. https://doi.org/10.22153/kej.2018.06.004
22. Knowles AJ et al (2014) Microstructure and mechanical properties of 6061 Al alloy based composites with SiC nanoparticles. J Alloy Compd 615:401–405. https://doi.org/10.1016/j.jallcom.2014.01.134
23. Ferkel H, Mordike B (2001) Magnesium strengthened by SiC nanoparticles. Mater Sci Eng A 298(1–2):193–199. https://doi.org/10.1016/S0921-5093(00)01283-1
24. Li X et al (2018) Precipitation behaviors and properties of solution-aging Al-Zn-Mg-Cu alloy refined with TiN nanoparticles. J Alloys Compd 746:462–470. https://doi.org/10.1016/j.jallcom.2018.02.271
25. Jeje SO et al (2021) Properties improvement of spark plasma sintered Ti-7Al-1Mo ternary alloy by TiN nanoparticles addition. Int J Eng Res Afr. Trans Tech Publications
26. Akbari MK, Baharvandi H, Shirvanimoghaddam K (2015) Tensile and fracture behavior of nano/micro TiB2 particle reinforced casting A356 aluminum alloy composites. Mater Des 66:150–161. https://doi.org/10.1016/j.matdes.2014.10.048
27. Kun L, Nabawy A, Chen X (2017) Influence of TiB2 nanoparticles on elevated-temperature properties of Al-Mn-Mg 3004 alloy. Trans Nonferrous Metals Soc China 27(4):771–778. https://doi.org/10.1016/S1003-6326(17)60088-8

28. Byra Reddy B, Bharathesh TP (2021) Influence of B4C nano particles on microstructure and mechanical properties of Al6063 alloy composites. In: AIP conference proceedings, vol 2327(1), p 020017. https://doi.org/10.1063/5.0039502
29. Manohar G, Pandey K, Maity S (2021) Effect of sintering mechanisms on mechanical properties of AA7075/B4C composite fabricated by powder metallurgy techniques. Ceram Int 47(11):15147–15154. https://doi.org/10.1016/j.ceramint.2021.02.073
30. Xie Y et al (2020) Sustaining strength–ductility synergy of CoCrFeNiMn high entropy alloy by a multilevel heterogeneity associated with nanoparticles. Scripta Mater 187:390–394. https://doi.org/10.1016/j.scriptamat.2020.06.054
31. Krupp U (2007) Basic concepts of metal fatigue and fracture in the engineering design process. In: Fatigue crack propagation in metals and alloys, pp 3–37. https://doi.org/10.1002/9783527610686.ch2
32. Krupp U (2007) Fatigue crack propagation in metals and alloys: microstructural aspects and modelling concepts. Wiley. https://doi.org/10.1002/9783527610686
33. Campbell FC (2008) Elements of metallurgy and engineering alloys. ASM International
34. Li X et al (2020) Mechanical properties and deformation mechanisms of gradient nanostructured metals and alloys. Nat Rev Mater 5(9):706–723. https://doi.org/10.1038/s41578-020-0212-2
35. Long J et al (2019) Improved fatigue resistance of gradient nanograined Cu. Acta Mater 166:56–66. https://doi.org/10.1016/j.actamat.2018.12.018
36. Lei Y, Xu J, Wang Z (2021) Controllable martensite transformation and strain-controlled fatigue behavior of a gradient nanostructured austenite stainless steel. Nanomaterials 11(8):1870. https://doi.org/10.3390/nano11081870
37. Leitner T et al (2017) Simultaneous enhancement of strength and fatigue crack growth behavior of nanocrystalline steels by annealing. Scripta Mater 139:39–43. https://doi.org/10.1016/j.scriptamat.2017.05.051
38. Al-Shammari MA et al (2020) Fatigue behavior of steel beam coated with nanoparticles under high temperature. J Mech Eng Res Dev 43(4):287–298
39. Höppel H et al (2006) An overview: fatigue behaviour of ultrafine-grained metals and alloys. Int J Fatigue 28(9):1001–1010. https://doi.org/10.1016/j.ijfatigue.2005.08.014
40. Valiev RZ et al (2008) Nanostructured titanium for biomedical applications. Adv Eng Mater 10(8):B15–B17. https://doi.org/10.1002/adem.200800026
41. Valiev RZ et al (2020) Developing nanostructured Ti alloys for innovative implantable medical devices. Materials 13(4):967. https://doi.org/10.3390/ma13040967
42. Hua P et al (2021) Nanocomposite NiTi shape memory alloy with high strength and fatigue resistance. Nat Nanotechnol 16(4):409–413. https://doi.org/10.6084/m9.figshare.13116578
43. Rahma NM, Eweed KM, Mohammed AA (2018) Investigation and improvement the properties of 7075 AL/T6 alloy using TiO2 nanomaterials. In: IOP conference series: materials science and engineering. IOP Publishing
44. Abed RM, Khenyab AY, Alalkawi HJM (2021) Development in mechanical and fatigue properties of AA6061/AL2O3 nanocomposites under stirring temperature (ST). East-Eur J Enterp Technol 4(12):112. https://doi.org/10.15587/1729-4061.2021.238588
45. Badwe N et al (2021) High-temperature mechanical properties and fatigue of nanocrystalline nickel-cobalt-phosphorous (NiCoP) alloy. Materialia 2021:101136. https://doi.org/10.1016/j.mtla.2021.101136
46. An X et al (2015) Improved fatigue strengths of nanocrystalline Cu and Cu–Al alloys. Mater Res Lett 3(3):135–141. https://doi.org/10.1080/21663831.2015.1029645
47. Bhaduri A (2018) Mechanical properties and working of metals and alloys, vol 264. Springer. https://doi.org/10.1007/978-981-10-7209-3
48. Tirupataiah Y, Sundararajan G (1987) A comprehensive analysis of the static indentation process. Mater Sci Eng 91:169–180. https://doi.org/10.1016/0025-5416(87)90295-3
49. Larsson P-L (2001) Investigation of sharp contact at rigid–plastic conditions. Int J Mech Sci 43(4):895–920. https://doi.org/10.1016/S0020-7403(00)00056-4

50. Torres H, Varga M, Ripoll MR (2016) High temperature hardness of steels and iron-based alloys. Mater Sci Eng A 671:170–181. https://doi.org/10.1016/j.msea.2016.06.058
51. Martin A et al (2021) Effect of the heat treatment on the microstructure and hardness evolution of a AlSi10MgCu alloy designed for laser powder bed fusion. Mater Sci Eng A 819:141487. https://doi.org/10.1016/j.msea.2021.141487
52. Hung CJ et al (2020) Novel Al-X alloys with improved hardness. Mater Des 192:108699. https://doi.org/10.1016/j.matdes.2020.108699
53. Al-Jaafari MA (2021) Study the effects of titanium dioxide nano particles reinforcement on the mechanical properties of aluminum alloys composite. In: IOP conference series: materials science and engineering. IOP Publishing
54. Liu H et al (2021) Effect of TiN concentration on microstructure and properties of Ni/W-TiN composites obtained by pulse current electrodeposition. Ceram Int. https://doi.org/10.1016/j.ceramint.2021.05.145
55. Govindarajan V et al (2022) Effect of tungsten carbide addition on the microstructure and mechanical behavior of titanium matrix developed by powder metallurgy route. Adv Mater Sci Eng 2022:2266951. https://doi.org/10.1155/2022/2266951
56. Borodianskiy K, Zinigrad M (2016) Nanomaterials applications in modern metallurgical processes. In: Diffusion foundations. Trans Tech Publications
57. Li X et al (2020) Microstructure and properties of mechanical alloying Al-Zr coating by high current pulsed electron beam irradiation. Nanomaterials 10(12):2398. https://doi.org/10.3390/nano10122398
58. Zhang L et al (2019) Nanocrystalline Cr-Ni alloying layer induced by high-current pulsed electron beam. Nanomaterials 9(1):74. https://doi.org/10.3390/nano9010074
59. Khanna V, Kumar V, Bansal SA (2021) Effect of carbonaceous nanomaterials' reinforcement on mechanical properties of aluminium metal-based nanocomposite: a review. Mater Today Proc 38:289–295. https://doi.org/10.1016/j.matpr.2020.07.221
60. Bakir M, Jasiuk I (2017) Novel metal-carbon nanomaterials: a review on covetics. Adv Mater Lett 8(884):10.5185. https://doi.org/10.5185/amlett.2017.1598
61. Callister WD, Rethwisch DG (2011) Materials science and engineering, vol 5. Wiley, New York
62. Cui X et al (2018) Polydimethylsiloxane-titania nanocomposite coating: fabrication and corrosion resistance. Polymer 138:203–210. https://doi.org/10.1016/j.polymer.2018.01.063
63. Hemanth J (2009) Development and property evaluation of aluminum alloy reinforced with nano-ZrO2 metal matrix composites (NMMCs). Mater Sci Eng A 507(1–2):110–113. https://doi.org/10.1016/j.msea.2008.11.039
64. Taha MA, Zawrah MF (2020) Fabrication of Al2O3-ZrO2-Ni composites with improved toughness using nano powders prepared by mechanical alloying. Ceram Int 46(11):19519–19529. https://doi.org/10.1016/j.ceramint.2020.05.002
65. Nagaral M et al (2019) Investigations on mechanical and wear behavior of nano Al2O3 particulates reinforced AA7475 alloy composites. J Mech Eng Sci 13(1):4623–4635. https://doi.org/10.15282/jmes.13.1.2019.19.0389
66. Huang B-C, Hung F-Y (2021) Al2O3 particle erosion induced phase transformation: structure, mechanical property, and impact toughness of an SLM Al-10Si-Mg alloy. Nanomaterials 11(8). https://doi.org/10.3390/nano11082131
67. Li W et al (2016) Effect of heat treatment on AlSi10Mg alloy fabricated by selective laser melting: microstructure evolution, mechanical properties and fracture mechanism. Mater Sci Eng A 663:116–125. https://doi.org/10.1016/j.msea.2016.03.088
68. Song X et al (2021) Fracture of magnesium matrix nanocomposites-a review. Int J Lightweight Mater Manuf 4(1):67–98. https://doi.org/10.1016/j.ijlmm.2020.07.002
69. Khalajhedayati A, Pan Z, Rupert TJ (2016) Manipulating the interfacial structure of nanomaterials to achieve a unique combination of strength and ductility. Nat Commun 7(1):1–8. https://doi.org/10.1038/ncomms10802
70. Liu L et al (2020) Making ultrastrong steel tough by grain-boundary delamination. Science 368(6497):1347–1352. https://doi.org/10.1126/science.aba9413

71. Malaki M et al (2019) Advanced metal matrix nanocomposites. Metals 9. https://doi.org/10.3390/met9030330
72. Pezzato L et al (2018) Tribological and corrosion behavior of PEO coatings with graphite nanoparticles on AZ91 and AZ80 magnesium alloys. Trans Nonferrous Metals Soc China 28(2):259–272. https://doi.org/10.1016/S1003-6326(18)64659-X
73. Lou B-S et al (2017) Plasma electrolytic oxidation coatings on AZ31 magnesium alloys with Si3N4 nanoparticle additives. Surf Coat Technol 332:358–367. https://doi.org/10.1016/j.surfcoat.2017.05.094
74. Baghani M et al (2018) Mechanical alloying of CuFe-alumina nanocomposite: study of microstructure, corrosion, and wear properties. Sci Eng Compos Mater 25(6):1085–1094. https://doi.org/10.1515/secm-2016-0313
75. Majd MT, Shahrabi T, Ramezanzadeh B (2019) Low carbon steel surface modification by an effective corrosion protective nanocomposite film based on neodymium-polyacrylic acid-benzimidazole. J Alloys Compd 783:952–968. https://doi.org/10.1016/j.jallcom.2018.12.367
76. ur Rahman O et al (2018) Lignin nanoparticles: synthesis, characterization and corrosion protection performance. New J Chem 42(5):3415–3425. https://doi.org/10.1039/C7NJ04103A
77. Asaad MA et al (2018) Enhanced corrosion resistance of reinforced concrete: role of emerging eco-friendly Elaeis guineensis/silver nanoparticles inhibitor. Constr Build Mater 188:555–568. https://doi.org/10.1016/j.conbuildmat.2018.08.140
78. Liu Z et al (2020) Integrated dual-functional ORMOSIL coatings with AgNPs@ rGO nanocomposite for corrosion resistance and antifouling applications. ACS Sustain Chem Eng 8(17):6786–6797. https://doi.org/10.1021/acssuschemeng.0c01294
79. Wen J-G et al (2019) Improvement of corrosion resistance of waterborne polyurethane coatings by covalent and noncovalent grafted graphene oxide nanosheets. ACS Omega 4(23):20265–20274. https://doi.org/10.1021/acsomega.9b02687
80. Hammad AE, Ibrahiem AA (2017) Enhancing the microstructure and tensile creep resistance of Sn-3.0Ag-0.5Cu solder alloy by reinforcing nano-sized ZnO particles. Microelectron Reliab 75:187–194. https://doi.org/10.1016/j.microrel.2017.07.034
81. Ganguly S, Mondal AK (2018) Influence of SiC nanoparticles addition on microstructure and creep behavior of squeeze-cast AZ91-Ca-Sb magnesium alloy. Mater Sci Eng A 718:377–389. https://doi.org/10.1016/j.msea.2018.01.131
82. Yang H et al (2019) Enhancing the creep resistance of AlN/Al nanoparticles reinforced Mg-2.85Nd-0.92Gd-0.41Zr-0.29Zn alloy by a high shear dispersion technique. Mater Sci Eng A 755:18–27. https://doi.org/10.1016/j.msea.2019.03.131
83. Yang C et al (2020) Nanostructural Sc-based hierarchy to improve the creep resistance of Al–Cu alloys. Mater Des 186:108309. https://doi.org/10.1016/j.matdes.2019.108309
84. Reihanian M, Ranjbar K, Rashno S (2012) Microstructure and impression creep behavior of Al–7Si–0.3Mg alloy with Zr addition. Metals Mater Int 27(8):2530–2540. https://doi.org/10.1007/s12540-020-00628-6
85. Rashno S, Reihanian M, Ranjbar K (2021) Effect of rare earth er on microstructure and creep behavior of Al–7Si–0.3Mg alloy. Metals Mater Int 27(6):1448–1457. https://doi.org/10.1007/s12540-019-00562-2
86. Gain AK, Zhang L (2020) Nanoindentation creep, elastic properties, and shear strength correlated with the structure of Sn-9Zn-0.5nano-Ag alloy for advanced green electronics. Metals 10(9). https://doi.org/10.3390/met10091137
87. Yang G, Park S-J (2019) Deformation of single crystals, polycrystalline materials, and thin films: a review. Materials 12(12). https://doi.org/10.3390/ma12122003
88. Zhang Z et al (2021) Toward the development of Mg alloys with simultaneously improved strength and ductility by refining grain size via the deformation process. Int J Miner Metall Mater 28(1):30–45. https://doi.org/10.1007/s12613-020-2190-1
89. Duan M, Luo L, Liu Y (2020) Microstructural evolution of AZ31 Mg alloy with surface mechanical attrition treatment: grain and texture gradient. J Alloy Compd 823:153691. https://doi.org/10.1016/j.jallcom.2020.153691

90. Meng X et al (2017) The deformation behavior of AZ31 Mg alloy with surface mechanical attrition treatment. Mater Sci Eng A 707:636–646. https://doi.org/10.1016/j.msea.2017.09.094
91. Deb S, Panigrahi SK, Weiss M (2019) The effect of annealing treatment on the evolution of the microstructure, the mechanical properties and the texture of nano SiC reinforced aluminium matrix alloys with ultrafine grained structure. Mater Charact 154:80–93. https://doi.org/10.1016/j.matchar.2019.05.023
92. Yang ZN et al (2017) Effect of aluminum alloying on the hot deformation behavior of nano-bainite bearing steel. J Mater Eng Perform 26(12):5954–5962. https://doi.org/10.1007/s11665-017-3018-7
93. Gruber B et al (2020) Mechanism of low temperature deformation in aluminium alloys. Mater Sci Eng, A 795:139935. https://doi.org/10.1016/j.msea.2020.139935
94. Wang H et al (2018) Improvement of interfacial interaction and mechanical properties in copper matrix composites reinforced with copper coated carbon nanotubes. Mater Sci Eng A 715:163–173. https://doi.org/10.1016/j.msea.2018.01.005
95. Han SZ et al (2016) Increasing strength and conductivity of Cu alloy through abnormal plastic deformation of an intermetallic compound. Sci Rep 6(1):30907. https://doi.org/10.1038/srep30907
96. Hung Y-W, Tran D-P, Chen C (2021) Effect of Cu ion concentration on microstructures and mechanical properties of nanotwinned Cu foils fabricated by rotary electroplating. Nanomaterials 11(8). https://doi.org/10.3390/nano11082135

Metal-Based Nanoparticles: Synthesis and Biomedical Applications

Amandeep Singh, Sovan Lal Banerjee, Aparesh Gantait, Kamlesh Kumari, and Patit Paban Kundu

Abstract Nanotechnology deals with the possibility to exploit materials at nanoscale with an aim to get desirable properties. Nanotechnology aims to enhance the properties of materials by taking advantage of their nanoscale properties. The scope of nanotechnology covers a wide variety of fields including electronics, catalysis, sensing, automobiles, and medical. In recent years, particular interest in the scientific community has been drawn to the biomedical applications of nanomaterials. They differ significantly from their bulk counterparts due to their peculiar and improved characteristics. Nanomaterials, including metal nanoparticles, are being increasingly used for different biological and medical applications. Metal nanoparticles are potential candidates for novel biocides and antibiotic treatments. The novel properties of nanoparticles, especially the physical properties of size and shape, enable them to penetrate all living organisms easily. Several methods are involved in synthesizing metallic nanoparticles, and in general, it can be categorized into either bottom-up or top-down approaches. The top-down method consists of cutting down the bulk materials into nano-sized particles through physical, chemical, or mechanical treatments, whereas, in a bottom-up approach, nanoparticles are formed by joining individual atoms or molecules. The top-down approach produces metallic nanoparticles in naked form, which can further agglomerate and are not suitable for biomedical applications. The bottom-up approach involves solid-state, liquid-state, gas-phase, biological, microfluidic technology-based, and other methods. Chemical reduction in the bottom-up approach is the most common method of metallic nanoparticle synthesis, which is flexible, simple, inexpensive, and produces particle

A. Singh (✉) · S. L. Banerjee · A. Gantait · P. P. Kundu
Department of Polymer Science and Technology, University of Calcutta, Kolkata, India
e-mail: adsingh.chemical@gmail.com; aspst_rs@caluniv.ac.in

A. Gantait
HPCL Mittal Energy Ltd, Noida, India

K. Kumari
Department of Chemical Engineering, SLIET, Longowal, India

P. P. Kundu
Department of Chemical Engineering, Indian Institute of Technology, Roorkee, India

S. K. Tiwari et al. (eds.), *Nanoparticles Reinforced Metal Nanocomposites*,
https://doi.org/10.1007/978-981-19-9729-7_13

in a homogenous shape. Biosynthesis of nanoparticles has recently gained popularity due to its toxic-free nature, affordability, sustainability, and eco-friendliness.

List of Abbreviations

NPs	Nanoparticles
QDs	Quantum dots
IONPs	Iron oxide nanoparticles
PNCs	Polymer nanocomposites
PTT	Photothermal therapy
OA	Optoacoustic contrast
LOT	Localization OA tomography
SERS	Surface-enhanced Raman scattering
MRI	Magnetic resonance imaging
CT	Computed tomography
PLA	Pulsed laser ablation
BSA	Bovine serum albumin
PBS	Phosphate-buffered saline
PDMS	Polydimethylsiloxane
GO	Graphene oxide
PVD	Physical vapor deposition
CVD	Chemical vapor deposition

1 Introduction

Nanotechnology is anticipated to transform the domains related to the biomedical advances by giving most sensitive and specific imaging systems, nanodevices, and nanorobotics in order to detect the biochemical changes at earliest, guided and less toxic targeted drug delivery, and regenerative treatment. As per the US National Nanotechnology Initiative, nanotechnology is a branch of understanding as well as control of matter at measurements nearly from 1 to 100 nm which facilitate size-dependent distinctive characteristics [1, 2]. This size range may resemble to the singular particles for polymers as well as other macromolecules but may also comprise higher-degree associations to the NPs. But for the angstrom-sized smaller atoms/molecules, the range of size (1–100 nm) may involve small aggregations or clusters. As far as biological systems are concerned, the range of size in nm is considered to be perfect in order to enabling NPs to circulate in blood stream, then traverse the tissues, and ultimately enter into the cells.

Metal-based nanoparticles have numerous applications in scientific fields as well as in the industries. Nanoparticles (NPs) show reasonably different characteristics than that of native precursor. Usually, it may be because of that NPs expose a greater proportion of the surface of their atoms as compared to the bulk material. For instance, NPs of size 3 nm (approximately 1000 atoms) can expose approximately 40% of thc total atoms on to the surface. Apart from the surface exposure concept, quantum confinement effect is also important which causes an alteration in the properties of the metals. The larger ratio of surfacc area to volume (surface area/volume) leads to bring the variations in the chemical reactivity as well as quantum confinement effects. It has been observed that the electronic structure of NPs gets changed from that of native bulk metal. For instances, ferromagnetic materials-derived NPs can be paramagnetic and conducting bulk material-derived NPs can be poor conductors. Since change in the properties of NPs depends upon the size, therefore, the properties of NPs are said to be tunable. It gives a free hand for tailor-made preparation of nanoparticles. Moreover, the quantum effect makes metal NPs unique and suitable for plasmon absorption, IR photoluminescence, and superparamagnetism. The multifunctional IONPs shows several applications such as biosensing and pathogen detection, magnetofection and gene therapy, MRI, drug delivery and targeted cell killing, biomarker tracking, bioimaging and therapeutics, hyperthermia and chemotherapy, microbe targeting as well as destruction, stem cell detection and therapy, immune system activation and nanovaccines, tissue engineering and organ transplant, etc. [3]. Furthermore, the uses of polymer nanocomposites (PNCs) consisting metal-based NPs as nanofillers are ever developing. The notable characteristic of NPs is matched with the polymer matrix and developed a hybrid structure comprising several advanced properties. However, the efficacies of metal-based nanofillers matter the end-use applications in various fields such hydrogels, drug delivery, shape memory PNCs, bioimaging, tumor therapy, bio-implants [4].

This chapter explains about various types of metal-based nanoparticles, methods of synthesis, and their applications in various biomedical applications.

1.1 Metal-Based Nanoparticles

Metal-based NPs have attracted the attention of scientists from across the globe for over time span of a century. As of now, such NPs are comprehensively used in different biomedical purposes. Metal-based nanoparticles are in the focus of attention due to their massive prospective in the nanotechnology [5–8]. Magnetic NPs such as iron oxide (Fe_3O_4), gold (Au) NPs, silver (Au) NPs are widely used for the diagnostic imaging and therapy of cancer, drug carriers, and as bioimaging agents. Furthermore, NPs like quantum dots (QD) [9], Au NPs [10], and magnetic NPs [11] have potential to be used in therapy as bioimaging agents or as drug carriers [12–14]. Thereafter, plasmonic NPs such as Au NPs are commonly used in photothermal therapy (PTT) to demolish the brain tumor cells [15]. The QD and magnetic iron NPs are used for the bioimaging purposes [16, 17]. Magnetoferritin, ferritin conjugated to iron oxide magnetic NP, is widely used to penetrate to blood–brain barricade, thus used in treatment of brain cancers [18, 19]. Apart from the above mentioned ones, several other NPs such as cadmium sulfide, bimetallic nanoparticles (such as iron platinum (Fe-Pt) and iron cobalt (Fe-Co)), and metal oxides (such as silica (SiO_2), titanium dioxide (TiO_2), zinc oxide (ZnO), cerium dioxide (CeO_2)) too have various biomedical applications as reported in the coming sections of this chapter.

Metal-based nanoparticles can be classified further into several categories as mentioned below.

1.2 Elemental Nanoparticles

The nanoparticles derived directly from the elements are called elemental nanoparticles. It consists of C, Al, Ti, Fe, Cu, Co, Ag, Ce, Pt, Au, etc.

1.3 Magnetic Nanoparticles

Magnetic nanoparticles are a class of nanoparticle that can be manipulated using magnetic fields such as ferrites.

1.4 Metal Oxide Nanoparticles

Metal oxide nanoparticles are one of the most used nanoparticles in biomedical application, for instances, TiO_2, ZnO, Al_2O_3, CrO_2, CeO_2, etc.

1.5 Bimetallic or Alloy Nanoparticles

Bimetallic nanoparticles are composed of two different metals and exhibit a number of distinct and improved properties. A bimetallic nanomaterial may take the form of an alloy, a core–shell structure, or a contact aggregate.

2 Synthesis of Metal-Based Nanoparticles

Till date, a number of methods are known to synthesize the metal-based NPs. Still, to develop NPs in the large amount with certain shapes, sizes, as well as crystalline properties is a challenging task and needs extra advancements. In a broader term, the preparation of metal-based NPs is categorized into two groups on the basis of precursor being used: one is top-down method and another is bottom-up method. As far as top-down method of NPs' synthesis is concerned, a bulk metal material acts as precursor. On the other hand, in thc bottom-up method, individual nuclei/atoms serve as precursors.

2.1 *Top-Down Methods*

As stated earlier, it includes downsizing of bulk material into desired nano-sized material by using any of method, i.e., chemical, physical, or mechanical method of treatments. As per the opted kind of method, the top-down method includes mechanical milling, laser ablation, and sputtering techniques. All these methods are described below in detail.

2.1.1 Mechanical Milling

Mechanical milling method consists reduction in the particle size with the help of great energy ball milling also called mechano-chemical milling. First, the bulk metallic material is taken in vessel, and heavy balls are added to them. Thereafter, high mechanical energy is created with the help of a rotating ball at very high speed. In the result of that, size of particles gets reduced in required form. The NPs synthesized by this method consist of excellent physical properties, for instance, enhanced solubility. However, this method have some drawbacks, i.e., high energy requirement, time consumption, chance of contamination of powder, and not suitable for synthesis of sensitive NPs, whereas, in mechano-chemical based ball milling method, bulk metallic substance is taken with proper stoichiometry and then milled. During milling, the taken material undergoes frequent distortion, welding, and fracture. Also, some chemical reactions may occur at the interface of substrate and reagent. Though this method seems to be simple one, the prepared NPs are very reactive toward grinding parameters and a prolonged milling is required to synthesize possible smallest sized NPs. In a recent study, Shojaei et al. [20] have synthesized and characterized $CuAlS_2$ NPs using mechanical milling. In this method, $CuAlS_2$ chalcopyrite nanocrystalline structures were prepared using combination of high-energy mechanical milling and homogenization annealing. The stoichiometric mixture of precursor elements, i.e., Cu, Al, and S, was alloyed together mechanically. The process was completed in argon atmosphere to avoid oxidation of NPs. In the same order, Velasquez et al. [21] have synthesized magnetite–maghemite NPs with the help of mechanical milling of sub-micrometric hematite in the presence of a stabilizing medium (polyethylene glycol). Bonding of stabilizing medium with surface of nanoparticles permits to obtain the particles with comparatively less aggregation. It has been found that the use of stabilizing medium narrows down the particle size distribution.

2.1.2 Laser Ablation

In laser ablation method, the irradiation of laser is required to prepare the NPs from the bulk substance. Fragmentations of precursor bulk solid particles into NPs take place by irradiation of laser energy. Thereafter, obtained NPs remain suspended in a liquid and form colloidal solution. It has been found in this method that the amount of ablated atoms and produced NPs relies upon energy of laser as well as pulse duration. Also, ablation efficiency and characteristics of obtained NPs depend upon pulse duration, wavelength of laser being used, laser fluence, as well as on ablation time. Laser ablation method shows a number of benefits over counter methods, i.e., easiness and efficiency, properties of NPs are tunable, and NPs are synthesized without using any surfactant. Nevertheless, one main disadvantage is that the exposure of laser for prolonged time can decrease the ablation rate.

2.1.3 Sputtering

In sputtering method of NPs' synthesis, a bulk material is vaporized using inert gas ions. The vaporization process is performed in vacuum chamber by supplying the sputtering gas under pressure. Thereafter, free electrons are generated using high voltage in cathode. These electrons flow in a spiral path, where free electrons ram into the atoms of sputter gas and result to the ionization of gas. Cations move to the target and get continuously impact. This process gets repeated over and over. Sputtering method is found to be most suitable to synthesize alloy NPs [22].

2.2 *Bottom-Up Methods*

Bottom-up method is a term to be given to a method if in it the precursors of NPs are atoms or molecules. The NPs are developed from such precursors using chemical reactions and various techniques. The techniques used in bottom-up methods include gas-phase synthesis, liquid-phase synthesis, solid-phase synthesis, microfluidic-based synthesis, biological method, etc. Out of all the methods, initially building block precursor is generated, and thereafter, their assembling takes place to synthesize required sized and shaped NPs.

2.2.1 Solid-Phase Method

In this method, precursor material is deposited onto the surface in form of a thin film using either physical vapor deposition or chemical vapor deposition method. A laser ablation method is used in PVD technique that leads to the development of plasma of ablated target, which is deposited onto the substrate in order to develop a thin film. Generally, the metal NPs are deposited onto the carbon nanotubes. However, said method seems to be simple but not able to produce the material in higher quantity as well as the procedure is quite expensive. In CVD, a chemical reaction between the gaseous molecules/atoms in order to develop thin film is used to deposit onto the substrate. Basically, three types of CVD are developed, namely plasma-enhanced chemical vapor deposition, thermally active chemical vapor deposition, and photo initiated chemical vapor deposition. In case of several volatile molecules, the target material is used to be deposited onto the surface of substrate. However, for temperature-sensitive material, for instance polymer substrate, TACVD method cannot be used. For PECVD technique, the plasma is produced inside a void vessel using inductively induced electric current and microwave. The surface morphology as well as crystal structure can be precisely managed in CVD than other methods discussed above. Moreover, the coating of thin film shows good durability, and also, it is very easy to further scale-up [23]. Both of the techniques of solid-state method, PVD and CVD, are typically used to form thin films onto the substrate. Thereafter, corresponding NPs are prepared. It has also been observed that the composites consisting Ag NPs developed through solid-state methods were found appropriate for the biomedical uses attributed to their inbuilt non-toxic and antimicrobial nature.

2.2.2 Gas-Phase Method

This method includes flame pyrolysis, laser pyrolysis, and spray pyrolysis. The spray pyrolysis is carried out using an apparatus having three segments: one unit nebulizer for spraying vapor form precursor in the hot reactor, a vessel reactor having temperature setting between 120 and 200 °C, and one unit of precipitator to collect NPs. However, an apparatus can be, if required, further reformed by exchanging the nebulizer unit with ultrasound unit. The benefits of using spray pyrolysis include simplicity, low-cost, easiness in controlling the size of particles, and the reproducibility. This method is significantly suitable for synthesizing the metal oxide as well as mixed metal oxide-based NPs. Furthermore, the laser energy is utilized in order to activate the homogeneous nucleation reaction after exposing the laser to the precursor material in laser pyrolysis technique. Widely used laser energy is IR CO_2 laser energy that is simply absorbed by the inert photosensitizers such as sulfur hexafluoride. After reaching to adequate quantity of supersaturation of product in vapor phase, CO_2 pyrolysis process gets started. The synthesis of NPs with uniform size distribution can be achieved using this method. Particle size can be tuned by regulating the flow rate of reagents by pyrolysis reaction. However, in the flame pyrolysis method, liquid precursor is directly sprayed in flame that results to the supply of precursors in vapor form. Therefore, this method is said to be beneficial for less volatile precursor materials. The flame pyrolysis method is said to be a favorable one to develop metal oxide-based NPs.

2.2.3 Liquid-Phase Method

This method of NPs' synthesis consists of chemical reduction, sol–gel, solvothermal, and hydrothermal method. Sol–gel method comprises the development of linkages of gelatine and colloidal suspension (sol) in the liquid phase which is accomplished in three means: first, the development of colloids of metal oxide and mix them with sol containing matrix-forming species resulting to form a gel; second, the direct mixing of metal and metal oxide within a pre-hydrolyzed silica sol; and third, development of compound of metal with silone followed by the reduction of the metal before hydrolysis process. In general, the colloids are prepared from metal alkoxides and alkoxysilanes ions. For instances, silica gel is prepared ordinarily from the tetramethoxysilane and tetraethoxysilane. The development of sol–gel consists of several steps, namely hydrolysis, condensation, particle growth, and followed by agglomeration. This method offers the liberty to control the morphology as well as particle size. Liquid-phase method is usually employed to prepare zinc peroxide nanostructures, thin films, and metal oxide-based NPs such as iron oxide, nickel oxide. [24].

The hydrothermal method consists of a chemical reaction between the solid material and solution vapor at elevated temperature and pressure. The cations get precipitated in form of polymeric hydroxide dihydrate which accelerates the formation of metal oxide crystals. On addition of a base to solid material, another cation is developed. Such cations are important to control the formation of particles by inhibiting the formation of complex hydroxide. However, the particles with high crystallinity are able to prepare using this technique, but the overall processes are quite complicated to be controlled. This method is said to be most suitable to synthesize the NPs in powdered form.

Further, the solvothermal-based method of NPs' formation includes the synthesis of NPs in liquid-phase solvents, i.e., water, methanol, polyol, ethanol, etc. Such solvents are able to get heated over to their boiling points in a closed vessel. This method needs a metallic precursor, a reducing agent, a solvent medium, and specific capping agents for controlled synthesis. Narrow-sized distribution of monodispersed crystals of NPs can be obtained through solvothermal method. Various metal-based NPs such as Pd, Rh, Au, Pt, Ni, Co, Ru, and Ag NPs can promptly be prepared using solvothermal method.

2.3 Biology-Sourced Methods

Till date, the biological techniques are emerging approaches to synthesize the metal-based NPs. Biological methods overwhelmed the restrictions related with conventional methods such as safety issues, reaction complications, and high cost. Also, the temperature, high pressure, toxic chemicals, and energy are not used in this method. Therefore, biological sourced synthesis of NPs is regarded as green or biomimetic synthesis. This method of NPs' synthesis includes the applications of microorganisms and their enzymes, plant extracts, plant products, etc. Generally, this broader method can be grouped into two types: first one is the bioreduction and second one is the biosorption. Briefly, in bioreduction method, various metal ions are reduced to a comparatively stable biological form using microorganisms and their enzymes. Thereafter, obtained NPs are separated from the specimen. Obtained NPs were fond stable and chemically inert, whereas in biosorption process, the metal cations get joint to cell wall of organism in the aqueous media. Reaction between peptide and cell wall forms quite stable NPs [25].

2.3.1 From Fungi

The metal salts can be reduced using protein and enzymes produced by the fungi. For this purpose, most common used fungi are *Aspergillus fumigatus*, *Trichoderma reesei*, and *Fusarium oxysporum*. Out of all the NPs, Ag NPs can easily be synthesized by this method because Ag NPs can easily bind with cytoplasmic membrane. Therefore, Ag nuclei can be developed by reducing the metal ions which further get aggregate to form NPs. The location of NPs synthesis is either intracellular or extracellular. However, the Pt NPs can also be prepared using fungi, i.e., *Neurospora crassa* and *Fusarium oxysporum*. The *Neurospora crassa*-based NPs preparation contains bioreduction of proteins which results in single-crystalline, quasi-spherical, and round nanoaggregates. The bioreduction of various enzymes within the *Fusarium oxysporum* results to crystalline-natured Pt NPs with spherical shape. Contrary, bioreduction of hydrogenase enzyme results to rectangular, triangular, and monodispersed spherical NPs. However, the pentagonal, hexagonal, square, spherical shaped, and circular Pt NPs can also be prepared using peptides and bioprecipitation of enzymes in *Fusarium oxysporum* [26]. Recently, Mohamed et al. [27] have given a detailed explanation about the synthesis of metal NPs from entophytic fungi and their possible biological applications. In recent years, the endophytic fungi have attained substantial curiosity for synthesizing the NPs. There is several unique properties of endophytic fungi such as metal tolerance, metal absorption, and accumulation capabilities. The fungi are excellent alternatives for synthesis of metallic NPs as compared to other sources. This method provides several benefits including: easy isolation from soil or plants, fungi secrete a variety of metabolites/extracellular enzymes which help in the reduction of metal ions into NPs, and due to their quick growth, fungi are simple to scale-up. Since most of the fungi grow in a wide range of pH, temperature, and NaCl concentrations, therefore, it is simple to alter the culture conditions to form different NPs.

2.3.2 From Bacteria

In recent time, Ag, Au, Pd, and Pt NPs have been effectively prepared using bacteria. The most common used bacteria for this purpose are Lactobacillus species (i.e., *Klebsiella pneumoniae* and *Enterobacter cloacae*), Bacillus species for producing Ag NPs, Rhodococcus species for Au, and *E. coli* for Pd, Pt, and Ag NPs. Because of the ability to survive in the environment and its richness in nature, the prokaryotes are receiving more consideration for the synthesis of metal NPs. However, the productivity is lower as compared to fungi because bacteria secretes less amount of protein as compared to the fungi. Furthermore, the spherical shaped Ag NPs can be prepared from endophytic bacterium *Bacillus siamensis* strain C1 extracted from the plant *Coriandrum sativum*. Recently, Wang et al. [28] have showed that the combination of nanotechnological method and bacterial therapeutic method may increase the safety and efficacy of focused ultrasonic ablation surgery. Researchers aimed to develop a new biological targeting system consisting of genetically engineered bacteria (GVs-*E. coli*) and multifunctional NPs. In order to escape the influence of excessive particle size, after the combination of GVs-*E. coli* with multifunctional NPs, separate delivery and self-assembly in vivo approach were used [29]. The GVs-*E. coli* synthesized from genetic engineering was injected into the tumor-bearing mice to infiltrate the tumor target areas by taking advantage of their tumor-targeting capability.

2.3.3 From Plant Parts

Generally, metal NPs such as Au, Ag, Pt, Cu, and Zn are effortlessly synthesized using plant extracts. Primary and secondary metabolites are found in the redox reactions of plant metabolism. The plant extracts are found to be a source of secondary metabolites such as phenolic acid, flavonoids, alkaloid, and terpenoid. These metabolites specifically reduce the metal ions which results to the formation of NPs. Such metabolites also act as stabilizing/capping agent in the synthesis process of NPs. The photoremediation of several heavy metals is possible using some specific plants, i.e., *Brassica juncea* (brown mustard), *Clethra barbinervis* (Japanese clethra), *Sesbania drummondii* (rattlebox), and *Acanthopanax sciadophylloides*. Thereafter, plant extraction is mixed with an aqueous solution of metal precursor properly. Numerous spontaneous chemical reactions take place at room temperature and is resulting to the formation of nanostructures. However, in order to fasten the synthesis process, small amount of Cd can be added to the reaction mixture. The plant-based synthesis of NPs is favored over other available biological-based technique because of its ease of availability, bulk production, and due to production of eco-friendly products [30]. Several plant parts such as leaf extract, fruit, herb extract, dried leaves, crusts, and tuber extract are being utilized to prepare Pt NPs having different sizes and shapes. Most usual plants being used for synthesizing the Pt NPs are date palm (*Phoenix dactylifera*), pomegranate (*Punica granatum*), tulsi (*Ocimum sanctum*), and vajradanti (*Barleria prionitis*).

2.4 Microfluidic Method

Microfluidics method is a part of micro-electromechanical technique that includes the movement of 10^{-9} to 10^{-18} liters fluid through micron-sized vessels. Because the dimensions of channel are small, the fluid flowing by this microfluidic channel exhibits several interesting properties different from any macro system. These properties include laminar flow, electro-osmotic flow for the charged elements, and ability to control the water inside the channel. Furthermore, the main property of microfluidic device is high ratio of surface area to volume which permits rapid transfer of heat and mass which results in rapid cooling and heating of the reagents. The intake of the reagent in microfluidics is least which reduces the synthesis cost. The development of reactions in spatial and temporal field is potential in the microfluidics. The reaction can be altered by controlling the flow rate of the reactant. The production of the products can be increased by employing a parallel setup at the same time. Furthermore, it is very easy to incorporate the reagents and simple to quench the on-going reaction after the formation of the product. The NPs in the microfluidics are prepared with the help of a single solvent (for single-phase) or multiple miscible solvents (for various phases). In order to synthesize metal NPs, two types of flow along with different mixing geometries are employed. The diffusion mixing of the reagents occurs in the continuous flow microfluidic. However, the segmented flow uses gas–liquid flow or liquid–liquid flow. The gas phase is used due to the inlet air or the air which is produced during the reaction, whereas the liquid phase is used due to the injection of immiscible liquid [31]. The shear force and interfacial tension within the two immiscible liquids produce the droplets, and it is accomplished using several geometries, i.e., flow focusing, T-junction, and co-flow. The winding and distributed mixing modules are generally employed in the laminar flow mixing. T-junction as well as Y-junction geometries are comparatively modest and easy to use, where the particle size can be tuned by controlling the channel width and flow rate. The geometries and the flow condition regulate the particle size in case of cross-channel geometry microfluidics [32]. Out of all the methods defined above, NPs synthesized using microfluidics technology showed excellent properties in the targeted cancer imaging, enhanced therapeutic efficiency, low sensing detection limit, and higher catalytic activity [33, 34]. A smart construction as well as design of microfluidic device is needed so that metal NPs with required shape, size, and crystallinity can be accomplished. Several parameters, i.e., concentration ratio of reactants, flow rate, seed quantity, seed aging time, residence time, temperature, heating time, and micro-channel wall thickness impact the synthesis of metal-based NPs.

3 Biomedical Applications

3.1 Gold Nanoparticles

3.1.1 Biomedical Imaging, PTT, and Drug Delivery

Gold and its compounds are used for several medical applications since its existence about 5000 years in the past. Among the metallic NPs, Au NPs are widely explored in detail across the globe and have delivered new comprehensions, particularly in the field of biomedical domain. Gold metal is regarded as noble one. Au NPs show exceptional characteristics like physicochemical catalytic, gets easily functionalized, non-toxicity, as well as several biological features. Attributed to the above mentioned unique properties, Au NPs are used for various applications such as therapeutics, biomedical imaging, drug delivery, sensing, analytical science, catalysis, and medical diagnosis [35–37]. Au NPs possess unique electronic and optical features along with the chemical inertness. These NPs have ability to get surface functionalization that is attributed to the presence of negative charge onto the surface of Au NPs. Due to the unique electronic and optical properties, Au NPs are being utilized in biosensors, PTT, and bioimaging. The ability of gold to get easily functionalized using organic compounds permits them to get conjugated with the ligands, antibodies, and drug molecules to be used for the active/passive drug delivery. The Au NPs show excellent biocompatibility in both of setups, in vitro and in vivo, due to their chemical inertness. Gold nanoparticles (Au NPs) have great potential for the biomedical applications and are being used since a decade as shown in Table 1.

3.1.2 Cancer Treatment

Au NPs and its compounds are being used for cancer treatment since a long time. Recently, Meir et al. [53] have developed a technique to be used in tracking tumor-specific T-cell via non-invasive longitudinal and quantitatively. This method is a combination of an imaging modality and labeling agents Au NPs. The results obtained from in vitro cytokine release as well as proliferation assays showed about the Au NP labeling being not obstructing the functioning of T-cells. The labeled cells then shifted into mice bearing melanoma. The CT imaging showed the circulation, migration, as well as perseverance of the cells inside tumorous tissue. T-cell tracking process as well as aggregation of directed T-cells in tumor is shown in Fig. 1. Prior to the injection, small tumor was not observable using CT scans (Fig. 1 IIA). A clear CT scan image can be seen (Fig. 1 IIB) after 24 h of the inoculation T-cells labeled with Au NP in tumorous tissue. This can be attributed due to the amount of Au NPs within T-cells which keep migrating toward the tumorous tissue. It has been observed that post 48 h, the signal gets strengthened which indicates a rise in concentration of targeted T-cells around the tumor spot (Fig. 1 IIC). Thereafter, signal gets decreased (Fig. 1 IID).

Table 1 Various biomedical applications of Au nanoparticles

Method	Observation	Applications	Refs.
• Au NP (size 16 nm) was prepared by a modified-Turkevich method • CNT-Au NP-silica was prepared by layer-by-layer self-assembly method	• CNT-Au NP-silica microsphere shows strong absorption in visible and NIR regions • High optoacoustic contrast (OA) or photoacoustics and Raman scattering on illuminating with lasers of $\lambda = 532$ and 785 nm • Amplifies the signal in OA flow cytometry at laser of $\lambda = 1064$ nm • Ex vivo brain tissue was done using a portable Raman spectrometer and imaging with the Raster-scanning OA mesoscopy technique	• Localization OA tomography (LOT) • OA flow cytometry • Multiplex SERS detection	[38]
• Au NPs were prepared by reduction method from $NaBH_4$ and $HAuCl_4$ • Fe_3O_4 NPs were prepared by coprecipitation, electrochemical, hydrothermal, microemulsions and reverse micelles, sol–gel, solvothermal, sonication, and thermal decomposition method • Fe_3O_4/Au core/shell NCs were prepared • Inorganic and organic coating of Fe_3O_4/Au core/shell	• Fe_3O_4/Au core/shell and Fe_3O_4/glue/Au core/glue/Au • NCs' structures have reduced synthesis problems • Modifications have reduced the physiochemical properties of core material of Fe_3O_4/Au and Fe_3O_4/glue/Au NCs • Intrinsic modularity enables quick interaction between NCs and metal/metals oxide to meet the necessities like surface area, specific functions, and morphology	• MRI • CT scan • Drug delivery • Biosensors • Therapeutic device • Hyperthermia applications	[39]

(continued)

Table 1 (continued)

Method	Observation	Applications	Refs.
• High-resolution ICP-MS is used for NPs' qualification tool • Novel Au NPs stabilized by N-heterocyclic carbenes	• Reactivity for serum biomolecules or resistance toward action of human serum milieu is consistently measured by recording the signals of Au or S isotopes • If apply on screening stage, they may be beneficial for shortening the timelines and reducing the cost for selection and initial testing of medicinal NPs		[40]
• Polypyrrole microcapsules with a hydrophobic liquid core and Au NPs • Oxidative polymerization of pyrrole by encapsulation of liquid phase and Au NPs • Microcapsules loaded with organic solvents (toluene, hexane) or 2-oxoheptyl isothiocyanate (anticancer agent) as liquid cores and stable or radioactive Au Nps (Au-197 or Au-198 isotopes)	• Resulting microcapsules have been demonstrated as promising agents for medical applications shown by in vitro on cancer and normal cell cultures	• CT scan • Gamma imaging • Drug delivery	[41]
• Green chemical approach for Au NPs synthesis from *Caulerpa racemosa*	• Cr@Au NPs showed a surface plasmon resonance (SPR) peak at 528 nm in UV-visible spectrum • Cr@Au NPs effectively controlled the growth of human colon adenocarcinoma (HT-29) cells, and it exhibited IC50 at 20.84 μg/ml • Cr@Au NPs showed non-toxic effects on Artemia nauplii, even at high concentration as 100 μg/ml	• Antibacterial uses • Drug development for cancer	[42]

(continued)

Table 1 (continued)

Method	Observation	Applications	Refs.
• *In situ* biosynthesize Au NPs in living platelets by ultrasound energy • In ultrasound exposure, biocompatible $HAuCl_4$ was permeated within platelet cytoplasm • Reducing agent ($NaBH_4$ and sodium citrate), platelet enzyme, and synthesized Au NPs were reacted *in situ*	• Atomic absorption spectrometry (AAS) showed the synthesized amount of Au is $(12.7 \pm 2.4) \times 10^{-3}$ pg per one platelet • GNPs in platelets can produce Raman enhancement effect and further be probed for both dark-field microscopy (DFM)-based imaging and computed tomography (CT) imaging • Mimicking GNPs' platelets with *in situ* GNPs' components remains inherent platelet bioactivity which will find potential theranostic implications with unique GNPs' properties	• Immune system	[43]
• Au NPs core/shell prepared was stabilized with a hydrophilic polymer, poly(3-dimethylammonium-1-propyne hydrochloride) (PDMPAHCl) • Au@PDMPAHCl is pH responsive	• Used for immobilize bovine serum amine oxidase (BSAO) • BSAO enzyme immobilize onto Au@PDMPAHCl • Activity tests on Au@PDMPAHCl-BSAO bioconjugates confirm 40% rise in enzymatic activity	• Inhibition to tumor growth • Au@PDMPAHCl-BSAO used in anticancer therapeutic system	[44]

(continued)

Table 1 (continued)

Method	Observation	Applications	Refs.
• Green synthesis of Au NPs • Endophytic strain *Fusarium solani* ATLOY-8 isolated from plant *Chonemorpha fragrans* was used	• Au NPs exhibited cytotoxicity on cervical cancer cells (HeLa) • Against human breast cancer cells (MCF-7) also • Au NPs showed dose-dependent cytotoxic effect • IC_{50} value was 0.8 ± 0.5 µg/mL on MCF-7 cell line and was found to be 1.3 ± 0.5 µg/mL on HeLa cell lines • NPs induced apoptosis on both cancer cell lines • Accumulation of apoptotic cells get declined in sub G0 and G1 phase of cell cycle in MCF-7 cancer cells • Found to be 55.13%, 52.11%, and 51.10% post 12 h contact with various concentrations	• Anticancer • Safer chemotherapeutic agent with little systemic toxicity	[45]
• Au NPs were synthesized from $HAuCl_4$ using $NaBH_4$ as a reducing agent	• In vitro targeting efficacy tested against three pancreatic cancer cell lines PANC-1, AsPC-1, and MIA Paca2 • Along with variable epidermal growth factor receptor (EGFR) expression • Results showed that Au uptake is correlated with EGFR expression	• Delivery carrier for pancreatic cancer (using cetuximab as a targeting agent and gemcitabine as an anticancer drug) • Inhibition of pancreatic tumor cell proliferation • Orthotopic pancreatic tumor growth *in vivo*	[46, 47]
• Au NPs were synthesized by the reduction of chloroauric acid ($HAuCl_4$) and sodium borohydride ($NaBH_4$)		• Pancreatic cancer	[48]

(continued)

Table 1 (continued)

Method	Observation	Applications	Refs.
• Au NPs were modified by aminoalkanethiol	• Developed for lectin biosensor • Polycrystalline gold modified by an aminoalkanethiol linker layer	• Label-free detection of glycoproteins	[49]
• Au NPs decorated reduced graphene oxide	• Supersandwich-type electrochemical biosensor • Capture DNA labeled with thiol was immobilized onto Au NPs' surface	• Sequence-specific DNA detection	[50]
• Amperometric biosensor based on Au NPs • Reduction method using tetrachloroauric (III) acid ($HAuCl_4$) and sodium borohydride ($NaBH_4$)	• Au NPs were anchored onto reduced graphene oxide (RGO-Au NPs) and L-lactate dehydrogenase (LDH)	• Sensing of L-lactate	[51]
• Biocompatible glycol chitosan-coated Au NPs	• Surface modified with biocompatible glycol chitosan (GC) • *In vivo* biodistribution of GC-Au NPs was studied through computed tomography (CT)	• Tumor-targeting CT imaging	[52]
• Au NPs	• Tracking cancer-specific T-cells *in vivo*	• Nanomedicine for Cancer • Immunotherapy • CT imaging	[53]
• pH-sensitive Au NPs	• NPs-based probe with "turn-on" application of theragnostic agent for simultaneous Raman imaging/diagnosis and PTT	• Selective SERS • PTT of cancer	[54]
• pH-induced aggregation of "smart" Au NPs	• Hydrolysis-susceptible citraconic amide surface assists to aggregate in mild acidic intracellular environments	• PTT of cancer	[55]

(continued)

Table 1 (continued)

Method	Observation	Applications	Refs.
• Biodegradable liposome Au NPs (LiposAu)	• Pharmacokinetic analysis of LiposAu NPs was performed in small animal model • Results showed *in situ* degradation in hepatocytes and further get cleared through hepatobiliary and renal route	• PTT of cancer	[56]
• Multifunctional Au-coated thermosensitive liposomes were prepared	• Au-coated liposomes showed excellent biocompatibility • High efficiency to kill cancer cells through photothermal transduction (PTT)	• Multimodal imaging and PTT of breast cancer cells • Imaging using specific NIR dyes	[57]
• Doxorubicin-tethered responsive Au NPs (DOX-Hyd@Au NPs)	• Significantly increased cytotoxicity of doxorubicin and induced elevated apoptosis of MCF-7/ADR cancer cells	• Intracellular drug delivery for overcoming multidrug resistance in cancer cells	[58]
• Au NPs-based drug delivery system in cancer therapy	• Au NPs easily penetrate blood vessels and tissue barriers into tumor foci. • Indicates that AU NPs are more effective drug carrier with great merits in reducing cytotoxicity	• Multiple functions in therapeutics, imaging, and surface modification • Effective antitumor drug carriers	[59]
• Doxorubicin-loaded magnetic Au NPs (MGNPs) were prepared	• MGNPs were functionalized with thiol-terminated polyethylene glycol (PEG) • Loaded with anticancer drug doxorubicin (DOX) • Aspartate aminotransferase (AST), alanine transaminase (ALT), lactate dehydrogenase (LDH), creatine kinase MB (CK-MB), urea, uric acid, and creatinine were measured to assess *in vivo* toxic effect of MGNPs' formulations	• *In vivo* magnetic targeted drug delivery	[60]

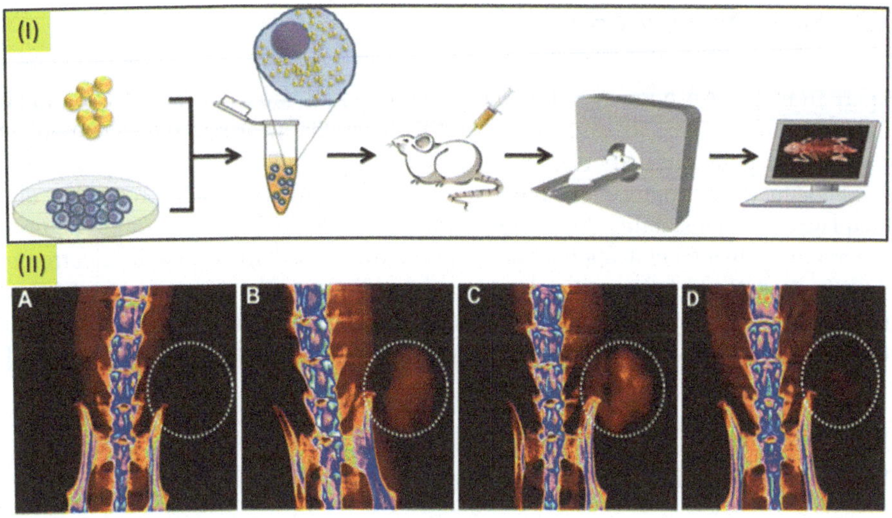

Fig. 1 **I** T-cell tracking process: T-cells were labeled with Au NPs in vitro; the cells were then injected to mice and tracked in vivo using CT imaging. **II** Time-dependent accumulation of targeted T-cells at the tumor. **a** Before T-cell injection, **b** 24 h post injection, **c** 48 h post injection, and **d** 72 h post injection. [Reused with due permission from [53]

In the same order, Abdulateef et al. [61] have used pulsed laser ablation (PLA) method for synthesizing the Au NPs. In this method, bovine serum albumin was used along with simulated body fluid. In this research, BSA was used as a stabilizer, reducing, and capping agent to develop Au NPs of size in the range of 3–10 nm. Agglomeration as well as precipitation of obtained Au NPs was confirmed by occurrence of BSA in the reaction solution by electrostatic repulsion among the Au NPs and BSA. Results showed the stability of synthesized Au NPs-BSA conjugates gets enhanced with addition of BSA. Also, the obtained Au NPs show high biocompatibility and low toxicity and thus are capable to use for drug delivery. This research has presented an environmental-friendly, effective, and a simplistic method for synthesizing the multifunctional Au NPs through PLA method, applicable for potential and efficient therapeutic cure of cervical cancer.

3.1.3 Tissue Engineering

The Au NPs have an excellent potential to be used in tissue engineering application. Cutroneo et al. [62] have developed PDMS sponge having 50–900 μm size with the help of sugar template. Thereafter, obtained sponges of PDMS were adsorbed with the GO and Au NPs developed by laser ablation method in liquid. The outcomes of this research exhibit that PDMS composites as well as adopted processing were found to be biocompatible after confirming by analyzing the cytotoxicity of PDMS composites toward various cell cultures.

3.2 Silver Nanoparticles

Apart from the Au NPs, Ag NPs and its compounds are also used for biomedical applications since its discovery. Ag NPs have characteristic physicochemical features like good thermal conductivity, excellent catalytic activity, adequate electrical conductivity, appropriate antibacterial activity, greater optical properties, and adequate chemical stability. Such unique features of Ag NPs allow them to be used in photonic, disinfectant applications, antimicrobial activities, and in electronic. Usually, Ag NPs are being utilized in wound dressings, medical industry for fabrics, and coating of devices because of their extraordinary antimicrobial activity. Other applications of NPs consist of development of biosensors, drug delivery, and PTT [63–74].

3.2.1 Antimicrobial

Silver nanoparticles are well-known to have antimicrobial properties [75, 76] and luckily seem to be potent and efficient antimicrobial agents as compared to other noble metal nanomaterials, owing to their characteristic properties like large surface to volume ratio, toxicity, possible interaction with the sulfur and phosphorus compounds present in the cells [77], crystallographic characteristics, and so on, which enable them to be useful agents for treating various microbial infectious diseases and also can serve as useful agents to overcome the microbial resistance against the conventional drugs, either used alone or in conjugation with therapeutic antimicrobial formulations [78]. Silver nanoparticles have also been researched and proved to be efficient antifungal and antiviral agents.

Furthermore, the Ag NPs can be conjugated or loaded with antibiotics or antimicrobials for synergistic and enhanced antimicrobial effects, resulting to the development of efficient, potent, highly biocompatible, broad-spectrum, and least resistant antimicrobial formulations. One of the studies conducted by Bonde et al. has reported strong antimicrobial activity of Ag NPs (size 40–80 nm) synthesized from *Murraya koenigii* against human pathogens like *Pseudomonas aeruginosa, Escherichia coli,* and *Staphylococcus aureus.* The synergistic enhanced antibacterial activity of Ag NPs loaded with tetracycline and gentamycin was observed [79].

3.2.2 Cancer Treatment

In recently reported research work, Shejawal et al. synthesized Ag NPs using 1% aqueous extract of the carotenoid phytopigment named lycopene, isolated from tomato, extracted in benzene and observed their anticancer activity. The lycopene Ag NPs were tested on Hella, COLO320DM, H29 cancer cell lines, and it was reported through MTT assay with the percent inhibition of 40.9 ± 0.69, 41.41 ± 0.41, and 35.43 ± 0.67 against Hella, COLO320DM and H29 cancer cell lines, respectively

[80]. In another study conducted by Lin et al., it was found that Ag NPs act as anticancer agents by induction of autophagy of cancer cells through activation of the Ptdlns3K pathway. Furthermore, they observed that inhibition of autophagy by autophagic inhibitor like wortmannin results in enhanced cancer cell killing efficacy in the mouse melanoma cell model (B16 cell lines) [81]. Shejawal et al., in their research work, synthesized iron and Ag NPs ranging in size from 50 to 100 nm by a green synthesis method from a polyphenolic bioactive phytocompound "proanthocyanidin", isolated from grape seed, and successfully evaluated various biological activities of synthesized nanoparticles. They reported significant anticancer activity against different colon cancer cell lines (COLO320DM and H29) through SRB and MTT assays. According to the SRB assay, proanthocyanidin Ag NPs inhibited the growth of COLO320DM (inhibition: 71.61.97%) and H29 (inhibition: 69.211.86%) cell lines. MTT assay reveals that proanthocyanidin Ag NPs showed 64.27 ± 1.63 and $63.34 \pm 1.64\%$ inhibition against COLO320DM and H29 cell lines, respectively [82]. In another research work carried out by Mittal et al., biosynthesized silver nanoparticles from plant extracts of *Potentilla fulgens* caused cytotoxicity in a dose-dependent manner in the U-87 and MCF-7 cell lines. The IC50 value was found to be 8.23 and 4.91 μg/mL in U-87 and MCF-7 cell lines, respectively [83]. Researchers carry out the desired conjugations with Ag NPs to have enhanced cytotoxicity toward cancer cells. Preethi and Padma observed the enhanced anticancer activity of silver nanobioconjugates synthesized from the leaf extract of *Piper betle* and its active polyphenol compound "eugenol". The cytotoxicity to oral "KB" cancer cell lines was found to be elevated by the application of nanoconjugates in comparison to non-conjugates or crude parts of the plant [84]. Yuan et al. reported synergistic enhanced cytotoxicity and apoptosis of Hela cancer cells with combinatorial therapy of Ag NPs and camptothecin (CPT) [85]. Venkatesan et al. biosynthesized porous chitosan–alginate Ag NPs and revealed that the nanocomposite has potential anticancer and antimicrobial properties. The nanocomposite was tested in vitro on MD-MB-231 (breast cancer cells), and the IC50 value is reported to be 4.6 μg/ml [86]. Gomathi et al. synthesized Ag NPs by using the fruit shell of *Tamarindus indica*, and the resulting biosynthesized nanoparticles proved effective in a dose-dependent manner against the MCF-7 cell line (human breast cancer) [87].

In vitro anticancer activity of green synthesized silver nanoparticles from the aqueous extract of *D. ciliolata* seaweed extract was evaluated by Venkatesan et al. [88]. Silver nanoparticles exhibited 82% toxicity to A549 cells with an IC50 value of 5 μg/mL. The nanoparticles induced DNA fragmentation and also inhibited cell migration in A549 cells. Antiangiogenic activity was exerted by the synthesized silver nanoparticles, as evidenced by the inhibition of tertiary blood vessel formation in CAM assay. The developed silver nanoparticles exerted potential anticancer and anti-angiogenic activity on non-small cell lung cancer cells.

3.2.3 Bio-implants

Ti implants are highly biocompatible and allow orderly bone growth but, unfortunately, in the first five years after implantation, 5–10% of them fail due to poor osseointegration and to the presence of bacterial infections in prosthesis. Ag NPs have been described to damage bacterial cell via prolonged release of Ag^+ ions as a mode of action when immobilized on a surface. The deposition of these nanomaterials on porous Ti substrates was previously fabricated using the space-holder technique. After silver nanoparticles were deposited on the porous Ti substrates, microstructural characteristics and antibacterial behavior were evaluated against the proliferation of *Staphylococcus aureus* on the Ag NPs functionalized substrates. Finally, the preliminary qualitative analysis showed the presence of inhibitory halos, being more relevant in the substrates with larger pores [70].

Jung et al. [89] have developed chitosan-Ag NPs hybrid 3D porous structure as a SERS substrate. The SERS substrate with chitosan-Ag NPs (chitosan-Ag NPs) hybrid 3D porous structure was fabricated simply by a one-step method. SERS enhancement by the chitosan-Ag NPs substrate was experimentally verified using rhodamine B as an analyte. Thiolated single-stranded DNA was also measured for atopic dermatitis genetic markers (chemokines CCL17) at a low concentration of 5 pM. Chitosan-Ag NPs SERS substrate has a simple, low-cost, large area, and excellent biocompatibility.

3.3 *Iron Oxide Nanoparticles*

Iron oxide NPs are considered as one of the widely utilized magnetic nanoparticles in biomedical stream since they possesses several characteristic biological, chemical, as well as magnetic features. These NPs show adequate biocompatibility, greater magnetic susceptibility, high saturation magnetization, non-toxicity, and also greater chemical stability. The iron oxide shows various oxidation numbers, for instances, +3 in Fe_2O_3, +2 in FeO, and both +3 and +2 in Fe_3O_4. It has been found that Fe_2O_3 shows various crystalline polymorphs. These crystalline polymorphs are α-Fe_2O_3, β-Fe_2O_3, γ-Fe_2O_3, and ε-Fe_2O_3. The γ-Fe_2O_3 and Fe_3O_4 NPs were found to be very much biocompatible, but Fe_3O_4 NPs is widely utilized biomedicine applications. Though Fe_3O_4 NPs are sensitive to get oxidize, thus a coating with some biocompatible shell such as a suitable polymer, ceramic, or metal is compulsory. Coating with a shell imparts numerous benefits; for instances, it inhibits the agglomeration and aids the functionalization and conjugation with enzymes, proteins, anticancer drugs, and antibodies. Various iron-based NPs are explored and studied for the targeted drug delivery, magnetic hyperthermia treatment, contrast agents for MRI, and for some other biomedical uses. The magnetic features of such NPs are further enhanced by the means of doping of magnetically sensitive atoms, i.e., Ni, Co, and Mn. However, the ferrites doped by Co as well as Mn are widely used in biomedical applications. $CoFe_2O_4$ NPs were found to have high magnetocrystalline anisotropy, moderate magnetization saturation, high Curie temperature, chemically stable, and high coercivity. Therefore, such NPs are potentially being researched for to make them feasible for magnetic hyperthermia applications and as possible contrast agents in MRI. The $MnFe_2O_4$ NPs possess greater magnetization, good magnetic susceptibility with appropriate relativities, and biocompatibility. Therefore, $MnFe_2O_4$ NPs too are widely explored for magnetic hyperthermia applications as well as contrast agents in MRI.

3.3.1 MRI and Biomarker Tracking

Magnetic resonance imaging (MRI) is an efficient bioimaging method that gives both 3-D and cross-sectional images of delicate tissues without the use of radioactive radiations. Magnetic IONPs act as contrasting agent in MRI, and additionally, they are able to track the specific biomarker, cell, or organ by MRI in diseases like stroke, cancer, aortic or cerebral dysfunctions, etc. Therefore, magnetic IONPs are considered as multifunctional due to their synergistic properties of bioimaging and biomarker tracking [90]. Ferromagnetic and superparamagnetic IONPs are currently used in biomedical industry as they shorten T1 spin–lattice relaxation time and T2 spin–spin relaxation time. This results in brighter and sharp image. They can modify spin–spin relaxation effects by inducing local field inhomogenicity and thus shorten T1 and T2 relaxation times [91]. To improve biocompatibility, hydrophilicity, and degradability, magnetic IONPs are surface modified with polyethylene glycol (PEG) polymers and other functional entities.

3.3.2 Hyperthermia and Chemotherapy

Hyperthermia is considered as a supporting treatment for chemotherapy, surgery, or radiotherapy. It is based on the principle of heating tumor region by generating minimal damage to healthy cells. Cancer cells are comparatively more sensitive to temperature than normal ones. When IONPs are under influence of changing external magnetic field, heat production occurs due to the magnetic hysteresis loss and Brown and Neel relaxation that destroys cancer cells. This therapy can be implemented for brain, breast, prostate, melanoma, lymph node, glioblastoma, cervical cancers. IONPs are surface modified with several biocompatible molecules (dextran, polyethylene glycol, folate, etc.) to achieve specific targeting and destruction. Applications of external magnetic field and temperature for specific period of time promote the destruction of targeted cells. Outcome depends upon IONP particle size and their distribution, susceptibility of magnet, temperature control, and immune response activation. Magnetic guided drug targeting provides efficacy, reduces unwanted side effects, and therefore establishes potential of IONPs in the diagnosis and treatments of cancer. Major technical issue in hyperthermia is the damage causes to healthy tissues near tumor region. Nanomaterial benefits spatial absorption by cancer cells and inhibits specific target. Hence, scientists are emphasizing on combination therapy consisting of magnetic hyperthermia and chemotherapy. It may conquer individual

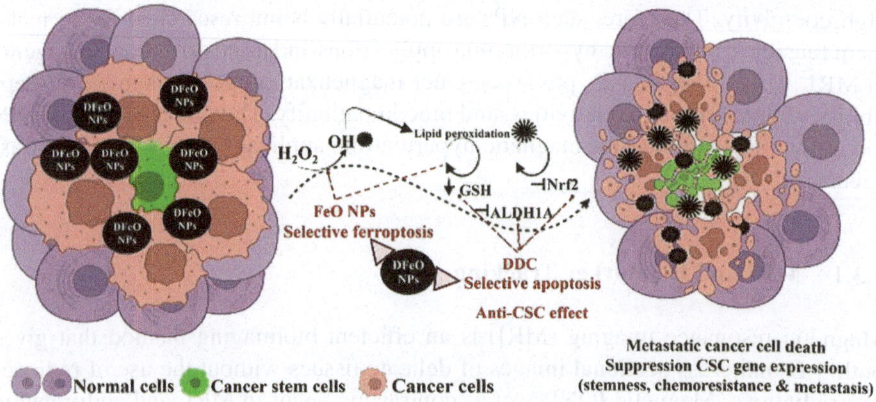

Fig. 2 Working mechanism of iron oxide NPs against CSCs. Reused with due permission from [92]

treatment problems like magnetic hyperthermia contract cancer cells and enlarges space between cells. This allows drug to spread at tumor site and demolish cancer cells. Heat generation due to IONPs triggers drug release. This enables consistent anticancer drug release by using IONPs; moreover, magnetic hyperthermia may destruct cancer cells by potentiating drug effect.

Mortality rate of metastatic breast cancer is linked to the cancer stem cells' (CSCs) aggressive features. Abu-serie et al. [92] The nanoformulations of DDC with green chemically synthesized ferrous oxide nanoparticles (FeO NPs) and ferric oxide (Fe_2O_3 NPs) were prepared. Due to nanoparticle characters and synergistic effect between iron oxide NPs and DDC, nanocomplexes (DFeO NPs and DFe_2O_3 NPs, respectively) exhibited the strongest antimetastatic cancer potency in vitro. Because of corresponding iron oxide nature, DFeO NPs demonstrated better therapeutic efficacy than DFe_4O_3 NPs, in mammary tumor liver metastasis-bearing mice, in terms of tumor size, histological analysis, immunostaining % of ki-67+ and caspase 3+, and gene expression of p53 and BCl2. The potent antitumor effect of DFeO nanocomplex is attributed to the maximum elevation of reactive oxygen species and lipid peroxidation (ferroptosis hall marker) with severe depletion of GSH and Nrf2 selectively in both tumor tissues, causing CSC eradication with halting metastatic activity. The mechanism is shown in Fig. 2.

3.3.3 Magnetofection and Gene Therapy

Magnetofection is associated with the non-viral gene vectors in the form of magnetic IONPs that transfects gene into the targeted region inside the body. Magnetic IONPs provide a novel tool for high-throughput gene delivery. Such delivered genes once reach to the target, either knock-down or knock-up the functionality of specific gene or translated protein. Such manipulations lead to the gene therapy in case of various genetic disorders and tumor destructions. Therefore, magnetic IONPs possess multifunctional approach with respect to magnetofection and gene therapy [93]. IONPs are incorporated for effective gene therapy to treat various diseases. One of the common applications of gene therapy is to insert a specific gene as a replacement of an abnormal gene. Likewise, other applications are swapping, repairing, or altering an abnormal gene for normal gene. Surface of nanomaterials must be functionalized with different components for efficient gene binding [94].

3.3.4 Nanovaccine and Immune System Activation

Furthermore, IONPs conjugated vaccines increase immunogenicity of the vaccine. It has been reported that IONPs promote immune cells' activation and thus cytokine production that induces strong humoral as well as cellular immune response. This established a potent vaccine delivery platform using nanovaccines [95]. For example, IONP-based nanovaccine has been designed as an adjuvant for Avian influenza virus (AIV) subtype H9N2 vaccine delivery and immune system activation [96].

3.3.5 Tissue Engineering

Tissue or organ replacement needs spatial and temporal regulation over biological circumstances and in situ monitoring. Targeted delivery of biomolecules like genes, growth factors, protein inhibitors, and contrast agents is crucial step in monitoring the engineered tissues. In such cases, it has been reported that nanoparticle-based scaffolds provide contrast for bioimaging as well as control for scaffold movements. As per the application, nanoparticles are synthesized and surface functionalized with different polymers, ceramics, metals, biomolecules, etc. [97]. Composites made up of combination of different biopolymers (like bovine serum albumin, chitosan, etc.), calcium phosphate (CP), and IONPs have been demonstrated as biocompatible relatively slow degraded scaffolds for in vivo tissue engineering studies [98]. It has been reported that composite of calcium phosphate and MNPs is used for osteoinduction of stem cells with potent biocompatibility. Composites accelerated biological organization of seeded stem cells that resulted into bone tissue regeneration [99]. Michelle et al. have reported the IONP-based nanoenzymes for effective tissue engineering and regenerative medicine. It has been reported that they possess high stability, catalytic action, and effective enzymatic nature that help in cardioprotection in the angiogenesis and heart-related therapies, wound healing, as well as bone tissue engineering.

3.3.6 Stem Cell Therapy and Regenerative Medicine

The current emergence of nanobiotechnology is expected to expedite the field of stem cell and regenerative medicine that covers damaged tissue or organ repair, replacement, and regeneration to restore their original function with nanoparticles. Specifically, IONPs with characteristic physicochemical properties, higher surface area, provide promising platform for advance diagnosis and therapeutic in the field of regenerative medicine. IONP-based applications cover drug delivery and release, tissue regeneration, cancer stem cells destruction, transplanted cell tracking, cell proliferation and differentiation regulation, channels and signal pathway, etc. IONPs bind to ion channels by specific and non-specific interactions to regulate in situ ion channels and manipulate the stem cell function with an externally induced magnetic field. This generates the mechanical force or heat that controls adhesion and differentiation of stem cells and intracellular tissue regeneration signaling. IONPs embedded in scaffolds tend to promote adhesion, proliferation, and growth of stem cells by regulating interactions between cell and scaffolds [100]. IONPs provide diagnostic as well as therapeutic effects when modified according to the need. Small interfering RNAs (siRNA) are considered as potent gene therapeutic agent for their gene knockout ability. However, due to their polyanionic and unstable nature, their entry inside the cells becomes difficult by passive diffusion. It has been reported that IONPs loaded with siRNA cross cell membrane and are transported to specific gene site with induced external magnetic field. It has been reported that specific biomolecule such as cancer stem cell (CSC) membrane conjugation with IONPs targets the tumor-specific site and shows potent antitumor in vivo activity [101]. IONP-mediated targeted destruction of CSCs is possible in combination with hyperthermia and chemotherapy that establishes potent cancer therapeutic platform [102, 103].

3.3.7 Drug Delivery

IONPs are one of the most popular theranostics agents in the field of nanomedicine for targeted and controlled drug delivery. IONPs are prominently being in focus as vital components in medical industry, biomedicine, and biotechnology. IONP-mediated cancer treatment was first performed in 1957. Due to small size and less cytotoxicity of IONPs, it is possible to perform in vivo studies by targeting specific sites. Stable suspensions of IONPs are in colloidal form dispersed in liquid organic and inorganic carriers. Bare IONPs possess hydrophobic surfaces containing high surface area to volume proportion, particle agglomeration, large cluster formation, and larger particle size. These properties limit superparamagnetic characteristics and initiate opsonization. Hence, surface functionalization is necessary for better activity of IONPs. Polymers such as polyethylene imide (PEI), polyethyleneglycol (PEG), poly-L-lysine (PLL) are coated on surfaces of IONPs to change surface of nanomaterial and thereby exist a new multifunctional IONP system [104, 105]. Magnetite provides comparatively better biocompatibility. Surface functionalization not only stabilizes IONP suspension but also lessens remnant magnetization. Functionalization provides biostability, biodegradability, and non-toxicity [106]. Natural and synthetic polymer-coated IONPs have biomedical drug delivery functions. Therefore, magnetic IONPs tend to deliver drugs at specific targeted region and destruct cell functioning by killing or inhibiting them. This provides the multifunctionality to the IONPs with synergistic biomedical potential.

3.4 Zirconium Nanoparticles

3.4.1 Antimicrobial and Antioxidant

In a recent work, scientist has synthesized a green and eco-friendly Zr NPs (size 20–60 nm) from fruit peels of *Punica granatum* (pomegranate). Antimicrobial activity of synthesized Zr NPs was tested against gram-positive strains (*Bacillus subtilis, Staphylococcus aureus*), gram-negative strains (*Escherichia coli, Klebsiella pneumoniae*), and fungi (*Aspergillus niger*) by agar well diffusion method. Zr NPs showed maximum zone of inhibition against *S. aureus* (19 mm) and *A. niger* (18 mm) at the maximum concentration of 200 μg/mL. The antioxidant scavenging activity of obtained Zr NPs was analyzed using DPPH radical scavenging activity, hydroxyl radical scavenging activity, ferric reducing antioxidant power, and hydrogen peroxide radical scavenging activity. Considering the tolerance of zirconium toward human body, it can also be used as antimicrobial coating material on human implants [107, 108].

3.4.2 Bone Tissue Engineering

To satisfy the stressed zone criterion, biomaterial scaffolds must have sufficient mechanical strength [109]. Additionally, in order to enable the development of osteoblasts, vasculature, and new bone, these scaffolds must possess specific osteoinductivity and cytocompatibility qualities as well as the connectivity structure. Through the replication process, researchers have created porous nano-ZrO_2 scaffolds [110]. Various materials are available for the functionalization of zirconium dioxide utilized in bone and tissue applications. The investigation's main goal is to use various functionalization techniques to promote the regeneration of broken bone or tissue, which will substantially impact the usage of ceramic materials in biomedical applications [111].

3.4.3 Bone Resorption

Wear debris and periprosthetic osteolysis of medical implants have a significant role in loosening prosthetic implants. Zirconium dioxide is typically used to manage debris and inflammatory reactions to wear products. Comparing osteoblast, fibroblast, and macrophage cell proliferation to CoCrMo alloy and zirconia particles perform better [112]. In osteoarthritis and rheumatoid synovial cells, Liagre et al. investigated the effects of zirconia or aluminum particles on the production of proinflammatory interleukin and the metabolism of arachidonic acid. In this study, Ramaswamy et al. investigated the attachment of human osteoclasts grown in vitro to resorbable bioceramics with various surface properties and chemical compositions to affect osteoclast resorption and the creation of resorption lacunae. Specifically, the researchers were interested in how these factors could affect the resorption of bone. Using a mouse coculture with osteoblast-like cells on bone slices, Sabokbar et al. showed how the addition of zirconium dioxide nanoparticles increased tartrate-resistant acid phosphate (TRAP) expression and bone resorption in the system [113].

In the same order, in another research, Imran et al. [114] have used the sol–gel method to synthesize of zirconia NPs as shown in Fig. 3. The $ZrOCl_2{\cdot}8H_2O$ is utilized as a precursor material, Fe_3O_4 NPs as a stabilizer, and deionized water as a solvent to synthesize Zr NPs. The Fe_3O_4 NPs prepared by sol–gel method stabilizes the Zr NPs. A maximum scavenging antioxidant activity and a weak hemolytic activity were perceived in the optimized parameters. Therefore, for conclusion, it states that the optimized NPs, i.e., tetragonal Zr NPs stabilized using basic Fe_3O_4 in 6 wt% amount, can be a potential material for therapeutical and pharmaceutical uses.

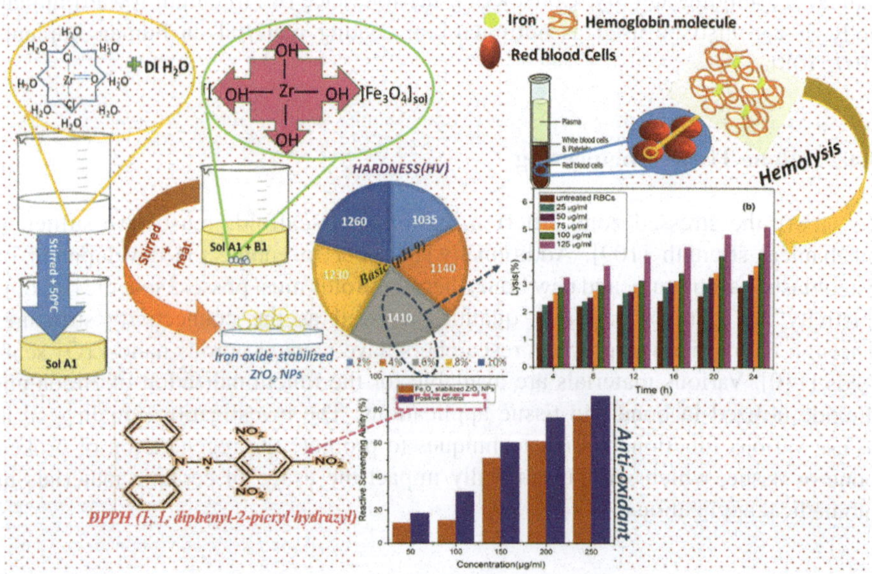

Fig. 3 Sol–gel route for the synthesis of zirconia NPs and their properties. Reused with permission from [114]

3.5 Titanium Oxide Nanoparticles

The metal oxide-based NPs, i.e., TiO_2 have a great impact on advancements in the nanotechnology. A number of research papers have been reported in past several years being focused onto metal oxide NPs due to their greater biological, chemical, as well as magnetic features such as non-toxicity, chemical stability, high saturation magnetization, excellent biocompatibility, as well as greater magnetic susceptibility. Due to such features, metal oxide NPs are being used in several biomedical applications, i.e., bioimaging, hyperthermia, drug delivery, cell labeling, and gene delivery. TiO_2 is one of the widely studied materials, and it is being used in several photocatalytic and photovoltaic devices. TiO_2 NPs are also used in food coloring, paint, cosmetics, and toothpaste. The TiO_2 NPs possess several characteristic features, i.e., chemical stability, biocompatibility, as well as optical properties. Due to above mentioned unique features, TiO_2 NPs have a great potential to be used for biomedical uses, i.e., bioimaging, photoablation therapy, drug delivery, and biosensors.

3.5.1 Protein Adsorption

Recently, this method was adopted by Pantaroto et al. [115] to synthesize various crystalline phases in sputtered TiO_2 films as shown in Fig. 4. These phases were custom-made to regulate their morphology as well as electrochemical characteristics, apatite layer formation on TiO_2 implant material, and protein adsorption. The deposition parameters of TiO_2 crystalline phases, anatase and rutile, are recognized. Thereafter, these phases were grown onto the commercially pure titanium (cpTi) using magnetron sputtering in order to prepare several groups, namely A-TiO_2 (anatase), M-TiO_2 (anatase and rutile mixture), and R-TiO_2 (rutile). The results exhibited that M-TiO_2 sample has reflected normal cell adhesion and morphology. It has been found that the amalgamation of anatase and rutile structures in order to develop TiO_2 layers is an encouraging approach to advance the properties of biomedical implants such as higher protein adsorption, greater corrosion protection, non-cytotoxicity effect, and bioactivity.

3.5.2 Antimicrobial

In the same manner, Song et al. [116] have used co-sputtering method to develop TiO_2 thin films consisting Ag NPs on commercially pure titanium substrates in order to maximize the bactericidal property along with the sustained biocompatibility. Correlation between the dispersion of Ag NPs throughout the film and their antibacterial efficiency was analyzed. The outcomes revealed that two major factors affect the bacterial inhibition addition to the surface of sample: surface structure and release kinetics of Ag ions. Furthermore, the MTT assay findings showed that no cytotoxicity toward fibroblast cells for any of the group was found. Therefore, magnetron

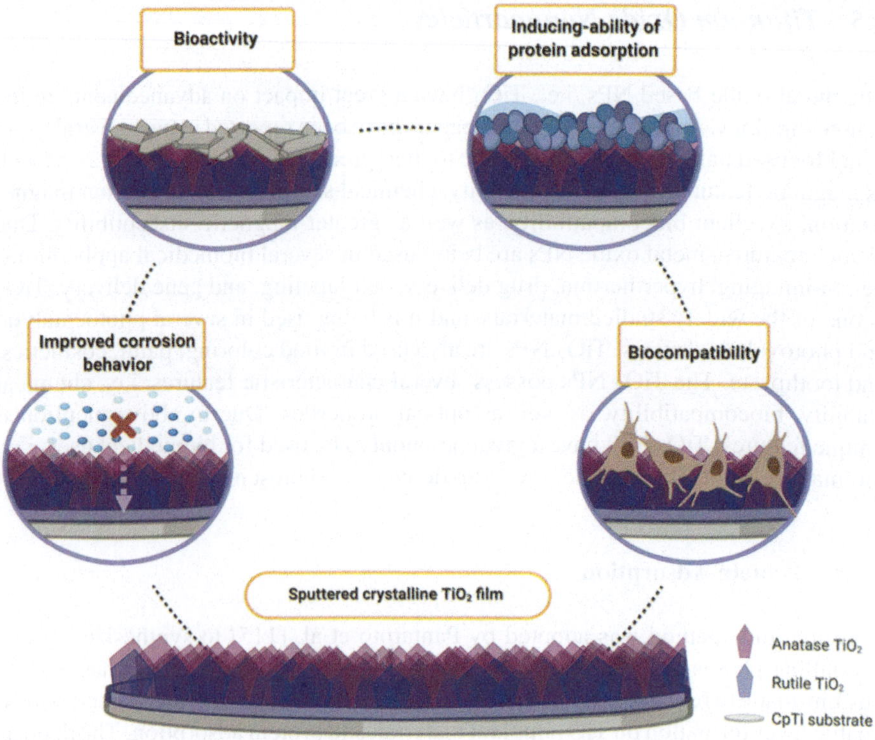

Fig. 4 Proposed mechanism of sputtered bi-phasic rutile and anatase TiO_2 film on the electrochemical and biological behavior of cpTi substrate. Reused with due permission from [115]

sputtered Ag NPs coated with TiO_2 are potential candidate as antibacterial agents along with sustained biocompatibility.

3.5.3 Implants

A method was developed to coat the nanomembrane using SiC–TiO_2–graphene with varying percentages of *Azadirachta indica* (neem) extract. The nanomembranes have been synthesized by electrospinning machine over aluminum foil paper using the raw materials PVA grain, SiC, TiO_2, graphene, and neem. No toxic effects were observed from the membrane during the cytotoxicity test. The concept of this type of nanofibrous–nanofiller membrane can be used in dental implants to mitigate the complication of stress shielding and implant-related infections and inadequate osseointegration problem [117]. The surface of Ti6Al4V alloy widely used for the fabrication of implants was modified by incorporating a novel biocompatible and antibacterial coating composed of multiple layers of hydroxyapatite (HA) and Ag deposited on a TiO_2/TiN/Ti trilayer. In vitro biological response was evaluated by cell viability tests with mouse mesenchymal stem cells, while bacterial viability was evaluated with *Staphylococcus aureus* strain. Meanwhile, there was a relief in the residual stresses of >50%, which generated increases of up to 74% in the adhesion of the coatings to the substrate. The biological tests showed that biocompatible coatings with high antibacterial potential were obtained. Such material can be used for implant applications [118].

3.6 CeO_2 Nanoparticles

Another metal oxide NPs' material that is gaining the attention for being used in biomedical stream is cerium oxide (CeO_2) NPs, also known as nanoceria. The nanoceria possesses a specific feature and that is its ability to get switched between various oxidation states. Generally, CeO_2 NPs have defects on the surface which act as oxygen vacancies. Thus, CeO_2 NPs have facility to shift the oxidation number between +4 and +3 and these oxidation states may also coexist. This property of CeO_2 NPs has taken them in the focus of research, especially as a possible biological antioxidant. CeO_2 NPs are being used in several biomedical streams including biosensor and cancer treatment. Recently, simple sol–gel combustion method was used by Elayakumar et al. [119] to prepare spinel copper ferrite ($CuFe_2O_4$) NPs doped by various amount of rare earth element cerium (Ce^{3+}) ions. The Ce^{3+} ions were doped within spinel lattice of $CuFe_2O_4$ NPs without further distortion. Magnetic behavior of Ce^{3+}-doped $CuFe_2O_4$ NPs was analyzed by VSM method at room temperature. Magnetic properties were changed after incorporation of Ce^{3+} ions into $CoFe_2O_4$ lattice. In application point of view, antibacterial potential of Ce^{3+}-doped $CuFe_2O_4$ NPs was examined and found to be increased with increase in Ce^{3+} amount.

Unarguably, the most important and highly explored property of CNPs is their inherent ability to act as antioxidant in an environment-dependent manner. The case is further strengthened by possession of "redox switch" to confer them auto-regenerating capabilities by automatically shifting between Ce^{4+} and Ce^{3+} oxidation states [120]. The nanoparticle form of cerium oxide is superior as compared to the bulk cerium because it can alter the ratio of different oxidation states in its nanoparticle form. In the nanoparticle form, cerium oxide exists as a mixture of Ce^{3+} and Ce^{4+} on the surface of CNPs. Further, with a decrease in the diameter of CNPs, there is an increase in number of surface oxygen vacancies and Ce^{3+}. To offset the oxygen loss from surface, the oxidation state of two atoms changes from Ce^{4+} to Ce^{3+} [121].

3.6.1 Cancer Treatment

The basal levels of ROS are already higher in cancer cells, and a further increase in ROS by CNPs due to their oxidant properties in acidic pH results in crossing of the threshold levels causing cellular death. Although a number of different redox active nanomaterials are useful to combat cancer, the uniqueness of CNPs lies with their ability to regenerate themselves, reduce the levels of increased ROS without affecting basal levels in normal cells and selective pro-oxidant activity in case of cancer cells ultimately leading to ROS-induced cell death in cancer cells. The pro-oxidant effects of CNPs selectively in cancer cells were demonstrated by Alili and associates, where they showed that CNPs induced cytotoxicity in melanoma cells by causing a two-fold increase in the intracellular ROS levels, without affecting ROS levels in fibroblasts [122]. The similar results were obtained by Pesic et al., on different cancer cell lines including adenocarcinoma DLD1, NCI-H460, 518A2, HT-29 and normal cells such as lung fetal fibroblasts, MRC-5, and keratinocytes, HaCaT cells [123].

However, CNPs modified with different ligands were also tried to improve targeting capabilities in cancer. Hijaz et al. used folic acid tagged CNPs to induce cytotoxicity in ovarian cancer [124]. Similarly, Sulthana et al. also used folate conjugated CNPs to actively target the NSCLC and used a combination of doxorubicin (DOX) and Hsp90 inhibitor, Ganetespib, to achieve significant cellular death [125]. Further, Vassie and group also decorated CNPs with folic acid to provide active targeting of folate receptor expressing ovarian cells [126]. The surface modification of CNPs with folic acid increased the uptake of CNPs by folate receptors which resulted in induced ROS to a greater extent in ovarian cells, in contrast to, colon carcinoma cells where folic acid-tagged CNPs actually reduced the ROS generation. Fernandez-Varo et al. demonstrated the efficacy of CNPs in diethylnitrosamine-induced hepatocellular carcinoma (HCC), where CNPs administration was associated with cellular apoptosis, decreased macrophage infiltration, reduced inflammatory gene overexpression, and increased survival in rats [127]. A schematic presentation of cellular environment-dependent activity of CNPs is shown in Fig. 5.

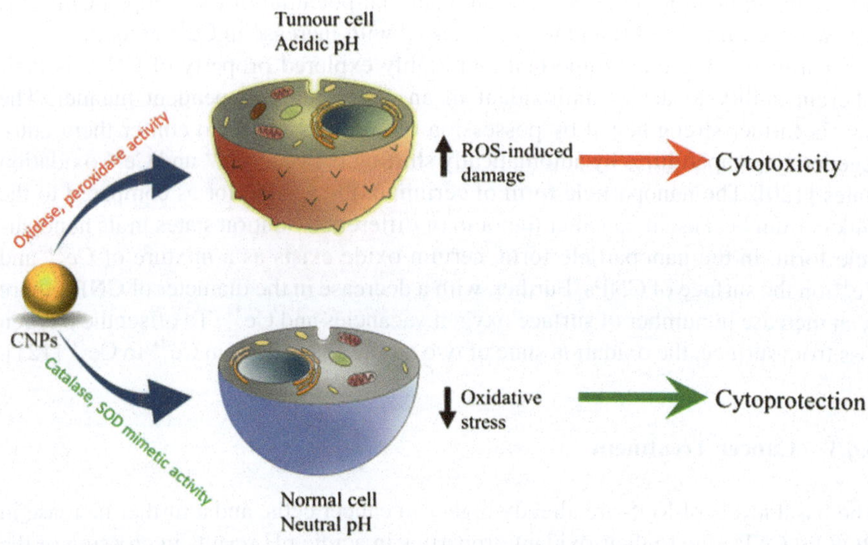

Fig. 5 Cellular environment-dependent activity of CNP. Reused with due permission from [128]

3.6.2 Wound Healing

The improvement in wound healing by CNPs could be well attributed to proliferation enhancing and cytoprotective activities on different cells such as fibroblasts and keratinocytes. Chigurupati and associates showed accelerated effects of topical application of water-soluble CNPs on cutaneous wound healing by accelerating migration and proliferation of skin keratinocytes and fibroblasts along with an enhanced tube formation by vascular endothelial cells (VECs). The findings were also supported by in vivo results, where CNPs accelerated wound healing in C57BL/6 mice and showed an increase in the immunoreactivity for α-smooth muscle actin (α-SMA), a marker for differentiated myofibroblasts [129]. Further, CNPs were reported to enhance the proliferation of embryonic fibroblasts in vitro by ROS reduction and increasing the expression of antioxidant genes. However, the authors also observed that high concentrations of CNPs significantly reduced ROS below the physiological levels. Lord and group coated hyaluronan on CNPs to target CD44 and ROS in human fibroblast and reported that HA-conjugated CNPs showed better cellular uptake, improved ROS reduction, and induced autophagy in the cells [130]. Another report demonstrated that CNPs preserved mitochondrial functions and improved the energy balance in basal as well as pro-oxidant conditions in primary cultured skin fibroblasts [131]. In addition, Naseri-Nosar et al. incorporated CNPs on the poly(ε-caprolactone)/gelatin films by electrospinning method to develop wound dressing material with enhanced wound closure activity [132]. Zgheib et al. conjugated CNPs with miRNA-146a, a "molecular break" for the inflammatory processes, and found that conjugated CNPs accelerated wound healing without inducing any pathologic abnormalities post healing. Further, it also decreased inflammation and increased angiogenesis for improved wound healing [133].

3.6.3 Antimicrobial

Babu et al. investigated antibacterial activity of gold-supported CNPs against different bacterial cultures including *E. coli*, *S. enteritidis*, *B. subtilis*, and *S. aureus* monoculture, though the inhibitory effect was found to be moderate modest on different bacteria [134]. Kasinathan et al. synthesized photocatalytically active cerium-doped TiO_2 nanocomposite using hydrothermal method and reported potent activity against gram-positive bacteria, *E. coli* at 10 μg/ml level [135]. Further, Alpaslan et al. showed antibacterial activity of CNPs against gram-positive as well as gram-negative bacteria depending on the pH as the CNPs were able to show much more effectiveness at basic pH of 9 as compared to acidic pH [136]. Mohamed and associates used a green method to synthesize CNPs using fruits of *Hyphaene thebaica* and found them to possess excellent biological properties including antioxidant, antimicrobial, and antiviral [137].

3.7 *Zinc Oxide Nanoparticles*

ZnO NPs are another metal oxide NPs being used for bioimaging and drug delivery. Though the surface of ZnO NPs is required to be altered in order to protect because these NPs are dissolvable in water and acidic solutions. ZnO [138–140], ZnO tetrapod materials [141, 142], etc., are being explored since a long duration of time for medical uses. Recently, El-Gendy et al. [143] have synthesized and characterized ZnO NPs with the help of laser ablation at several ablation times. By applying laser ablation method, it has been found that the synthesized ZnO NPs were found to be spherical shaped having the average size as 10.2 and 9.8 nm using ablation times as 20 and 30 min, respectively. In order to know their antimicrobial properties, synthesized ZnO NPs were subjected against two ophthalmological bacteria: one was *Pseudomonas aeruginosa* and second was methicillin-resistant *S. aureus*. A substantial decline was observed in bacterial growth when treated with synthesized ZnO NPs. Furthermore, the obtained ZnO NPs were analyzed for their antioxidant property as well as biocompatibility for the retinal epithelial cells and were found reasonable antioxidant too. However, the synthesized ZnO NPs using 20-min ablation time were found to be safer and most biocompatible as compared to synthesized using 30-min ablation time.

3.7.1 Antimicrobial

ZnO NPs can be selected as an antibacterial material because of its superior properties, such as high specific surface area and high activity to block a wide scope of pathogenic agents. But recently, the antibacterial activity of ZnO NPs is still scarcely known. Prior reports had suggested that the main antibacterial toxicity mechanisms of ZnO NPs were based on their ability to induce excess ROS generation, such as superoxide anion, hydroxyl radicals, and hydrogen peroxide production [144]. The antibacterial activity may involve the accumulation of ZnO NPs in the outer membrane or cytoplasm of bacterial cells and trigger Zn^{2+} release, which would cause bacterial cell membrane disintegration, membrane protein damage, and genomic instability, resulting in the death of bacterial cells [145, 146].

Epidemic disease *cholera*, a serious diarrheal disease caused by the intestinal infection of gram-negative bacterium *V. cholera*, mainly affects populations in the developing countries [147, 148]. Aiming at the development of nanomedicine against *cholera*, Sarwar et al. carried out a detailed study about ZnO NPs against *Vibrio cholerae* (two biotypes of cholera bacteria (classical and El Tor)). ZnO NPs were observed to be more effective in hindering the growth of El Tor (N16961) biotype of *V. cholera*, which was closely associated with ROS production. These results would damage bacterial membrane, increase permeabilization, and substantially modify their morphology [149]. They also detected the antibacterial activity of the ZnO NPs in cholera toxin (CT) mouse models. It was found that ZnO

NPs induce the collapsing of secondary structures by interrupting the CT binding with GM1 anglioside receptor [150].

3.7.2 Bioimaging

ZnO NPs exhibit efficient blue emissions and near-UV emissions, which have green or yellow luminescence related to oxygen vacancies, therefore further extending its application into bioimaging field [151]. Using a simple sol–gel method, Xiong et al. prepared stable aqueous ZnO@polymer core–shell nanoparticles (ZnO@poly(MAA-co-PEGMEMA)) for the first time. The ZnO@polymer core–shell nanoparticles exhibited high quantum yield and very stable broad photoluminescence in aqueous solutions. As shown in Fig. 3, within human hepatoma cells, ZnO-1 (derived from LiOH) with an average size of 3 nm showed green fluorescence, while ZnO-2 (derived from NaOH) with an average size of 4 nm appeared yellow. It was worth to note that these nanoparticles did not show any remarkable toxicity for human hepatoma cells when their concentrations were less than 0.2 mg/mL. Furthermore, the luminescence was very stable during cell culturing, and the cells were alive at 45 min of exposure. So, as a type of safe and cheap luminescent labels, the ZnO@polymer core–shell nanoparticles can be used as fluorescent probes for cell imaging in vitro [152].

3.8 Bimetallic Nanoparticles

The bimetallic NPs, iron cobalt (Fe–Co), possess excellent magnetic features, i.e., elevated Curie temperature, superparamagnetism property, as well as greater saturation magnetization. But, their ability is to being get oxidize easily and their biocompatibility limits their applications. In order to resolve this issue, a coating of some biocompatible material over Fe–Co NPs is required. Since Fe–Co NPs have high saturation magnetization, thus are used as magnetic hyperthermia as well as contrast agents for the MRI. Further, iron nickel (Fe–Ni) NPs too show high saturation magnetism feature along with elevated Curie temperature and thus are primarily utilized in MRI as contrast agents. Thereafter, the copper nickel (Cu–Ni) NPs exhibit magnetic features and elevated Curie temperatures and thus are chiefly being used in magnetic hyperthermia. Iron platinum (Fe–Pt) NPs own characteristic magnetic and chemical properties, i.e., elevated Curie temperature, high X-ray absorption, chemical stability, superparamagnetization, and greater saturation magnetization. Such unique features allow these NPs to be used in MRI, hyperthermia treatment, biosensors, and drug delivery. Apart from these, Ag–Fe_3O_4 [153], Au–Ni [154], Ag–Au [155–157], Pd–Pt [158], Ag–Pd core-shell [159], Ag–Ni [160], PEG–Fe–Au alloy [161], Ag–Pt-modified TiO_2 [162], (Pd–Pt)/SiO_2 [163], Pt–Co [164], Ni–Pd@MIL-101 [165] are biomedical relevant bimetallic NPs. Bimetallic NPs such as iron cobalt

(Fe–Co) [166–168] as well as iron platinum (Fe–Pt) [169, 170] are also very much clinically relevant and are being used since a long time.

4 Conclusion

Synthesis of nanoparticles using several known methods fulfills maximum demand of the nanomaterial to be used in biomedical sector. However, each method is most suitable to some of the specific type of nanoparticles and also depends upon the end-use applications. The progress in nanotechnology is facilitating us to resolve the obstacles faced in our basis routine life. However, numerous approaches are still at the clinical stage or pilot stage, and these methods are required to be established at the ground level. The healthcare workers including doctors, nurses, and scientists get blessed with nanotechnology for the specific drug target delivery, healthier diagnosis, and the utilization of requirement-based compatible implants. Various nanomaterials assist the different health associated dimensions like treatment, therapy, surgery, implants, tissue engineering, etc. In summary, nanoparticles have great potential to offer numerous multimodal diagnostic stages and therapeutic uses for human health. The efforts are presently in the process to interpret such scientific advancements for the clinic uses, and definitely, it will bring new paradigm in the biomedicine world [171].

5 Challenges and Future Aspects

Synthesis of nanoparticles has several challenges, especially, as far as their end-use applications are considered. The challenges for nanotechnology in biomedical applications are added on daily basis. New and new NPs are being discovered to counter novel diseases. However, there are some basic challenges those need to be effectively resolved prior to the remarkable prospective of nanotechnology for the biomedical uses which is completely comprehended. Though unique characteristics of various NPs are used for several medical science applications, NPs should also be scrutinized for their possible potential dangerous side effects to ensure the health safety. However, the agglomeration of nanotechnology with medicine domain gives hope to improve the human health, but their toxicological aspects should also be analyzed thorough in vitro and in vivo manner. The main challenge in diagnostics is to make affordable and robust point-of-care set of systems to detect human diseases at their early stage with highest sensitivity and specificity. In current time, cancer diagnosis and treatment are in main focus. However, one of the major challenges is to spread the advancements and technologies to the mass of people to aware them about infectious diseases those are considered to be the biggest killers. Preferably, the in vitro diagnostic platforms with the help of nanotechnology should be developed for communicable diseases. In addition to that, as far as the circular economy concept is considered, the cost of raw material and processing, their availability, their sustainability toward end-use, as well as recyclability is must to take under consideration [172–176].

References

1. Roco MC (2011) The long view of nanotechnology development: the National Nanotechnology Initiative at 10 years. Nanotechnol Res Dir Societal Needs. Springer, Dordrecht, 1–28
2. Kumar V, Singh A (2013) Polypropylene clay nanocomposites. Rev Chem Eng 29(6):439–448
3. Gambhir RP et al (2022) Multifunctional surface functionalized magnetic iron oxide nanoparticles for biomedical applications: a review. Appl Surf Sci Adv 11:100303
4. Bokka S, Chowdhury A (2022) Evolving trends of nanotechnology for medical and biomedical applications: a review. Encycl Mater Plast Polym 4:734–745
5. Kumar P et al (2015) Incorporation of nano-Al_2O_3 within the blend of sulfonated-PVdF-co-HFP and Nafion for high temperature application in DMFCs. RSC Adv 5(78):63465–63472
6. Singh A et al (2017) Extrusion and evaluation of chitosan assisted AgNPs immobilized film derived from waste polyethylene terephthalate for food packaging applications. J Packag Technol Res 1(3):165–180
7. Singh A et al (2018) Microbial, physicochemical, and sensory analyses-based shelf life appraisal of white fresh cheese packaged into PET waste-based active packaging film. J Packag Technol Res 2(2):125–147
8. Pattanayak P et al (2021) Performance evaluation of poly (aniline-co-pyrrole) wrapped titanium dioxide nanocomposite as an air-cathode catalyst material for microbial fuel cell. Mater Sci Eng, C 118:111492
9. Mertens H et al (2006) Polarization-selective plasmon-enhanced silicon quantum-dot luminescence. Nano Lett 6(11):2622–2625
10. Erik DC et al (2012) Size matters: gold nanoparticles in targeted cancer drug delivery. Ther Deliv 3(4):457–478
11. Thomsen LB et al (2015) Targeted drug delivery to the brain using magnetic nanoparticles. Ther Deliv 6(10):1145–1155
12. Sharma HS et al (2007) Drug delivery to the spinal cord tagged with nanowire enhances neuroprotective efficacy and functional recovery following trauma to the rat spinal cord. Ann N Y Acad Sci 1122(1):197–218
13. Williams D (2006) Quantum dots in medical technology. Med Dev Technol 17(4):8–9
14. Fang J (2006) Nano-or submicron-sized liposomes as carriers for drug delivery. Chang Gung Med J 29(4):358
15. Jain S et al (2012) Gold nanoparticles as novel agents for cancer therapy. Br J Radiol 85(1010):101–113
16. Wang Y et al (2013) Functionalized quantum dots for biosensing and bioimaging and concerns on toxicity. ACS Appl Mater Interfaces 5(8):2786–2799
17. Qiao R et al (2012) Receptor-mediated delivery of magnetic nanoparticles across the blood–brain barrier. ACS Nano 6(4):3304–3310
18. Fan K et al (2012) Magnetoferritin nanoparticles for targeting and visualizing tumour tissues. Nat Nanotechnol 7(7):459–464
19. Wang T et al (2019) Bioengineered magnetoferritin nanozymes for pathological identification of high-risk and ruptured atherosclerotic plaques in humans. Nano Res 12(4):863–868
20. Shojaei M et al (2021) Synthesis and characterization of $CuAlS_2$ nanoparticles by mechanical milling. Mater Today Commun 27:102243

21. Velásquez AA, Urquijo JP (2021) Synthesis and characterization of magnetite-maghemite nanoparticles in presence of polyethylene glycol obtained by mechanical milling. Mater Sci Eng, B 263:114873
22. Jamkhande PG et al (2019) Metal nanoparticles synthesis: an overview on methods of preparation, advantages and disadvantages, and applications. J Drug Deliv Sci Technol 53:101174
23. Piszczek P, Radtke A (2018) Silver nanoparticles fabricated using chemical vapor deposition and atomic layer deposition techniques: properties, applications and perspectives: review. Noble Precious Metals; Seehra, MS, Bristow, AD, Eds, 187–213
24. Owens GJ et al (2016) Sol–gel based materials for biomedical applications. Prog Mater Sci 77:1–79
25. Santra TS et al (2014) Biosynthesis of silver and gold nanoparticles for potential biomedical applications—a brief review. J Nanopharmaceutics Drug Deliv 2(4):249–265
26. Puja P, Kumar P (2019) A perspective on biogenic synthesis of platinum nanoparticles and their biomedical applications. Spectrochim Acta Part A Mol Biomol Spectrosc 211:94–99
27. Mohamed HI et al (2023) Endophytic fungi-derived biogenic nanoparticles: mechanisms and applications. In: Fungal cell factories for sustainable nanomaterials productions and agricultural applications. Elsevier, pp 361–391
28. Wang Y et al (2022) Genetically engineered bacteria-mediated multi-functional nanoparticles for synergistic tumor-targeting therapy. Acta Biomater 150:337–352
29. Chen C et al (2020) Bifidobacterium-mediated high-intensity focused ultrasound for solid tumor therapy: comparison of two nanoparticle delivery methods. Int J Hyperth 37(1):870–878
30. Iravani S (2011) Green synthesis of metal nanoparticles using plants. Green Chem 13(10):2638–2650
31. Cacho-Bailo F et al (2018) Microfluidic synthesis of MOFs and MOF-based membranes. Microfluid Fundament Dev Appl Fundament Appl 479–515
32. Hao N et al (2018) Microfluidic synthesis of functional inorganic micro-/nanoparticles and applications in biomedical engineering. Int Mater Rev 63(8):461–487
33. Illath K et al (2020) Microfluidics-based metallic nanoparticle synthesis and applications. In: Microfluidics and bio-MEMS: devices and applications. Jenny Stanford Publishing, pp 429–501
34. Kumar A (2020) Microfluidic technologies for cell manipulation, therapeutics, and analysis. In: Microfluidics and bio-MEMS: devices and applications. Jenny Stanford Publishing, pp 1–48
35. Santra TS et al (2020) Near-infrared nanosecond-pulsed laser-activated highly efficient intracellular delivery mediated by nano-corrugated mushroom-shaped gold-coated polystyrene nanoparticles. Nanoscale 12(22):12057–12067
36. Huang H et al (2018) Continuous flow synthesis of ultrasmall gold nanoparticles in a microreactor using trisodium citrate and their SERS performance. Chem Eng Sci 189:422–430
37. Shinde P et al (2018) Current trends of microfluidic single-cell technologies. Int J Mol Sci 19(10):3143
38. Nozdriukhin D et al (2021) Gold nanoparticle-carbon nanotube multilayers on silica microspheres: Optoacoustic-Raman enhancement and potential biomedical applications. Mater Sci Eng C 120:111736
39. Dheyab MA et al (2020) Synthesis and coating methods of biocompatible iron oxide/gold nanoparticle and nanocomposite for biomedical applications. Chin J Phys 64:305–325
40. Kuznetsova OV et al (2020) An ICP-MS-based assay for characterization of gold nanoparticles with potential biomedical use. Anal Biochem 611:114003
41. Krug P et al (2019) Polypyrrole microcapsules loaded with gold nanoparticles: perspectives for biomedical imaging. Synth Met 248:27–34
42. Manikandakrishnan M et al (2019) Facile green route synthesis of gold nanoparticles using Caulerpa racemosa for biomedical applications. J Drug Deliv Sci Technol 54:101345
43. Jin J et al (2018) Rapid in situ biosynthesis of gold nanoparticles in living platelets for multimodal biomedical imaging. Colloids Surf B 163:385–393

44. Venditti I et al (2015) Bioconjugation of gold-polymer core–shell nanoparticles with bovine serum amine oxidase for biomedical applications. Colloids Surf, B 134:314–321
45. Clarance P et al (2020) Green synthesis and characterization of gold nanoparticles using endophytic fungi Fusarium solani and its in-vitro anticancer and biomedical applications. Saudi J Biol Sci 27(2):706–712
46. Patra CR et al (2008) Targeted delivery of gemcitabine to pancreatic adenocarcinoma using cetuximab as a targeting agent. Can Res 68(6):1970–1978
47. Bhattacharya R et al (2007) Gold nanoparticles inhibit the proliferation of multiple myeloma cells. Adv Mater 19(5):711–716
48. Patra CR et al (2010) Fabrication of gold nanoparticles for targeted therapy in pancreatic cancer. Adv Drug Deliv Rev 62(3):346–361
49. Bertok T et al (2013) Label-free detection of glycoproteins by the lectin biosensor down to attomolar level using gold nanoparticles. Talanta 108:11–18
50. Wang J et al (2015) An ultrasensitive supersandwich electrochemical DNA biosensor based on gold nanoparticles decorated reduced graphene oxide. Anal Biochem 469:71–75
51. Azzouzi et al (2015) A novel amperometric biosensor based on gold nanoparticles anchored on reduced graphene oxide for sensitive detection of l-lactate tumor biomarker. Biosens Bioelectron 69:280–286
52. Sun I et al (2014) Biocompatible glycol chitosan-coated gold nanoparticles for tumor-targeting CT imaging. Pharm Res 31(6):1418–1425
53. Meir R et al (2015) Nanomedicine for cancer immunotherapy: tracking cancer-specific T-cells in vivo with gold nanoparticles and CT imaging. ACS Nano 9(6):6363–6372
54. Jung S et al (2013) Theragnostic pH-sensitive gold nanoparticles for the selective surface enhanced Raman scattering and photothermal cancer therapy. Anal Chem 85(16):7674–7681
55. Nam J et al (2009) pH-induced aggregation of gold nanoparticles for photothermal cancer therapy. J Am Chem Soc 131(38):13639–13645
56. Rengan AK et al (2015) In vivo analysis of biodegradable liposome gold nanoparticles as efficient agents for photothermal therapy of cancer. Nano Lett 15(2):842–848
57. Rengan AK et al (2014) Multifunctional gold coated thermo-sensitive liposomes for multimodal imaging and photo-thermal therapy of breast cancer cells. Nanoscale 6(2):916–923
58. Wang F et al (2011) Doxorubicin-tethered responsive gold nanoparticles facilitate intracellular drug delivery for overcoming multidrug resistance in cancer cells. ACS Nano 5(5):3679–3692
59. Ajnai G et al (2014) Trends of gold nanoparticle-based drug delivery system in cancer therapy. J Exp Clin Med 6(6):172–178
60. Elbialy NS et al (2015) Doxorubicin loaded magnetic gold nanoparticles for in vivo targeted drug delivery. Int J Pharm 490(1–2):190–199
61. Abdulateef SA et al (2022) Rapid synthesis of bovine serum albumin-conjugated gold nanoparticles using pulsed laser ablation and their anticancer activity on hela cells. Arab J Chem 104395
62. Cutroneo M et al (2022) Porous polydimethylsiloxane composite filled with graphene oxide and gold nanoparticles produced by laser ablation in liquids. Vacuum 199:110951
63. Malik M et al (2022) Biosynthesis of silver nanoparticles for biomedical applications: a mini review. Inorgan Chem Commun 109980
64. Garg R et al (2022) Biomedical and catalytic applications of agri-based biosynthesized silver nanoparticles. Environ Pollut 119830
65. Maddalozzo AED et al (2022) Development and characterization of natural rubber latex wound dressings enriched with hydroxyapatite and silver nanoparticles for biomedical uses. React Funct Polym 105316
66. Subha V et al (2022) Functionalization of spray coated cellulose nanofiber sheet with montmorillonite (MMT) and silver nanoparticles (AgNPs) to biomedical nanocomposite as wound regeneration scaffold. Prog Org Coat 166:106782
67. Suriyakala G et al (2021) Plumeria pudica Jacq. flower extract-mediated silver nanoparticles: characterization and evaluation of biomedical applications. Inorgan Chem Commun 126:108470

68. Narciso AM et al (2021) Antimicrobial green silver nanoparticles in bone grafts functionalization for biomedical applications. Biocatal Agric Biotechnol 35:102074
69. Ramalingam V (2022) Silver nanoparticles for biomedical applications. In: Nanoparticle therapeutics. Academic Press, pp 359–375
70. Gaviria J et al (2021) Synthesis and deposition of silver nanoparticles on porous titanium substrates for biomedical applications. Surf Coat Technol 406:126667
71. Deka R (2020) Highly stable silver nanoparticles containing guar gum modified dual network hydrogel for catalytic and biomedical applications. Carbohyd Polym 248:116786
72. Sanjivkumar M et al (2019) Investigation on characterization and biomedical properties of silver nanoparticles synthesized by an actinobacterium Streptomyces olivaceus (MSU3). Biocatal Agric Biotechnol 17:151–159
73. Vale AC et al (2019) Optimization of silver-containing bioglass nanoparticles envisaging biomedical applications. Mater Sci Eng, C 94:161–168
74. Singh A et al (2020) Fabrication of calcium hydroxyapatite incorporated polyurethane-graphene oxide nanocomposite porous scaffolds from poly (ethylene terephthalate) waste: a green route toward bone tissue engineering. Polymer 195:122436
75. Marambio-Jones C, Hoek E (2010) A review of the antibacterial effects of silver nanomaterials and potential implications for human health and the environment. J Nanopart Res 12(5):1531–1551
76. Wang Z et al (2020) Embedding ultrasmall Ag nanoclusters in Luria-Bertani extract via light irradiation for enhanced antibacterial activity. Nano Res 13(1):203–208
77. Yin J et al (2013) Attachment of silver nanoparticles (AgNPs) onto thin-film composite (TFC) membranes through covalent bonding to reduce membrane biofouling. J Membr Sci 441:73–82
78. Morones JR et al (2005) The bactericidal effect of silver nanoparticles. Nanotechnology 16(10):2346
79. Bonde SR et al (2012) Murraya koenigii-mediated synthesis of silver nanoparticles and its activity against three human pathogenic bacteria. Nanosci Methods 1(1):25–36
80. Shejawal KP et al (2021) Green synthesis of silver, iron and gold nanoparticles of lycopene extracted from tomato: their characterization and cytotoxicity against COLO320DM, HT29 and Hella cell. J Mater Sci Mater Med 32(2):1–12
81. Lin J et al (2014) Inhibition of autophagy enhances the anticancer activity of silver nanoparticles. Autophagy 10(11):2006–2020
82. Shejawal KP et al (2020) Green synthesis of silver and iron nanoparticles of isolated proanthocyanidin: its characterization, antioxidant, antimicrobial, and cytotoxic activities against COLO320DM and HT29. J Genet Eng Biotechnol 18(1):1–11
83. Mittal AK et al (2015) Bio-synthesis of silver nanoparticles using Potentilla fulgens Wall. ex Hook. and its therapeutic evaluation as anticancer and antimicrobial agent. Mater Sci Eng, C 53:120–127
84. Preethi R, Padma PR (2016) Green synthesis of silver nanobioconjugates from Piper betle leaves and its anticancer activity on A549 cells. Asian J Pharm Clin Res 9(1):252–257
85. Yuan YG et al (2018) Silver nanoparticles potentiates cytotoxicity and apoptotic potential of camptothecin in human cervical cancer cells. Oxidative Med Cell longevity
86. Venkatesan J et al (2017) Antimicrobial and anticancer activities of porous chitosan-alginate biosynthesized silver nanoparticles. Int J Biol Macromol 98:515–525
87. Gomathi AC et al (2020) Anticancer activity of silver nanoparticles synthesized using aqueous fruit shell extract of Tamarindus indica on MCF-7 human breast cancer cell line. J Drug Deliv Sci Technol 55:101376
88. Pavan SR et al (2022) Anticancer activity of silver nanoparticles from the aqueous extract of Dictyota ciliolata on non-small cell lung cancer cells. J Drug Deliv Sci Technol 74:103525
89. Jung GB et al (2013) Fabrication of chitosan-silver nanoparticle hybrid 3D porous structure as a SERS substrate for biomedical applications. Appl Surf Sci 273:179–183
90. Korchinski DJ et al (2015) Iron oxide as an MRI contrast agent for cell tracking: supplementary issue. Magn Reason Insights 8:MRI-S23557

91. Salehipour M et al (2021) Recent advances in polymer-coated iron oxide nanoparticles as magnetic resonance imaging contrast agents. J Nanopart Res 23(2):1–35
92. Abu-Serie MM, Abdelfattah EZA (2022) Anti-metastatic breast cancer potential of novel nanocomplexes of diethyldithiocarbamate and green chemically synthesized iron oxide nanoparticles. Int J Pharm 627:122208
93. Plank C et al (2003) Magnetofection: enhancing and targeting gene delivery with superparamagnetic nanoparticles and magnetic fields. J Liposome Res 13(1):29–32
94. Wang Z et al (2018) Shape-controlled magnetic mesoporous silica nanoparticles for magnetically-mediated suicide gene therapy of hepatocellular carcinoma. Biomaterials 154:147–157
95. Zhao Y et al (2018) Iron oxide nanoparticles-based vaccine delivery for cancer treatment. Mol Pharm 15(5):1791–1799
96. Jonas M et al (2018) Identification of avian influenza virus subtype H9N2 in chicken farms in Indonesia. Prev Vet Med 159:99–105
97. Fathi-Achachelouei M et al (2019) Use of nanoparticles in tissue engineering and regenerative medicine. Front Bioeng Biotechnol 7:113
98. Cojocaru FD et al (2019) Biopolymers-Calcium phosphates composites with inclusions of magnetic nanoparticles for bone tissue engineering. Int J Biol Macromol 125:612–620
99. Mu X et al (2018) siRNA delivery with stem cell membrane-coated magnetic nanoparticles for imaging-guided photothermal therapy and gene therapy. ACS Biomater Sci Eng 4(11):3895–3905
100. Liu XL et al (2019) Magnetic nanomaterials for advanced regenerative medicine: the promise and challenges. Adv Mater 31(45):1804922
101. Bu LL et al (2019) Cancer stem cell-platelet hybrid membrane-coated magnetic nanoparticles for enhanced photothermal therapy of head and neck squamous cell carcinoma. Adv Func Mater 29(10):1807733
102. Liu D et al (2020) Targeted destruction of cancer stem cells using multifunctional magnetic nanoparticles that enable combined hyperthermia and chemotherapy. Theranostics 10(3):1181
103. Su Z et al (2019) CD44-targeted magnetic nanoparticles kill head and neck squamous cell carcinoma stem cells in an alternating magnetic field. Int J Nanomed 14:7549
104. Ziarani GM et al (2019) The role of hollow magnetic nanoparticles in drug delivery. RSC Adv 9(43):25094–25106
105. Ayubi M et al (2019) Magnetic nanoparticles decorated with PEGylated curcumin as dual targeted drug delivery: synthesis, toxicity and biocompatibility study. Mater Sci Eng, C 104:109810
106. Mandal K et al (2019) AIEgen-conjugated magnetic nanoparticles as magnetic–fluorescent bioimaging probes. ACS Appl Nano Mater 2(5):3292–3299
107. Chau TP et al (2022) Green synthesis of Zirconium nanoparticles using Punica granatum (pomegranate) peel extract and their antimicrobial and antioxidant potency. Environ Res 209:112771
108. Tabassum N et al (2021) Zirconium oxide (ZrO_2) nanoparticles from antibacterial activity to cytotoxicity: a next-generation of multifunctional nanoparticles. Mater Today Commun 26:102156
109. Zhu Y et al (2015) In vitro cell proliferation evaluation of porous nano-zirconia scaffolds with different porosity for bone tissue engineering. Biomed Mater 10(5):055009
110. Hasan A et al (2014) Electrospun scaffolds for tissue engineering of vascular grafts. Acta Biomater 10(1):11–25
111. Kim HW et al (2003) Porous ZrO_2 bone scaffold coated with hydroxyapatite with fluorapatite intermediate layer. Biomaterials 24(19):3277–3284
112. Dalal A et al (2012) Orthopedic implant cobalt-alloy particles produce greater toxicity and inflammatory cytokines than titanium alloy and zirconium alloy-based particles in vitro, in human osteoblasts, fibroblasts, and macrophages. J Biomed Mater Res, Part A 100(8):2147–2158

113. Liagre B et al (2002) Effects of alumina and zirconium dioxide particles on arachidonic acid metabolism and proinflammatory interleukin production in osteoarthritis and rheumatoid synovial cells. J Bone Joint Surg Br Vol 84(6):920–930
114. Imran M et al (2020) In-vitro hemolytic activity and free radical scavenging by sol-gel synthesized Fe_3O_4 stabilized ZrO_2 nanoparticles. Arab J Chem 13(11):7598–7608
115. Pantaroto HN et al (2021) Sputtered crystalline TiO_2 film drives improved surface properties of titanium-based biomedical implants. Mater Sci Eng, C 119:111638
116. Song DH et al (2012) Synthesis of titanium oxide thin films containing antibacterial silver nanoparticles by a reactive magnetron co-sputtering system for application in biomedical implants. Mater Res Bull 47(10):2994–2998
117. Chowdhury MA et al (2022) Development of SiC–TiO_2-Graphene neem extracted antimicrobial nano membrane for enhancement of multiphysical properties and future prospect in dental implant applications. Heliyon 8(9):e10603
118. Lenis JA et al (2022) Mechanical, structural, and biological evaluation of multilayer HA-Ag/TiO_2/TiN/Ti coatings on Ti_6Al_4V obtained by magnetron sputtering for implant applications. Surf Coat Technol 449:128925
119. Elayakumar K et al (2019) Enhanced magnetic property and antibacterial biomedical activity of Ce^{3+} doped $CuFe_2O_4$ spinel nanoparticles synthesized by sol-gel method. J Magn Magn Mater 478:140–147
120. Das M et al (2007) Auto-catalytic ceria nanoparticles offer neuroprotection to adult rat spinal cord neurons. Biomaterials 28(10):1918–1925
121. Deshpande S et al (2005) Size dependency variation in lattice parameter and valency states in nanocrystalline cerium oxide. Appl Phys Lett 87(13):133113
122. Alili L et al (2013) Downregulation of tumor growth and invasion by redox-active nanoparticles. Antioxid Redox Signal 19(8):765–778
123. Pešić M et al (2015) Anti-cancer effects of cerium oxide nanoparticles and its intracellular redox activity. Chem Biol Interact 232:85–93
124. Hijaz M et al (2016) Folic acid tagged nanoceria as a novel therapeutic agent in ovarian cancer. BMC Cancer 16(1):1–14
125. Sulthana S et al (2017) Combination therapy of NSCLC using Hsp90 inhibitor and doxorubicin carrying functional nanoceria. Mol Pharm 14(3):875–884
126. Vassie JA et al (2018) Targeted delivery and redox activity of folic acid-functionalized nanoceria in tumor cells. Mol Pharm 15(3):994–1004
127. Fernández-Varo G et al (2020) Bespoken nanoceria: An effective treatment in experimental hepatocellular carcinoma. Hepatology 72(4):1267–1282
128. Saifi MA et al (2021) Nanoceria, the versatile nanoparticles: promising biomedical applications. J Control Release 338:164–189
129. Chigurupati S et al (2013) Effects of cerium oxide nanoparticles on the growth of keratinocytes, fibroblasts and vascular endothelial cells in cutaneous wound healing. Biomaterials 34(9):2194–2201
130. Lord MS et al (2016) Hyaluronan coated cerium oxide nanoparticles modulate CD44 and reactive oxygen species expression in human fibroblasts. J Biomed Mater Res, Part A 104(7):1736–1746
131. Pezzini I et al (2017) Cerium oxide nanoparticles: the regenerative redox machine in bioenergetic imbalance. Nanomedicine 12(4):403–416
132. Naseri-Nosar M et al (2017) Cerium oxide nanoparticle-containing poly (ε-caprolactone)/gelatin electrospun film as a potential wound dressing material: in vitro and in vivo evaluation. Mater Sci Eng, C 81:366–372
133. Zgheib C et al (2019) Use of cerium oxide nanoparticles conjugated with microRNA-146a to correct the diabetic wound healing impairment. J Am Coll Surg 228(1):107–115
134. Babu KS et al (2014) Cytotoxicity and antibacterial activity of gold-supported cerium oxide nanoparticles. Int J Nanomed 9:5515
135. Kasinathan K et al (2016) Photodegradation of organic pollutants RhB dye using UV simulated sunlight on ceria based TiO_2 nanomaterials for antibacterial applications. Sci Rep 6(1):1–12

136. Alpaslan E et al (2017) pH-controlled cerium oxide nanoparticle inhibition of both gram-positive and gram-negative bacteria growth. Sci Rep 7(1):1–12
137. Mohamed HEA et al (2020) Promising antiviral, antimicrobial and therapeutic properties of green nanoceria. Nanomedicine 15(05):467–488
138. Mehrvarz A et al (2022) The effect of ZnO nanoparticles on nanomechanical behavior of Hydroxyapatite electrodeposited on NiTi biomedical alloy. Ceram Int 48(23):35039–35049
139. Riyas ZM et al (2022) Green synthesis and biomedical behavior of Mg-doped ZnO nanoparticle using leaf extract of Ficus religiosa. Ceram Int
140. AlSalem HS et al (2022) Physico-chemical and biological responses for hydroxyapatite/ZnO/graphene oxide nanocomposite for biomedical utilization. Mater Chem Phys 283:125988
141. Mishra YK et al (2015) Direct growth of freestanding ZnO tetrapod networks for multifunctional applications in photocatalysis, UV photodetection, and gas sensing. ACS Appl Mater Interfaces 7(26):14303–14316
142. Sonntag SR et al (2022) Zinc oxide tetrapods modulate wound healing and cytokine release in vitro—a new antiproliferative substance in glaucoma filtering surgery. Life 12(11):1691
143. El-Gendy AO et al (2022) Preparation of zinc oxide nanoparticles using laser-ablation technique: retinal epithelial cell (ARPE-19) biocompatibility and antimicrobial activity when activated with femtosecond laser. J Photochem Photobiol, B 234:112540
144. Zhang ZY, Xiong HM (2015) Photoluminescent ZnO nanoparticles and their biological applications. Materials 8(6):3101–3127
145. Jiang Y et al (2016) Role of physical and chemical interactions in the antibacterial behavior of ZnO nanoparticles against E. coli. Mater Sci Eng C 69:1361–1366
146. Dutta RK et al (2013) Antibacterial effect of chronic exposure of low concentration ZnO nanoparticles on E. coli. J Environ Sci Health Part A 48(8):871–878
147. Chatterjee T et al (2010) The effect of zinc oxide nanoparticles on the structure of the periplasmic domain of the Vibrio cholerae ToxR protein. FEBS J 277(20):4184–4194
148. Salem W et al (2015) Antibacterial activity of silver and zinc nanoparticles against Vibrio cholerae and enterotoxic Escherichia coli. Int J Med Microbiol 305(1):85–95
149. Sarwar S et al. (2016). The antimicrobial activity of ZnO nanoparticles against Vibrio cholerae: variation in response depends on biotype. Nanomed Nanotechnol Biol Med 12(6):1499–1509
150. Sarwar S et al (2017) Zinc oxide nanoparticles provide anti-cholera activity by disrupting the interaction of cholera toxin with the human GM1 receptor. J Biol Chem 292(44):18303–18311
151. Zhu P et al (2016) Biomedical applications of functionalized ZnO nanomaterials: from biosensors to bioimaging. Adv Mater Interfaces 3(1):1500494
152. Xiong HM et al (2008) Stable aqueous ZnO@ polymer core–shell nanoparticles with tunable photoluminescence and their application in cell imaging. J Am Chem Soc 130(24):7522–7523
153. Peng S et al (2011) Plasmonic/magnetic bifunctional nanoparticles. Angew Chem Int Ed 50(14):3158–3163
154. Wang X et al (2014) Preparation and catalytic activity of PVP-protected Au/Ni bimetallic nanoparticles for hydrogen generation from hydrolysis of basic NaBH4 solution. Int J Hydrogen Energy 39(2):905–916
155. Mukha I et al (2017) Anticancer effect of Ag, Au, and Ag/Au bimetallic nanoparticles prepared in the presence of tryptophan. J Nanosci Nanotechnol 17(12):8987–8994
156. Shmarakov IO et al (2014) Tryptophan-assisted synthesis reduces bimetallic gold/silver nanoparticle cytotoxicity and improves biological activity. Nanobiomedicine 1:1–6
157. Pal A et al (2007) Preparation of silver, gold and silver–gold bimetallic nanoparticles in w/o microemulsion containing TritonX-100. Colloids Surf A 302(1–3):483–487
158. Nakamura T, Sato S (2015) Green and facile synthesis of Pd–Pt alloy nanoparticles by laser irradiation of aqueous solution. J Nanosci Nanotechnol 15(1):426–432
159. Mottaghi N et al (2014) Ag/Pd core-shell nanoparticles by a successive method: pulsed laser ablation of Ag in water and reduction reaction of $PdCl_2$. Appl Surf Sci 292:892–897
160. Xiao Q et al (2011) Synthesis and characterization of Ag–Ni bimetallic nanoparticles by laser-induced plasma. Thin Solid Films 519(20):7116–7119

161. Amendola V et al (2013) Coexistence of plasmonic and magnetic properties in $Au_{89}Fe_{11}$ nanoalloys. Nanoscale 5(12):5611–5619
162. Zielińska-Jurek A, Zaleska A (2014) Ag/Pt-modified TiO_2 nanoparticles for toluene photooxidation in the gas phase. Catal Today 230:104–111
163. Hierso JC et al (1998) Metal-organic chemical vapor deposition in a fluidized bed as a versatile method to prepare layered bimetallic nanoparticles. J Mol Catal A: Chem 135(3):321–325
164. Choi DS et al (2016) Low-temperature chemical vapor deposition synthesis of Pt–Co alloyed nanoparticles with enhanced oxygen reduction reaction catalysis. Adv Mater 28(33):7115–7122
165. Hermannsdörfer J et al (2012) Ni/Pd@ MIL-101: synergistic catalysis with cavity-conform Ni/Pd nanoparticles. Angew Chem Int Ed 51(46):11473–11477
166. Manjula N et al (2021) Synthesis and characterization of iron-cobalt oxide/polypyrrole nanocomposite: an electrochemical sensing platform of anti-prostate cancer drug flutamide in human urine and serum samples. Colloids Surf A 628:127367
167. Alonso J et al (2015) FeCo nanowires with enhanced heating powers and controllable dimensions for magnetic hyperthermia. J Appl Phys 117(17):17D113
168. Shokuhfar A, Afghahi SSS (2014) Size controlled synthesis of FeCo alloy nanoparticles and study of the particle size and distribution effects on magnetic properties. Adv Mater Sci Eng 295390
169. Shi Y et al (2015) Recent advances in FePt nanoparticles for biomedicine. J Nanomater 2:2
170. Liang S et al (2015) Water-soluble L-cysteine-coated FePt nanoparticles as dual MRI/CT imaging contrast agent for glioma. Int J Nanomed 10:2325
171. Bharathala S, Sharma P (2019) Biomedical applications of nanoparticles. In: Nanotechnology in modern animal biotechnology. Elsevier, pp 113–132
172. Singh A et al (2018) Adaptation of 3D printing techniques in bone tissue engineering: an assessment of its need, reliability, validity, sustainability, and future scope. 3D Print Technol. CRI, New Delhi
173. Singh A et al (2022) Surface functionalizations of nanocellulose for wastewater treatment. Handbook of nanocelluloses: classification, properties, fabrication, and emerging applications. Springer International Publishing, Cham, pp 1–48
174. Singh A et al (2021) Nanocellulose biocomposites for bone tissue engineering. Handbook of nanocelluloses: classification, properties, fabrication, and emerging applications. Springer International Publishing, Cham, pp 1–51
175. Singh A et al (2022) Recent innovations in chemical recycling of polyethylene terephthalate waste: a circular economy approach toward sustainability. In: Handbook of solid waste management: sustainability through circular economy. Springer Singapore, Singapore, pp 1149–1176
176. Singh A et al (2022) Polyurethane nanocomposites for bone tissue engineering. In: Engineered nanomaterials for innovative therapies and biomedicine. Springer, Cham, pp 373–403

The manufacturer's authorised representative in the EU is Springer Nature Customer Service Centre GmbH, Europaplatz 3, 69115 Heidelberg, Germany. If you have any concerns regarding our products, please contact ProductSafety@springernature.com

Printed and bound by CPI Group (UK) Ltd, Croydon, CR0 4YY
30/09/2026
02225244-0013